AF410869

Carbon Emission Reduction—Carbon Tax, Carbon Trading, and Carbon Offset

Carbon Emission Reduction—Carbon Tax, Carbon Trading, and Carbon Offset

Editor

Wen-Hsien Tsai

MDPI • Basel • Beijing • Wuhan • Barcelona • Belgrade • Manchester • Tokyo • Cluj • Tianjin

Editor
Wen-Hsien Tsai
National Central University
Taiwan

Editorial Office
MDPI
St. Alban-Anlage 66
4052 Basel, Switzerland

This is a reprint of articles from the Special Issue published online in the open access journal *Energies* (ISSN 1996-1073) (available at: https://www.mdpi.com/journal/energies/special_issues/carbon_emission_reduction).

For citation purposes, cite each article independently as indicated on the article page online and as indicated below:

LastName, A.A.; LastName, B.B.; LastName, C.C. Article Title. *Journal Name* **Year**, *Volume Number*, Page Range.

ISBN 978-3-0365-0160-4 (Hbk)
ISBN 978-3-0365-0161-1 (PDF)

Cover image courtesy of Feng-Ying Ho.

Contents

About the Editor

Wen-Hsien Tsai is a distinguished professor of accounting and information systems in the Department of Business Administration at National Central University, Taiwan. He has served as a Guest Editor for Special Issues of the journals Energies and Sustainability and as an associate editor of the journal Decision Support Systems. He is also a certified consultant of SAP financial modules. He received his Ph.D. degree in industrial management from the National Taiwan Science and Technology University. He received his MBA degree and his MS degree in industrial engineering from the National Taiwan University and National Tsing-Hwa University, respectively. His research interests include Industry 4.0; carbon emissions; carbon tax; activity-based costing (ABC); ERP implementation and auditing; green production and optimization decision; and the International Financial Reporting Standards (IFRS). He has published several papers in high-quality international journals, such as *Energies, Sustainability, Decision Support Systems, European Journal of Operational Research, Omega, Transportation Science, Industrial Marketing Management, Journal of the Operational Research Society, Computers and Operations Research, Journal of Cleaner Production, International Journal of Production Economics, Computers and Industrial Engineering*, and the *International Journal of Production Research*.

Preface to "Carbon Emission Reduction—Carbon Tax, Carbon Trading, and Carbon Offset"

The World Bank stated that there are some incentives which have been created to encourage carbon emission reduction, such as the removal of fossil fuels subsidies, the introduction of carbon pricing, the increase in energy efficiency standards, and the implementation of auctions for the lowest-cost renewable energy. Among these, carbon pricing refers to charges for those who emit carbon dioxide (CO2) from their emissions, including carbon taxes, emissions trading systems (ETSs), offset mechanisms, and results-based climate finance (RBCF). This Special Issue collects 19 carbon emissions-related papers (including five that are carbon tax-related) and five energy-related papers using various methods or models, such as the LMDI (Logarithmic Mean Divisia Index) decomposition method, panel data model, ordered weighted regression (OWA), geographically-and-temporally-weighted regression (GTWR), and the expanded stirpat* model. The research studies come from China, Taiwan, Thailand, Czech Republic, Pakistan, Sweden, Norway, and United States. These studies involved various industries such as agricultural industry, transportation industry, electric-power industry, electronic industry, paper industry, iron and steel industry, and the oil and gas industry. Although this Special Issue does not fully solve our concerns, it still provides abundant material for implementing energy conservation and carbon emissions reduction. However, there are still many issues regarding the problems caused by global warming that require research. Finally, I am grateful to MDPI for the invitation to act as the Guest Editor of this Special Issue and I am indebted to the editorial office of Energies for their kind cooperation, patience, and committed engagement. I would like to thank the authors for submitting their excellent contributions to this Special Issue. My thanks are extended to the reviewers for evaluating the manuscripts and providing helpful suggestions. Sincere thanks also go to the editorial team of MDPI and *Energies* for providing the opportunity to publish this book and helping in all possible ways.

Wen-Hsien Tsai
Editor

Editorial

Carbon Emission Reduction—Carbon Tax, Carbon Trading, and Carbon Offset

Wen-Hsien Tsai

Department of Business Administration, National Central University, Jhongli, Taoyuan 32001, Taiwan; whtsai@mgt.ncu.edu.tw; Tel.: +886-3-426-7247

Received: 29 October 2020; Accepted: 19 November 2020; Published: 23 November 2020

---

1. Introduction

The Paris Agreement was signed by 195 nations in December 2015 to strengthen the global response to the threat of climate change following the 1992 United Nations Framework Convention on Climate Change (UNFCC) and the 1997 Kyoto Protocol. In Article 2 of the Paris Agreement, the increase in the global average temperature is anticipated to be held to well below 2 °C above pre-industrial levels, and efforts are being employed to limit the temperature increase to 1.5 °C. The United States Environmental Protection Agency (EPA) provides information on emissions of the main greenhouse gases. It shows that about 81% of the totally emitted greenhouse gases were carbon dioxide (CO_2), 10% methane, and 7% nitrous oxide in 2018. Therefore, carbon dioxide (CO_2) emissions (or carbon emissions) are the most important cause of global warming. The United Nations has made efforts to reduce greenhouse gas emissions or mitigate their effect. In Article 6 of the Paris Agreement, three cooperative approaches that countries can take in attaining the goal of their carbon emission reduction are described, including direct bilateral cooperation, new sustainable development mechanisms, and non-market-based approaches.

The World Bank stated that there are some incentives that have been created to encourage carbon emission reduction, such as the removal of fossil fuels subsidies, the introduction of carbon pricing, the increase of energy efficiency standards, and the implementation of auctions for the lowest-cost renewable energy. Among these, carbon pricing refers to charging those who emit carbon dioxide (CO_2) for their emissions, including carbon taxes, emissions trading systems (ETSs), offset mechanisms, results-based climate finance (RBCF), and so on. In view of the urgent need for carbon emission reduction, this special issue collects 19 related papers concerning carbon emission reduction by using various models and methods.

2. Summary Information of 19 Papers in the Special Issue

Table 1 shows the summary information of 19 papers in this special issue, including Research Topic, Papers' Author, Method/Model, Research Object, and Industry/Field. From Table 1, we can see that this special issue has 5 papers for investigating the influencing factors of carbon emissions; 2 papers for exploring the relationship among carbon emissions, economic growth, and agricultural production; and 6 papers for discussing various tools of carbon emission reduction such as carbon tax, carbon trading, carbon offset, carbon storage, and carbon footprint. In additions, there are 6 papers involving the related issues for carbon emission reduction.

Table 1. Summary information of 19 papers in this special issue.

Topic			Paper/Author	Method/Model	Research Object	Industry/Field
1. Influencing Factors of Carbon Emissions	1.1	Transportation Carbon Emissions	Zhu, Gao [1]	Panel Data Model	57 'the Belt and Road Initiative' (BIT) countries	Transportation
			Zhu, Du [2]	LMDI (Logarithmic Mean Divisia Index) Decomposition Method	Six Asia-Pacific countries	Transportation
			Zhu, Wang, Yang [3]	LMDI (Logarithmic Mean Divisia Index) Decomposition Method	Regions in China	Transportation
	1.2	Agricultural Carbon Emissions	Chen, Li, Su, Li [4]	Ordered Weighted Regression (OWA); Geographically-and-Temporally-Weighted Regression (GTWR)	Fujian, China	Agricultural Industry
	1.3	Carbon Emissions from Energy Consumption	Li, Li, Shao [5]	Expanded STIRPAT * Model	China	National level
2. Relationship among Carbon Emissions, Economic Growth, and Agricultural Production			Ali, Ying, Shah, Tariq, Chandio, Ali [6]	Autoregressive Distributed Lag (ARDL) Model; Pairwise Granger Causality Test	Pakistan	Agricultural Industry
			Ali, Gucheng, Ying, Ishaq, Shah [7]	Autoregressive Distributed Lag (ARDL) Model; Kwiatkowski–Phillips–Schmidt–Shin (KPSS) Test	Pakistan	Agricultural Industry
3. Carbon Tax			Che, Zhang, Lang [8]	Mixed Integer Nonlinear Programming Model	Numerical Experiments	Electric-power Industry
			Hsieh, Tsai, Chang [9]	Mixed Integer Linear Programming (MILP) Model	Taiwan	Paper Industry
4. Carbon Trading			Duan, Han, Mu, Yang, Li [10]	Two-stage Game Theory Model	China	Iron and Steel Industry
5. Carbon Offset			Krasovskii, Khabarov, Lubowski, Obersteiner [11]	Two-stage Stochastic Technological Portfolio Optimization Model	Not specified	Power Industry
6. Carbon Storage			Cao, Liu, Hou, Mehmood, Liao, Feng [12]	Literature Review	Review Paper	Not specified
7. Carbon Footprint			Quintana-Pedraza, Viera-Agudelo, Muñoz-Galeano [13]	Cradle-to-Grave Multi-Pronged Methodology	Sweden and China	Electronic Industry
8. Others		Carbon Leakage	Fan, Zhang, Gao, Chen, Li, Miao [14]	Single-Region Input–Output Model	China	Industrial Sector
		Gas Leaks	Log, Pedersen [15]	Differential Absorption Lidar (DIAL) Technique	Norway	Oil and Gas Industry
		CO_2 Efficiency Break Point	Bat'a, Fuka, Lešáková, Heckenbergerová [16]	Modified Life Cycle Assessment (LCA)	Czech Republic	Transportation
		Efficiency of Sustainable Development Policy	Sutthichaimethee, Naluang [17]	SEM-VARIMAX ** Model	Thailand	National Level
		Optimization in the Stripping Process of CO_2 Gas	Chen, Lai [18]	Taguchi Method	Not specified	Not specified
		Low Emission Taxiing Path Optimization	Li, Sun, Yu, Li, Zhang, Tsai [19]	Path Optimization Model	China	Airport

<Note> * STIRPAT: Stochastic Impacts by Regression on Population, Affluence, and Technology. ** SEM-VARIMAX: Structural Equation Modeling/Vector Autoregressive Model with Exogenous Variables.

3. Review of the Special Issue

3.1. Influencing Factors of Carbon Emissions

3.1.1. Transportation Carbon Emissions

Zhu and Gao [1] used the Panel Data Model to investigate the factors affecting the carbon emissions of the transportation industry in 57 of 'the Belt and Road Initiative' (BRI) countries. The research results indicate that the positive factors influencing carbon emissions of the transportation industry are capita

GDP, urbanization level, and energy consumption structure, while the negative factors are technology level and trade openness. Zhu and Du [2] applied the Logarithmic Mean Divisia Index (LMDI) decomposition method to research the driving factors for carbon emissions of the road transportation industry in six Asia-Pacific countries from 1990 to 2016. This research found that the economic output and the population size have positive driving influences on carbon emissions for the road transportation industry; the energy intensity and the transportation intensity have different influences on driving carbon emissions for the road transportation industry for these six Asia-Pacific Countries. In addition, the carbon emissions coefficient has a relatively small influence. Based on data from 1997 to 2017, Zhu, Wang, and Yang [3] adopted the Logarithmic Mean Divisia Index (LMDI) decomposition method to analyze the influencing degree of several major factors on the carbon emissions of transportation industry in different regions, and they put forward some suggestions according to local conditions and provided references for the carbon reduction of Chinese transportation industry. Their results showed that economic output effect is the most important factor to promote the carbon emissions of transportation industry in various regions. The regional energy intensity effect in most stages inhibited carbon emissions of the transportation industry.

3.1.2. Agricultural Carbon Emissions

Based on the carbon emission sources of five main aspects in agricultural production, Chen, Li, Su, and Li [4] applied the latest carbon emission coefficients released by Intergovernmental Panel on Climate Change of the UN (IPCC) and World Resources Institute (WRI), and then used the ordered weighted aggregation (OWA) operator to remeasure agricultural carbon emissions in Fujian from 2008–2017. The research results showed that the regression coefficients of each selected factor in the cities were positive or negative, which indicated that the impacts on agricultural carbon emission had the characteristics of geospatial nonstationarity.

3.1.3. Carbon Emissions from Energy Consumption

Li, Li, and Shao [5] used the IPCC (The Intergovernmental Panel on Climate Change) method to calculate the carbon emissions of energy consumption in China from 1996 to 2016, and used it as a dependent variable to analyze the influencing factors. Their results showed that the order of impact on carbon emissions from high to low is total population, per capita GDP, technology level, industrial structure, primary energy consumption structure, and urbanization level.

3.2. Relationship among Carbon Emissions, Economic Growth, and Agricultural Production

Ali, Ying, Shah, Tariq, Chandio, and Ali [6] employed augmented Dickey–Fuller (ADF) and Phillips–Perron (PP) tests to check the stationarity of the variables. The results of the analysis revealed that there is both a short- and long-run association between agricultural production, economic growth, and carbon dioxide emissions in the country.

Ali, Gucheng, Ying, Ishaq, and Shah [7] aimed to explore the casual relationship between agricultural production, economic growth, and carbon dioxide emissions in Pakistan. An autoregressive distributed lag (ARDL) model was applied to examine the relationship between agricultural production, economic growth, and carbon dioxide emissions using time series data from 1960 to 2014. The results showed both short-run and long-run relationships between agricultural production, gross domestic product (GDP), and carbon dioxide emissions in Pakistan.

3.3. Carbon Tax

Carbon tax is a tax on energy sources that emit carbon dioxide. It is a pollution tax and a form of carbon pricing. The objective of a carbon tax is to reduce the harmful and unfavorable levels of carbon dioxide emissions, thereby decelerating climate change and its negative effects on the environment and human health. A Carbon tax also can prompt companies to find more efficient ways to manufacture their products or deliver their services. Generally, a carbon tax is determined by the carbon tax rate and the quantity of carbon emissions that a company processes in its manufacturing, and it is represented as the amount paid for every ton of greenhouse gas released into the atmosphere. However, carbon tax also will have some disadvantages, such as imposing expensive administration costs for businesses, prompting them to move their operations to "pollution havens," and so on.

Che, Zhang, and Lang [8] formulated the generation self-scheduling problem under the proposed carbon-tax policy as a mixed integer nonlinear programming model. Numerical results demonstrated that the proposed decomposition algorithm can solve the considered problem in a reasonable time and indicated that the proposed carbon-tax policy can enhance the incentive for generation companies to invest in a low-carbon generation capacity.

Using mathematical programming with activity-based costing (ABC) and based on the theory of constraints (TOC), Hsieh, Tsai, and Chang [9] proposed a green production model for the traditional paper industry to achieve the purpose of energy saving and carbon emission reduction. A numerical example was used to demonstrate how to apply the model presented in their paper.

3.4. Carbon Trading

Carbon trading is another form of carbon pricing under cap-and-trade systems. Cap-and-trade is one method for regulating and ultimately reducing the amount of carbon emissions. The government sets a cap on carbon emissions for the whole country, and then limits the amount of carbon dioxide that companies are allowed to release. A company that can more efficiently reduce carbon emissions can sell any extra permits in the emission market to companies that cannot easily afford to reduce carbon emission. Thus, carbon trading markets are set up. The number of emissions trading systems around the world is increasing. In addition to the EU emissions trading system (EU ETS), national or subnational systems are already in operation or under development in Canada, China, Japan, New Zealand, South Korea, Switzerland, and the United States.

To study the emission reduction policies' impact on the production and economic level of the steel industry, Duan, Han, Mu, Yang, and Li [10] constructed a two-stage dynamic game model and analyzed various emission reduction policies' impact on the steel industry and enterprises. The research results indicated that with the increasing emission reduction target (15–30%) and carbon quota trading price (12.65–137.59 Yuan), social welfare and producer surplus show an increasing trend and emission macro losses show a decreasing trend.

3.5. Carbon Offset

A carbon offset is a reduction in emissions of carbon dioxide or greenhouse gases made in order to compensate for or to offset an emission made elsewhere. One ton of carbon offset represents the reduction of one ton of carbon dioxide or its equivalent in other greenhouse gases. There are two markets for carbon offsets: (1) The larger compliance market, where companies, governments, or other entities buy carbon offsets in order to comply with caps on the total amount of carbon dioxide they are allowed to emit; and (2) the smaller voluntary market, where individuals, companies, or governments purchase carbon offsets to mitigate their own greenhouse gas emissions from transportation, electricity use, and other sources. Carbon offset usually supports projects that reduce the emission of greenhouse gases in the short- or long-term. A common project type is renewable energy, such as wind farms, biomass energy, or hydroelectric dams. Others include energy efficiency projects, the destruction of industrial pollutants or agricultural byproducts, the destruction of landfill methane, LULUCF (land use, land-use change, and forestry), REDD (reducing emissions from deforestation and forest degradation), and so on.

Krasovskii, Khabarov, Lubowski, and Obersteiner [11] were motivated by the risks associated with the future CO_2 price uncertainty in the context of the offsetting of carbon emissions by regulated entities. They asked whether it is possible to reduce these financial risks. In this study, authors considered the bilateral interaction of a REDD supplier and a greenhouse gas (GHG)-emitting energy producer in an incomplete emission offsets market. Their results showed that flobsion's flexibility had advantages compared to a standard option, which can help GHG-emitting energy producers with managing their compliance risks, while at the same time facilitating the development of REDD programs.

3.6. Carbon Storage

Cao, Liu, Hou, Mehmood, Liao, and Feng [12] aim to provide the latest developments of CO_2 storage from the perspective of improving safety and economics. This review demonstrates that CO_2 storage in depleted oil and gas reservoirs could play an important role in reducing CO_2 emission in the near future and CO_2 storage in saline aquifers may make the biggest contribution due to its huge storage capacity. Comparing the various available strategies, CO_2—enhanced oil recovery (CO_2—EOR) operations are supposed to play the most important role for CO_2 mitigation in the next few years, followed by CO_2—enhanced gas recovery (CO_2—EGR).

3.7. Carbon Footprint

Quintana-Pedraza, Viera-Agudelo, and Muñoz-Galeano [13] propose the application of a cradle-to-grave multi-pronged methodology to obtain a more realistic carbon footprint (CF) estimation of electro-intensive power electronic (EIPE) products. The proposed methodology is applied in a cradle-to-grave scenario, being composed of two approaches of LCA. Results show that D-STATCOM considerably decreases the CF and saves emissions taken place during the usage stage. A comparison was made between Sweden and China to establish the environmental impact of D-STATCOM in electrical networks, showing that saved emissions in the life cycle of D-STATCOM were 5.88 and 391.04 ton CO_2eq in Sweden and China, respectively.

3.8. Others

3.8.1. Carbon Leakage

On the basis of trade data for China's 20 industrial sectors, Fan, Zhang, Gao, Chen, Li, and Miao [14] built a panel data model to test the effect of trade on carbon dioxide emissions and the presence of carbon leakage for all industrial sectors. They derived a single-region input–output model for open economies based on the industrial sectors' diversity and carbon dioxide emissions, and performed an empirical test. The results show that higher trade openness leads to a reduction in the intensity of CO_2 emissions and gross emissions and that there are obvious structural differences in different sectors with different carbon emission intensity. The coefficient of trade openness for LCSs is -0.073 and is statistically significant at the 1% level, so higher trade openness for LCSs leads to a reduction in the CO_2 emissions intensity.

3.8.2. Gas Leaks

Log and Pedersen [15] developed a simple logarithmic table based on an existing consequence matrix for safety related incidents extended to include non-safety related fugitive emissions. An evaluation sheet was also developed as a guide for immediate risk evaluations when new leaks are identified. The leak rate table and evaluation guide were tested in the field at five land-based oil and gas facilities during Optical Gas Inspection (OGI) campaigns. It is demonstrated how the suggested concept can be used for presenting and analyzing detected leaks to assist in Leak Detection and Repair (LDAR) programs.

3.8.3. CO_2 Efficiency Break Point

Bat'a, Fuka, Lešáková, and Heckenbergerová [16] aim to deal with CO_2 emissions in energy production process in an original way, based on calculations of total specific CO_2 emission and, depending on the type of fuel and the transport distance. It is based on a modified life cycle assessment (LCA), supplemented with a system of equations. Their key finding is the break point for associated processes at a distance of 1779.64 km, since it is better to burn brown coal than wood in terms of total CO_2 emissions. The research can conclude that, in some cases, it is more efficient to use coal instead of wood as fuel in terms of CO_2 emissions, particularly in regard to transport distance and type of transport.

3.8.4. Efficiency of Sustainable Development Policy

Sutthichaimethee and Naluang [17] aim to predict the efficiency of the Sustainable Development Policy for Energy Consumption under Environmental Law in Thailand for the next 17 years (2020–2036), and aim to analyze the relationships among causal factors by applying a structural equation modeling/vector autoregressive model with exogenous variables (SEM-VARIMAX Model). With the implementation of the Sustainable Development Policy for Energy Consumption under Environmental Law (S.D.EL), the forecast results derived from the SEM-VARIMAX Model indicate a continuously high change in energy consumption from 2020 to 2036, and the change exceeds the rate determined by the government.

3.8.5. Optimization in the Stripping Process of CO_2 Gas

Chen and Lai [18] aimed to explore the effects of variables on the heat of regeneration, the stripping efficiency, the stripping rate, the steam generation rate, and the stripping factor. The results showed that the stripping efficiency was in the range of 20.98–55.69%, the stripping rate was in the range of 5.57×10^{-5}–4.03×10^{-4} kg/s, and heat of regeneration was in the range of 5.52–18.94 GJ/t.

3.8.6. Low Emission Taxiing Path Optimization

Li, Sun, Yu, Li, Zhang, and Tsai [19] considered the aircraft's taxiing distance, the number of large steering times, and collision avoidance in the taxi, and established a path optimization model for aircraft taxiing at airport surface with the shortest total taxi time as the target. The experimental results show that the total fuel consumption and emissions of the aircraft are reduced by 35% and 46%, respectively, before optimization, and the taxi time is greatly reduced, which effectively avoids the taxiing conflict and reduces the pollutant emissions during the taxiing phase.

4. Concluding Remarks

The World Bank stated that there are some incentives that have been created to encourage carbon emission reduction, such as the removal of fossil fuels subsidies, the introduction of carbon pricing, the increase of energy efficiency standards, and the implementation of auctions for the lowest-cost renewable energy. Among these, carbon pricing refers to charging those who emit carbon dioxide (CO_2) for their emissions, including carbon taxes, emissions trading systems (ETSs), offset mechanisms, results-based climate finance (RBCF), and so on. This Special Issue collects 19 carbon emissions-related papers (including 5 that are carbon tax-related) and 5 energy-related papers using various methods or models. Although this special issue did not fully satisfy our needs, it still provides abundant related material for energy conservation and carbon emissions reduction. However, there still are many research topics waiting our efforts to study to solve the problems of global warming.

Funding: This research received no external funding.

Acknowledgments: Wen-Hsien Tsai is grateful to the MDPI Publisher for the invitation to act as guest editor of this special issue and is indebted to the editorial staff of "Energies" for the kind cooperation, patience and committed engagement, especially the senior editor, Julyn Li. The guest editor would also like to thank the authors for submitting their excellent contributions to this special issue. Thanks are also extended to the reviewers for evaluating the manuscripts and providing helpful suggestions.

Conflicts of Interest: The author declares no conflict of interest.

References

1. Zhu, C.; Gao, D. A Research on the Factors Influencing Carbon Emission of Transportation Industry in "the Belt and Road Initiative" Countries Based on Panel Data. *Energies* **2019**, *12*, 2405.
2. Zhu, C.; Du, W. A Research on Driving Factors of Carbon Emissions of Road Transportation Industry in Six Asia-Pacific Countries Based on the LMDI Decomposition Method. *Energies* **2019**, *12*, 4152.
3. Zhu, C.; Wang, M.; Yang, Y. Analysis of the Influencing Factors of Regional Carbon Emissions in the Chinese Transportation Industry. *Energies* **2020**, *13*, 1100.
4. Chen, Y.; Li, M.; Su, K.; Li, X. Spatial-Temporal Characteristics of the Driving Factors of Agricultural Carbon Emissions: Empirical Evidence from Fujian, China. *Energies* **2019**, *12*, 3102.
5. Li, Z.; Li, Y.; Shao, S. Analysis of Influencing Factors and Trend Forecast of Carbon Emission from Energy Consumption in China Based on Expanded STIRPAT Model. *Energies* **2019**, *12*, 3054.
6. Ali, S.; Ying, L.; Shah, T.; Tariq, A.; Chandio, A.A.; Ali, I. Analysis of the Nexus of CO_2 Emissions, Economic Growth, Land under Cereal Crops and Agriculture Value-Added in Pakistan Using an ARDL Approach. *Energies* **2019**, *12*, 4590.
7. Ali, S.; Gucheng, L.; Ying, L.; Ishaq, M.; Shah, T. The Relationship between Carbon Dioxide Emissions, Economic Growth and Agricultural Production in Pakistan: An Autoregressive Distributed Lag Analysis. *Energies* **2019**, *12*, 4644.
8. Che, P.; Zhang, Y.; Lang, J. Emission-Intensity-Based Carbon Tax and Its Impact on Generation Self-Scheduling. *Energies* **2019**, *12*, 777.
9. Hsieh, C.-L.; Tsai, W.-H.; Chang, Y.-C. Green Activity-Based Costing Production Decision Model for Recycled Paper. *Energies* **2020**, *13*, 2413.
10. Duan, Y.; Han, Z.; Mu, H.; Yang, J.; Li, Y. Research on the Impact of Various Emission Reduction Policies on China's Iron and Steel Industry Production and Economic Level under the Carbon Trading Mechanism. *Energies* **2019**, *12*, 1624.
11. Krasovskii, A.; Khabarov, N.; Lubowski, R.; Obersteiner, M. Flexible Options for Greenhouse Gas-Emitting Energy Producer. *Energies* **2019**, *12*, 3792.
12. Cao, C.; Liu, H.; Hou, Z.; Mehmod, F.; Liao, J.; Feng, W. A Review of CO_2 Storage in View of Safety and Cost-Effectiveness. *Energies* **2020**, *13*, 600.
13. Quintana-Pedraza, G.A.; Vieira-Agudelo, S.C.; Muñoz-Galeano, N. A Cradle-to-Grave Multi-Pronged Methodology to Obtain the Carbon Footprint of Electro-Intensive Power Electronic Products. *Energies* **2019**, *12*, 3347.
14. Fan, B.; Zhang, Y.; Li, X.; Miao, X. Trade Openness and Carbon Leakage: Empirical Evidence from China's Industrial Sector. *Energies* **2019**, *12*, 1011.
15. Log, T.; Pedersen, W.B.B. A Common Risk Classification Concept for Safety Related Gas Leaks and Fugitive Emissions? *Energies* **2019**, *12*, 4063.
16. Baťa, R.; Fuka, J.; Lešáková, P.; Heckenbergerová, J. CO_2 Efficiency Break Points for Processes Associated to Wood and Coal Transport and Heating. *Energies* **2019**, *12*, 3864.
17. Sutthichaimethee, P.; Naluang, S. The Efficiency of the Sustainable Development Policy for Energy Consumption under Environmental Law in Thailand: Adapting the SEM-VARIMAX Model. *Energies* **2019**, *12*, 3092.
18. Chen, P.C.; Lai, Y.-L. Optimization in the Stripping Process of CO_2 Gas Using Mixed Amines. *Energies* **2019**, *12*, 2202.
19. Li, N.; Sun, Y.; Yu, Y.; Yu, J.; Li, J.-C.; Zhang, H.-F.; Tsai, S. An Empirical Study on Low Emission Taxiing Path Optimization of Aircrafts on Airport Surfaces from the Perspective of Reducing Carbon Emissions. *Energies* **2019**, *12*, 1649.

Publisher's Note: MDPI stays neutral with regard to jurisdictional claims in published maps and institutional affiliations.

Article

A Research on the Factors Influencing Carbon Emission of Transportation Industry in "the Belt and Road Initiative" Countries Based on Panel Data

Changzheng Zhu * and Dawei Gao

School of Modern Post, Xi'an University of Posts and Telecommunications, Xi'an 710061, China; g1997dw@outlook.com
* Correspondence: zhuchangzheng@xupt.edu.cn; Tel.: +86-182-9207-8891

Received: 13 April 2019; Accepted: 19 June 2019; Published: 22 June 2019

Abstract: Carbon emissions in countries in the "Belt and Road Initiative (BRI)" account for more than half of the world's total volume. According to the international energy agency report, the world transportation industry carbon emissions in 2015 came second on the list for the proportion of global carbon emissions across all industries, accounting for 23.96% of the total. Along with the advancement of the BRI construction, transportation industry carbon emissions will continue their rapid growth. Therefore, studying the factors affecting the carbon emissions of the transportation industry in countries in the BRI is conducive to the formulation of policies to control carbon emissions. In this paper, the CO_2 emissions of the transportation industry in countries in the BRI line from 2005 to 2015 were measured, and then the influencing factors of 57 countries in the BRI were analyzed by using the panel data model. The results show that per capita GDP, urbanization level, and energy consumption structure have positive effects on the carbon emissions of transportation industry, while technology level and trade openness have negative effects on carbon emissions of the transportation industry. Therefore, in order to effectively control the carbon emissions of the transportation industry in the BRI countries, it is necessary to reasonably control the transportation industry carbon emissions caused by urbanization, optimize the energy consumption structure of the transportation industry, optimize the structure of the transportation industry, and improve the openness of trade and the technical level of the BRI countries.

Keywords: carbon emissions; transportation industry; influencing factors; the "Belt and Road Initiative"

---

1. Introduction

As an important part of modern service industry, the transportation industry is a high energy consumption industry and a large carbon emitter in the national economic system. According to the international standard industry classification (ISIC Rev 4.0), the transportation industry is divided into road transportation, railway transportation, navigation transportation, aviation transportation, and pipeline transportation, and the transportation industry studied in this paper is specified as follows: road transportation, railway transportation, domestic aviation transportation, pipeline transport, and domestic navigation transportation. According to the "carbon dioxide emissions from fuel combustion" report released by the International Energy Agency (IEA) in 2017, the world's transportation industry accounted for the second largest share of global carbon emissions in 2017, accounting for 23.96% of the total [1]. Therefore, the development of low-carbon transportation is not only the requirement of the transportation industry to save energy and to reduce emissions and the impact on the environment, but also the requirement of the era for a low-carbon economy. It is the most effective direction of transportation development to deal with the serious energy consumption in society and prevent global warming.

The phrase the "Belt and Road Initiative"(BRI), which stands for the "silk road economic belt" and "marine silk road in the 21st century", was successively put forward by Chinese President Xi Jinping during his visit to Central Asia and Southeast Asian countries in September and October 2013. Then China's national development and reform commission, ministries of foreign affairs, and commerce jointly issued the report called "the vision and action of pushing to build the silk road economic belt and the maritime silk road into the 21st century". BRI involves 66 countries in Asia, Europe, and Africa. An in-depth study of the factors affecting the carbon emissions of the transport industry for these countries will help formulate relevant policies to control their carbon emissions of the transport industry and make contributions to the control of global warming.

At present, research on the carbon emission of the transportation industry in the BRI countries mainly focuses on the factors influencing the carbon emission and the low-carbon transportation development measures of specific countries and regions.

Regarding the research on influencing factors, Danish et al. studied the relationship between transport energy consumption, economic growth, foreign direct investment, and carbon dioxide emissions in urban transport sector in Pakistan [2]. Liang et al. analyzed the influence of energy structure, energy efficiency, transportation mode, transportation development, economic development, and population size on the CO_2 emissions of China's transportation sector from 2001 to 2014 [3]. Xu et al. analyzed the driving force of carbon dioxide emissions in China's transportation industry by vector autoregressive model and found that energy efficiency plays a leading role in reducing carbon dioxide emissions [4]. Song et al. built super-efficiency non-expected models, made evaluation analysis on transportation energy efficiency among 30 provinces of the China's eastern, central, and western parts, and found that the Chinese transportation industry's overall energy efficiency presents the anti-N type market decline, and the large discreteness maintains in the eastern and western provinces, and the strong consistency while the overall level is low in the northeast provinces [5]. Li et al. discussed the influencing factors of CO_2 emissions from road freight in China from 1985 to 2007, and concluded that economic growth is the most important factor for the increase of CO_2 emissions, while industrial added value and market concentration levels have a significant impact on the reduction of CO_2 emissions [6]. Wang et al. analyzed the driving factors of carbon emissions in China's transportation industry from 2000 to 2015 by using the generalized partition index method, and found that the added value, energy consumption and per capita carbon emissions of the transportation industry were the main factors leading to the increase of carbon emissions, and energy carbon emission intensity was the key factor to reduce carbon emissions [7]. Lin et al. used quantile analysis to investigate the impact of China's GDP, energy intensity, carbon intensity and urbanization on the carbon emissions of the transportation industry from 1980 to 2010, and found that GDP, energy intensity and carbon intensity had a greater impact on carbon emissions than urbanization [8]. Liu et al. investigated the impact of urbanization on road traffic energy use by using the data of 386 cities in Norway from 2006 to 2009, and found that the urban density level had a significant negative impact on per capita road energy use [9]. Fameli et al., after calculating the traffic carbon emissions in Attica, Greece, concluded that vehicle composition, popularity of diesel cars, urban speed, and vehicle renewal are the most effective parameters for formulating carbon emissions reduction policies [10]. Hassan et al. empirically analyzed the influence of aircraft technology, operational improvement, and sustainable biofuels on future carbon dioxide emissions from the perspective of aviation transportation industry [11]. Wu et al. used data from the years from 1949 to 2012, and applied vector autoregressive model to discuss the dynamic relationship among the interaction among China's transportation, economic growth and carbon emissions [12]. Xiao et al. introduced factors such as the proportion of added value of transportation industry, energy structure, and development level of transportation to analyze the direct and indirect impacts of various factors on carbon emissions of the transportation industry [13]. Lu et al. calculated the regional differences in CO_2 emissions of the transportation industry in nine provinces and two cities in the Yangtze river economic belt from three aspects: per capita carbon emissions, carbon emissions per unit of added value, and carbon emissions per unit of converted turnover [14].

Du et al., based on the panel data of carbon emissions of China's transportation industry in the past 20 years, concluded that economic level, transportation intensity, and energy intensity are the main factors affecting carbon emissions of the transportation industry [15]. Ou et al. analyzed the impact path of transportation technology progress on carbon dioxide emissions based on the multiple equilibrium theory and Stochastic Impacts by Regression on Population, Affluence, and Technology model [16]. Li et al. selected carbon emissions, per capita carbon emissions, and carbon intensity indexes from the spatial–temporal perspective to analyze the spatial agglomeration characteristics of carbon emission of China's transportation industry and the convergence characteristics with the change of time axis [17].

Regarding low-carbon transport development measures, Bakker et al. made a comparative analysis of the approaches and status quo of sustainable and low-carbon transport policies in Association of Southeast Asian Nations (ASEAN) countries based on the strategy component developed by Howlett and Cashore [18]. Selvakkumaran et al. adopted a recursive dynamic optimization model based on bottom-up modeling principle to analyze the energy utilization potential and emission reduction potential of Thailand's transportation sector under two scenarios of low-carbon social measures and emission tax clusters [19]. Fungtammasan et al. found that Thailand could achieve the national autonomous contribution (INDC) target by improving energy efficiency (especially in the transportation sector) and deploying the renewable energy and de-carbonization of power sectors [20]. Lah et al. believe that in order to minimize the rebound effect, a balanced and integrated policy approach is needed to significantly curb greenhouse gas emissions from the transportation industry by combining vehicle efficiency standards, fuel tax and differentiated vehicle tax with mode selection and compact urban design [21]. Shukla et al. showed that carbon dioxide emission reduction in transportation system can be achieved more effectively by combining emission reduction policies with measures to change traffic structure [22]. Rashidi et al. selected three low-carbon investment cases of environmental protection departments and transportation departments in Nairobi (Kenya), Balikpapan (Indonesia), and Colombo (Sri Lanka) to illustrate how to promote low-carbon investment in transportation and waste treatment in developing countries [23]. Lu et al. analyzed and calculated the total factor productivity and carbon emission efficiency of the transportation industry in Eastern China, and believed that the transportation industry in most provinces and cities still has a large space for energy conservation and emission reduction [24]. Wang et al. discussed the carbon emission characteristics of Beijing's commuting from different perspectives (macro and micro) and believed that reducing carbon dioxide emissions related to transfer is of great significance and necessity for the low-carbon emission of China's urban transportation development. In order to avoid the increasing carbon dioxide emissions related to commuting, the use of cars should be restricted and public transportation is a priority [25]. Dong et al. adopted life cycle assessment method to quantitatively analyze the environmental impact of Shenzhen urban public transport system (including buses and subways) from 2005 to 2015, in order to realize the low-carbon transformation of Shenzhen's urban transportation sector [26]. Gambhir et al. found that passenger cars and heavy trucks would constitute the majority of the carbon dioxide emission saving potential between 2010 and 2050 [27]. Diaz-ruiz-Navamuel et al. empirically analyzed the values of the reduction in emissions obtained and the advantages of installing Automatic Mooring Systems and operating RoRo/Pax terminals in commercial ports [28]. Wu et al. compared the environmental impact of the use of electric vehicles in developing and developed countries by taking Bric countries as an example and found that the electric car's impact on the environment is closely related to the power structure and transmission efficiency in the involved countries and regions because of the high proportion of thermal power in developing countries. Therefore, considered that in a sense, the development of the electric car industry in developing countries is transferring environmental pollutant emissions from vehicles to process of electricity production [29]. Shi et al. analyzed the emission reduction factors affecting electric vehicles, and also believed that the power generation energy structure and the power supply route of coal power technology play a decisive role in the carbon emission reduction space of the fuel life cycle of electric vehicles [30].

As for the research on country's development and carbon emissions of BRI, scholars outside China pay little attention to it, while Chinese scholars have conducted research from the fields of economy and culture. Lei et al. believed that in order to avoid repeating the mistake of sharp growth of carbon emissions after China's accession to the WTO, China must balance the relationship between economic growth and carbon emission reduction. Therefore, the relationship between China's economic growth and carbon emission reduction was studied from the perspective of synergy [31]. Xu et al. empirically investigated the influence of cultural dimensions on carbon emissions of 42 sample countries in the BRI from 2000 to 2013 under the framework of environmental Kuznets curve [32]. In the context of BRI, Zhang et al. studied the threshold effect of economic growth on carbon emissions by using panel data of 30 provinces (regions) in China from 2006 to 2015 and non-dynamic threshold panel estimation method [33]. Chen et al. discussed how the trade opening, green investment and energy cooperation under BRI would contribute to the low-carbon development of the countries in the BRI based on the current reality and the academic research results [34]. Xiao et al. selected the inter-provincial panel data of China from 2004 to 2015, and investigated the influence of the carbon emissions on provinces in the BRI by constructing the simultaneous equation model [35].

A comprehensive review of the existing studies shows that countries in the BRI pay great attention to low-carbon development, and study the relationship between relevant factors and carbon emissions from different time spans, regional scope, and professional fields, providing many references and ideas for the realization of global carbon emission reduction. However, most scholars study carbon emission from one country or region. Since BRI was proposed by the Chinese government, Chinese scholars pay more attention to BRI. However, Chinese scholars mainly study national economic pattern and strategic development issues from the perspective of theoretical framework and strategy, lacking quantitative research. A small number of studies are conducted from the economic and cultural fields. With the advance of the BRI construction, the carbon emission in the field of transportation will grow rapidly, and how to realize the carbon emission reduction in the field of transportation is extremely urgent. This paper intends to use the transnational panel data to calculate the carbon emissions of the transportation industry of the countries in the BRI, and use the econometric model to analyze the influencing factors of the carbon emissions of the countries in the BRI, and finally put forward suggestions to control the carbon emissions of the transportation industry.

2. Research Methods and Data Sources

2.1. Model Assumption

Panel Data is the data combining time series and cross section, in which time series data can reflect the dynamic changes of individuals, while cross section data can better reflect the differences between individuals. The panel data model combines the advantages of time series and cross-sectional data, and the application of panel data model for empirical research can provide more information, more changes, less collinearity, more degrees of freedom and efficiency, making up for the deficiency of two-dimensional data information in the classical linear econometric model. Panel data models allow for the construction and validation of more complex behavioral models than cross-sectional data or time series.

The general form of the panel data model is as follows:

$$y_{it} = \sum_{k=1}^{k} \beta_{ki} x_{kit} + \mu_{it}, \tag{1}$$

where, $i = 1, 2, \cdots N$, N is the number of individuals; $t = 1, 2, \cdots, T$, T is the number of known time points. y_{it} is the observed value of the explained variable to the individual i at the time t; x_{kit} is the observation value of the kth non-random explanatory variable for the individual i at the time t, β_{ki} is the parameter to be estimated; μ_{it} is the random error term.

2.2. Data Source and Data Processing

BRI includes 66 countries. In Asia, except Japan, South Korea, and North Korea, the remaining 46 countries have all joined BRI, including 19 countries in Europe, and Egypt in Africa. Through consulting the World Bank database and the International Energy Agency database, this paper obtained relevant data. Variable data in Laos, Syria, Palestine, Oman, Afghanistan, Iraq, Maldives, Bhutan, and Serbia was excluded for its overmuch missing. In this paper, after a small amount of missing data were processed by interpolation method, the variable data of 57 countries (see Table 1) in the BRI were obtained from 2005 to 2015.

Table 1. Countries in the "Belt and Road Initiative (BRI)".

Category	Region	Member Countries
High-income countries 19	Asia 8	Singapore, Brunei, Israel, Saudi Arabia, UAE, Qatar, Kuwait, Bahrain
	Europe 11	Greece, Cyprus, Poland, Lithuania, Estonia, Latvia, Czech Republic, Slovakia, Hungary, Slovenia, Croatia
Middle- and high-income countries 19	Asia 9	China, Malaysia, Thailand, Iran, Turkey, Jordan, Lebanon, Kazakhstan, Turkmenistan
	Europe 10	Russia, Belarus, Georgia, Azerbaijan, Bosnia and Herzegovina, Montenegro, Albania, Romania, Bulgaria, Macedonia
Low- and middle-income countries 19	Asia 16	Indonesia, Vietnam, Myanmar, Cambodia, Philippines, Mongolia, Yemen, Egypt, India, Pakistan, Bangladesh, Sri Lanka, Nepal, Uzbekistan, Tajikistan, Kyrgyzstan
	Europe 3	Ukraine, Armenia, Moldova

Note: data were obtained from the World Bank database in 2015.

2.2.1. Explained Variables and Their Data Sources

Carbon emissions is the general term of greenhouse gas emissions, expressed in carbon dioxide equivalent (CO_2eq), including carbon dioxide, nitrous oxide, freon, and methane, of which carbon dioxide emissions account for 60% of the total greenhouse gas emissions, so it is the main greenhouse gas causing global warming, therefore having been studied by most scholars. In this paper, total carbon dioxide emissions are used to measure the carbon emissions level of the transportation industry of the BRI countries. Due to the incompleteness of global carbon emissions data, most scholars collect energy consumptions and calculate formulas to obtain carbon emissions data. Therefore, the specific expression formula of carbon emissions adopted in this paper is as follows:

$$CE = \sum_{i}^{n} CO_{2i} = \sum_{i}^{n} E_i \times \delta_i = \sum_{i}^{n} E_i \times NCV_i \times CEF_i \times COF_i \times \frac{44}{12}. \tag{2}$$

In which, CE stands for carbon dioxide emissions from transportation industry; i is the type of fossil fuel, The IEA database classifies fuels consumed by the transportation industry into five categories: coal, petroleum products, biomass energy, natural gas and electricity. E_i refers to energy consumption of fossil fuel i; δ_i is the carbon dioxide emission coefficient of carbon energy i; NCV_i is the average low calorific value of energy i; CEF_i is the carbon emission coefficient of energy i, namely, the carbon content per unit of heat; COF_i is the carbon oxidation factor, that is, the carbon oxidation rate during energy combustion. 44 and 12 are the molecular weights of carbon dioxide and carbon.

According to the national greenhouse gas inventory guidelines of Intergovernmental Panel on Climate Change (IPCC), the carbon emission coefficients of various energy sources are shown in Table 2.

Table 2. Carbon emission coefficient of transportation and energy.

Types of Energy	Average Low Calorific Value (KJ/kg, m^3)	Carbon Oxidation Rate	CO_2 Emission Factor (kgCO$_2$/GJ)	Carbon Emission Factor (kg C/GJ)
Coal	20,908	1	94.6	25.8
Oil products	43,070	1	72.35	19.7
Biomass energy	42,338	1	75.18	20.5
Natural gas	38,931	1	56.1	15.3
Electric power	—	—	—	—

Note: data source: Intergovernmental Panel on Climate Change (IPCC) 2006 edition.

Description: since electricity belongs to secondary energy, the method to calculate the carbon emission of electricity in this paper is to convert the energy consumption of electricity into equivalent standard coal, and then use the carbon emission of standard coal to represent the carbon emission of electricity.

According to the date from Organization for Economic Co-operation and Development (OECD) database, the carbon emission changes of the 57 countries in the BRI from 2005 to 2015 is shown in Figure 1, the 57 countries accounted for about 50.06% of the world's carbon emissions in 2015, and China alone accounted for 28.01%, far more than the sum of the carbon emissions of the other 56 countries. Therefore, reducing the carbon emissions of the BRI countries is of great significance to the global carbon emission reduction.

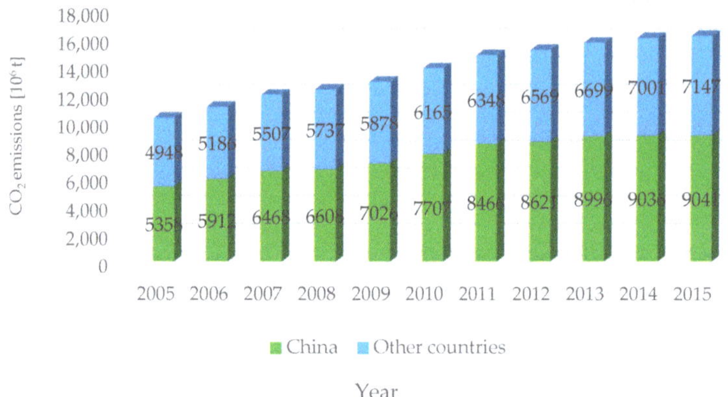

Figure 1. 2005–2015 BRI national carbon emissions.

Figure 2 shows the changes in the top 20 carbon emissions of the BRI countries from 2005 to 2015. In terms of growth rate, China and India, the two countries in the BRI, saw the largest growth rate of carbon emissions in their transportation industry, doubling in 11 years, which was related to their rapid economic growth. In other countries, the changing trend of the transportation industry carbon emission is stable. In terms of total carbon emission, China, Russia, and India are significantly higher than other countries. China's total carbon emission of the transportation industry in 2015 was 871.56 million tons, roughly equivalent to the total carbon emission of the transportation industry of the second to fourth countries. Russia's carbon emissions from its transportation industry has remained at around 3.3 million tons over the years, while India, a fast-growing economy, has seen a significant increase in its total carbon emissions.

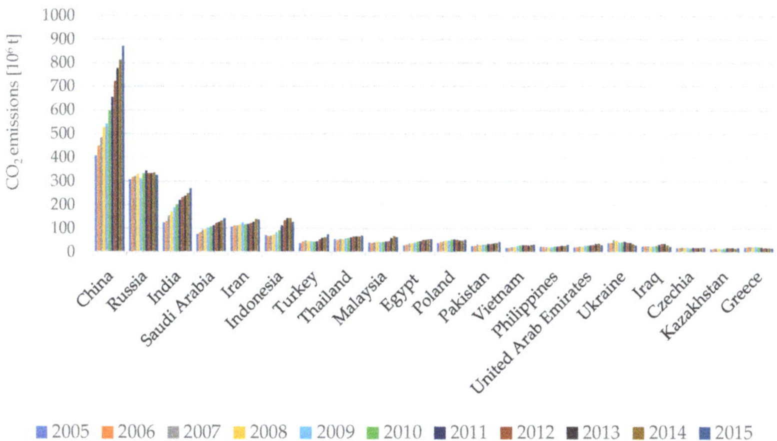

Figure 2. 2005–2015 BRI national transportation carbon emission.

2.2.2. Explanatory Variables and Their Data Sources

According to the above literatures, the explanatory variables selected in the carbon emission research of the transportation industry are usually as follows: energy indicators like energy consumption, energy structure, energy efficiency, energy carbon emission intensity, per capita carbon emission, carbon intensity, and energy intensity; economic indicators such as: economic growth, economic development, GDP, economic level, foreign direct investment; indicators of the transportation industry include transportation mode, transportation intensity, transportation development, added value of transportation industry, transportation development level, and technological progress of transportation; Macro indicators such as population size and urbanization.

Some explanatory variables share some certain similarities, such as economic indicators: economic growth, economic development, GDP, economic level; energy indexes: energy and carbon intensity, per capita carbon emissions, carbon intensity, energy intensity and traffic development level index: transportation development, added value of transportation industry, and traffic development level, while it is so microcosmic for modes of transportation, the intensity of transportation with indicators of technological progress in transportation that they are not suitable for the macroscopical study of carbon emissions of BRI transportation industry.

Therefore, for the research on the influencing factors of carbon emissions of the BRI national transportation industry, this paper preliminarily selected per capita GDP, urbanization level, technology level, energy consumption structure, net inflow of foreign capital, industry proportion, and trade openness as explanatory variables, the explanatory variables selected cover most of the explanatory variables used in the references, and the technical level and energy consumption structure were newly-added as explanatory variables.

Since there would likely be multicollinearity problems among model variables, stepwise regression was conducted to determine the optimal explanatory variables before data processing. The specific method was to gradually introduce other explanatory variables under the initial model. After introducing variables, the determination coefficient $R2$ of the model was gradually improved on the basis of 0.963. However, t test and p values of foreign capital inflow (FDI) and carbon emissions proportion of the transportation industry did not pass the test at the significance level of 10%, so they were excluded.

Therefore, per capita GDP, urbanization level, technological level, energy consumption structure and trade openness are selected as explanatory variables in this paper. All data are from the World Bank database.

Per capita GDP is the ratio of GDP to the total population, which is used to reflect the size of the economy. Due to the different economic sizes of countries in the world, it is not scientific to simply consider GDP, therefore, per capita GDP (2010 constant dollar) was selected as the explanatory variable in this paper. Figure 3 shows the per capita GDP of the BRI countries in 2015. It can be seen from the figure that the countries with higher per capita GDP are mainly distributed: the oil producing countries in West Asia, Brunei and Singapore in Asia, and the EU members in Eastern Europe, while countries with low per capita GDP mainly distributed in Southeast Asia, South Asia, and Central Asia.

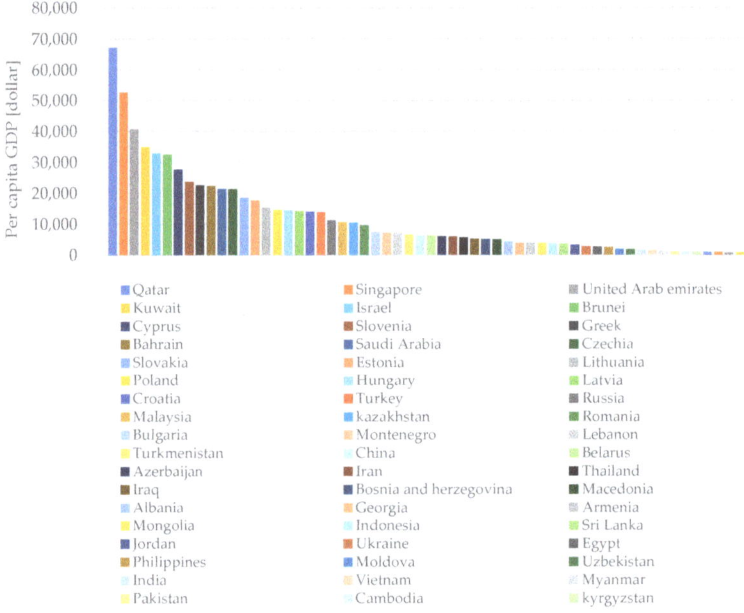

Figure 3. Per capita GDP of some countries in the BRI in 2015.

The urbanization level is represented by the proportion of urban population in the total population. Figure 4 shows the urbanization level of regions in the BRI in 2015, which reflects strong regional characteristics: the urbanization level of West Asia was close to 80%, then Central and Eastern Europe, East Asia, and cis countries (the national league built up by several Former Soviet Republics consists of 9 countries such as Russia and Belarus), and general South Asia, with the lowest urbanization level less than 30%.

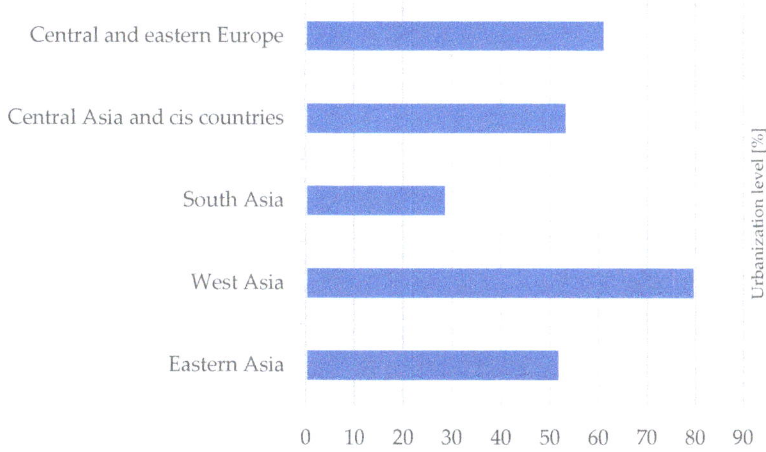

Figure 4. Urbanization rate in the BRI in 2015.

The technical level is represented by the percentage of high-tech export in the export of manufactured goods. High-technology exports are products with high research and development intensity, such as in aerospace, computers, pharmaceuticals, scientific instruments, and motors, reflecting the manufacturing level of a country's high-tech products. Figure 5 shows the technical level ranking of the BRI countries in 2015. The high-ranking countries were mainly concentrated in two regions: one in South Asia and Southeast Asia, and the other in Eastern Europe. The Philippines in Southeast Asia ranked the highest, followed by Singapore and Malaysia. The reason is that with superior geographical conditions, these countries have developed export processing industries. While the Eastern European region is located in the economic circle of the EU with its own high scientific and technological level and strong high-tech R&D and production capacity, so high-tech exported products account for a high proportion.

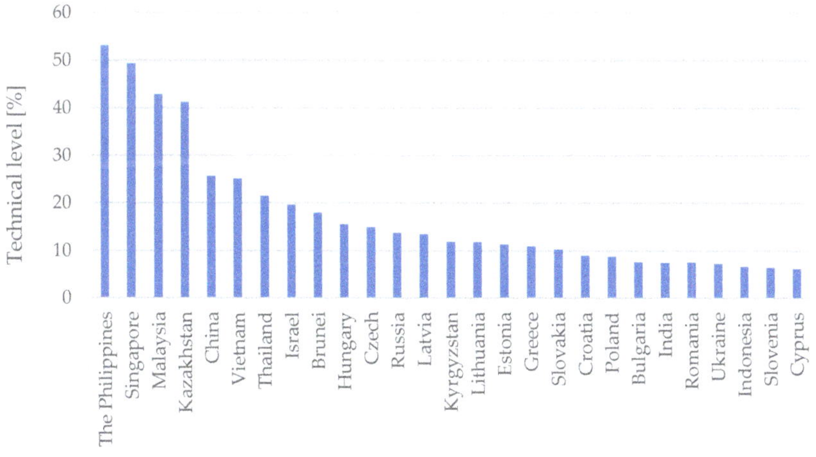

Figure 5. High-tech exports rank of the BRI countries in 2015

The energy consumption structure is expressed as the percentage of fossil energy consumed by the transportation industry in the total energy consumption. Figure 6 is the change trend of energy consumption structure of the transportation industry in the BRI countries. From 2005 to 2015, the proportion of fossil energy consumption of the BRI national transportation industry showed a slight decline trend, but the average proportion of fossil energy consumption was still as high as 97%, and the use of clean energy was very low.

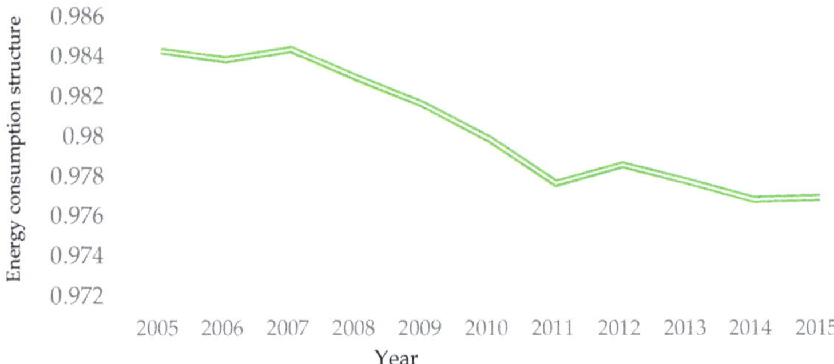

Figure 6. Changing trend of the BRI national transportation industry energy consumption structure from 2005 to 2015.

Trade openness is expressed as the ratio of trade volume to GDP. The trade volume is the sum of imports and exports of goods and services. Figure 7 is a part of the BRI national trade openness, containing the countries whose import and export trades are more than 100% and less than 50% of the total. It can be seen from the figure, countries with high trade openness are mainly distributed within two economic circles of the ASEAN and the European Union, which show that the facilitation of regional trade can promote the growth of trade. The top countries in terms of GDP, such as China, India, Russia, Indonesia, and Turkey, are relatively low in terms of trade as a share of GDP.

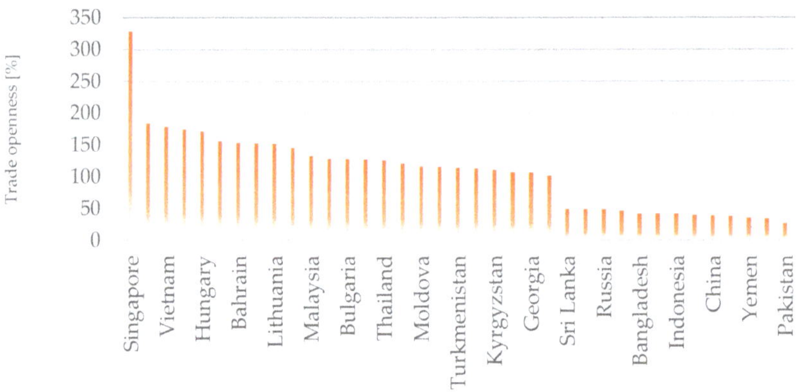

Figure 7. BRI countries' trade openness in 2015.

3. Model specification

Introduce the following notations, CE represents carbon emissions, PGDP represents per capita GDP, UL represents urbanization level, HE represents technical level, ECS represents energy consumption structure, TO represents trade openness.

3.1. Descriptive Statistics

Based on the statistics, we list the descriptive statistics of variables in Table 3.

Table 3. Descriptive statistics of variables.

Statistics	CE	PGDP	UL	HE	ECS	TO
Mean	37.8998	11,546.1500	58.3852	9.0348	98.0092	99.5843
Median	8.1200	6029.3970	55.5370	4.3997	99.0553	94.0826
Maximum	871.5600	72,670.9600	100.0000	98.7291	100.0000	441.6038
Minimum	0.1500	502.2394	15.1830	0.0000	87.7193	0.1674
Std. Dev.	97.0310	13,769.6000	20.9816	13.2503	2.3993	54.5934
Kurtosis	34.2811	7.3023	2.2997	12.7772	3.4220	14.8253

Note: all data are based on the World Bank database. CE: carbon emissions; PGDP: per capita GDP; UL: urbanization level; HE: technical level; ECS: energy consumption structure; TO: trade openness.

3.2. Panel Data Unit Root Test Results

In order to ensure the effectiveness of parameter estimation and avoid the occurrence of "pseudo-regression", the stationarity of data should be tested before establishing the model. In order to ensure the reliability of the test results, this paper applies the Levin–Lin–Chu (LLC) in the same root case and the Fisher-ADF and Fisher-PP methods in the different root cases to conduct the unit root test. The co-integration test can only be carried out when the variables are single integrals of the same order.

Multiple test results show that the horizontal value of 6 variables cannot reject the null hypothesis, that is, the horizontal value of variables has unit root and is a non-stationary sequence. However, the first-order difference of the six variables all rejected the null hypothesis at the significance level of 1%. Therefore, the variables were all first-order single integral sequences, and the co-integration analysis could be continued.

3.3. Co-Integration Test Results of Panel Data

In order to test whether there is a long-term stable relationship between variables and carbon emissions, the co-integration test of panel data is required. Using Kao test [36], the null hypothesis is that there is no co-integration relationship. The results are shown in Table 4. The test results reject the null hypothesis at the significance level of 5%, and there is a long-term stable relationship between carbon emission and all variables.

Table 4. Kao Test.

Test Method	Testing Hypothesis	Statistic Name	t-Statistic	Prob.
Kao test	H0: no co-integration relationship	ADF	3.0521	0.0011

3.4. Panel Model Setting

Because of the panel data's dimensionality, if the model set is not correct, and the resulting parameter estimation method is undeserved. This article examines the types of models in two steps. Firstly, we carried out a Redundant test of Fixed Effect-likelihood ratio [37], which is also called Fixed effect redundancy test (constraint test), to determine whether there is an effect and the number of effects. The test results are shown in Table 5:

Table 5. Redundant Fixed Effects-likelihood ratio test.

Effects Test	Statistic	Prob.
Cross-section F	208.5271 ***	0.0000
Cross-section chi-square	1939.2385 ***	0.0000
Period F	0.1091	0.9997
Period Chi-square	1.2318	0.9996
Cross-Section/Period F	177.5094 ***	0.0000
Cross-Section/Period Chi-square	1941.1888 ***	0.0000

Note: p value in brackets; *, ** and *** represent the significance levels of 10%, 5%, and 1% respectively.

From the test results, the p value of time fixed effects regression model test was >0.05, therefore, we accept the null hypothesis—there was no fixed effect at the time point. The individual test $p < 0.05$, so, reject the null hypothesis that the individual has a fixed effect; therefore, the model exists an individual single factor effect.

According to the different ways of the individual influence, models can be divided into two kinds: fixed effect model and random effect model. This paper uses the Hausman test method [38] to determine the model affect the form, and the null hypothesis of the model is that a random effect model should be established. Table 6 indicates the result the p value is less than 0.01, therefore, reject the null hypothesis and adopted fixed effect model.

Table 6. Hausman test of panel data.

Test Summary	Chi-Sq. Statistic	Chi-Sq. d.f.	Prob.
Cross-section random	93.6754	5	0.0000

Based on the comprehensive analysis, the regression model of fixed individual (i.e., variable intercept) should be established between the explanatory variable and the explained variable of the BRI national transportation industry carbon emission from 2005 to 2015. In addition, from the perspective of actual research needs, the research object is the transportation carbon emissions and other variables of 57 countries, and there is no problem with random sampling from the population. Therefore, compared with the random effect model, it is appropriate to establish the individual fixed effect model too.

3.5. Panel Model Regression Result

According to the test results above, the individual fixed effect model should be established in this paper. The model and parameter estimation results are as Table 7:

Table 7. Regression result.

Variable	Coefficient	Std. Error	t-Statistic	Prob.
C	−535.3166	83.1595	−6.4372	0.0000
PGDP	0.0011	0.0002	4.7791	0.0000
UL	7.7284	0.3067	25.1961	0.0000
HE	−0.2182	0.0795	−2.7467	0.0062
ECS	1.1906	0.6861	1.7345	0.0334
TO	−0.0481	0.0099	−4.8707	0.0000
R-squared	0.9636	Log likelihood	−2719.2810	
Adjusted R-squared	0.9596	F-statistic	244.9956	
S.E. of regression	19.4938	Prob(F-statistic)	0.0000	
Sum squared residuals	214,703.6000			

The regression result shows that what the adjusted R^2 very close to 1 indicates a good regression fitting effect. F value is greater than the critical value of 5% significance level, indicating that the linear relationship of the model is significant. T value of each parameter was greater than the critical value at significance level = 0.05, indicating that each explanatory variable had significant influence on the equation.

4. Empirical Analysis

According to the regression results, every 1% is increased in per capita GDP, the carbon emissions of the transportation industry increased by 0.0011%. For every 1% increase in urbanization level, the carbon emission of the transportation industry will rise by 7.7284%. For every 1% increase in the proportion of fossil energy consumption in the transportation industry, the carbon emissions of the transportation industry will grow by 1.1901%. For every 1% increase in technology level, carbon emissions of the transportation industry will be reduced by 0.2182%. Trade openness was increased by 1%, and carbon emissions from transportation industry were reduced by 0.0481%.

The change of urbanization level has a great influence on the carbon emissions of the transportation industry in the countries in the BRI. According to the analysis results in Figure 4, by 2015, the average urbanization rate of the countries in the BRI is 59.13%, among which half of the countries are lower than 50%, and one-third of the countries are lower than 35%, while in the same period, the urbanization level of major developed countries has reached over 80%. Most of the BRI countries are at the initial stage of urbanization, and the urbanization level will continue to increase, which will lead to the continuous increase of carbon emissions of the transportation industry in these countries.

The energy consumption structure of the transportation industry is an important driving factor of transportation carbon emission in countries in the BRI. According to the analysis results in Figure 6, by 2015, the fossil energy consumption of the transportation industry in countries in the BRI was as high as 97.69%, and the use of clean energy was very low. In some major developed countries, for example, in 2015 the United States, 93.98% of the total energy consumption is from the transportation industry, Germany (93.53%), and France (91.59%), so there still exists a large space for optimizing transportation energy consumption structure in the BRI countries, which makes the energy structure optimization of the transportation industry become an important driving factor of reducing carbon emissions.

It cannot be underestimated that per capita GDP holds a lot of weight for carbon emissions of the transportation industry in the BRI countries. BRI countries account for 62.2% of the world's population and 30.9% of the world's GDP. Per capita GDP is 1/2 of the world average and about 1/3 of that of the European Union. In recent years, per capita GDP of 47 countries in the BRI has shown an upward trend, and the growth of per capita GDP has become an important factor driving the growth of carbon emissions.

Improving the technical level has an obviously inhibitory effect on the carbon emissions of the transportation industry in the BRI countries. From the development trend in recent years, only individual countries such as Vietnam, Israel, and Kazakhstan, with high-tech exports accounting for the obvious rising trend in the proportion of manufactured goods exported, while most countries witnessed a volatile downward trend. Therefore, the BRI countries should increase the R&D of high-tech products, promoting the industry from low-end to high-end. Impulse optimization and upgrading of industrial structure transformation will make it become a significant driving factor of reducing carbon emissions.

The improvement of trade openness has an inhibitory effect on the carbon emissions of the transportation industry in the BRI countries. According to the data collected, the overall level of trade openness of the BRI countries is greatly affected by the world political and economic situation. In 2008, with the acceleration of the globalization of the world economy, the trade openness level of the BRI countries witnessed a steady rise, however, because of the influence of the U.S. financial crisis, their trade openness plunged by 13.28% in 2009. From 2009 to 2012, the world economic situation improved, and trade openness level increased year by year, while from 2013 to 2015, trade openness level showed a downward trend again for the rise of trade protectionism, but the globalization trend is irreversible, and improving trade openness will still be an important factor to curb carbon emissions.

The individual effects intercept items of the model are shown in Table 8. The meaning of the term "intercept" is the level of carbon emission when all explanatory variables are equal to zero.

Table 8. Fixed effects (cross) intercept item.

Country	Intercept	Country	Intercept	Country	Intercept
China	665.7295	Qatar	−402.8491	Georgia	17.1848
Singapore	−359.1432	Kuwait	−370.7331	Azerbaijan	11.9394
Malaysia	−72.8431	Bahrain	−280.0096	Armenia	−70.2872
Indonesia	134.9449	Greece	−172.5202	Moldova	76.8339
Myanmar	176.4387	Cyprus	−123.5516	Poland	−6.8274
Thailand	146.9235	Egypt	128.0329	Lithuania	−94.7458
Cambodia	270.8803	India	380.5904	Estonia	−114.0947
Vietnam	217.3133	Pakistan	167.9745	Latvia	−106.6167
Brunei	−195.1263	Bangladesh	189.0075	Czech	−134.5726
Philippines	107.6275	Sri Lanka	281.5551	Slovakia	0.6559
Mongolia	−99.1237	Nepal	289.5090	Hungary	−97.9231
Iran	−11.0146	Kazakhstan	24.7058	Slovenia	21.1239
Turkey	−87.0962	Uzbekistan	151.4763	Croatia	−29.0579
Jordan	−212.3560	Turkmenistan	56.7199	Bosnia and Herzegovina	116.1757
Lebanon	−255.3461	Tajikistan	218.1378	Montenegro	−71.3814
Israel	−303.1303	Kyrgyzstan	152.0016	Algeria	16.3055
Saudi Arabia	−125.9317	Russia	176.2817	Armenia	18.1747
Yemen	179.3969	Ukraine	−62.2023	Bulgaria	−130.3784
United Arab Emirates	−239.8403	Belarus	−142.6494	Macedonia	−22.2893

This reflects the influence of the neglected variables representing the difference of cross section in the model. According to the cross-sectional data in Table 8, the value of 29 countries including China, India, Nepal, Bangladesh, Indonesia, Thailand, and Vietnam, is positive: it indicates that the neglected variables in the model have a greater impact on the carbon emission of the transportation industry in these countries. While 28 countries, including Singapore, Israel, Turkey, and Brunei, have negative values, it indicates that the neglected variables in the model have little impact on the carbon emission of the transportation industry in these countries.

5. Conclusions and Suggestions

5.1. Conclusions

Based on the 2005–2015 data of 57 BRI countries, this paper, using panel data model, coming to the following main conclusions: there exists a long-term and stable relationship among transport carbon emissions, the level of urbanization, transportation energy consumption structure, per capita GDP, technical level, and trade openness in the BRI countries. For every 1% increase in per capita GDP, 0.0011% of the transportation industry carbon emissions are added. For every 1% increase in urbanization level, the carbon emissions of the transportation industry will increase by 7.728%. For every 1% increase in the proportion of fossil energy consumption in the transportation industry, the carbon emissions of which will increase by 1.1901%. For every 1% increase in technology level, carbon emissions of the transportation industry will be reduced by 0.2182%. Trade openness was increased by 1%, and carbon emissions from transportation industry were reduced by 0.0481%.

Therefore, for West Asia and Eastern Europe with their high urbanization level and per capita GDP, the main way to reduce the carbon emissions of the transportation industry is to enrich the energy consumption structure, such as the use of biomass energy, natural gas, solar energy, and other clean energy, and gradually reduce the proportion of fossil energy in the energy consumption of the transportation industry. For South Asia, with its low urbanization level, special attention should be paid to the negative impact of urban development on traffic carbon emissions in the process of urbanization; not only should the energy consumption structure be optimized, but also the excessive growth of carbon emissions be reasonably controlled, which is caused by fossil energy consumption in the process of urbanization. East Asia and Southeast Asia, which have developed rapidly, should further promote the reduction of carbon emissions by improving the level of foreign trade. On the whole, countries in the BRI are different in resource endowment, economic, and technological level. In the process of BRI development, increasing the proportion of clean energy consumption, expanding trade openness, and improving the level of science and technology will curb the carbon emissions of the BRI transportation industry.

5.2. Policy Suggestions

First, traffic carbon emissions caused by urbanization should be properly controlled. From the research conclusion, urbanization has a great impact on carbon emissions. For countries in the BRI, urbanization is an irreversible process, so reasonable measures should be taken to control the increase of carbon emissions caused by the increase of urbanization level. The strategies include vigorously developing public transport, establishing a low-carbon transport mode dominated by public transport, and building a comprehensive public transport system with rail transit as the skeleton, conventional transport as the meridian, taxi as the supplement, and slow traffic as the extension, guiding citizens to choose "walk + public transport" and "bike + public transport" travel modes, reducing the use frequency of cars.

Second, the share of fossil energy consumption in the transportation industry should be reduced. At present, the average proportion of fossil energy consumption in the transportation industry of the BRI countries is about 97%. In the past 10 years, the proportion of fossil energy in the transportation industry has not decreased much, the transportation industry need optimizing energy consumption structure, reducing carbon emissions. Therefore, BRI countries need to promote new energy vehicles such as hybrid, pure electric, and fuel cells, to improve the electrification rate of railway construction, to encourage the use of green ships based on new technologies and new energy sources, to actively promote the use of aviation biofuels.

Third, trade openness should be increased. Trade protectionism is on the rise in the modern international community. From the perspective of reducing carbon emissions of the transportation industry, trade openness needs to be enhanced. Countries in the BRI should further reduce tariffs and non-tariff barriers through bilateral or multilateral economic cooperation mechanisms, and reduce market access barriers to commodity flows, carry out cooperation in customs clearance, and actively promote the integration of regions and customs clearance. Making full use of modern information, network, and communication technologies should improve the efficiency of customs clearance from the perspectives of optimizing the process of customs clearance and improving the efficiency of key links.

Fourth, the technical level of the BRI countries should be improved. In today's world, a new round of scientific and technological revolution and industrial transformation is looming, becoming a new driving force for world economic growth. Driving the optimization and upgrading of industrial structure with high-tech industry as the driving force and modern service industry and modern manufacturing industry should be the direction of development.

Author Contributions: C.Z. and D.G. wrote original draft.

Funding: This research was funded by Shaanxi natural science foundation project (Grant No. 2019JQ-533), and Xi'an science and technology plan (Grant No. 2017111SF/RK005-5).

Acknowledgments: The authors thank Wenbo Du for his help in statistical analysis and participation in data collection.

References

1. International Energy Agency (IEA). *CO$_2$ Emissions from Fuel Combustion Highlights 2017*; International Energy Agency: Paris, France, 1 October 2017; ISBN 9789264278196.
2. Danish; Baloch, M.A.; Suad, S. Modeling the impact of transport energy consumption on CO_2 emission in Pakistan: Evidence from ARDL approach. *Environ. Sci. Pollut. Res.* **2018**, *25*, 9461–9473. [CrossRef] [PubMed]
3. Liang, Y.; Niu, D.X.; Wang, H.C.; Li, Y. Factors Affecting Transportation Sector CO_2 Emissions Growth in China: An LMDI Decomposition Analysis. *Sustainability* **2017**, *9*, 1730. [CrossRef]
4. Xu, B.; Lin, B.Q. Carbon dioxide emissions reduction in China's transport sector: A dynamic VAR (vector autoregression) approach. *Energy* **2015**, *83*, 486–495. [CrossRef]
5. Song, Z.; Cong, L. Study on energy efficiency of China's transportation industry under environmental constraints. *J. Transp. Syst. Eng. Inf. Technol.* **2016**, *16*, 39–45.
6. Li, H.Q.; Lu, Y.; Zhang, J.; Wang, T.Y. Trends in road freight transportation carbon dioxide emissions and policies in China. *Energy Policy* **2013**, *57*, 99–106. [CrossRef]
7. Wang, Y.; Zhou, Y.; Zhu, L.; Zhang, F. Influencing Factors and Decoupling Elasticity of China's Transportation Carbon Emissions. *Energies* **2018**, *11*, 1157. [CrossRef]
8. Lin, B.Q.; Benjamin, N.I. Influencing factors on carbon emissions in China transport industry. A new evidence from quantile regression analysis. *J. Clean. Prod.* **2017**, *150*, 175–187. [CrossRef]
9. Liu, Y.P.; Huang, L.Z.; Kaloudis, A. Does urbanization lead to less energy use on road transport? Evidence from municipalities in Norway. *Transp. Res. Part D Transp. Environ.* **2017**, *57*, 363–377. [CrossRef]
10. Fameli, K.M.; Assimakopoulos, V.D. Development of a road transport emission inventory for Greece and the Greater Athens Area: Effects of important parameters. *Sci. Total Environ.* **2015**, *505*, 770–786. [CrossRef]
11. Hassan, M.; Pfaender, H.; Mavris, D. Probabilistic assessment of aviation CO_2 emission targets. *Transp. Res. Part D Transp. Environ.* **2018**, *63*, 362–376. [CrossRef]
12. Wu, J.G.; Ye, A.Z. A study on the dynamic relationship among transport, economic growth and carbon emissions—based on empirical analysis of data from 1949 to 2012. *J. Transp. Syst. Eng. Inf. Technol.* **2015**, *15*, 10–17.
13. Xiao, Z.W.; Yuan, C.W.; Yun, H. Path analysis model of carbon emission of China's transportation industry. *Stat. Dec.* **2016**, *15*, 25–29.
14. Lu, S.R.; Jiang, H.Y.; Liu, Y. Regional differences and influencing factors of CO_2 emission in transportation industry. *J. Transp. Syst. Eng. Inf. Technol.* **2017**, *17*, 32–39.
15. Du, Q.; Su, Q.; Yang, Q.; Feng, X.Y.; Yang, J. Path analysis method for carbon emission drivers in China's transportation industry. *J. Traff. Transp. Eng.* **2017**, *17*, 143–150.
16. Ou, G.L.; Wang, Y. The impact of technological progress in transportation on carbon dioxide emissions—Based on China's provincial panel data. *Ecol. Econ.* **2018**, *34*, 64–71.
17. Li, W.; Sun, W. Spatial and temporal distribution characteristics of carbon emission of provincial transportation industry. *Syst. Eng.* **2016**, *34*, 30–38.
18. Bakker, S.; Contreras, K.D.; Kappiantari, M.; Tuan, N.A.; Guillen, M.D.; Gunthawong, G.; Zuidgeest, M.; Liefferink, D.; van Maarseveen, M. Low-Carbon Transport Policy in Four ASEAN Countries: Developments in Indonesia, the Philippines, Thailand and Vietnam. *Sustainability* **2017**, *9*, 1217. [CrossRef]
19. Selvakkumaran, S.; Limmeechokchai, B. Low carbon society scenario analysis of transport sector of an emerging economy-The AIM/Enduse modelling approach. *Energy Policy* **2015**, *81*, 199–214. [CrossRef]
20. Fungtammasan, B.; Tippichai, A.; Otsuki, T.; Tam, C. Transition pathways for a sustainable low-carbon energy system in Thailand. *J. Renew. Sustain. Energy* **2017**, *9*, 021405. [CrossRef]
21. Lah, O. The barriers to low-carbon land-transport and policies to overcome them. *Eur. Transp. Res. Rev.* **2015**, *7*, 5. [CrossRef]
22. Shukla, P.R.; Dhar, S. Energy policies for low carbon sustainable transport in Asia. *Energy Policy* **2015**, *81*, 170–175. [CrossRef]
23. Rashidi, K.; Stadelmann, M.; Patt, A. Valuing co-benefits to make low-carbon investments in cities bankable: the case of waste and transportation projects. *Sustain. Cities Soc.* **2017**, *34*, 69–78. [CrossRef]
24. Lu, X.Y.; Ma, X.M.; Xiong, S.Q. Analysis on environmental efficiency of transportation industry in east China. *Mod. Manag.* **2017**, *3*, 88–91.

25. Wang, H.H.; Zeng, W.H. Revealing Urban Carbon Dioxide (CO_2) Emission Characteristics and Influencing Mechanisms from the Perspective of Commuting. *Sustainability* **2019**, *11*, 385. [CrossRef]
26. Dong, D.; Duan, H.B.; Mao, R.C.; Song, Q.B.; Zuo, J.; Zhu, J.S.; Wang, G.; Hu, M.W.; Dong, B.Q.; Liu, G. Towards a low carbon transition of urban public transport in megacities: A case study of Shenzhen, China. *Resour. Conserv. Recycl.* **2018**, *134*, 149–155. [CrossRef]
27. Gambhir, A.; Tse, L.K.C.; Tong, D.L.; Martinez-Botas, R. Reducing China's road transport sector CO_2 emissions to 2050: Technologies, costs and decomposition analysis. *Appl. Energy* **2015**, *157*, 905–917. [CrossRef]
28. Diaz-Ruiz-Navamuel, E.; Piris, A.O.; Perez-Labajos, C.A. Reduction in CO_2 emissions in RoRo/Pax ports equipped with automatic mooring systems. *Environ. Pollut.* **2018**, *241*, 879–886. [CrossRef] [PubMed]
29. Wu, Y.; Zhang, L. Can the development of electric vehicles reduce the emission of air pollutants and greenhouse gases in developing countries? *Transp. Res. Part D Trans. Environ.* **2017**, *51*, 129–145. [CrossRef]
30. Shi, X.Q.; Li, X.N.; Yang, J.X. Analysis on carbon emission reduction potential of low-carbon transportation electric vehicles and its influencing factors. *Environ. Sci.* **2013**, *34*, 385–394.
31. Lei, Y.; Zhang, W.L.; Qu, L. Coordinated research on China's economic growth and carbon emission reduction under the "One Belt One Road" strategy. *J. Hohai Univ. (Philos. Soc. Sci.)* **2016**, *18*, 23–29.
32. Xu, R.N.; Wu, Y.M. How does culture affect carbon emissions in countries along the "One Belt One Road"?—Empirical study under the framework of environmental Kuznets curve. *Ecol. Econ.* **2018**, *34*, 14–19.
33. Zhang, Q.R.; Wang, H.B.; Lu, Y.Y. Research on threshold effect of China's economic growth and carbon emission under "One Belt One Road". *Chin. Coal* **2018**, *44*, 26–31.
34. Chen, Z. Research on the significance and path of realizing low-carbon development strategy of countries along the "One Belt One Road". *Mod. Manag. Sci.* **2019**, *3*, 27–29.
35. Xiao, D.; Chen, W. Study on carbon emission effect of overseas direct investment in provinces along "One Belt One Road". *J. Hubei Univ. (Philos. Soc. Sci.)* **2018**, *45*, 118–125.
36. Kao, C. Spurious regression and residual-based tests for cointegration in panel date. *J. Econom.* **1999**, *90*, 1–44. [CrossRef]
37. Liu, M.Z.; Liu, X.X. Factors affecting Chinese urban household energy consumption and spatial differences based on static panel data modelling for eight regions. *Resour. Sci.* **2016**, *12*, 2295–2306.
38. Hausman, J.A.; Taylor, W.E. Panel date and unobservable individual effects. *Econometrica* **1981**, *49*, 1377–1398. [CrossRef]

Article

A Research on Driving Factors of Carbon Emissions of Road Transportation Industry in Six Asia-Pacific Countries Based on the LMDI Decomposition Method

Changzheng Zhu * and Wenbo Du

School of Modern Post, Xi'an University of Posts & Telecommunications, Xi'an 710061, China; dwb694845437@outlook.com

* Correspondence: zhuchangzheng@xupt.edu.cn; Tel.: +86-182-9207-8891

Received: 2 August 2019; Accepted: 29 October 2019; Published: 31 October 2019

Abstract: The transportation industry is the second largest industry of carbon emissions in the world, and the road transportation industry accounts for a large proportion of this in the global transportation industry. The carbon emissions of the road transportation industry in six Asia-Pacific countries (Australia, Canada, China, India, Russia, and the United States) accounts for more than 50% of this in the global transportation industry. Therefore, it is of great significance to study driving factors of carbon emissions of the road transportation industry in six Asia-Pacific countries for controlling global carbon emissions. In this paper, the Logarithmic Mean Divisia Index (LMDI) decomposition method is adopted to analyze driving factors on carbon emissions of the road transportation industry in six Asia-Pacific countries from 1990 to 2016. The results show that carbon emissions of the road transportation industry in these six Asia-Pacific countries was 2961.37 million tons in 2016, with an increase of 84.43% compared with those in 1990. The economic output effect and the population size effect have positive driving influences on carbon emissions of the road transportation industry, in which the economic output effect is still the most important driving factor. The energy intensity effect and the transportation intensity effect have different influences on driving carbon emissions of the road transportation industry for these six Asia-Pacific Countries. Furthermore, the carbon emissions coefficient effect has a relatively small influence. Hence, in order to effectively control carbon emissions of the road transportation industry in these six Asia-Pacific countries, it is necessary to control the impact of economic developments on the environment, to reduce energy intensity by promoting the conversion of road transportation to rail and water transportation, and to lower the carbon emissions coefficient by continuously improving vehicle emission standards and fuel quality.

Keywords: logarithmic mean Divisia index; road transportation industry; carbon emissions; driving factors

---

1. Introduction

Global warming has become one of the most important challenges for human beings, and the essential cause is excess emissions of greenhouse gases such as carbon dioxide, etc. According to the statistics of International Energy Agency Data (IEA Data), the transportation industry accounted for 23.96% of the 32.5804 billion tons of global carbon dioxide emissions in 2017, making it the second largest industry of carbon emissions. Among them, carbon emissions in sub-industries of road transportation accounted for the highest proportion of the transportation industry, with more than 70%. Furthermore, according to a rough calculation in this paper, the carbon emissions of the road transportation industry in six Asia-Pacific countries (Australia, Canada, China, India, Russia, and the United States) are about 50.81% of the global total volumes of that industry. Therefore, it is of great significance to study the driving factors of carbon emissions of the road transportation industry in these six Asia-Pacific countries for controlling global carbon emissions.

Currently, relevant researches on carbon emissions of the road transportation industry mainly involve two aspects: One is the relationship between carbon emissions and economic growth, and the other is the influence factors of carbon emissions. In terms of the relationship between carbon emissions and economic growth, Grossman et al. [1] firstly proposed an inverted U-shaped relationship between environmental quality and economic development based on Kuznets Curve (Kuznets [2]). Panayotou [3] called it the Environmental Kuznets Curve (EKC). Based on the EKC theory, Kwon [4] proposed, for the first time, that whether British road transportation was fit for the turning point of EKC should be verified. Abdallah et al. [5] verified, for the first time, that carbon emissions of road transportation in Tunisia conform to the law of EKC. With the same method, Kharbach et al. [6], Alshehry et al. [7], and Azlina et al. [8], respectively, verified the applicability of EKC in the road transportation industry in the United States, Saudi Arabia, and Malaysia. However, some scholars believe that some countries cannot verify that there is an EKC relationship between environment and carbon emission (Huang [9]).

In addition, some scholars have also verified the relationship between economy and environment through Decoupling Theory. The Organization for Economic Cooperation and Development (OECD) introduced "Decoupling Theory" and created the decoupling model. Lu [10] used this model to verify the decoupling relationship between the economic development and carbon emissions of the road transportation industry for Taiwan, Germany, Japan, and South Korea. The study revealed that Taiwan shows a decoupling relationship, while Korea, Germany, and Japan show a relative decoupling relationship. Tapio [11] optimized the basic decoupling model by introducing the elastic concept of economics and established the Tapio Decoupling Model to study the decoupling state of dynamic data. With this method, Sorrell et al. [12] analyzed the energy consumption of the road freight industry for Britain from 1989 to 2004 with the decoupling analysis method. The research results showed that the United Kingdom (UK) has been more successful than most European Union (EU) countries in decoupling the environmental influences of road freight transportation from GDP. Tapio [11] created a theoretical framework of decoupling to analyze carbon emissions of road transportation for the European Union from 1970 to 2001. The result indicated that the freight of the European Union transforms the relations from weak decoupling to expansive negative decoupling. In the 1990s, there existed a weak decoupling relation between freight transportation and carbon dioxide emissions in the UK, Sweden, and Finland, while a strong decoupling relation between road traffic volume and carbon dioxide emissions from the road transportation industry in Finland from 1990 to 2001. Kveiborg et al. [13] combined the Divisia Index Decomposition Method and Tapio Decoupling Model to analyze the carbon emissions of road freight from 1981 to 1997. The research results presented an obvious decoupling relationship in road freight from 1989 to 1997.

The decoupling model mainly calculates the decoupling index and the decoupling factor (OECD reference) to determine whether there is decoupling relationship between the environment and the economy; however, it cannot explain the specific reasons of decoupling. Therefore, some scholars have begun to study which factors have an influence on carbon emissions. At present, the research mainly focuses on using the factor decomposition method or the econometric model to analyze the influencing factors of carbon emissions. The factor decomposition method is mainly divided into the Laspeyres Index Decomposition method and the Divisia Index Decomposition method. (The Laspeyres Index Decomposition method follows the Laspeyres price and quantity indices in economics analysis.) Hankinson et al. [14], Reitler et al. [15], Howarth et al. [16], Howarth et al. [17], Park [18], Park et al. [19], and Lin [20] all used this method to analyze carbon emissions of different countries and regions. Due to the defect that the Factor-Reversal Test and the Time-Reversal Test cannot pass the tests in the Laspeyres Index method, Sobrino et al. [21] adopted the improved Laspeyres Index method to analyze driving factors for carbon emissions of the road transportation industry in Spain from 1990 to 2010. The conclusion showed that economic growth reveals a close relationship with the rise of carbon emissions, and improved energy efficiency has been a powerful contributor to the carbon emissions decrease.

Because relatively large residual errors in the calculated results in the Divisia Index Decomposition method exist, and it cannot solve the problem of zero value, Ang et al. [22] proposed the Logarithmic Mean Divisia Index (LMDI) in 1998. It effectively solves the above problems and acquires a wide range of applications. M'raihi et al. [23] adopted this method to study the influencing factors of carbon emissions of the road transportation industry in Tunisia. The research results showed that economic growth is the main reason for the increase of carbon dioxide emissions. Effects of fossil fuel share, fossil fuel intensity, and road freight transport intensity are all found as secondary factors responsible for CO_2 emission changes, while Timilsina et al. [24] considered that the economic activity effect and the transportation energy intensity effect are found to be the main driver of CO_2 emissions of road transportation in Latin American and Caribbean countries. Liu et al. [25], Howarthetal et al. [16], Paul et al. [26], and Lise [27] all used this method to analyze the relationship between energy consumption and carbon emissions.

Econometric models can effectively analyze time series data. Wei [28] used the impulse response function and the factor decomposition method to study the carbon emissions of China's road transportation industry. The research results showed that traffic structure and carbon emissions had long-term influences and that dynamic interactive mechanisms exited in China from 1989 to 2009. Wang et al. [29] used a combined research model, including co-integration analysis, the error correction model, and the dynamic model, to study the influences of different factors on energy consumptions in China and OECD countries. However, the paper only showed the strong and weak relationship of each factor—it did not quantify their influence degrees. Liimatainen et al. [30] firstly proposed the "road freight–economy" relationship analysis framework (McKinnon et al. [31]) for McKinnon's improvement, and introduced three indexes of CO_2 intensity, transport intensity, and energy efficiency. He used a joint analysis method for comparison to analyze carbon dioxide emissions and energy efficiencies of the road transportation industry for the four countries of Denmark, Finland, Norway, and Sweden in northern Europe in 2010. It indicated that transportation intensity and energy efficiency have significant influences on carbon dioxide emissions. Puliafito et al. [32] calculated the carbon emissions data of Argentina's road transportation industry from 1960 to 2010 and predicted the data from 2011 to 2050, and Monte Carlo sensitivity analysis and scenario analysis methods were applied to analyze the relations between energy demand and greenhouse gas emissions. Melo [33] applied both the spatial and non-spatial panel data models and introduced ten influence factors, such as urbanization, vehicle ownership, and income levels, etc., to analyze the causal relationship between demand-led, as well as supply-led, factors and carbon emissions of the road transportation industry. The multi-factor and multi-angle analysis strategy provided in the paper can provide a basis for future researches on causality and influence factors. Hasan et al. [34] used a multiple regression model to determine the main driving factors of transportation emissions of passenger vehicles in New Zealand. The results showed that there is a significant causal relationship between fuel economy and transportation emissions. The present study can provide reference values for future studies in different effect factors, and might offer further policy implications for other countries. Sundo [35], adopting a new mathematical original-destination (O-D) approach of estimating CO_2 emissions, made a comparison among five different low-carbon scenarios. The results showed that increasing the proportion of clean energy can effectively reduce the carbon emissions of the road transportation industry.

Seen from the above references, scholars at home and abroad have conducted in-depth researches on the carbon emissions of the road transportation industry, but several problems also exist, as follows: (1) The expansion of Kaya identity is a little simpler when the factor decomposition method is used to analyze carbon emissions of transportation industry; and (2) currently, only a few scholars conduct comparative studies among countries, while other scholars take only one country as the research object, failing to fully explain the differences of carbon emissions among countries. This paper takes six Asia-Pacific countries as the research object, and expands Kaya identity by introducing transportation turnover and other indexes, so as to analyze the influence of more factors on the carbon emissions of the road transportation industry. The LMDI decomposition method is used to emphatically discuss the driving factors of carbon emissions of road transportation, and comparative studies among the six countries are conducted to analyze the influence mode and degree of various factors on carbon emissions of the road transportation industry in these six countries.

2. Research Method

2.1. Expansion of Kaya Identity

Kaya identity, firstly proposed by Japanese professor Yoichi Kaya at the seminar of Intergovernmental Panel on Climate Change (IPCC) in 1989 [36], establishes a relationship between carbon dioxide emissions and economic, policy, as well as population factors, etc. It can decompose driving factors for carbon dioxide emissions and quantify the contribution rate of each influencing factor accurately. Its expression is as follow:

$$C = \frac{GDP}{POP} \times \frac{PE}{GDP} \times \frac{C}{PE} \times POP \tag{1}$$

In Formula (1), C, POP, GDP, and PE respectively represent the volume of carbon dioxide emissions, the whole population of a country, gross domestic product, and total energy consumption.

Kaya identity has been widely used in the fields of energy, environment, and economy. However, due to the limited numbers of examined variables, the results obtained are basically confined to the quantitative relationships between carbon dioxide emissions and energy, economy, and population at the macro level. In recent years, when studying influencing factors for carbon emissions of road transportation, most scholars have mainly selected population size, GDP per capita, and the carbon emissions coefficient of energy [37–39]. However, since carbon emissions are not only connected to these factors, but also relatively closely related to factors of transportation intensity and energy intensity, etc., the index of road transportation turnover is added in this paper, and the Kaya identity is extended. The expression of the expanded Kaya identity is as follows:

$$C = \frac{GDP}{POP} \times \frac{TRS}{GDP} \times \frac{PE}{TRS} \times \frac{C}{PE} \times POP \tag{2}$$

In Formula (2), GDP and POP have the same meaning as Formula (1); C represents total carbon emissions of a country's road transportation industry; TRS says road transportation turnover of a country; and PE indicates energy consumptions of a country's road transportation industry.

Let:

$$G = \frac{GDP}{POP}; \quad R = \frac{TRS}{GDP}; \quad P = \frac{PE}{TRS}; \quad S = \frac{C}{PE}; \quad O = POP \tag{3}$$

Formula (2) can be simplified into Formula (4) by applying Formula (3):

$$C = G \times R \times P \times S \times O \tag{4}$$

In Formula (4), G, R, P, S, and O respectively represent economic output, transportation intensity, energy intensity, and the carbon emissions coefficient of energy, as well as population size.

2.2. The LMDI Decomposition Method Based on Extended Kaya Identity

The factor decomposition method is a further extension of Kaya identity, mainly including the Laspeyres Index decomposition method, the Logarithmic Mean Divisia Index (LMDI) decomposition method, and the Fisher's Ideal Index method, etc. Among them, the LMDI decomposition method, proposed by Ang. B.W. etc. in 1998, solved the problems of inherent salvage value and zero value for the index decomposition method. It witnesses an advantage of complete decomposition and the results' uniqueness [22,40]. Therefore, the LMDI decomposition method has become a mainstream research tool in the field of energy and environment.

The LMDI decomposition method includes the two specific methods of additive decomposition and multiplication decomposition [41]. Because decomposition results of the two methods can be converted to each other, and their converted results are consistent, this paper adopts the additive decomposition method to decompose the model shown in Formula (4). The specific formula is shown in Formula (5).

$$\Delta C = DG + DR + DP + DS + DO \tag{5}$$

In Formula (5), DG represents economic output effect, DR represents transportation intensity effect, DP represents energy intensity effect [42], DS represents carbon emissions coefficient effect of energy, and DO represents population size effect. Hence, the formulas for calculating the effects of various factors influencing carbon emissions are shown in Formulas (6)–(10), and the detailed calculation process is included in the Appendix A.

$$DG = \frac{C^t - C^0}{\ln C^t - \ln C^0} \cdot \ln\left(\frac{G^t}{G^0}\right) \tag{6}$$

$$DR = \frac{C^t - C^0}{\ln C^t - \ln C^0} \cdot \ln\left(\frac{T^t}{T^0}\right) \tag{7}$$

$$DP = \frac{C^t - C^0}{\ln C^t - \ln C^0} \cdot \ln\left(\frac{P^t}{P^0}\right) \tag{8}$$

$$DS = \frac{C^t - C^0}{\ln C^t - \ln C^0} \cdot \ln\left(\frac{S^t}{S^0}\right). \tag{9}$$

$$DO = \frac{C^t - C^0}{\ln C^t - \ln C^0} \cdot \ln\left(\frac{O^t}{O^0}\right) \tag{10}$$

Here, among Formulas (6)–(10), C^0 indicates the baseline year value of carbon emissions for one country's road transportation industry; C^t represents carbon emissions of a country's road transportation industry in year T; G^t, R^t, P^t, S^t, and O^t respectively show the economic output, transport intensity, energy intensity, the carbon emissions coefficient of energy, and population size in the Tth year of a country's road transportation industry; and G^0, R^0, P^0, S^0, and O^0 respectively show a baseline year's economic output, transport intensity, energy intensity, energy coefficient of carbon emissions, and population size of a country's road transportation industry.

3. Description of Variables and Data

This paper selects China and India in Asia, and the United States, Canada, Australia, and Russia in the Pacific Rim, with a total of six countries. The selected countries are characterized by the following commonalities: That all of the six countries respectively have a large territory area, and their carbon emissions from the road transportation industry account for a large proportion of that from the transportation industry. In addition, all of them, being members of the World Trade Organization (WTO), have a sound multilateral trading system and all of their total economic aggregates rank among the top in the world. Therefore, meaningful research conclusions can be obtained by comparing and analyzing the factors influencing carbon emissions of the road transportation industry in these six countries. The research interval of this paper is from 1990 to 2016, and the data in this paper come from the International Energy Agency database (IEA database), the United Nations database (UN database), and the World Bank database.

3.1. Decomposed Variables and Their Database Sources

Carbon emissions refer to the general term of greenhouse gases, expressed by carbon dioxide equivalent (CO_2eq). It mainly includes carbon dioxide, methane, nitrous oxide, and other carbon oxides, among which carbon dioxide emissions account for more than 60% of greenhouse gas emissions. Due to a lack of comprehensive statistics of global carbon emissions at present, most scholars adopt methods provided by IPCC national guidelines for inventory calculations of greenhouse gas [42], and use energy consumption data to calculate carbon emissions. Therefore, the specific expression formula of carbon emissions adopted in this paper is as follows:

$$C = \sum_{i}^{n} CO_{2i} = \sum_{i}^{n} E_i \times \delta_i = \sum_{i}^{n} E_i \times NCV_i \times CEF_i \times COF_i \times \frac{44}{12} \tag{11}$$

where CE stands for carbon dioxide emissions from road transportation industry; i is the type of fossil fuel (the IEA database classifies fuels consumed by the road transportation industry into five categories: Coal, petroleum products, biomass energy, natural gas, and electricity); E_i refers to energy consumption of fossil fuel I; δ_i is the carbon dioxide emission coefficient of carbon energy i; NCV_i is the average low calorific value of energy i; CEF_i is the carbon emissions coefficient of energy i, namely, the carbon content per unit of heat; COF_i is the carbon oxidation factor, that is, the carbon oxidation rate during energy combustion; and 44 and 12 are the molecular weights of carbon dioxide and carbon [43].

According to the glosses of International Energy Agency database [44], the unit for all energy consumption is oil equivalent, and the carbon emissions coefficients of various energies are shown in Table 1.

Table 1. Carbon emissions coefficients of transportation and energy.

Types of Energy	Conversion Factor (KJ/toe)	Carbon Oxidation Rate	CO_2 Emission Factor (kgCO$_2$/GJ)
Coal	41,868	1	94.6
Oil products	41,868	1	72.35
Biomass energy	41,868	1	75.18
Natural gas	41,868	1	56.1
Electric power	-	-	-

Note: Data source: Intergovernmental Panel on Climate Change (IPCC) 2006 edition.

Since electricity is a secondary energy, and carbon emissions from electricity of the road transportation industry in the six Asia-Pacific countries in 2016 only account for 1.50% of total carbon emissions of all energy consumptions, the method to calculate carbon emissions of electricity in this paper is to convert the energy consumption of electricity into equivalent standard coal, and then use the carbon emissions of standard coal to represent the carbon emissions of electricity.

3.2. Decomposition Variables and Their Data Sources

The driving factors of the decomposition model for carbon emissions based on LMDI mainly include economic output, transportation intensity, energy intensity, the carbon emissions coefficient, and population size. The data of the six countries' GDP and population are derived from the UN database, in which the GDP of the six countries is calculated by constant 2010 prices in US Dollars. Road transportation turnover comes from the World Bank database. Energy consumptions of road transportation based on the energy consumptions of the IEA database are converted to standard coal by the method of "oil equivalent–calorific value–standard coal". The detailed forms are shown in Tables 2 and 3.

Table 2. The element's description and data sources.

Elements	Description	Data Resource
GDP	Gross Domestic Production at constant 2010 prices in US Dollars	UN database
POP	Population	UN database
TRS	Total road turnover	World Bank database
PE	Total energy consumption of road transportation	IEA database
C	Total CO_2 emissions of road transportation	Estimate by Formula (11)

Table 3. The driving factors of the decomposition model.

Driving Factors	Description	Symbols
G	Economic output	G = GDP/POP
R	Transportation intensity	R = TRS/GDP
P	Energy intensity	P = PE/TRS
S	Carbon emissions coefficient	S = C/PE
O	Population size	O = POP

3.2.1. Economic Output

According to Figure 1, China's per capita GDP in 2016 increased by 857.20% compared with that in 1990, ranking first among the six countries. The per capita GDP of Russia in 2016 increased by 19.00% compared with 1990, being last among the six countries. In 2016, China's per capita GDP reached $6,770 per person, Russia's $11,500 per person, while the per capita GDP of Australia, Canada, and the United States exceeded $50,000 per person.

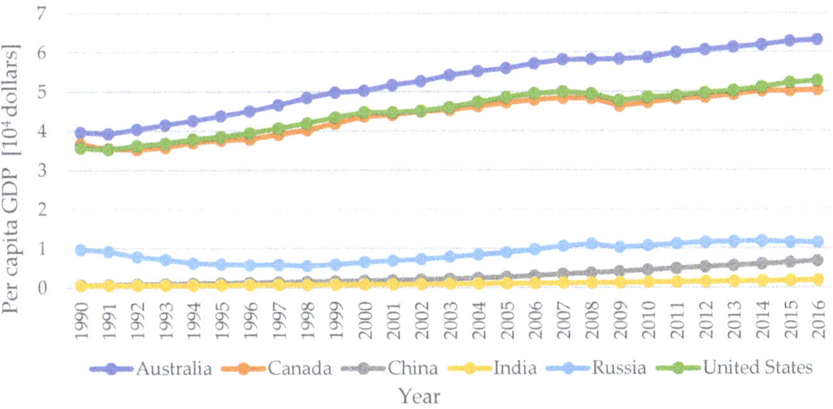

Figure 1. Per capita GDP of six Asia-Pacific countries from 1990 to 2016. Note: Data source: United Nations (UN) database.

3.2.2. Transportation Intensity

The transportation intensity of the six countries is shown in Figure 2. Indian transportation intensity of the road transportation industry is highest in 2016, reaching 1620.96 million tonne-kilometer/$billions, secondary in China at 653.77 million tonne-kilometer/$billions, and lowest in Canada at 121.96 million tonne-kilometer/$billions. In addition, India's road transportation industry presents the largest increase in transport intensity, with 269.04% growth in 2016 compared with 1990, ranked first among all countries, followed by Canada with an increase of 28.94%. In the study range, the transportation intensity of the United States and Russia show a decreasing trend. The transportation intensity of the United States decreased by 23.01% in 2016 compared with 1990, while that of Russia decreased by 35.04%. The reason for the high transportation intensity of India lies in its relatively high proportion of manufacturing and agriculture, and relatively high proportion of road transportation in the five transportation modes. China witnessed a high transportation intensity, which is also due to its relatively high proportion of manufacturing industry in its national economic industry.

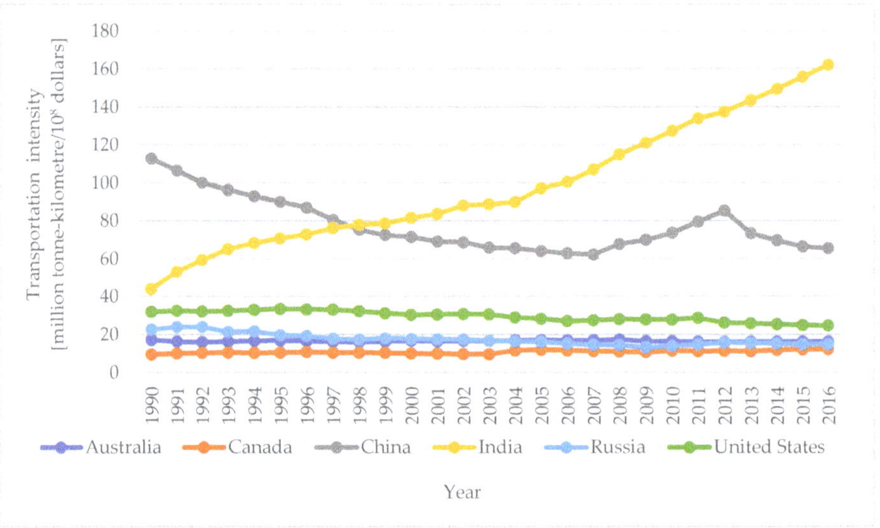

Figure 2. Transportation intensities of six Asia-Pacific countries from 1990 to 2016. Note: Data source: Organization for Economic Cooperation and Development (OECD) Database.

3.2.3. Energy Intensity

The changes of energy intensity for the road transportation industry in the six Asia-Pacific countries are shown in Figure 3. While energy intensities of the road transportation industry in Australia, Canada, United States, and India show decreasing trends within the research range, energy intensities of China and Russia increase by 45.26% and 20.78% respectively. In 2016, Canada's road transportation industry showed the highest energy intensity, reaching 3.09 tons per million tonne-kilometers; Russia's was second at 2.88 tons per million tonne-kilometers; and the United States' third, with 1.81 tons per million tonne-kilometers. The reason for India's low energy intensity is that motorcycles account for nearly 80% of all motor vehicles in India, while trucks and lorries account for only 5.3%. The energy consumption of motorcycles is far less than that of vehicles with four wheels or above. The reason for China's low energy intensity is that the statistics of China's road transport industry cover operating vehicles, excluding private cars, while the statistics of the other five countries cover private cars.

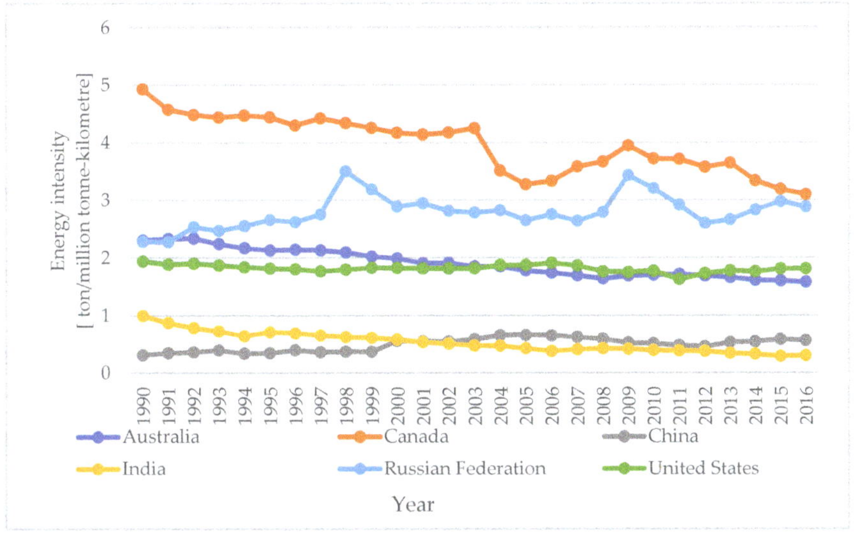

Figure 3. Energy intensities of the six Asia-Pacific countries from 1990 to 2016. Note: Data source: International Energy Agency (IEA) database.

3.2.4. Population Size

The population sizes of the six Asia-Pacific countries are shown in Figure 4. In the study range, the population sizes of the countries, except Russia, have increased to some extent. Among them, India's population in 2016 increased by 52.18% compared with 1990, ranking first among the six countries, while Russia's decreased by 2.44% year-on-year, ranking last of countries. In 2016, China and India, respectively, had a population of 1.403 billion and 1.324 billion, accounting for 43.13% and 40.69% of the total population of the six Asia-Pacific countries.

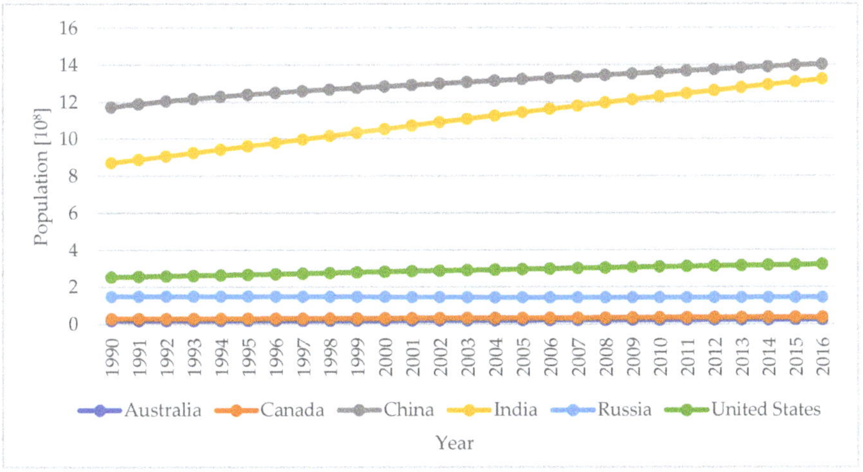

Figure 4. Population sizes of the six Asia-Pacific countries from 1990 to 2016. Note: Data source: United Nations (UN) database.

3.2.5. Carbon Emissions Coefficient

The carbon emissions coefficients of the six Asia-Pacific countries are shown in Figure 5. Within the research range, the carbon emissions coefficients of the six Asia-Pacific countries show downward trends, but the decline is relatively small.

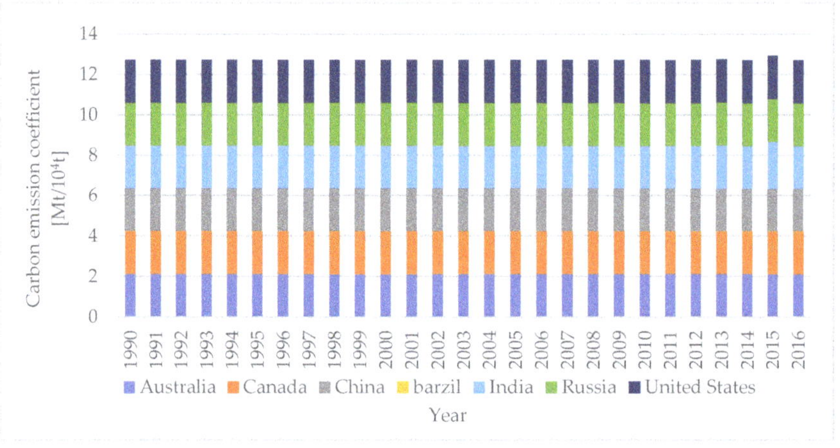

Figure 5. Carbon emissions coefficients of the six Asia-Pacific countries from 1990 to 2016. Note: Data source: International Energy Agency (IEA) database.

4. Results and Discussions

4.1. Analysis on Total Carbon Emissions

The results of the calculation for carbon emissions of the road transportation industry in the six Asia-Pacific countries from 1990 to 2016 are shown in Figure 6 and Table 4. The total carbon emissions from the road transportation industry of the six countries increased from 1605.73 million tons in 1990 to 2961.37 million tons in 2016. Among them, in 2016, the combined carbon emissions from the road transportation industry of the United States and China accounted for 53.73% of the total volume of the six countries. On the whole, carbon emissions of the road transportation industry in the six countries in the study range increased rapidly, among which the average annual growth rate of China's carbon emissions is 9.65%, far higher than those of other countries; India ranks second with 6.36%, and Russia last with a rate of −0.17%.

On the other hand, in the United States and Canada, appeared turning points appeared in carbon emissions in 2007 and 2011, respectively. In 2016, the per capita carbon emissions of China and the United States far exceeded those of the other countries, reaching 4.04 tons and 4.99 tons per capita, respectively.

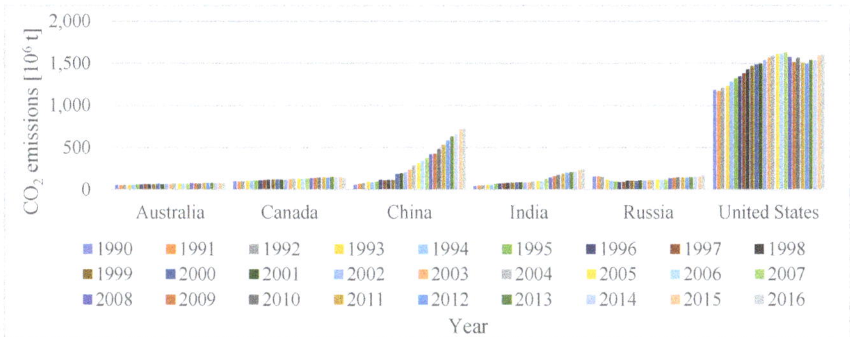

Figure 6. Carbon emissions of road transportation in the six Asia-Pacific countries from 1990 to 2016.

Table 4. Carbon emissions from road transportation in the six Asia-Pacific countries.

Country	Average Annual Growth Rate (%)	Carbon Emissions in 1990 (million tons)	Carbon Emissions in 2016 (million tons)	1990–2016 Increasing Multiples of Carbon Emissions	Turning Points' Year for Carbon Emissions	Per capita Carbon Emissions in 2016 (ton/person)
Australia	1.46%	56.68	82.29	0.45	-	3.41
Canada	1.49%	100.36	146.60	0.46	2011	4.04
China	10.67%	60.58	728.49	11.03	-	0.52
India	6.76%	46.95	247.94	4.28	-	0.19
Russia	0.78%	156.77	177.94	0.14	-	1.24
United states	1.20%	1184.39	1606.24	0.36	2007	4.99
Total	22.36%	1605.73	2989.51	16.71	-	14.38

Note: Data source: The International Energy Agency database.

4.2. Analysis on Main Driving Effect

While calculated with Formulas (5)–(10), the years from 1990 to 2016 are divided into nine time periods at intervals of three years. The effect value and contribution rate of every factor driving carbon emissions of road transportation industry are calculated separately in each time period. This paper uses the LMDI decomposition method to decompose carbon emissions of the road transportation industry in the six Asia-Pacific countries. They are mainly decomposed into economic output effect, transportation intensity effect, energy intensity effect, energy carbon emissions coefficient effect, and population size effect, and the calculated effect value and contribution rate of each factor. The contribution rate of each influence factor is the ratio of effect value of the influencing factor to the total effect value of carbon emissions, e.g., $B_{DG} = DG/\Delta C$, whose results are shown in Tables 5 and 6. Seen from the decomposition results, the economic output effect presents the largest contribution rate, the population size effect is the second, and the carbon emissions coefficient the smallest. Therefore, the economic output and population size effects are the main driving factors for the growth of carbon emissions from the road transportation industry in the six Asia-Pacific countries.

Table 5. LMDI decomposition results of carbon emissions from road transportation in the six Asia-Pacific countries (unit: million tons).

Australia	DG	DR	DP	DS	DO	ΔC	China	DG	DR	DP	DS	DO	ΔC
1990–1992	1.01	−4.00	0.72	−0.01	1.45	−0.82	1990–1992	13.51	−7.16	9.97	0.02	1.84	18.18
1993–1995	3.14	1.83	−3.01	0.00	1.28	3.36	1993–1995	19.84	−7.37	−11.22	0.01	1.77	3.47
1996–1998	4.58	−2.27	−1.48	−0.01	1.35	2.24	1996–1998	17.26	−7.41	−17.50	−0.01	1.60	−5.29
1999–2001	2.45	0.41	−3.92	−0.03	1.40	0.34	1999–2001	22.58	−8.45	62.81	−0.01	1.83	79.73
2002–2004	3.32	1.14	−2.29	0.00	1.71	3.70	2002–2004	44.58	−12.40	46.61	−0.59	2.91	82.39
2005–2007	2.84	−0.27	−3.61	0.01	2.53	0.54	2005–2007	83.42	3.20	−33.09	0.33	3.95	59.15
2008–2010	0.66	−4.49	2.87	−0.01	2.71	0.65	2008–2010	81.02	37.72	−63.60	0.03	5.12	61.41
2011–2013	1.74	0.00	−2.72	0.00	2.34	0.71	2011–2013	80.71	−46.93	58.87	−1.18	6.46	98.94
2014–2016	1.62	0.02	−1.74	−0.02	2.22	1.60	2014–2016	84.51	−42.65	23.44	0.34	6.63	74.19
1990–2016	31.96	−4.03	−26.15	−0.03	23.88	25.62	1990–2016	606.64	−146.15	161.82	−2.71	48.31	667.91

Canada	DG	DR	DP	DS	DO	ΔC	India	DG	DR	DP	DS	DO	ΔC
1990–1992	−3.72	9.17	−9.42	0.00	2.46	−1.51	1990–1992	0.83	15.02	−11.80	0.00	2.02	6.06
1993–1995	5.16	−0.15	0.02	0.00	2.24	7.50	1993–1995	6.69	5.30	−1.68	0.00	2.42	12.79
1996–1998	6.97	−4.14	1.11	0.01	2.11	6.50	1996–1998	4.98	5.31	−7.70	0.00	2.86	5.61
1999–2001	5.91	−5.46	−3.28	0.07	2.22	−0.21	1999–2001	4.69	5.16	−11.32	−0.12	2.98	1.69
2002–2004	3.53	22.70	−21.42	−0.01	2.47	7.33	2002–2004	11.86	1.89	−7.15	−0.10	3.06	10.06
2005–2007	3.16	−7.48	11.97	0.05	2.97	10.26	2005–2007	13.86	11.39	−4.63	0.04	3.51	24.98
2008–2010	−3.17	5.26	1.96	0.04	3.23	6.74	2008–2010	20.77	16.61	−12.92	−0.20	4.47	30.23
2011–2013	3.16	−0.19	−2.69	0.02	3.06	3.01	2011–2013	17.88	13.51	−24.44	4.26	4.92	18.45
2014–2016	0.73	4.57	−10.73	−0.02	2.82	−2.62	2014–2016	28.67	18.78	−20.26	0.02	5.36	35.49
1990–2016	38.91	31.02	−56.80	0.13	32.99	46.25	1990–2016	140.19	157.71	−146.96	−0.66	50.72	200.99

Russia	DG	DR	DP	DS	DO	ΔC	United States	DG	DR	DP	DS	DO	ΔC
1990–1992	−32.59	8.10	16.17	−0.04	0.78	−7.57	1990–1992	17.00	4.78	−23.65	0.00	23.08	21.20
1993–1995	−19.14	−8.89	7.89	0.15	−0.10	−19.70	1993–1995	56.71	42.59	−38.76	0.34	27.32	85.83
1996–1998	−3.60	−10.16	28.73	−0.05	−0.46	15.30	1996–1998	86.64	−36.00	−6.14	0.10	34.28	72.74
1999–2001	16.55	−2.88	−8.61	0.14	−0.82	5.39	1999–2001	43.15	−31.96	−12.38	−0.01	31.49	30.59
2002–2004	16.90	−7.20	0.30	0.00	−0.91	9.74	2002–2004	75.27	−97.33	49.48	0.33	27.13	63.41
2005–2007	19.63	−11.95	−0.33	−0.07	−0.39	6.87	2005–2007	45.98	−46.45	−8.04	0.70	29.72	31.21
2008–2010	−5.34	−6.18	19.62	0.00	0.07	7.45	2008–2010	−27.64	−15.78	10.31	0.36	27.04	6.16
2011–2013	7.55	7.46	−13.53	0.07	0.34	0.70	2011–2013	39.93	−106.80	78.42	1.00	21.83	49.78
2014–2016	−4.78	−4.29	2.71	−0.07	0.22	−6.22	2014–2016	47.14	−44.00	44.78	0.31	21.90	84.54
1990–2016	26.67	−66.12	35.69	0.59	−3.78	−6.96	1990–2016	540.90	−362.01	−97.61	3.30	337.27	421.84

Table 6. Contribution rates of carbon emission drivers for road transportation in the six Asia-Pacific countries (unit: %).

Australia	B_{DG}	B_{DR}	B_{DP}	B_{DS}	B_{DO}	B_{DC}	China	B_{DG}	B_{DR}	B_{DP}	B_{DS}	B_{DO}	B_{DC}
1990–1992	−122.52	485.53	−87.95	0.75	−175.82	100	1990–1992	74.33	−39.37	54.83	0.09	10.12	100
1993–1995	93.56	54.60	−89.70	0.02	38.02	100	1993–1995	571.10	−212.22	−323.04	0.36	51.00	100
1996–1998	204.57	−101.31	−66.10	−0.55	60.17	100	1996–1998	−326.31	140.06	330.70	0.24	−30.17	100
1999–2001	729.70	122.35	−1169.14	−10.11	416.27	100	1999–2001	28.32	−10.59	78.77	−0.01	2.29	100
2002–2004	89.81	30.75	−61.91	−0.09	46.24	100	2002–2004	54.11	−15.06	56.57	−0.71	3.53	100
2005–2007	525.30	−50.31	−667.91	2.03	467.42	100	2005–2007	141.03	5.41	−55.95	0.56	6.67	100
2008–2010	101.61	−687.50	439.48	−0.87	415.45	100	2008–2010	131.92	61.41	−103.56	0.04	8.34	100
2011–2013	245.41	0.66	−383.39	−0.70	329.49	100	2011–2013	81.57	−47.43	59.50	−1.19	6.52	100
2014–2016	101.12	1.09	−108.66	−1.48	139.00	100	2014–2016	113.91	−57.49	31.60	0.46	8.94	100
1990–2016	124.76	−15.74	−102.10	−0.14	93.22	100	1990–2016	90.83	−21.88	24.23	−0.41	7.23	100

Canada	B_{DG}	B_{DR}	B_{DP}	B_{DS}	B_{DO}	B_{DC}	India	B_{DG}	B_{DR}	B_{DP}	B_{DS}	B_{DO}	B_{DC}
1990–1992	245.81	−605.44	622.05	−0.13	−162.29	100	1990–1992	13.67	247.88	−194.86	−0.01	33.30	100
1993–1995	68.76	−1.94	0.28	−0.02	29.89	100	1993–1995	52.27	41.46	−13.16	0.00	18.89	100
1996–1998	107.34	−63.70	17.12	0.15	32.51	100	1996–1998	88.88	94.73	−137.35	0.01	51.01	100
1999–2001	−2823.09	2606.35	1565.11	−32.56	−1058.92	100	1999–2001	276.95	304.54	−667.93	−7.23	175.75	100
2002–2004	48.15	309.83	−292.37	−0.08	33.70	100	2002–2004	117.91	18.82	−71.09	−1.02	30.37	100
2005–2007	30.83	−72.95	116.69	0.47	28.91	100	2005–2007	55.48	45.58	−18.53	0.15	14.07	100
2008–2010	−47.01	78.05	29.02	0.54	47.92	100	2008–2010	68.70	54.94	−42.72	−0.67	14.80	100
2011–2013	104.77	−6.21	−89.25	0.54	101.48	100	2011–2013	96.92	73.22	−132.46	23.06	26.65	100
2014–2016	−27.90	−174.25	409.06	0.67	−107.58	100	2014–2016	80.77	52.91	−57.10	0.06	15.09	100
1990–2016	84.14	67.07	−122.83	0.29	71.34	100	1990–2016	69.75	78.47	−73.12	−0.33	25.23	100

Russia	B_{DG}	B_{DR}	B_{DP}	B_{DS}	B_{DO}	B_{DC}	United States	B_{DG}	B_{DR}	B_{DP}	B_{DS}	B_{DO}	B_{DC}
1990–1992	430.34	−106.93	−213.59	0.48	−10.29	100	1990–1992	80.15	22.55	−111.54	0.00	108.84	100
1993–1995	97.15	45.15	−40.07	−0.75	0.53	100	1993–1995	66.07	49.62	−45.16	0.40	31.83	100
1996–1998	−23.54	−66.43	187.78	−0.34	−3.02	100	1996–1998	119.11	−49.49	−8.44	0.14	47.12	100
1999–2001	306.86	−53.49	−159.74	2.58	−15.15	100	1999–2001	141.08	−104.47	−40.49	−0.03	102.94	100
2002–2004	173.56	−73.98	3.09	0.01	−9.35	100	2002–2004	118.70	−153.49	78.02	0.52	42.79	100
2005–2007	285.82	−173.92	−4.79	−0.96	−5.70	100	2005–2007	147.30	−148.80	−25.77	2.25	95.22	100
2008–2010	−71.66	−82.98	263.47	0.04	0.93	100	2008–2010	−449.03	−256.36	167.53	5.89	439.32	100
2011–2013	1085.39	1072.53	−1945.79	9.77	49.18	100	2011–2013	80.20	−214.52	157.53	2.02	43.85	100
2014–2016	76.92	68.91	−43.51	1.15	−3.47	100	2014–2016	55.77	−52.05	52.97	0.37	25.91	100
1990–2016	−383.12	950.05	−512.81	−8.49	54.38	100	1990–2016	128.22	−85.81	−23.14	0.78	79.95	100

4.2.1. Economic Output Effect

As can be seen from Table 5, most of the economic output effect values of the six countries in the research interval are positive, and only a few years witness small negative values of absolute values. This indicates that the economic output effect plays a positive driving role in the carbon emissions of the road transportation industry.

Seen from Figure 7, the values of the economic output effect for the United States, Canada, Russia, and Australia present an overall inverted "U" pattern, which conforms to the development law of the environmental Kuznets curve. China and India are both developing countries; their national economy and industrialization are in a stage of rapid development, and they are still at the left end of the environmental Kuznets curve, having no turning points yet. Therefore, the influence of economic output on carbon emissions of the road transportation industry continues to increase.

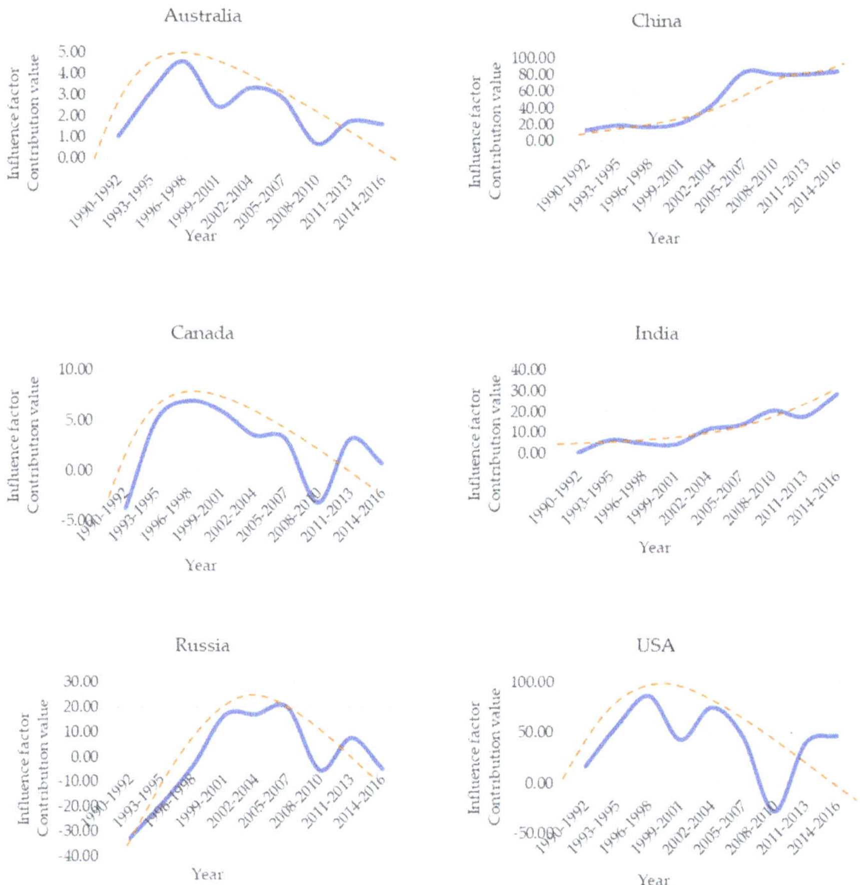

Figure 7. Effect values of the economic output effective for the six Asia-Pacific countries.

From the overall results, the value of the economic output effect (DG, Table 5) and its contribution rate (B_{GD}, Table 6) in each period of the six countries from 1990 to 2016 are greater than the other effects. This indicates that the economic output effect is the main influencing factor for the growth of the road transportation industry's carbon emissions. According to the calculation, the values of the economic output effect for China, Australia, and the United States from 1990 to 2016 are 606.64, 31.96, and 540.90, respectively, whose respective contribution rates are 90.83%, 127.76%, and 128.22%. The results indicate that the economic output effects of China, Australia, and the United States have greater influences on carbon emissions of the road transportation industry than other countries. This reveals that the economic output effects of the three countries still have a relatively large space to decline.

4.2.2. Transportation Intensity Effect

Most of the effect values (DR, Table 5) are negative in the transportation intensity effects from China, Russia, and the United States to carbon emissions of road transportation industry in each stage. This indicates that the transportation intensity effects of the three countries present negative driving forces of carbon emissions of the road transportation industry. India's has a positive driving effect. The reason is that the development of India's manufacturing industry and the improvement of people's living standard leads to increasing traffic. Thus, the growth rate of road freight volume and passenger volume exceeds that of GDP, bringing an increase of transportation intensity. However, Australia's and Canada's have no significant influence, which shows that for China, Russia, and the United States, the transportation intensity effect is an important factor to curb the growth of carbon emissions, while for India, measures need to be taken to reduce the transportation intensity and the influences of transportation intensity effect on the growth of carbon emissions.

4.2.3. Energy Intensity Effect

There are some differences in the influences of energy intensity effect (DP, Table 5) on carbon emissions of the road transportation industry in the six countries. Overall, Australia, the United States, India, and Canada play negative roles in their energy intensity effects driving carbon emissions of the road transportation industry, while China and Russia act as positive driving roles. Since the 1990s, the slow promotion of Chinese vehicle energy saving technology and the increasing requirements of enterprises on the speed, as well as efficiency of transportation, make a gradual rise in energy consumption intensity of Chinese operating road transportation vehicles. Hence, this finally leads to the rise of energy intensity for China's road transportation industry. Therefore, it shows that China's energy intensity effect has a positive influence on carbon emissions of the road transportation industry.

During the study period, the contribution rates in 1990–2016 (B_{DP}, Table 6) of the energy intensity effect in Canada and Russia are, respectively, −122.83% and −512.81%. This indicates that the energy intensity effect of the two countries is the main influencing factor in reducing carbon emissions of the road transportation industry. China's contributes at a rate of 24.23% to carbon emissions of the road transportation industry, ranking first among all countries. This indicates that the energy intensity effect on carbon emissions has a positive effect.

4.2.4. Carbon Emissions Coefficient Effect

It can be seen from Tables 5 and 6 that the values of the carbon emissions coefficient effect (B_{DS}, Table 5) for the six countries are all less than 4, and the contribution rates of the carbon emissions coefficient effect are less than 9%. So, the carbon emissions coefficient effect has a relatively small influence on the carbon emissions of the road transportation industry for the six countries.

4.2.5. Population Size Effect

It can be seen from Table 6 that, except for Russia, most values of the population size effect (DO, Table 5) for the other five countries in each period are positive. This indicates that the population size effect plays a positive driving role in the carbon emissions of the road transportation industry. The effect value of the Russian population size effect is negative, mainly due to the decline of their population in the research period—and the decline is obvious in that their population decreased by 2.44% in 2016 compared with that in 1990, with a total of 3.59 million people.

5. Conclusions and Suggestions

5.1. Conclusions

This paper uses the LMDI decomposition model to analyze the carbon emissions and driving factors of the road transportation industry for six Asia-Pacific countries. It comes to the following main conclusion that an overall rise is seen in the carbon emissions of the road transportation industry of the six Asia-Pacific countries from 1990 to 2016. In 2016, the total carbon emissions of the road transportation industry in the six Asia-Pacific countries reached 2961.37 million tons compared with that in 1990, with a growth of 84.42%. Among them, the carbon emissions of the road transportation industry for the United States accounted for 54.24% of the total volume of that in the six Asia-Pacific countries in 2016, ranking highest among them; China was second with 24.60%, and Australia came in last with only 2.78%. Among the driving factors, the economic output and population size effects play positive driving roles in the road transportation industry; the economic output effect is the main factor for their increasing carbon emissions. However, the transportation intensity effect and the energy intensity effect, being divergent to some degree in different countries, have negative driving effects on the carbon emissions of the road transportation industry in most countries. Where the transportation intensity effect of India plays a positive driving role in the carbon emissions of the road transportation industry, contributing with a rate of 78.47%, the energy intensity effects of China and Russia also play positive driving roles in the carbon emissions. The carbon emission coefficient effect has a relatively small influence on the carbon emissions of the road transportation industry in the six countries—and, except Russia, the population size effect of the other five countries plays a positive role in driving the carbon emissions of the road transportation industry.

5.2. Policy Suggestions

The United States, which accounts for more than half of the total carbon emissions from the six Asia-Pacific countries, withdrew from Paris Agreement in 2017. This signaled that the United States did not want to fulfill its international obligations to reduce carbon emissions. The international community should exert pressure on them to change their attitude towards carbon emissions control. China is second only to the United States in carbon emissions of the road transportation industry, and has made some progress in carbon emissions in recent years. However, the rising energy intensity restricts its achievement of carbon emissions control. China should make policies, such as accelerating the promotion of new energy vehicles and improving vehicle emissions standards, etc., to reduce the energy intensity of road transportation industry. The carbon emissions of the Indian road transportation industry are growing relatively fast. Therefore, India should reduce its transportation intensity and control its carbon emissions growth by controlling the carbon emissions of transportation in the process of developing manufacturing industry.

Compared with 1990, Russia is the only one of the six Asia-Pacific countries whose carbon emissions from the road transportation industry decreased in 2016. However, the age of Russian road operating vehicles is generally older, which leads to its rising energy intensity in the road transportation industry since 1990. Russia should reduce this energy intensity and promote its carbon emissions to fall faster by formulating policies such as accelerating the elimination of old cars. Since 2005, the carbon emissions of the road transportation industry in Canada have begun to slow down, and since 2014, its carbon emissions have shown a negative growth. So, Canada can lower its carbon emissions by reducing its transportation intensity. Australia witnesses a relatively slow growth in its carbon emissions of the road transportation industry, and a relatively good result in its carbon emissions control. Therefore, it can be improved from the aspect of reducing transportation intensity next.

Author Contributions: C.Z. and W.D. wrote the original draft.

Funding: This research was funded by the National Social Science Foundation in China. (Grant No. 19BJY175).

Acknowledgments: The authors thank Dawei Gao and Meng Wang for their help in participation in data collection, and thank Lijiao Qin for her help in the English editor.

Conflicts of Interest: The authors declare no conflicts of interest.

Appendix A

In this section, we give the calculation process of the LMDI method. To this end, it follows from Formula (4) that:

$$\ln C = \ln G + \ln R + \ln P + \ln S + \ln O$$

Therefore:

$$\ln C^t - \ln C^0 = \ln \frac{G^t}{G^0} + \ln \frac{R^t}{R^0} + \ln \frac{P^t}{P^0} + \ln \frac{S^t}{S^0} + \ln \frac{O^t}{O^0} \tag{A1}$$

Let $\Delta C = C^t - C^0$.

Then, we can rewrite ΔC as follows:

$$\Delta C = \frac{C^t - C^0}{\ln C^t - \ln C^0} \left(\ln C^t - \ln C^0 \right)$$

Furthermore, by Formula (A1), we have:

$$\Delta C = \frac{C^t - C^0}{\ln C^t - \ln C^0} \cdot \left[\ln \frac{G^t}{G^0} + \ln \frac{R^t}{R^0} + \ln \frac{P^t}{P^0} + \ln \frac{S^t}{S^0} + \ln \frac{O^t}{O^0} \right]$$

Let:

$$DG = \frac{C^t - C^0}{\ln C^t - \ln C^0} \cdot \ln \frac{G^t}{G^0}; DR = \frac{C^t - C^0}{\ln C^t - \ln C^0} \cdot \ln \frac{R^t}{R^0};$$
$$DP = \frac{C^t - C^0}{\ln C^t - \ln C^0} \cdot \ln \frac{P^t}{P^0}; DS = \frac{C^t - C^0}{\ln C^t - \ln C^0} \cdot \ln \frac{S^t}{S^0};$$
$$DO = \frac{C^t - C^0}{\ln C^t - \ln C^0} \cdot \ln \frac{O^t}{O^0}$$

Then:

$$\Delta C = C^t - C^0 = DG + DR + DP + DS + DO$$

References

1. Grossman, G.M.; Krueger, A.B. Environmental impacts of a north American free trade agreement. *Soc. Sci. Electron. Pub.* **1991**, *8*, 223–250.
2. Kuznets, P.; Simon, P. Economic growth and income inequality. *Am. Econ. Rev.* **1955**, *45*, 1–28.
3. Panayotou, T. *Empirical Tests and Policy Analysis of Environmental Degradation at Different Stages of Economic Development*; ILO Working Papers; International Labour Office: Geneva, Switzerland, 1993; p. 238.
4. Kwon, T.H. Decomposition of factors determining the trend of CO_2 emissions from car travel in Great Britain (1970–2000). *Ecol. Econ.* **2005**, *53*, 261–275. [CrossRef]
5. Abdallah, K.B.; Belloumi, M.; Wolf, D.D. Indicators for sustainable energy development: A multivariate cointegration and causality analysis from Tunisian road transport sector. *Renew. Sustain. Energy Rev.* **2013**, *25*, 34–43. [CrossRef]
6. Kharbach, M.; Chfadi, T. CO_2 Emissions in Moroccan Road Transport sector: Divisia, Cointegration, and EKC analyses. *Sustain. Cities Soc.* **2017**, *35*, 396–401. [CrossRef]
7. Alshehry, A.S.; Belloumi, M. Study of the environmental Kuznets curve for transport carbon dioxide emissions in Saudi Arabia. *Renew. Sustain. Energy Rev.* **2017**, *75*, 1339–1347. [CrossRef]
8. Azlina, A.A.; Law, S.H.; Nik Mustapha, N.H. Dynamic linkages among transport energy consumption, income and CO2 emission in Malaysia. *Energy Policy* **2014**, *73*, 598–606. [CrossRef]
9. Huang, W.M.; Lee, G.W.M.; Wu, C.C. GHG emissions, GDP growth and the Kyoto Protocol: A revisit of Environmental Kuznets Curve hypothesis. *Energy Policy* **2008**, *36*, 239–247. [CrossRef]
10. Lu, I.J.; Lin, S.J.; Lewis, C. Decomposition and decoupling effects of carbon dioxide emission from highway transportation in Taiwan, Germany, Japan and South Korea. *Energy Policy* **2007**, *35*, 3226–3235. [CrossRef]

11. Tapio, P. Towards a theory of decoupling: Degrees of decoupling in the EU and the case of road traffic in Finland between 1970 and 2001. *Transp. Policy* **2005**, *12*, 137–151. [CrossRef]

12. Sorrell, S.; Lehtonen, M.; Stapleton, L.; Pujol, J.; Champion, T. Decoupling of road freight energy use from economic growth in the United Kingdom. *Energy Policy* **2012**, *41*, 84–97. [CrossRef]

13. Kveiborg, O.; Fosgerau, M. Decomposing the decoupling of Danish road freight traffic growth and economic growth. *Transp. Policy* **2007**, *14*, 39–48. [CrossRef]

14. Hankinson, G.A.; Rhys, J.M.N. Electricity consumption, electricity and industrial structure. *Energy Econ.* **1983**, *5*, 146–152. [CrossRef]

15. Reitler, W.; Rudolph, M.; Schaefer, M. Analysis of the factors influencing energy consumption in industry: A revised method. *Energy Econ.* **1987**, *9*, 145–148. [CrossRef]

16. Howarth, R.B.; Richard, B. Energy use in U.S. manufacturing: The impacts of the energy shocks on sectoral output, industry structure, and energy intensity. *J. Energy Dev.* **1991**, *14*, 175–191.

17. Howarth, R.B.; Schipper, L. Manufacturing energy use in eight OECD countries: Trends through 1988. *Energy J.* **1992**, *12*, 15–40. [CrossRef]

18. Park, S.H. Decomposition of industrial energy consumption. *Energy Econ.* **1992**, *13*, 265–270. [CrossRef]

19. Park, S.H.; Dissmann, B.; Nam, K.Y.A. Cross-country decomposition analysis of manufacturing energy consumption. *Energy* **1993**, *18*, 843–858. [CrossRef]

20. Lin, J.; Zhou, N.; Levine, M.; Fridley, D. Taking out 1 billion ton of CO2: The magic of China's 11th five-year plan. *Energy Policy* **2008**, *36*, 954–970. [CrossRef]

21. Sobrino, N.; Monzon, A. The impact of the economic crisis and policy actions on GHG emissions from road transport in Spain. *Energy Policy* **2014**, *74*, 486–498. [CrossRef]

22. Ang, B.W.; Zhang, F.Q.; Choi, K.H. Factorizing Changes in Energy and Environmental Indicators through Decomposition. *Energy* **1998**, *23*, 489–495. [CrossRef]

23. M'raihi, R.; Mraihi, T.; Harizi, R.; Bouzidi, M.T. Carbon emissions growth and road freight: Analysis of the influencing factors in Tunisia. *Transp. Policy* **2015**, *42*, 121–129. [CrossRef]

24. Timilsina, G.R.; Shrestha, A. Factors affecting transport sector CO2 emissions growth in Latin American and Caribbean countries: An LMDI decomposition analysis. *Int. J. Energ. Res.* **2009**, *33*, 396–414. [CrossRef]

25. Liu, N.; Ang, B. Factors shaping aggregate energy intensity trend for industry: Energy intensity versus product mix. *Energy Econ.* **2007**, *29*, 609–635. [CrossRef]

26. Paul, S.; Bhattacharya, R.N. Causality between energy consumption and economic growth in India: A note on conflicting results. *Energy Econ.* **2004**, *26*, 977–983. [CrossRef]

27. Lise, W. Decomposition of CO2 emissions over 1980–2003 in Turkey. *Energy Policy* **2006**, *34*, 1841–1852. [CrossRef]

28. Wei, Q.; Zhao, S.; Xiao, W.A. Quantitative Analysis of Carbon Emissions Reduction Ability of Transportation Structure Optimization in China. *J. Transp. Syst. Energy Inf. Technol.* **2013**, *13*, 10–17.

29. Wang, T.; Lin, B. Fuel consumption in road transport: A comparative study of China and OECD countries. *J. Clean. Prod.* **2019**, *206*, 156–170. [CrossRef]

30. Liimatainen, H.; Arvidsson, N.; Hovi, I.B.; Jensen, T.C.; Nykänen, L.; Kallionpää, E. Road freight energy efficiency and CO2 emissions in the Nordic countries. *Res. Transp. Bus. Manag.* **2014**, *2*, 11–19. [CrossRef]

31. McKinnon, A.; Woodburn, A. Logistical restructuring and road freight traffic growth: An empirical assessment. *Transportation* **1996**, *23*, 141–161. [CrossRef]

32. Puliafito, S.E.; Allende, D.; Pinto, S.; Castesana, P. High resolution inventory of GHG emissions of the road transport sector in Argentina. *Atmos. Environ.* **2014**, *101*, 303–311. [CrossRef]

33. Melo, P.C. Driving down road transport CO2 emissions in Scotland. *Int. J. Sustain. Transp.* **2016**, *10*, 906–916. [CrossRef]

34. Hasan, M.A.; Frame, D.J.; Chapman, R.; Archie, K.M. Emissions from the road transport sector of New Zealand: Key drivers and challenges. *Environ. Sci. Pollut. Res.* **2019**, *23*, 23937–23957. [CrossRef] [PubMed]

35. Sundo, M.B.; Vergel, K.N.; Sigua, R.G.; Regidor, J.R.F. Methods of estimating energy demand and CO2 emissions for inter-regional road transport. *Int. J. GEOMATE* **2016**, *11*, 2182–2187. [CrossRef]

36. Kaya, Y. *Impact of Carbon Dioxide Emission Control on GNP Growth: Interpretation of Proposed Scenarios*; IPCC Energy and Industry Subgroup, Response Strategies Working Group: Paris, France, 1990.

37. Jie, Y.U.; Da, Y.B.; OuYang, B. Analysis of Carbon Emission Changes in China's Transportation Industry Based on LMDI Decomposition Method. *China J. Highw. Transp.* **2015**, *28*, 112–119.

38. Liang, D.P.; Liu, T.S.; Li, Y.J. Comparative study of BRICS' CO_2 emission cost and its influential factors based on LMDI model. *Resour. Sci.* **2015**, *37*, 2319–2329.

39. Scholl, L.; Schipper, L.; Kiang, N. CO_2 Emissions from Passenger Transport: A Comparison of International Tends from 1973 to 1992. *Energy Policy* **1996**, *24*, 17–30. [CrossRef]

40. Greening, L.A.; Ting, M.; Davis, W.B. Decomposition of aggregate carbon intensity for freight: Trends from 10 OECD countries for the period 1971–1993. *Energy Econ.* **1999**, *21*, 331–361. [CrossRef]

41. Ang, B.W. Decomposition analysis for policymaking in energy: Which is the preferred method. *Energy Policy* **2004**, *32*, 1131–1139. [CrossRef]

42. Wang, W.W.; Zhang, M.; Zhou, M. Using LMDI method to analyze transport sector CO_2 emissions in China. *Energy* **2011**, *36*, 5909–5915. [CrossRef]

43. Intergovernmental Panel on Climate Change (IPCC). *IPCC Guidelines for National Greenhouse Gas Inventories 2006 Volume 2 Energy*; Intergovernmental Panel on Climate Change: Kanagawa, Japan, April 2007; ISBN 4-88788-032-4.

44. IEA Sankey Diagram. Available online: https://www.ipcc.ch/data/ (accessed on 3 May 2019).

Article

Analysis of the Influencing Factors of Regional Carbon Emissions in the Chinese Transportation Industry

Changzheng Zhu *, Meng Wang and Yarong Yang

School of Modern Post, Xi'an University of Posts & Telecommunications, Xi'an 710061, China; wm853364635@outlook.com (M.W.); yyr240267861@outlook.com (Y.Y.)
* Correspondence: zhuchangzheng@xupt.edu.cn; Tel.: +86-182-9207-8891

Received: 29 January 2020; Accepted: 28 February 2020; Published: 2 March 2020

Abstract: Global warming caused by excessive emissions of CO_2 and other greenhouse gases is one of the greatest challenges for mankind in the 21st century. China is the world's largest carbon emitter and its transportation industry is one of the fastest growing sectors for carbon emissions. However, China is a vast country with different levels of carbon emissions in the transportation industry. Therefore, it is helpful for the Chinese government to formulate a reasonable policy of regional carbon emissions control by studying the factors influencing the carbon emissions of the Chinese transportation industry at the regional level. Based on data from 1997 to 2017, this paper adopts the logarithmic mean divisia index (LMDI) decomposition method to analyze the influencing degree of several major factors on the carbon emissions of transportation industry in different regions, puts forward some suggestions according to local conditions, and provides references for the carbon reduction of Chinese transportation industry. The results show that (1) in 2017, the total carbon emissions of the Chinese transportation industry were 714.58 million tons, being 5.59 times of those in 1997, with an average annual growth rate of 9.89%. Among them, the carbon emissions on the Eastern Coast were rising linearly and higher than those in other regions. The carbon emissions in the Great Northwest were always lower than those in other regions, with only 38.75 million tons in 2017. (2) Economic output effect is the most important factor to promote the carbon emissions of transportation industry in various regions. Among them, the contribution values of economic output effect to carbon emissions on the Eastern Coast, the Southern Coast and the Great Northwest continued to rise, while the contribution values of economic output effect to carbon emissions in the other five regions decreased in the fourth stage. (3) The population size effect promoted the carbon emissions of the transportation industry in various regions, but the population size effect of the Northeast had a significant inhibitory influence on the carbon emissions in the fourth stage. (4) The regional energy intensity effect in most stages inhibited carbon emissions of the transportation industry. Among them, the energy intensity effects of the North Coast and the Southern Coast in the two stages had obvious inhibitory influences on carbon emissions of the transportation industry, but the contribution values of the energy intensity effect in the Great Northwest and the Northeast were positive in the fourth stage. (5) Except for the Great Southwest, the industry-scale effects of other regions had inhibited the carbon emissions of transportation industry in all regions. (6) The influences of the carbon emissions coefficient effect on carbon emissions in different regions were not significant and their inhibitory effects were relatively small.

Keywords: transportation industry; carbon emissions; regional; influencing factor; LMDI

---

1. Introduction

Global climate change is a major challenge in the field of sustainable development for the world. Global CO_2 emissions, influenced by rising energy demand, rose 1.7% in 2018 to a record 33.1 billion tons, according to the research report of "carbon emissions from fuel combustion" released by the International Energy Agency (IEA) [1]. In 2017, the carbon emissions of the transportation sector accounted for nearly a quarter of the global total, reaching 8.04 billion tons. According to IEA data, in 2007, Chinese CO_2 emissions surpassed those of the United States and China became the world's largest CO_2 emitter [2]. The rapid growth of Chinese carbon emissions has attracted global attention. Based on IEA data, this paper calculates that Chinese carbon emissions reached 9.302 billion tons in 2017, accounting for 28.33% of the total in the world. Among them, the transportation industry accounts for 9.5% of Chinese total carbon emissions, becoming one of the industries with the fastest growth of carbon emissions in China [3].

In recent years, Chinese carbon reduction has also received extensive attention [4]. Furthermore, as one of the important sources of Chinese energy consumption and carbon emissions, still in a stage of rapid growth, the transportation industry is bound to become a key industry to achieve targets of carbon reduction in the future [5]. China is a vast country with different industrial structures between provinces and regions [6], as well as different levels of carbon emissions in the transportation sector. Therefore, it is of positive guiding significance for the Chinese government to scientifically identify the influence degree of major factors on the carbon emissions of Chinese regional transportation industry to formulate the policies of total carbon emissions control and distribution as well as carbon reduction.

Schipper of Lawrence Berkeley national laboratory was the first scholar to study the carbon emissions of the transportation industry [7]. Later, Chinese scholars Zhu et al. and Su et al. also carried out relevant studies [8,9]. At present, studies at home and abroad mainly focus on applying the STIRPAT (Stochastic Impacts by Regression on Population, Affluence, and Technology) model, Kaya identity and its extension model, etc., applying the Laspeyres index method or the logarithmic mean divisia index (LMDI) method to analyze the degree of key factors influencing on carbon emissions and using econometric analysis and other methods to study the influencing factors of carbon emissions. Timilsina and Shrestha (2009) used the LMDI method to study CO_2 emissions of transportation sector in some Asian countries. The results showed that the change of GDP (gross domestic product) per capita, population growth and transport energy intensity were the main factors for the growth of carbon emissions; fiscal policy, fuel economy policy, and the policy of encouraging transferring into clean energy and energy-saving vehicles also played positive roles to curb carbon emissions [10,11]. Based on the Laspeyres index decomposition method, Zhang et al. (2017) selected China and other 6 countries as the research objects and constructed a secondary decomposition model for CO_2 emissions of the roads and railways. The study found that the growth of GDP per capita is the most important reason for the growth of road and railway turnover; the improvement of energy intensity and energy structure can slow down the growth of CO_2 emissions [12]. Du et al. (2017) analyzed the influencing factors of carbon emissions in the Chinese transportation industry and proposed a path analysis method based on multiple regression analysis. The results showed that economic level, transportation intensity, and energy intensity were the main factors influencing the carbon emissions of the transportation industry [13]. Talbi (2017) used the vector autoregressive model to study the relationship between CO_2 emissions and energy consumption, energy intensity, economic growth and fuel efficiency in the road transportation sector in Tunisia [14]. Based on the panel data of nine provinces and two cities in the Yangtze River economic belt from 2005 to 2014 and combined with the extended Kaya identity, Lu et al. (2017) analyzed the influencing factors of CO_2 emissions in the transportation industry through the LMDI decomposition method. The results showed that energy structure, energy consumption of per unit added value and turnover of per unit GDP inhibited CO_2 emissions, while added value of per unit turnover, per capita GDP and population promoted CO_2 emissions [15]. Based on the improved STIRPAT model, Fan et al. (2019) used ridge regression to explore the influences of passenger-freight turnover, per capita GDP, energy intensity, urbanization rate and private car ownership on traffic carbon emissions in the five Great Northwestern provinces. The result of ridge regression analysis showed that, except for Gansu Province, the contribution degrees of influence factors in the Great Northwest were all energy intensity > per capita GDP > urbanization rate > private car ownership > passenger-freight turnover [16]. In addition, in order to analyze the carbon emissions features of different regions, some scholars have also conducted studies on specific provinces and cities, such as Henan Province [17], Jiangsu Province [18] and Beijing [19].

To sum up, scholars at home and abroad further studied the influencing factors of carbon emissions in the transportation industry combining with different models from different perspectives. However, there are also the following problems: (1) the vast majority of scholars only focused on the national level [10–14], specific regions or provinces [15–19], ignoring the differences of CO_2 emissions in the transportation industry between regions of a country [20]; (2) few scholars pay attention to the degree of the same factor influencing carbon emissions in transportation industry in different regions of a country. This paper divides 30 Chinese provinces into 8 regions, calculates and analyzes the current situation of carbon emissions of different regions in the transportation industry from 1997 to 2017, and uses the LMDI decomposition method to analyze the influencing factors on carbon emissions in the transportation industry of different regions. Finally, the paper puts forward some suggestions according to local conditions to provide references for the carbon reduction of the Chinese transportation industry.

2. Methods

2.1. Carbon Emissions Calculation Method

Carbon emissions indicate "the general term of greenhouse gases, which is presented by CO_2 equivalent (CO_2 eq)", mainly including carbon dioxide, methane, nitrous oxide, and other carbon oxides. Among them, the proportion of CO_2 emissions is more than 60% of that of greenhouse gas emissions [21]. For lacking comprehensive statistics of the latest global carbon emissions, most scholars apply the methods, provided by *IPCC* (Intergovernmental Panel on Climate Change) *Guidelines on National Greenhouse Gas Inventories*, and adopt the data of energy consumption to calculate carbon emissions. In the aspect of carbon emissions measurement of transportation industry, the common methods are "top-down" and "bottom-up" methods. The "top-down" method is based on the conversion factors of energy consumptions and energy carbon emissions coefficient of vehicles to calculate the carbon emissions of the transportation industry. The "bottom-up" method is to calculate the energy consumption of the transportation industry based on the data of different types of vehicles and fuel consumptions of traveled mileage and per unit of traveled mileage, etc., thus calculating the carbon emissions. Yang et al. (2014) and Ma et al. (2017) both used the "top-down" method to calculate the carbon emissions of the transportation industry in China and the Chinese Beijing-Tianjin-Hebei Region [22,23]. As for the "bottom-up" measurement method, Zhang et al. (2009) calculated CO_2 emissions of different transportation modes for residents in Shanghai from 2002 to 2006 [24]. Wang et al. (2019) divided the Chinese comprehensive transportation system into four modes—road, railway, domestic water transportation and domestic civil aviation—and analyzed the factors of influencing carbon emissions change from 2003 to 2015 [25]. However, the "bottom-up" method needs to take into account such factors as the types of vehicles, influencing distance, energy consumption of per unit mileage, etc. At present, the relevant provincial and regional statistics in China are not yet perfect and moreover, the uncertainty is relatively large in the calculation [26]. Therefore, this paper adopts a "top-down" calculation method.

This paper, based on the data of terminal energy consumptions of industry, calculates the carbon emissions of the transportation industry in 30 provinces of China and selects 12 main types of energies according to the energy classification of IPCC. The energies include raw coal, cleaned coal, briquette coal, carbon coke, crude oil, gasoline, kerosene, diesel, fuel oil, liquefied petroleum gas and natural gas. The calculation formula of carbon emissions of fuel combustion recommended by IPCC is as follows:

$$C = \sum_{ij} C_{ij} = \sum_{ij} E_{ij} \times f_i = \sum_{ij} \left(E_{ij} \times NCV_i \times CC_i \times COF_i \times \frac{44}{12} \right) \tag{1}$$

In the formula, C represents the total CO_2 emissions of the transportation industry in China or its certain region; C_{ij} represents CO_2 emissions of energy i in province j; i is the type of the fossil fuels ($i = 1, 2, 3, \cdots, 12$); E_{ij} stands for the terminal consumption of fossil fuel i in province j; f_i indicates for the CO_2 emissions coefficient of carbon energy i; NCV_i shows the mean low calorific value of energy i; CC_i refers to the carbon per calorific value of energy i, that is, the carbon content per unit of heat; COF_i is the carbon oxidation factor, namely, the carbon oxidation rate during energy combustion; and 44 and 12 are the molecular weights of CO_2 and carbon [27].

The specific folding standard coal coefficients and carbon emissions coefficients for all energies are shown in Table 1.

Table 1. Carbon emissions coefficient of transportation and energy.

Names of Energies	Mean Low Calorific Value [NCV_i] (KJ/kg or KJ/m^3)	Carbon Per Calorific Value [CC_i] (kg-C/GJ)	Carbon Oxidation Rate [COF_i] (%)	Carbon Dioxide Emission Coefficients [f_i] (kg-CO$_2$/kg or kg-CO$_2$/m^3)
Raw coal	20,908	26.37	0.94	1.9003
Cleaned coal	26,344	25.41	0.93	2.2829
Briquette coal	15,910	33.56	0.90	1.7622
Carbon coke	28,435	29.42	0.93	2.8604
Coke oven gas	17,354	13.58	0.98	0.8469
Crude oil	41,816	20.08	0.98	3.0202
Gasoline	43,070	18.90	0.98	2.9251
Kerosene	43,070	19.60	0.98	3.0179
Diesel	42,652	20.20	0.98	3.0959
Fuel oil	41,816	21.10	0.98	3.1705
Liquefied petroleum gas	50,179	17.20	0.98	3.1013
Natural gas	38,931	15.32	0.99	2.1622

Note: data are from the General Principles for Calculating Comprehensive Energy Consumption (GB/T 2589-2008) and the Guide to National Greenhouse Gas Emission Inventory.

2.2. Kaya Identity Extension

Firstly proposed by Yoichi Kaya in 1989 [28], the Kaya identity establishes the relationship between economic and demographic factors and CO_2 emissions, decomposes the influencing factors of CO_2 emissions and accurately quantifies the contribution degree of various influencing factors.

Kaya identity is simple in structure and easy to operate. Although it has been widely used in such fields as energy, environment and economy, etc., the identity is limited in the number of focused variables. In recent years, when studying influencing factors on carbon emissions of transportation industry, some scholars have found that the change of carbon emissions is also related to energy intensity, energy structure, added value of the transportation industry, etc. In view of this, this paper extends the Kaya identity to introduce the scale and energy intensity of the transportation industry. The specific expression is:

$$C = \frac{C}{E} \times \frac{E}{ADV} \times \frac{ADV}{GDP} \times \frac{GDP}{POP} \times POP \tag{2}$$

In the formula, C represents carbon emissions of transportation industry, E represents energy consumption of transportation industry, ADV represents added value of transportation industry, GDP represents gross regional product, and POP represents regional population size. Among them, $\frac{C}{E}$ is the energy carbon emissions coefficients of the transportation industry, represented by G; $\frac{E}{ADV}$ is the factor of energy intensity, namely, the energy consumption of per unit added value for the transportation industry, represented by S; $\frac{ADV}{GDP}$ is the industry scale, that is, the proportion of the added value of the transportation industry in the gross regional product, represented by A; $\frac{GDP}{POP}$ is the factor of economic output, that is, regional GDP per capita, expressed by R; POP represents the factor of population size, expressed by P.

Thus, Equation (2) can be expressed as Equation (3):

$$C = G \times S \times A \times R \times P \tag{3}$$

In Equation (3), C, G, S, A, R, and P indicate the relationships between carbon emissions in transportation and factors of energy carbon emissions coefficients, energy intensity, industry size, economic output, and population size, etc.

2.3. Logarithmic Mean Divisia Index (LMDI) Decomposition Method Based on Extended Kaya Identity

In recent years, scholars have made great achievements in studying the relationships between energy consumptions and carbon emissions by means of decomposition analysis. Currently, the two relatively popular methods of decomposition are structural decomposition analysis (SDA) and index decomposition analysis (IDA). Based on the input-output table, SDA conducts a detailed analysis on various influencing factors [29]. In 1991, IDA expanded from the field of energy consumption to the study of carbon emissions related to energy consumption for the first time. Based on the aggregate data of departments and its time series analysis, IDA can make a meaningful decomposition for the industry and find out the deep factors that indirectly affect the total index [29,30].

Based on the comprehensive comparison of SDA and IDA methods, Ang et al. (1998) suggested that the Laspeyres index and Divisia index of the IDA method should be adopted when studying the decomposition analyses on factors of energy consumption and gas emissions [31]. However, the Laspeyres decomposition method will produce relatively large residuals during the decomposition process, which will have an influence on the results of decomposition analyses. On the contrary, LMDI in the Divisia index decomposition method solves the residual value and zero value problems inherent in the index decomposition method, which has the advantage of complete decomposition and a unique result, making the results more convincing [32,33]. Therefore, the LMDI decomposition method is finally adopted in this paper to analyze the carbon emissions of the Chinese regional transportation industry.

The LMDI decomposition method includes two forms, addition and multiplication. Their decomposition results can be inter-converted and are consistent for the two methods [32]. Therefore, this paper applies the method of "additive decomposition" to decompose the model in Equation (3). The specific formula is presented in Equation (4):

$$\Delta C = DG + DS + DA + DR + DP \tag{4}$$

In Equation (4), DG indicates for the carbon emissions coefficient effect of energy, DS indicates for the energy intensity effect, DA indicates for the industry scale effect, DR indicates for the economic output effect, and DP indicates for the population size effect. Then, the expressions of each factor, which influences on carbon emissions, are respectively Equations (5)–(9). Similarly with those in reference [34], they are used for the detailed derivation process.

$$DG = \frac{C^T - C^0}{\ln C^T - \ln C^0} \cdot \ln\left(\frac{G^T}{G^0}\right) \tag{5}$$

$$DS = \frac{C^T - C^0}{\ln C^T - \ln C^0} \cdot \ln\left(\frac{S^T}{S^0}\right) \tag{6}$$

$$DA = \frac{C^T - C^0}{\ln C^T - \ln C^0} \cdot \ln\left(\frac{A^T}{A^0}\right) \tag{7}$$

$$DR = \frac{C^T - C^0}{\ln C^T - \ln C^0} \cdot \ln(\frac{R^T}{R^0}) \tag{8}$$

$$DP = \frac{C^T - C^0}{\ln C^T - \ln C^0} \cdot \ln(\frac{P^T}{P^0}) \tag{9}$$

In Equations (5)–(9), C^0 represents the base year's carbon emissions of the transportation industry in China or its certain region; C^T indicates for the carbon emissions of the transportation industry in year T in China or its certain region. G^0, S^0, A^0, R^0 and P^0 respectively denote the base year's carbon emissions coefficient of energy, energy intensity, industry scale, economic output and population size of the transportation industry; G^T, S^T, A^T, R^T and P^T respectively indicate the energy carbon emissions coefficient, energy intensity, industry scale, economic output and population size of the transportation industry in year T.

2.4. Data Sources

The data range of this paper is from 1997 to 2017. Considering the completeness and availability of data, this study covers 30 provinces (autonomous regions and municipalities) in China, excluding Taiwan, Hong Kong, Macao and Tibet. Among them, the reference coefficients of conversion standard coal for energy and the energy consumption data of terminal transportation industry in each year of Chinese each province was derived from the *China Energy Statistical Yearbook* from 1998 to 2018. The factors of carbon emissions for the energies were derived from the *General Principles for Calculation of Comprehensive Energy Consumption* (GB/T 2589-2008) and the 2006 edition of the *IPCC Guidelines for National Greenhouse Gas Emission Inventory*. Population, GDP and ADV (added value of transportation industry) are all from the *China Statistical Yearbook* from 1998 to 2018. GDP and the added value of the transportation industry are constant price based on the base year 1997. In addition, individual missing data are obtained by interpolation, assuming the same annual growth rate.

In order to facilitate the research, this paper divides China into eight comprehensive regions according to the concept of regional division of comprehensive economy, proposed by the development research center of the state council of China [35]. They are respectively the Northeast (i.e., Liaoning, Jilin and Heilongjiang Provinces), the North Coast (i.e., Beijing, Tianjin, Hebei and Shandong), the Eastern Coast (i.e., Shanghai, Jiangsu and Zhejiang), the Southern Coast (i.e., Fujian, Guangdong and Hainan), the middle reaches of the Yellow River (i.e., Shaanxi, Shanxi, Henan, Inner Mongolia), the middle reaches of the Yangtze River (i.e., Hubei, Hunan, Jiangxi, Anhui), the Great Southwest (i.e., Yunnan, Guizhou, Sichuan, Chongqing, Guangxi), and the Great Northwest (i.e., Gansu, Qinghai, Ningxia, Xinjiang).

3. Variable Description and Analysis

3.1. Carbon Emissions Status of Chinese Transportation Industry

The "top-down" calculation method is used to calculate the total carbon emissions of Chinese transportation industry from 1997 to 2017 by using Equation (1), as shown in Figure 1.

From Figure 1, the total CO_2 emissions in the Chinese transportation industry are on the rise year by year, that is, from 108.43 million tons in 1997 to 714.58 million tons in 2017, increasing 5.59 times, with an average annual growth rate of 9.89%. From 1997 to 2012, the average annual growth rate was as high as 12.39%, and then from 2012 to 2017, the growth rate slowed down to only 2.71%. This is mainly because the economic development of China grew rapidly from 1997 to 2012, with the average annual GDP growth rate of 11.61%. As a result, the rigid demand for transportation continued to increase and the total carbon emissions of transportation continued to grow at a high speed. After 2012, Chinese economic development turned into the new normal, from the pursuit of speedy growth to quality growth. Therefore, economic growth was slowing down. From 2012 to 2017, GDP grew by an average of 8.03% a year. The demand for transportation decreased due to the optimization of the Chinese industrial structure, the decline of traditional manufacturing, and the increasing proportions of the dominating service industry of financial, information, etc. Therefore, carbon emissions growth slowed down for the Chinese transportation industry from 2012 to 2017.

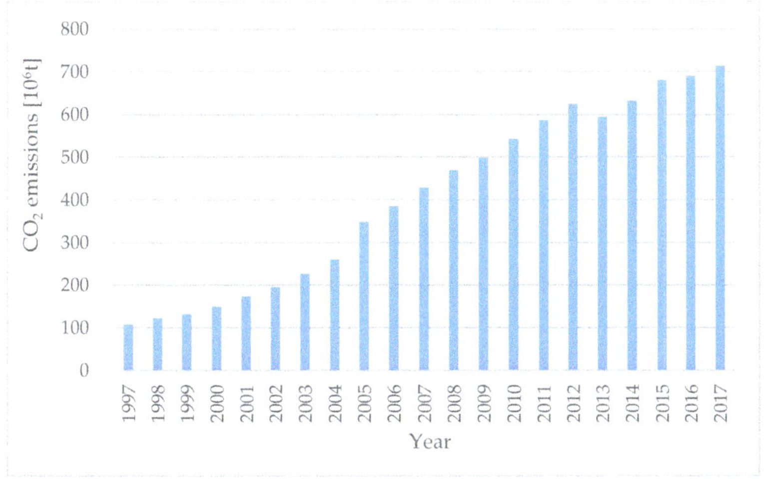

Figure 1. Carbon emissions of Chinese transportation industry from 1997 to 2017.

CO_2 emissions from transportation industry in the eight regions of China are shown in Figure 2 and Table 2. Overall, carbon emissions in all the regions were on the rising trend. Among them, the carbon emissions on the Eastern Coast rose sharply and were higher than that in other regions. Its carbon emissions increased from 19.56 million tons in 1997 to 123.36 million tons in 2017, with an average annual growth rate of 9.64%. The Eastern Coast is located in a favorable geographical position. Its economic development has always been at a high level and its demand for transportation continues to increase, so its carbon emissions are the highest. The emissions of the Great Northwest were always lower than those of other regions, with only 38.75 million tons in 2017. The reason is that the Great Northwest of China has always been an economically underdeveloped region. Its agriculture and animal husbandry have been in relatively high proportions for a long time, compared with other regions; while its industry and commerce account for relatively low proportions. Therefore, the capital attraction is relatively weak, resulting in insufficient demand for freight. In addition, the Great Northwest has a small population and relatively less demand for transportation, so it is low for the carbon emissions of the transportation industry. Carbon emissions in the Great Southwest grew the fastest during the study period, with an average annual growth rate of 12.18% and an 8.96-fold increase from 1997 to 2017. Although the economic development level of the Great Southwest is not as high as that of the Southeastern Coast, it is geographically close to the southeast, carrying on the industrial transfer of the Southeastern Coast; therefore, the demand for freight is large. At the same time, the Great Southwest is a place where the population gathers, with a total population of 246 million, accounting for 17.7% of the total population of China, therefore, it shows a large demand for passenger transportation. The Great Southwest has a more favorable traffic condition. Compared with the Great Northwest, although the railway mileage of the Great Southwest is comparable, its road transportation is more developed. Its traffic mileage is more than 94.37% of the Great Northwest, especially with the convenience of shipping in the Yangtze River of Sichuan and ports in Guangxi, the inland waterway navigation mileage being 16.7 times that of the Great Northwest. Therefore, with the rapid growth of the transportation industry, the carbon emissions in the Great Southwest increase rapidly.

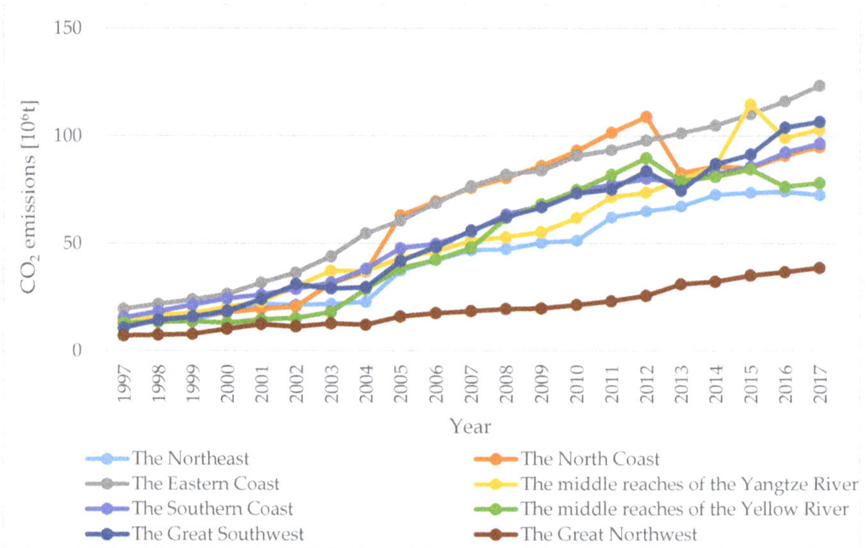

Figure 2. CO$_2$ emissions in the transportation industry in the eight regions of China from 1997 to 2017.

Table 2. Carbon emissions of the transportation industry in eight Chinese regions.

Region	Average Annual Growth Rate (%)	Carbon Emissions in 1997 (Million Tons)	Carbon Emissions in 2017 (Million Tons)	1997–2017 Increasing Multiples of Carbon Emissions
The Northeast	8.56	14.07	72.73	4.17
The North Coast	9.83	14.57	95.00	5.52
The Eastern Coast	9.64	19.56	123.36	5.31
The middle reaches of the Yangtze River	10.69	13.53	103.11	6.62
The Southern Coast	9.51	15.71	96.63	5.15
The middle reaches of the Yellow River	9.37	13.04	78.29	5.00
The Great Southwest	12.18	10.71	106.72	8.96
The Great Northwest	8.75	7.24	38.75	4.35
Total	9.89	108.43	714.58	5.59

3.2. Decomposition Variables

3.2.1. Carbon Emissions Coefficients

The carbon emissions coefficients of energy in eight regions of China are shown in Figure 3. Within the study range, the carbon emissions coefficients show a slow downward trend, with a relatively small decline. This is because in 2017, compared with 1997, the proportion of clean energy such as natural gas increased. In addition, the improvements of oil quality and fuel efficiency of vehicles reduced the CO$_2$ emissions under the same energy consumption, which lowered the carbon emissions coefficients of energy. However, most of the vehicles in China still rely on traditional petroleum energy, so the carbon emissions coefficients of energies in each region are relatively small.

3.2.2. Energy Intensity

In this paper, the energy intensity is the ratio between the energy consumption of transportation and the added value of transportation industry (at the constant price in 1997) and its unit is 10,000 tons of standard coal per 100 million yuan. The change of energy intensities of the transportation industry in the Chinese regions are shown in Figure 4. In general, the energy intensities of the transportation industry in most regions present an "inverted U-shaped" trend within the research range, that is, it rises first and then slowly declines. Among them, the middle reaches of the Yangtze River presented the fastest growth in energy intensity, rising 51.07% in 2017 compared with that in 1997, with an average annual growth rate of 2.08%. In 2017, the energy intensity in the Great Northwest was the highest, with 23,200 tons of standard coal/100 million yuan. Because the proportion of road transportation in the region continued to increase, its energy consumption was higher than those of the other regions. The energy intensity of the North Coast is the lowest, which was only 8500 tons of standard coal/100 million yuan in 2017. The energy intensity of the region decreased rapidly after 2012. It is mainly because the relatively strict energy conservation policies for vehicles have been adopted to reduce energy consumption. Thus, its energy intensity is the lowest.

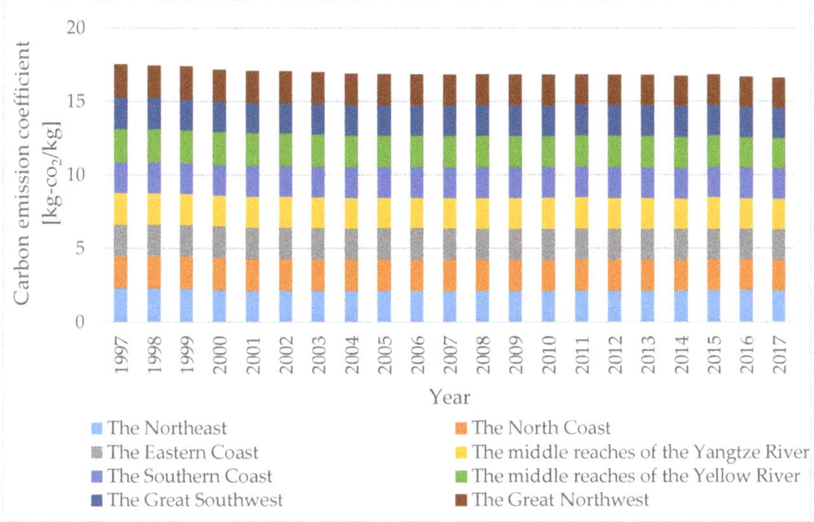

Figure 3. Carbon emissions coefficients of different regions in China from 1997 to 2017.

3.2.3. Industry Scale

Industry scale is the proportion of added value of transportation industry in gross regional product. The change of industry scale in each region is shown in Figure 5. During the study range, all regions show an "inverted U-shaped" trend, which is first in a slow rise and then a fluctuating decrease. After 20 years of reform and opening up, the Chinese market economy has realized great development and domestic manufacturing industry has begun to make progress. At the same time, due to the cheap labor and various preferential policies, large numbers of processing and manufacturing were shifted to China, resulting in growing demands for transportation. Therefore, the scale of the regional industry from 1997 to 2002 had different degrees of increase. After 2002, the proportion of services represented by finance, tourism and information in each region increased, while the proportion of traditional industries, mainly manufacturing, decreased and the demand for transportation slowed down, thus slowing down the growth rate of added value of transportation industry.

Among them, the decline rates of the scale of industry in the Southern and the Eastern Coast are relatively fast, with an average annual decline rate of 2.91% and 2.18% respectively. This is because the Southern and Eastern Coast are forefronts of reform and opening-up in China. Their industrial transformation and upgrading paces are faster. Whether information technology or the services industry, like financial, are all at the advanced level. In addition, while a variety of emerging industries are influencing regional economic level, they slow down the transportation demand, making industry scale reduce quickly. In contrast, the industrial scale in the Great Southwest did not decline, but grew at an average annual rate of 0.08%. This is because, firstly, the industrial transformation and upgrading in the Great Southwest is relatively slow and the proportion of the traditional secondary industry is not substantially optimized. Secondly, there are many basins and plateaus in the Great Southwest, with rugged terrain and inconvenient transportation, which makes it difficult to reduce transportation costs.

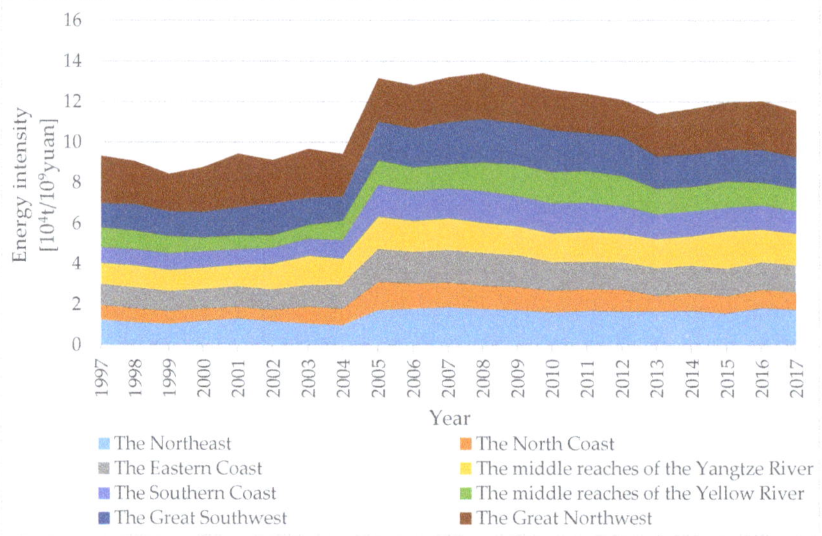

Figure 4. Regional energy intensities in China from 1997 to 2017.

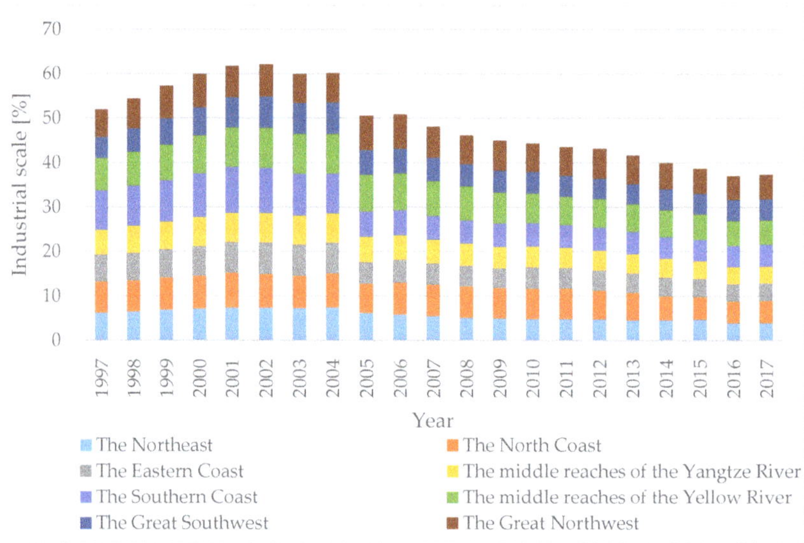

Figure 5. Industry scale in various regions of China from 1997 to 2017.

3.2.4. Economic Output

Economic output is all expressed by GDP per capita in this paper. From 1997 to 2017, Chinese economic output continued to grow at an average annual rate of 10.01%. GDP per capita in 2017 was 42,407.71 yuan, 5.73 times higher than that in 1997. As shown in Figure 6, the economic output of each region continued to rise during the study range. Among them, the GDP per capita of the Eastern Coast has always been higher than that of other regions, with an average annual growth rate of 9.60%, reaching 70,347.40 yuan by 2017. Due to the superior geographical location, the Eastern Coast is the earliest to implement the reform and opening up in China and supported by national policies, so its economic output keeps rising. The economic output of the Southern Coast ranking second in 2017 was mainly because that the Southern Coast has had a strong business atmosphere and is also an area that firstly implemented the reform and open policy. With the locative advantage of being adjacent to Hong Kong, Macao and Taiwan and with the help of the good shipping conditions, it attracted the inflow of large amounts of foreign capital, making the change in industrial structure, proportional increase in manufacturing industry and rapid development of foreign trade, so it reached a relatively higher economic development level. The economic output in the Great Northwest in 2017 was the lowest in all the regions. The reason is that the Great Northwest has poor natural conditions, with wide distribution of desert, Gobi, plateaus and mountains. It makes relatively high proportions of farming and animal husbandry, with relatively low proportions of commerce and industry. In addition, the development pattern of the priority given to the marine trade since modern times also causes the decline of inland commercial. Moreover, the factor of social instability in the Great Northwest also has a certain influence on the development of its economy.

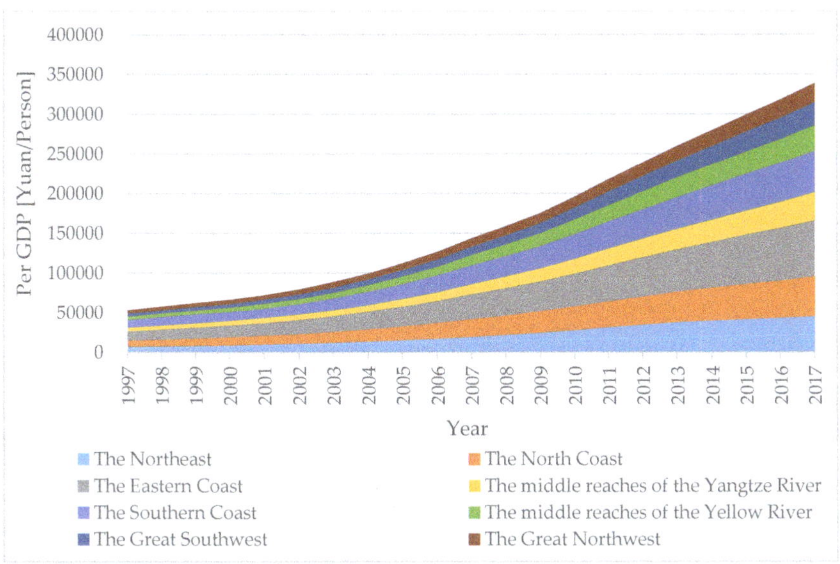

Figure 6. Regional economic output of China from 1997 to 2017.

3.2.5. Population Size

Compared with 1997, the Chinese population increased by 153.82 million at the end of 2017, reaching 1390.08 million. The population sizes of the eight regions are shown in Figure 7. On the whole, the population sizes of each region increased slowly to varying degrees. Among them, the Southern Coast was witnessed the highest growth rate, with an average annual growth rate of 1.86% from 1997 to 2017. It is followed by the Eastern Coast, with an average annual growth rate of 1.06%. The two regions with the slowest growth rates were the Northeast and the Great Southwest, both with average annual growth rates of 0.17%. The population of the Northeast decreased 0.31% annually from 2015 to 2017. It is mainly because the Northeast is an old industrial base with unbalanced industrial structure and slow economic growth, resulting in a large population outflow.

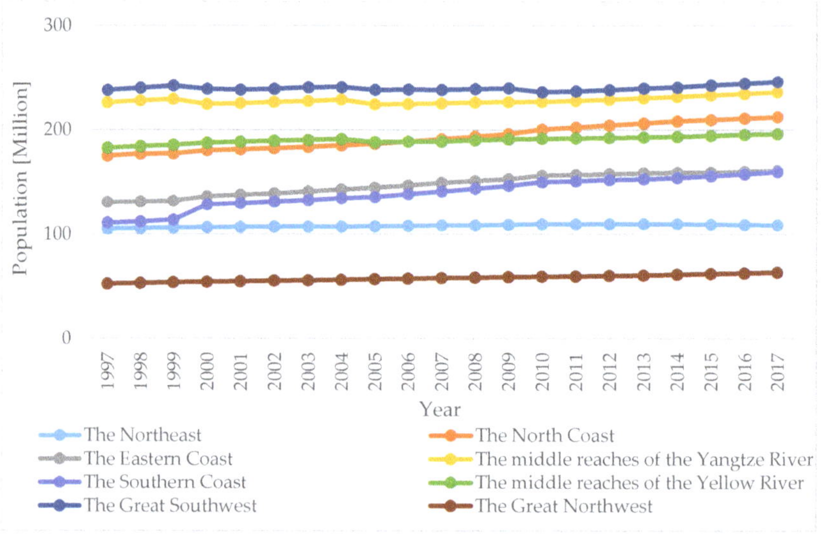

Figure 7. Regional population sizes of China from 1997 to 2017.

4. Results and Discussion

4.1. Decomposition Results of Total Carbon Emissions in the Chinese Transportation Industry

In this paper, the research range is from 1997 to 2017. Because the research range experienced financial volatilities of Asian financial crisis and the world financial crisis, etc., Chinese economic growth has begun to slow markedly since 2012. So this paper divides the research range into four stages: the first stage (1997–2002), the second stage (2002–2007), the third stage (2007–2012) and the fourth stage (2012–2017).

By using the LMDI decomposition method, this paper decomposes the factors, which influence on the carbon emissions of Chinese transportation industry, into economic output, industry scale, energy intensity, population size, and carbon emissions coefficients. Equations (5)–(9) are used to calculate the contribution values (D) and contribution rates (R) of each influencing factor. The results are shown in Table 3. Among them, the contribution rate is the ratio between the contribution value of each influencing factor and the sum of contribution values of the five influencing factors, e.g., $RG = DG/\Delta C$.

From 1997 to 2017, the contribution value and contribution rate of energy carbon emissions coefficients effect were −16.14 and −2.65%, respectively, to the carbon emissions of transportation industry. It can be seen that the carbon emissions coefficients effect is conducive to the suppression of the industry's carbon emissions, but the influence is relatively small. This is because the improvements of vehicle fuel efficiency and oil quality reduced the carbon emissions coefficients of energy, thus promoting carbon reduction in the transportation industry.

Energy intensity effect has different influences on carbon emissions of Chinese transportation industry in different stages. It has a specific and significant effect on curbing carbon emissions of Chinese transportation industry in the third and fourth stages. The main reasons are as follows: (1) since 2007 and 2012, the Chinese government has proposed and implemented standards for the national pollutant emissions of motor vehicles, standardized air pollution emissions from the policy perspective and improved the quality of oil products for motor vehicles. These measures are conducive to reducing carbon emissions. (2) The proportion of clean energy consumption has increased. The proportion of clean energy consumption in Chinese transportation industry has increased from 0.02% in 1997 to 4.76% in 2017. (3) The continuous improvement of Chinese informatization level has promoted the information technology development of platforms of online ride-hailing, taxi dispatching, freight dispatching, etc., effectively improved the motor vehicles' operation efficiencies of trucks, passenger cars, taxis and etc., and reduced empty influencing rate, thus lowering carbon emissions.

Table 3. Decomposed contribution value (unit: million tons) and contribution rate (unit: %) of influencing factors on carbon emissions of Chinese transportation.

Effect		1997–2002	2002–2007	2007–2012	2012–2017	1997–2017
Energy Carbon Emission Factor	DG	−4.06	−2.55	−0.34	−8.95	−16.14
	RG	−4.63	−1.09	−0.18	−10.03	−2.65
Energy Intensity	DS	−2.27	132.84	−38.01	−64.12	85.07
	RS	−2.59	56.85	−19.38	−71.80	13.95
Industrial Scale	DA	25.72	−83.88	−56.31	−95.89	−116.71
	RA	29.33	−35.89	−28.71	−107.37	−19.14
Economic Output	DR	62.35	180.22	273.56	238.58	616.80
	RR	71.10	77.12	139.47	267.15	101.14
Population size	DP	5.95	7.06	17.24	19.68	40.82
	RP	6.78	3.02	8.79	22.04	6.69
Total Effect	ΔC	87.69	233.68	196.15	89.31	609.84
	RC	100	100	100	100	100

The industry scale effect has a relatively obvious influence on the inhibition of carbon emissions in Chinese transportation industry, with the overall contribution value and contribution rate of −116.71% and −19.14%, respectively, in the study range. From different stages, except the first stage to promote the increase of carbon emissions, the last three stages all act as inhibitory effects. This is mainly because China has begun to actively seek the transformation and upgrading for industrial structure. The proportion of secondary industry in the national economy has declined, while that of the tertiary industry has been rising. As the transportation volume for per unit output value of the secondary industry is higher than that of the tertiary industry, the demand for transportation is gradually reduced and the industry scale is also continuously reduced, thus restraining the increase of carbon emissions.

As shown in Table 3, the contribution value of economic output effect to the carbon emissions of the transportation industry increased year by year. The contribution rate increased from 71.10% in the first stage to 267.15% in the fourth stage. It can be seen that economic output is the key factor for the rise of the total carbon emissions of Chinese transportation industry. Therefore, if China wants to reduce the carbon emissions of the transportation industry, it should slow down its speed of economic development.

The contribution value of the population size effect to the carbon emissions of the transportation industry continued to rise from 1997 to 2017, but the total contribution value was only 40.82. This shows that the population size effect has a certain promoting influence on the carbon emissions of the Chinese transportation industry, but its overall influence is limited. It is because Chinese population control policy made the population growth rate begin to slow down. Although the continuing growth of population promoted the increase of the demand for transportation, its increase rate is not large, so the carbon emissions of transportation do not increase much.

4.2. Analyses of the Key Influencing Factors for Regional Carbon Emissions

In this paper, the LMDI decomposition method is used to decompose the factors affecting the carbon emissions of transportation in eight regions of China. Their contribution values and contribution rates of each factor are calculated. The results are shown in Tables 4 and 5.

Table 4. Results of LMDI decomposition for carbon emissions from transportation industry in the 8 regions (unit: million tons).

Year	Factor	The Northeast	The North Coast	The Eastern Coast	The Middle Reaches of the Yangtze River	The Southern Coast	The Middle Reaches of the Yellow River	The Great Southwest	The Great Northwest
1997–2002	DG	−1.43	−0.03	−0.05	−0.47	−0.07	−0.43	−0.55	−0.43
	DS	−1.41	−3.86	−0.91	4.40	0.08	−6.72	5.32	−0.86
	DA	2.88	1.59	4.04	3.82	2.66	3.12	7.80	1.64
	DR	7.12	7.86	12.04	8.72	6.82	5.95	7.65	3.25
	DP	0.33	0.73	1.72	0.05	3.66	0.55	0.09	0.51
	ΔC	7.49	6.28	16.84	16.51	13.15	2.48	20.31	4.11
2002–2007	DG	−0.13	−1.23	−0.10	−0.36	−0.23	−1.05	0.28	−0.37
	DS	15.49	32.12	25.77	8.04	25.32	18.29	12.33	0.50
	DA	−9.31	−3.44	−20.89	−9.38	−25.16	−4.01	−12.42	−0.80
	DR	18.97	25.78	31.56	23.13	23.70	19.57	24.90	7.28
	DP	0.41	1.95	3.93	−0.22	2.98	−0.10	−0.10	0.70
	ΔC	25.44	55.18	40.26	21.20	26.61	32.71	24.98	7.31
2007–2012	DG	0.32	−0.31	−1.15	0.75	0.35	0.24	−0.35	−0.28
	DS	−7.01	−9.19	−15.61	−6.02	−5.75	14.72	−5.87	−3.92
	DA	−7.88	−7.10	−5.31	−10.41	−4.38	−12.15	−9.56	−0.60
	DR	32.00	43.30	38.44	37.30	29.51	37.71	43.54	11.01
	DP	0.62	6.27	4.72	1.01	5.09	1.30	−0.03	0.84
	ΔC	18.04	32.97	21.08	22.63	24.82	41.82	27.73	7.05
2012–2017	DG	−0.05	−1.21	−0.64	−1.27	−0.39	−3.39	−1.37	−0.55
	DS	3.94	−25.71	−0.13	8.30	−15.60	−25.98	−21.10	7.52
	DA	−13.76	−25.67	−15.71	−15.58	−2.78	−13.83	3.87	−6.85
	DR	18.18	34.70	39.83	35.03	30.71	30.02	38.47	11.27
	DP	−0.62	3.90	2.26	2.75	4.39	1.58	3.12	1.64
	ΔC	7.69	−13.99	25.61	29.24	16.33	−11.60	22.99	13.03
1997–2017	DG	−2.88	−1.98	−1.29	−1.51	−0.36	−3.89	−1.75	−1.91
	DS	11.61	7.70	14.78	18.20	16.17	4.02	10.95	−0.03
	DA	−16.53	−13.71	−24.88	−17.55	−26.34	−10.71	0.71	−2.21
	DR	65.26	80.10	103.29	88.48	75.05	76.38	84.66	32.04
	DP	1.20	8.33	11.90	1.95	16.40	2.71	1.44	3.62
	ΔC	58.66	80.44	103.79	89.58	80.92	68.52	96.00	31.51

4.2.1. Carbon Emissions Coefficient Effect of Energy

From Table 4 (DG), the contribution values of carbon emissions coefficient effect of energy in the eight regions to the carbon emissions of the transportation industry are all nearly negative and their absolute values are small. It can be seen that the carbon emissions coefficient effect inhibits the carbon emissions of the transportation industry in different regions. However, the inhibition effect is small and the regional differences are not large. This is due to the fact that the fuel sources and the quality of oil products for vehicles in different regions of China are similar, with a low proportion of clean energy and a low degree of improvement in the quality of oil products.

4.2.2. Energy Intensity Effect

The energy intensity effect (DS, Table 4) has certain differences in influencing carbon emissions of the transportation industry in different regions. In general, except for the Great Northwest, the contribution values of the energy intensity effect to carbon emissions in the other seven regions were positive from 1997 to 2017.

Within the study range, the energy intensity effects of each region inhibited the carbon emissions of the transportation industry in most stages. Among them, the energy intensity effects of the Southern and the North Coast have significant inhibitory influences on the carbon emissions of the transportation industry in the third and fourth stages. This is mainly because these two regions adopted relatively strict energy-saving policies for transportation vehicles, which reduced energy consumption in the transportation process and restrained the increase of carbon emissions. In the fourth stage, the energy intensity effects in the Great Northwest and the Northeast regions became positive, and their energy intensities were still rising. This is because the two regions have relatively weak control over the energy saving of vehicles, and clean energy consumption accounts for a low proportion.

Table 5. Contribution rate of LMDI decomposition of carbon emissions in transportation industry in the 8 regions (unit: %).

Year	Factor	The Northeast	The North Coast	The Eastern Coast	The Middle Reaches of the Yangtze River	The Southern Coast	The Middle Reaches of the Yellow River	The Great Southwest	The Great Northwest
1997–2002	RG	−19.15	−0.52	−0.33	−2.87	−0.54	−17.33	−2.73	−10.40
	RS	−18.83	−61.51	−5.38	26.68	0.62	−271.14	26.19	−20.94
	RA	38.49	25.26	23.99	23.11	20.26	126.06	38.43	39.93
	RR	95.12	125.17	71.49	52.79	51.81	240.32	37.67	79.07
	RP	4.37	11.61	10.23	0.29	27.85	22.10	0.44	12.34
	RC	100	100	100	100	100	100	100	100
2002–2007	RG	−0.49	−2.24	−0.24	−1.68	−0.88	−3.20	1.11	−5.01
	RS	60.89	58.21	64.00	37.91	95.17	55.90	49.35	6.85
	RA	−36.60	−6.23	−51.90	−44.26	−94.55	−12.25	−49.73	−10.89
	RR	74.57	46.73	78.39	109.08	89.06	59.84	99.69	99.55
	RP	1.63	3.54	9.76	−1.05	11.20	−0.29	−0.42	9.50
	RC	100	100	100	100	100	100	100	100
2007–2012	RG	1.77	−0.95	−5.45	3.32	1.42	0.58	−1.26	−4.01
	RS	−38.87	−27.88	−74.04	−26.59	−23.16	35.21	−21.18	−55.55
	RA	−43.67	−21.54	−25.21	−45.99	−17.66	−29.06	−34.49	−8.58
	RR	177.35	131.35	182.32	164.81	118.90	90.17	157.04	156.23
	RP	3.41	19.01	22.38	4.45	20.50	3.10	−0.11	11.92
	RC	100	100	100	100	100	100	100	100
2012–2017	RG	−0.66	8.67	−2.49	−4.34	−2.37	29.25	−5.97	−4.23
	RS	51.22	183.80	−0.50	28.39	−95.54	223.91	−91.76	57.71
	RA	−179.01	183.54	−61.32	−53.28	−17.01	119.19	16.84	−52.57
	RR	236.47	−248.12	155.49	119.83	188.05	−258.77	167.34	86.48
	RP	−8.03	−27.90	8.81	9.39	26.87	−13.58	13.55	12.61
	RC	100	100	100	100	100	100	100	100
1997–2017	RG	−4.90	−2.46	−1.24	−1.68	−0.45	−5.68	−1.83	−6.07
	RS	19.79	9.58	14.24	20.32	19.98	5.86	11.40	−0.09
	RA	−28.17	−17.05	−23.97	−19.59	−32.55	−15.63	0.73	−7.01
	RR	111.25	99.58	99.52	98.78	92.75	111.48	88.18	101.69
	RP	2.04	10.35	11.46	2.18	20.27	3.96	1.51	11.48
	RC	100	100	100	100	100	100	100	100

4.2.3. Industry Scale Effect

From Table 4 (DA), except for the Great Southwest, the contribution values of industry scale effects in the other 7 regions were negative from 1997 to 2017, indicating that the industrial scale effect inhibited the carbon emissions of the transportation industry on the whole.

Among them, the absolute values of the contribution values of industry scale effects in the Southern and the Eastern Coast were relatively large, which played a great role in restraining the carbon emissions from the transportation industry. This is mainly caused by the change of regional industrial structure. The proportion of the secondary industry in the Southern Coast decreased from 47% in 1997 to 37% in 2017, but the tertiary industry increased from 37% in 1997 to 52% in 2017. On the Eastern Coast, the proportion of secondary industry decreased from 52% in 1997 to 39% in 2017, but the tertiary industry increased from 36% to 58%. The decreased proportion of secondary industry leads to a gradual decrease in the demand for transportation and the scale of the transportation industry, thus restraining the increase of carbon emissions. However, the industrial structure of the Great Southwest has not changed significantly. In 1997, its secondary industry accounted for 41%, and in 2017, it still accounted for 39%. The demand for transportation was still high, so the carbon emissions of transportation industry in this region were still rising.

4.2.4. Economic Output Effect

It can be seen that the contribution values of the economic output effect in each region within the research range (DR, Table 4) are always positive. Their contribution rates to the carbon emissions of the transportation industry are greater than other effects in the same period (RR, Table 5). This indicates that economic output effect is the most important factor influencing on the growth of the carbon emissions of transportation industry in each region.

Overall, the contribution values of the economic output effects of the Eastern Coast and the Southern Coast and the Great Northwest to the carbon emissions of the transportation industry continued to rise. This is mainly due to the fact that the Eastern and the Southern Coast are the first regions to implement reform and opening up in China, with strong state policy support, a high degree of foreign trade, superior geographical location and convenient land and sea transportation, resulting in its rapid economic development. The sustained and rapid economic growth increases the demand for transportation, so the economic output effect plays a great role in promoting the carbon emissions of the transportation industry in these two regions. For the Great Northwest, since China officially implemented the "western development strategy" in 2000, the state council has issued four five-year plans for the development of the western regions, focusing on the top-level design of industrial development, ecological and environmental protection, infrastructure construction and opening to the outside world, etc. The continuous progress of the "western development strategy" made the Great Northwest realize the transformation and upgrading of partly traditional industries, develop a series of characteristic industries, promoting the transformation from the resource advantage to economic advantage. Moreover, because of the low base of economic output in the Great Northwest, its economic output has grown rapidly, which in turn has increased the speed of demand for transportation, thus objectively bringing about the increase in carbon emissions.

The contribution values of economic output effect to carbon emissions in the Northeast, the North Coast, the middle reaches of Yangtze River, Yellow River and the Great Southwest increased continuously in the first three stages but decreased in the fourth stage. This is mainly due to the continuous growth of economic output in various regions since the reform and opening up, and these five regions are no exception. Therefore, the rigid demand for transportation keeps rising, which promotes the growth of carbon emissions. Consequently, the contribution values of economic output effect to the carbon emissions of the transportation industry in the first three stages continue to increase. Since 2012, Chinese economic development has slowed down and turned from high-speed growth to high-quality development. Due to the slow industrial transformation in these five regions, the economic growth slowed down, so that the contribution value of economic output effect to the carbon emissions of the transportation industry declined in the fourth stage.

4.2.5. Population Size Effect

From Tables 4 and 5, the contribution values of population size effects (DP, Table 4) of the eight regions in each stage are mostly positive, indicating that population size effect can promote the increase of carbon emissions in the transportation industry.

In general, from 1997 to 2017, the contribution values of population size effects to the carbon emissions of the transportation industry in the Southern and Eastern Coast were relatively large, 16.40 and 11.90, respectively, with contribution rates of 20.27% and 11.46%, respectively. This is mainly due to the fact that the Southern and Eastern Coast are the two regions with the most developed economy in China. The large population inflow resulted in the soaring demand for transportation, which promoted the increase of carbon emissions.

The contribution degree of population size effect to carbon emissions in the Northeast was relatively small in the first three stages. Even the contribution value and contribution rate were negative in the fourth stage. This is mainly because the population of the Northeast has continually decreased for three consecutive years. The population in 2017 was 0.92% lower than that in 2014. There are three main reasons for the decline. Firstly, the birth rate in the Northeast is far below that of

Chinese average and the population is ageing severely. Secondly, the government in the Northeast is less efficient than those of other economically developed regions, which makes it difficult to attract investment and thus absorbs less employment. Finally, as the traditional heavy industry in the Northeast has weakened over the years and light industry has not developed, it attracts fewer talents.

5. Conclusions and Suggestions

5.1. Conclusions

In this paper, 30 provinces of China are divided into 8 regions to calculate and analyze the current situation of regional carbon emissions in the transportation industry from 1997 to 2017. Based on the LMDI decomposition method, the contribution values and contribution rates of each influencing factor are analyzed, and the following main conclusions are drawn as follows.

(1) The total CO_2 emissions in the Chinese transportation industry show a ladder-type annual growth, from 108.43 million tons in 1997 to 714.58 million tons in 2017, with an average annual growth rate of 9.89%. Among them, the carbon emissions on the Eastern Coast are rising steeply and are higher than those in other regions as a whole. In 2017, the carbon emissions of the transportation industry in the Eastern Coast accounted for 17.27% of the Chinese total, but the total area only accounted for 2.28% of Chinese total. The emissions of the Great Northwest were always lower than those of other regions, with only 38.75 million tons in 2017. Carbon emissions in the Great Southwest grew the fastest during the study range, with an average annual growth rate of 12.18% from 1997 to 2017, an 8.96-fold increase.

(2) Based on the results of LMDI decomposition, the economic output effect is the most important factor to promote the carbon emissions of the transportation industry in various regions. Among them, the contribution values of economic output effect to carbon emissions of the Eastern Coast, the Southern Coast and the Great Northwest continued to rise, while the contribution values of the economic output effect to carbon emissions of the other five regions decreased to some degree in the fourth stage. The population size effect promoted the carbon emissions of the transportation industry in various regions, but the population size effect of the Northeast had a significant restraining effect on the carbon emissions in the fourth stage. Energy intensity effects of each region in most stages suppressed the carbon emissions of the transportation industry. Among them, the energy intensity effects of the North Coast and the Southern Coast in the two stages had obvious inhibitory influences on transportation carbon emissions, while the energy intensity effects of the Great Northwest and the Northeast still had positive contribution values in the fourth stage. Except the Great Southwest, the industry scale effects have inhibited the carbon emissions of the transportation industry in other regions. The carbon emissions coefficient effect to carbon emissions in different regions is not significant and the inhibitory effect is relatively small.

5.2. Policy Recommendations

In general, there are some differences in the factors influencing carbon emissions of the transportation industry in different regions of China. Therefore, based on the above research results, the following suggestions are proposed for policy makers.

(1) Change the pattern of economic growth and appropriately lower the speed of economic development. The results of this paper show that the economic output effect is the main factor leading to the increase of carbon emissions in the Chinese transportation industry. China should gradually change the pattern of economic growth, appropriately reduce the speed of economic development, and strive to achieve a coordinated development of economic growth and environmental protection. In the process of carbon reduction in the transportation industry, China should set differentiated emissions reduction targets. Economically developed regions such as the Eastern Coast and the Southern Coast regions can consider setting more stringent standards for industrial carbon reduction so as to realize the first transformation of the low-carbon economic growth mode. In the process of economic development, regions such as the Great Northwest, the Northeast and the middle reaches of the Yellow River can

gradually increase their responsibilities of carbon reduction and promote the steady transformation of economic growth pattern.

(2) Reduce the energy intensity of regional transportation industry. Energy intensity effect should have been an important factor for reducing carbon emissions of the transportation industry, but the research results show that energy intensity effect promotes carbon emissions of transportation industry on the whole. So China should actively promote the usages of new energy, clean energy vehicles and ships, increase the application of new energy and clean energy vehicles in the fields of city bus, taxi, express delivery, airports, railway freight yard, etc., and reduce the unit consumption of transportation industry and regional energy intensity to control the carbon emissions of the transportation industry. The regions of the Great Northwest and the Northeast, etc. with high energy intensities should formulate more reasonable policies of emissions reduction, vigorously promote the use of clean energy, and improve the efficiency of clean energy.

(3) Optimize the regional industrial structure. As a whole, the industry scale effect has a restraining influence on the carbon emissions of the Chinese transportation industry. Therefore, China can promote the optimization and upgrading of its industrial structure, develop strategic emerging industries and a modern service industry, and promote the industries to move towards the medium-high end and achieve high-quality development. The rapid transformation of the industrial structure in the Eastern Coast and Southern Coast regions has played a significant role in curbing the carbon emissions of the transportation industry. Therefore, China should continue to optimize the industrial structure, accelerate the expansion of tertiary industry, and encourage the development of new high-tech industries. For economically underdeveloped regions such as the Great Southwest, the Great Northwest and the middle reaches of the Yellow River, appropriate industrial transformation policies should be formulated to encourage them to constantly optimize their industrial structures, improve their capacities for scientific and technological innovation, and shift them from low-end traditional manufacturing to medium- and high-end manufacturing so as to reduce the demand for transportation and lower carbon emissions.

Author Contributions: C.Z. and M.W. conceived the study, wrote original draft and contributed to all aspects of this work, Y.Y. analyzed the data and gave some useful suggestions to this work. All authors have read and agreed to the published version of the manuscript.

Funding: This research was funded by the National Social Science Foundation in China. (Grant No. 19BJY175).

Acknowledgments: The authors thank Dawei Gao and Wenbo Du for their help in participation in data collection, and thank Lijiao Qin for her help in the English editing.

Conflicts of Interest: The authors declare no conflict of interest.

References

1. International Energy Agency. *CO₂ Emissions from Fuel Combustion 2019*; International Energy Agency: Paris, France, 2019.
2. International Energy Agency. *CO₂ Emissions from Fuel Combustion 2008*; International Energy Agency: Paris, France, 2008.
3. Xu, B.; Lin, B.Q. Differences in regional emissions in China's transport sector: Determinants and reduction strategies. *Energy* **2016**, *95*, 459–470. [CrossRef]
4. Duan, H.B.; Mo, J.L.; Fan, Y.; Wang, S.Y. Achieving China's energy and climate policy targets in 2030 under multiple uncertainties. *Energy Econ.* **2018**, *70*, 45–60. [CrossRef]
5. Yu, J.; Da, Y.B.; Ouyang, B. Analysis of carbon emission changes in China's transportation industry based on LMDI decomposition method. *China J. Highw. Transp.* **2015**, *28*, 112–119.
6. Lei, L.; Zhong, Y.Y.; Yuan, X.L. Decomposition Model and Empirical Study of Regional Carbon Emissions for China. *Mod. Econ. Sci.* **2011**, *33*, 59–65.
7. Shipper, L.; Scholl, L.; Price, L. Energy Use and Carbon Emissions from Freight in 10 Industrialized Countries: An Analysis of Trends from 1973 to 1992. *Transp. Res. D* **1997**, *2*, 57–76. [CrossRef]

8. Zhu, Y.Z. Analyses on Energy Development and Carbon Exhaustion According to Circumstances in Future China's Communications and Transportation. *China Ind. Econ.* **2001**, *12*, 30–35.
9. Su, T.Y.; Zhang, J.H.; Li, J.L.; Ni, Y. Influence Factors of Urban Traffic Carbon Emission: An Empirical Study with Panel Data of Big Four City of China. *Ind. Eng. Manag.* **2011**, *16*, 134–138.
10. Timilsina, G.R.; Shrestha, A. Transport Sector CO_2 Emission Growth in Asia: Underlying Factors and Policy Options. *Energy Policy* **2009**, *37*, 4523–4539. [CrossRef]
11. Timilsina, G.R.; Shrestha, A. Factors Affecting Transport Sector CO_2 Emissions Growth in Latin American and Caribbean Countries: An LMDI Decomposition Analysis. *Int. J. Energ. Res.* **2009**, *33*, 396–414. [CrossRef]
12. Zhang, H.J.; Wang, L.N.; Chen, W.Y. Decomposition analysis of CO_2 emissions from road and rail transport systems. *Tsinghua Univ. Sci. Technol.* **2017**, *57*, 443–448.
13. Du, Q.; Su, Q.; Yang, Q.; Feng, X.Y.; Yang, J. Path analysis method of influencing factors of carbon emissions for Chinese transportation industry. *J. Traffic Transp. Eng.* **2017**, *17*, 143–150.
14. Talbi, B. CO_2 emissions reduction in road transport sector in Tunisia. *Renew. Sust. Energ. Rev.* **2017**, *69*, 232–238. [CrossRef]
15. Lu, S.R.; Jiang, H.Y.; Liu, Y. Regional Disparities and Influencing Factors of CO_2 Emission in Transportation Industry. *J. Transp. Syst. Eng. Inf. Technol.* **2017**, *17*, 32–39.
16. Fan, Y.J.; Qu, J.S.; Zhang, H.F.; Xu, L.; Bai, J.; Wu, J.J. Study on the Current Situation and Influence Factors of Transportation Carbon Emissions in Five Great Northwest Provinces. *Ecol. Econ.* **2019**, *35*, 32–37.
17. Li, X.Y.; Li, D. Influencing factors and spatial pattern analysis of the transportation carbon emission in Henan Province. *J. Lanzhou Univ. Nat. Sci.* **2019**, *55*, 430–435.
18. Ouyang, B.; Feng, Z.H.; Li, Z.K.; Bi, Q.H.; Zhou, A.Y. Calculation and Evaluation Methodology of Transport Energy Consumption and Carbon Emission-The Case of Jiangsu Province. *Soft Sci.* **2015**, *29*, 139–144. [CrossRef]
19. Ke, S.F.; Wang, Y.; Chen, Y.G.; Liu, A.Y. Carbon Emissions and Reduction Scenarios of Transportation in Beijing. *China Popul. Resour. Environ.* **2015**, *25*, 81–87.
20. Zheng, J.L.; Mi, Z.F.; Coffman, D.; Milcheva, S.; Shan, Y.L.; Guan, D.B.; Wang, S.Y. Regional development and carbon emissions in China. *Energy Econ.* **2019**, *81*, 25–36. [CrossRef]
21. Zhu, C.Z.; Gao, D.W. A Research on the Factors Influencing Carbon Emission of Transportation Industry in 'the Belt and Road Initiative' Countries Based on Panel Data. *Energies* **2019**, *12*, 2405. [CrossRef]
22. Yang, Q.; Zhu, R.H.; Zhao, X.Q. Calculation decoupling analysis and scenario prediction of carbon emissions of transportation in China. *J. Chang'an Univ. Nat. Sci. Edit.* **2014**, *34*, 77–83.
23. Ma, H.T.; Kang, L. Spatial and temporal characteristics and prediction of carbon emissions from road traffic in the Beijing-Tianjin-Hebei Region. *Resour. Sci.* **2017**, *39*, 1361–1370.
24. Zhao, M.; Zhang, W.G.; Yu, L.Z. Resident Travel Modes and CO_2 Emissions by Traffic in Shanghai City. *Res. Environ. Sci.* **2009**, *22*, 747–752.
25. Wang, H.Y.; Wang, N. Research on Factors Affecting Carbon Emissions of China's Comprehensive Transportation System. *Log. Technol.* **2019**, *38*, 78–83.
26. Sun, J.K.; Zhang, J.H.; Tang, G.R.; Hu, H.; Chen, M. Review on Carbon Emissions by Tourism Transportation. *China Popul. Resour. Environ.* **2016**, *26*, 73–82.
27. Intergovernmental Panel on Climate Change (IPCC). *IPCC Guidelines for National Greenhouse Gas. Inventories 2006 Volume 2 Energy*; Intergovernmental Panel on Climate Change: Kanagawa, Japan, 2007.
28. Kaya, Y. *Impact of Carbon Dioxide Emission Control. on GNP Growth: Interpretation of Proposed Scenarios*; IPCC Energy and Industry Subgroup; Response Strategies Working Group: Paris, France, 1990.
29. Hoekstra, R.; Van den Bergh, J. Comparing structural decomposition analysis and index. *Energy Econ.* **2003**, *25*, 39–64. [CrossRef]
30. Zhang, M.; Mu, H.Y.; Song, Y.C. Decomposition of energy-related CO_2 emission over 1991–2006 in China. *Ecol. Econ.* **2009**, *68*, 2122–2128. [CrossRef]
31. Ang, B.W.; Zhang, F.Q.; Choi, K.H. Factorizing Changes in Energy and Environmental Indicators through Decomposition. *Energy* **1998**, *23*, 489–495. [CrossRef]
32. Ang, B.W. Decomposition analysis for policymaking in energy: Which is the preferred method. *Energy Policy* **2004**, *32*, 1131–1139. [CrossRef]
33. Wang, W.W.; Zhang, M.; Zhou, M. Using LMDI method to analyze transport sector CO_2 emissions in China. *Energy* **2011**, *36*, 5909–5915. [CrossRef]

34. Zhu, C.Z.; Du, W.B. A Research on Influencing Factors of Carbon Emissions of Road Transportation Industry in Six Asia-Pacific Countries Based on the LMDI Decomposition Method. *Energies* **2019**, *12*, 4152. [CrossRef]
35. Liu, B.S. A Research on the Economic Regionalizing of China. *China Soft Sci.* **2009**, *2*, 81–90.

Article

Spatial-Temporal Characteristics of the Driving Factors of Agricultural Carbon Emissions: Empirical Evidence from Fujian, China

Yihui Chen [1,2,3,*], **Minjie Li** [3,*], **Kai Su** [1] and **Xiaoyong Li** [4]

[1]   Anxi College of Tea Science, Fujian Agriculture and Forestry University, Fuzhou 350002, China
[2]   Anxi Cooperative Innovation Centre of Modern Agricultural Industrial Park, Quanzhou 362406, China
[3]   School of Economics and Management, Fuzhou University, Fuzhou 350116, China
[4]   School of International Trade and Economics, University of International Business and Economics, Beijing 100029, China
*   Correspondence: chenyihui@fafu.edu.cn (Y.C.); wcctxwb@hotmail.com (M.L.); Tel.: +86-595-2616-3105 (Y.C.)

Received: 12 July 2019; Accepted: 11 August 2019; Published: 13 August 2019

Abstract: With the development of agricultural modernization, the carbon emissions caused by the agricultural sector have attracted academic and practitioners' circles' attention. This research selected the typical agricultural development province in China, Fujian, as the research object. Based on the carbon emission sources of five main aspects in agricultural production, this paper applied the latest carbon emission coefficients released by Intergovernmental Panel on Climate Change of the UN (IPCC) and World Resources Institute (WRI), then used the ordered weighted aggregation (OWA) operator to remeasure agricultural carbon emissions in Fujian from 2008–2017. The results showed that the amount of agricultural carbon emissions in Fujian was 5541.95×10^3 tonnes by 2017, which means the average amount of agricultural carbon emissions in 2017 was 615.78×10^3 tonnes, with a decrease of 13.13% compared with that in 2008. In terms of spatial distribution, agricultural carbon emissions in the eastern coastal areas were less than those in the inland regions. Among them, the highest agricultural carbon emissions were in Zhangzhou, Nanping, and Sanming, while the lowest were in Xiamen, Putian, and Ningde. In addition, this paper selected six influencing variables, the research and development intensity, the proportion of agricultural labor force, the added value of agriculture, the agricultural industrial structure, the per capita disposable income of rural residents, and per capita arable land area, to clarify further the impacts on agricultural carbon emissions. Finally, geographically- and temporally-weighted regression (GTWR) was used to measure the direction and degree of the influences of factors on agricultural carbon emission. The conclusion showed that the regression coefficients of each selected factor in cities were positive or negative, which indicated that the impacts on agricultural carbon emission had the characteristics of geospatial nonstationarity.

Keywords: carbon emissions; agricultural sector; OWA aggregation operator; GTWR

---

1. Introduction

Since the 21st Century, global warming, which is mainly caused by the increase of the carbon dioxide concentration in the atmosphere, has attracted widespread attention. Although the total carbon emissions are mainly from the industrial and service sectors, agricultural carbon emissions cannot be underestimated. The main reason for this is that although the agricultural sector provides food for all mankind, it also needs a large number of inputs of agricultural machinery and equipment, fertilizers, pesticides, agricultural film, and other means of production, which may ultimately lead to high carbon dioxide emissions. In addition, according to the statistics, the agriculture, forestry, and other land use sectors are responsible for about 24% of anthropogenic carbon emissions worldwide and have become the second largest source of global greenhouse gas emissions, and the emissions are also increasing at a fast speed of approximately 1% per annum [1,2]. Therefore, under the background of increasingly severe global warming, carbon emission reduction in the agricultural sector is an indispensable link to improve the capability of agriculture to cope with climate change and also an inevitable choice to achieve economic growth, ecological environmental development, and sustainable agricultural development. In other words, it is necessary to pay attention to the research on agricultural carbon emissions.

As a large traditional agricultural country in the world, China's carbon emissions in the agricultural sector have a more noteworthy role in increasing global climate warming. Since 1978 and the progress of the reform and opening-up policy, China's agriculture has developed rapidly and become an important factor to promote economic development and social progress. However, these rapid developments are largely at the expense of high carbon emissions. According to the statistics, the agricultural sector in China accounts for approximately 17% of national carbon emissions [3,4]. Among them, the emissions of methane and nitrogen dioxide caused by agriculture account for 50% and 92% of the national total, respectively. Thus, under the circumstances that China pledged to peak its carbon dioxide emissions by around 2030 and make best efforts to peak early, reducing carbon emissions in the agricultural sector has become a hot issue of academics and the government. In order to reduce carbon emission in the agricultural sector, first of all, it is necessary to clarify the carbon emission sources, carbon emission quantities, and influencing factors on carbon emissions. To this end, it is necessary to measure carbon emissions in the agricultural sector and identify the factors driving these carbon emissions.

As a coastal province in southeastern China, Fujian has some special features that are different from other provinces. Fujian has many mountains and few farmland, while the cultivated land resources are scarce, even less than half of the national average level, which seriously restricts the development of agriculture. Therefore, in order to promote the development of agricultural production, this can only be done by adding the inputs of chemical fertilizers, pesticides, and agricultural film for Fujian to increase the outputs of grain and other cash crops. However, these measures have resulted in a large amount of carbon emissions, causing serious environmental pollution. Besides, in 2014, Fujian became the first national ecological civilization pilot zone in China, which was announced by the State Council. That is, low-carbon agricultural development will become an important way to realize ecological civilization in Fujian Province, so as to realize finally the coordinated development of agriculture, resources, and the environment. Hence, in order to effectively promote the reduction of carbon emissions in the agricultural sector and complete the construction of the ecological civilization pilot zone, it is of great significance to carry out research on Fujian's agricultural carbon emissions. Thus, this research takes the prefecture-level cities as the basic units to analyze the spatial and temporal pattern and the evolution process of agricultural carbon emissions in Fujian and then explores the influencing factors affecting agricultural carbon emissions, so as to provide scientific evidence for formulating agricultural carbon emission reduction policies and realizing low-carbon agriculture in Fujian.

Compared with the existing research, the innovative work of this research is mainly manifested in the following three aspects. Firstly, the study area of this paper is specific. More concretely, although some scholars' research involves the issue of agricultural carbon emissions, most of them stay at the macro level, that is the literature specific to a certain area is less [5]. Therefore, based on the data of nine prefecture-level cities in Fujian, this research remeasured the agricultural carbon emissions by using the latest emission coefficients released by the Intergovernmental Panel on Climate Change of UN (IPCC, Geneva, Switzerland) and World Resources Institute (WRI, Washington, DC, USA), so as to achieve more accurate calculation of carbon emissions in the agricultural sector. Secondly, a more scientific method was used to evaluate the agricultural carbon emissions of each prefecture-level city in the sample period. Different from other existing research using the simple arithmetic averaging method, this paper uses the ordered weighted averaging (OWA) aggregation operator to distribute weights in different years, so as to solve the problem of weighting the same indicators in different periods ignored in the calculation of carbon emissions, so as to realize the dynamic comprehensive evaluation of panel data. Moreover, agricultural carbon emissions are mainly the result of multivariate interaction, such as economic level, the infrastructure, and resource endowment. The mechanism of the above action is complex. Therefore, the magnitude and direction of the influencing factors are different under different in their temporal and spatial distributions. That is, traditional spatial econometric models will no longer meet the research requirements. The geographically- and temporally-weighted regression (GTWR) applied in this research is a local linear regression model that considers both geographical and temporal non-stationarity. On the whole, this study measures carbon emissions in the agricultural sector and uses the OWA aggregation operator to solve the problem of the dynamic comprehensive evaluation of panel data. Then, by adopting the GTWR model, this paper analyzes the spatial-temporal heterogeneity of the impact of factors on agricultural carbon emissions, aiming to establish an effective agricultural carbon emission reduction mechanism, and finally, aiding local sustainable development decision-making.

2. Literature Review

2.1. Measurement of Agricultural Carbon Emissions

At present, many existing research works have focused on the measurement of carbon emissions in the agricultural sector. It should be noted that different methods used to estimate carbon emissions will produce different results. For instance, Wang et al. followed the IPCC guidelines [6] released in 2006 to estimate the greenhouse gas emission intensity of rice, wheat, and maize yields in China from 1985–2010 [7]. According to the IPCC guidelines, Xiong et al. and Tian et al. estimated the carbon emissions of agricultural production in Hunan and Xinjiang, respectively [8,9]. In addition, Han et al. measured carbon emissions from the entire agricultural sector as a whole in China during the period from 1997–2015 [10]. However, the IPCC guidelines ignored soil emissions during agricultural land use change in its agricultural inventory [11,12] and are no longer fully suitable for current emissions. Therefore, some scholars have proposed novel methods to measure agricultural carbon emissions. For instance, Bell et al. used the Scottish Government's new method to calculate agricultural carbon emissions and compared it with the IPCC guidelines and national communications [12]. Wisniewski and Kistowski proposed a solution that enables local governments to estimate independently the carbon footprints and monitor the impacts of actions taken to reduce emissions [13]. Moreover, based on the national statistics, Yue et al. evaluated the carbon footprints of a range of 26 crop products and six livestock types [14]. Based on the above background, this paper applied the carbon emission coefficients released by IPCC and WRI to calculate the agricultural carbon emissions in Fujian from 2008–2017, which makes the measurement results more specific and accurate.

2.2. Influencing Factors of Agricultural Carbon Emissions

The driving and inhibiting factors of agricultural carbon emissions can be identified by studying the influencing factors of agricultural carbon emissions. Existing research on the influencing factors of agricultural carbon emissions mainly involves many aspects, including agricultural economic growth, technological progress, population size, income, and agricultural energy consumption. ACIL Tasman measured agricultural carbon emissions in the United States, Canada, India, the European Union, and New Zealand and demonstrated that the proportion of agricultural carbon emissions in total carbon emissions varies greatly, possibly due to different modes of agricultural production [5]. Ismael et al. also confirmed that agricultural production had a significant impact on carbon emissions [15]. For instance, the organic agricultural production mode had the function of restraining agricultural carbon emissions [16]. In addition, agricultural economic growth and the increase of the agricultural population has positive impacts on agricultural carbon emissions [11,17]. Moreover, the agricultural technology progress is also one of the important factors affecting agricultural carbon emissions. Gerlagh applied the endogenous technological progress model to study the influence of technological progress on carbon emission reduction [18] and confirmed that technological progress significantly reduced the cost of carbon emission reduction through the learning effect and increased the social benefits at the same time. Furthermore, agricultural land also affects the agricultural carbon emissions, such as per capita land use area [19], agricultural land use [20] and farmland conversion [21,22]. Besides, there also exists a close relationship between agricultural income and carbon emissions [23].

However, it should be noted that there is still no consensus on the causal relationship, direction, and extent between influencing factors and agricultural carbon emissions. Hence, when selecting the influencing factors, we should combine the relevant literature with the outstanding characteristics of Fujian in the process of agricultural development, so as to ensure that the factors are reasonable and scientific.

2.3. Methodologies of Agricultural Carbon Emissions

There exist many methodologies to explore the relationship between carbon emissions and their influencing factors in the agricultural sector. Among them, the autoregressive distribution lag model [24], the Granger causality test [25,26], and the vector error correction model [27] have been approved and applied by most scholars. Moreover, the logarithmic mean Divisia index [28,29] and the variance decomposition approach [15] mainly apply the exponential decomposition method to study the main factors causing the change of agricultural carbon emissions. Furthermore, other scholars have applied some other novel methodologies, including the denitrification-decomposition models [30], the spatial econometric models [25] and the fully-modified ordinary least squares [31,32].

However, when examining the degree and direction of the impact factors on agricultural carbon emissions, most of the literature only considers the time perspective, but ignores the spatial perspective. It should be noted that there exist great differences in the level of economic development, agricultural structure, resource endowment, and the agricultural production mode in each region, which will also lead to different degrees and directions of influencing factors at different times and in different regions. Hence, in this paper, the GTWR model is used to study the influences of factors affecting agricultural carbon emissions on each prefecture-level city from the perspective of time and space, so as to remedy the shortcomings of this research field.

3. Materials and Methodologies

3.1. Study Area

The study area covers Fujian on the southeast coast of China. As an important estuary for the Min River and also an important window for China's contacts with the world, Fujian encompasses a total land area of approximately 124,000 km^2 and a total maritime area of approximately 136,000 km^2. From the geographical perspective, Fujian is located approximately between longitudes 115°50′ E and 120°43′ E and between the latitudes 23°32′ N and 28°22′ N. As one of the provincial administrative regions in China, Fujian includes 9 prefectural-level cities: Fuzhou, Xiamen, Quanzhou, Zhangzhou, Sanming, Putian, Longyan, Nanping, and Ningde. According to the Fujian Statistical Yearbook, from 2010–2018, the gross output value of agriculture in Fujian increased from 136.367–237.982 billion Yuan at a rapid rate. Similarly, per capita disposable income of rural residents increased from 7426.86–17,821 Yuan. However, the rapid development of agriculture is at the expense of the environment. That is, at present, the agricultural development in Fujian is a typical chemical agriculture type, which relies heavily on high-carbon means of production such as chemical fertilizers and pesticides, which seriously affects the sustainable development of agriculture in the future. Thus, it is of great practical significance to study agricultural carbon emissions and their influencing factors in Fujian and to explore the way to realize the development of low-carbon agriculture. The location, the latitude and the longitude rage of the study area can be seen in Figure 1.

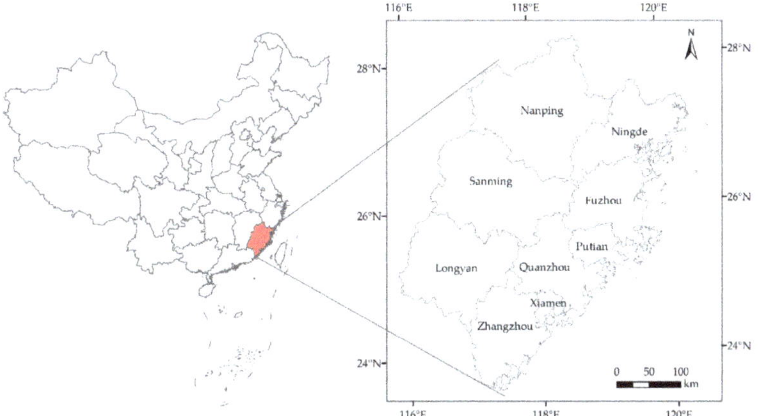

Figure 1. The location of the study area.

3.2. Selection of Measurement Indicators

In agriculture, there exist three main sources of carbon emissions, which can also be considered as sources of greenhouse gas emissions: agricultural land use, rice paddies and crop production, and livestock enteric fermentation and manure storage. In the United States, agricultural land use is the largest source of agricultural carbon emissions, mainly due to the large inputs of fertilizers, pesticides, and other agricultural materials and the loss of organic carbon caused by soil tillage; livestock enteric fermentation is the second largest source, and livestock manure storage is the third. Furthermore, the rice paddies produce fewer greenhouse gases than other agricultural productions because of the lesser rice planting area. Although Fujian's agricultural production situation differs from that of the United States, the composition of agricultural carbon emissions is consistent. Therefore, combining the above research with other references [17,23], this paper mainly calculates the agricultural carbon emissions according to the five types of carbon emission sources: (1) CO_2 emissions produced from agricultural land use; (2) CH_4 emissions caused by rice paddies; (3) CH_4 emissions caused by livestock breeding; (4) N_2O emissions triggered by crop production; (5) N_2O emissions triggered by livestock breeding. All carbon emission sources and their coefficients in agricultural sector in Fujian are listed in Table 1.

Table 1. Carbon emission sources and coefficients in the agricultural sector.

Sources	Detailed Sources	Units	Greenhouse Gases			References
			CO_2	CH_4	N_2O	
agricultural land use	chemical fertilizer	kg/kg	3.28	n/a	n/a	IPCC
	pesticide	kg/kg	18.09	n/a	n/a	IPCC
	plastic sheeting	kg/kg	19.00	n/a	n/a	IPCC
	diesel	kg/kg	3.17	n/a	n/a	IPCC
	tillage	kg/km^2	1146.31	n/a	n/a	IPCC
	irrigation	kg/ha	977.19	n/a	n/a	IPCC
rice paddies	early rice	kg/ha	n/a	77.39	n/a	IPCC
	late rice	kg/ha	n/a	525.95	n/a	IPCC
	in-season rice	kg/ha	n/a	434.66	n/a	IPCC

Table 1. *Cont.*

Sources	Detailed Sources	Units	Greenhouse Gases			References
			CO$_2$	CH$_4$	N$_2$O	
crop production	paddy rice	kg/ha	n/a	n/a	0.24	IPCC
	winter wheat	kg/ha	n/a	n/a	2.05	IPCC
	soybean	kg/ha	n/a	n/a	0.77	IPCC
	vegetable	kg/ha	n/a	n/a	4.21	IPCC
	maize	kg/ha	n/a	n/a	2.53	IPCC
	other dry crops	kg/ha	n/a	n/a	0.95	IPCC
livestock: manure storage	dairy	kg/head/year	n/a	8.33	2.07	WRI
	non-dairy	kg/head/year	n/a	3.31	0.85	WRI
	goat	kg/head/year	n/a	0.28	0.11	WRI
	pig	kg/head/year	n/a	5.08	0.18	WRI
	poultry	kg/head/year	n/a	0.02	0.01	WRI
	rabbit	kg/head/year	n/a	0.08	0.02	IPCC
livestock: enteric fermentation	dairy	kg/head/year	n/a	89.3	n/a	WRI
	non-dairy	kg/head/year	n/a	67.9	n/a	WRI
	goat	kg/head/year	n/a	9.4	n/a	WRI
	pig	kg/head/year	n/a	1	n/a	WRI
	poultry	kg/head/year	n/a	n/a	n/a	WRI
	rabbit	kg/head/year	n/a	0.25	n/a	IPCC

Note: Data of IPCC and WRI in the column of the "References" are from [6] and [33], respectively.

3.3. Selection of Influencing Factors

3.3.1. Research and Development Intensity

Agricultural technological progress is an important factor to promote the development of low-carbon agriculture and has the role of agricultural carbon emission reduction [34]. Hence, the absence of agricultural technology has become an important factor restricting the sustainable development of agriculture [35]. Therefore, it is necessary and effective for governments to increase the input intensity of research and development and promote the progress of agricultural technology to further improve the technical system of energy saving and emission reduction in agriculture [10,36]. It is noteworthy that the current role of science and technology input is not necessarily reflected in the current period, such as R & D investment [37]. R & D investment will take effect only after a later period of time. Thus, it is necessary to consider the lagged rank of R & D investment affecting agricultural carbon emissions in a certain period of time. The proportion of R & D investment to GDP is adopted in this research to measure the science and technology intensity and denoted as research and development intensity (RDI).

3.3.2. Proportion of Agricultural Labor Force

Agricultural labor force (ALF) can affect the carbon emissions from crop production [9], which in turn affects agricultural carbon emissions. Based on the Kaya model, Zhang and Fang decomposed factors affecting agricultural carbon emissions and found that reducing the proportion of agricultural labor can significantly limit the growth of carbon emissions [38]. According to the logarithmic mean Divisia index (LMDI) method, Yao et al. confirmed that the increase of agricultural labor force is an important factor for the sustained growth of carbon emissions produced from animal husbandry [39]. Similarly, Satterthwaite [40] and Al-Mulali et al. [41] analyzed the relationship between agricultural labor force and agricultural carbon emissions from the perspective of urbanization and confirmed that with the acceleration of urbanization, the proportion of agricultural labor force continued to decline and then had a positive impact on the reduction of carbon emissions. The proportion of agricultural labor force in Fujian decreased from 31.15–21.71% during the time period from 2008–2017, which inevitably had an influence on agricultural carbon emissions.

3.3.3. Added Value of Agriculture

This paper applies the proportion of added value of agriculture (AVA), forestry, animal husbandry, and fishery production to measure the level of agricultural economic development. By and large, the influence of agricultural added value on agricultural carbon emissions has regional characteristics. Tian et al. combined multiple linear regression with decoupling analysis to evaluate the influencing factors of agricultural carbon emission and found that there was a weak and unstable decoupling relationship between agricultural carbon emissions and the added value [9]. Murad also confirmed that there existed no Granger causality between agricultural output and carbon emissions in Bangladesh [42]. However, Jebli et al. [43] and Rafiq [44] confirmed that there was a two-way causal relationship between agricultural added value and carbon dioxide emissions. Besides, by using data from provinces from 2001–2013 in Iran, Alamdarlo demonstrated that there existed an inverted "U" relationship between agricultural value added and agricultural carbon emissions; however, the above conclusion was not suitable for all provinces, because of the heterogeneity of agricultural development in provinces, such as the disunity of agricultural infrastructure [45]. Hence, it is necessary to analyze the impact of the influencing factor based on the specific situation of agricultural development in Fujian.

3.3.4. Agricultural Industrial Structure

The percentage of the output value of the plant products industry to total agricultural output value is applied in this paper to measure the structure of agricultural industry. The plant products industry mainly relies on the input of agricultural materials such as pesticides and fertilizers to increase output, resulting in an increase in agricultural carbon emissions. Hence, with the optimization of the agricultural industrial structure (AIS), the decline of the proportion of plant products industry can reduce agricultural carbon emissions [29]. Nevertheless, Yao et al. obtained slightly different conclusions by studying the influencing factors of the agricultural carbon emission change in animal husbandry [39]. They found that the impact of the optimization of agricultural industrial structure on carbon emissions from animal husbandry changed from positive to negative, which is particularly evident in central and Eastern China. Therefore, empirical research is needed to analyze the direction and magnitude of the impacts of agricultural industrial structure on agricultural carbon emissions in Fujian.

3.3.5. Per Capita Disposable Income of Rural Residents

In 1993, Panayotou first introduced and named the relationship between economic growth and environmental conditions as the environmental Kuznets curve (EKC), indicating that per capita income had a strong inverted "U" curve relationship with the level of environmental pollution [46]. However, Liu and Xin confirmed that the evolutionary trend between economic growth and agricultural carbon emissions showed an "N" curve, indicating that agricultural carbon pollution becomes more serious than before as the economy continues to expand [47]. In addition, Tian et al. also demonstrated that when per capita income of agriculture increased by 1 unit, agricultural carbon emissions would increase by 0.354 units [9]. One explanation could be that while vigorously increasing the per capita income of agriculture, the extensive use of chemical fertilizers and pesticides into agriculture promoted the increase of carbon emissions [48]. Therefore, the increase of rural residents' income at this stage may lead to an increase in agricultural carbon emissions.

3.3.6. Per Capita Arable Land Area

Reducing per capita arable land area (ALA) will reduce the total agricultural energy demand per capita, such as chemical fertilizers, pesticides, and plastic film, thus restraining agricultural carbon emissions [19]. However, if the per capita arable land area were reduced through the transformation of cultivated land to industrial land, the greenhouse effect would be aggravated [49]. Thus, the degraded land can be restored by returning cultivated land to grassland. That is, reducing arable land area is conducive to reducing greenhouse gas emissions from agricultural activities. In brief, the land use and land use change in the agricultural sector are two important factors influencing agricultural carbon emissions [50].

3.4. Research Methodologies and Data Sources

3.4.1. Estimation of Agricultural Carbon Emissions

According to the IPCC guidelines, on the basis of the existing research about the carbon emission equation [20,28,29,36], in view of the current situation of agricultural development in Fujian Province, this paper chooses carbon emission sources and corresponding carbon emission coefficients to build a model for calculating agricultural carbon emissions. The specific formula is as follows:

$$E = \sum_{i=1}^{n} E_i = \sum_{i=1}^{n} T_i \cdot \mu_i \tag{1}$$

where E represents total agricultural carbon emissions; E_i denotes the carbon emission of the specific source i; T_i represents the amount of the specific source i; and μ_i denotes carbon emission coefficient of the specific source i. In accordance with the usual practice, it is necessary to convert CO_2, CH_4, and N_2O to standard carbon. By and large, the greenhouse effects caused by 1 tonne of CO_2, CH_4, and N_2O are equivalent to that produced by 0.2727, 6.8182, and 81.2727 tonnes of standard carbon, respectively.

3.4.2. Ordered Weighted Averaging Aggregation Operator

The ordered weighted averaging (OWA) aggregation operator, first proposed by Yager in 1988, is a novel time empowerment method. The basic idea of OWA is to reorder the data according to the numerical value and then determine the weight by the position of the data in the ranking [51]. In addition, the OWA aggregation operator determines the weights based on the data themselves; therefore, since it was introduced into the application, the fairness of its empowerment has been controversial. Scholars in various countries have constantly improved it. Hence, in this paper, an improved OWA aggregation operator proposed by Xu [52], that is a smooth and continuous normal distribution density function, is applied to determine the time weights of the panel data of agricultural carbon emissions in Fujian. The specific steps of the OWA aggregation operator are as follows:

(1) Assume that there exist m regions and n years; besides, E_{ij} denotes total carbon emissions in the agricultural sector in specific region i in specific year j. After summing up the value of E_{ij} in each year, the average value is as follows:

$$\overline{E}_j = \frac{1}{m} \sum_{i=1}^{m} E_{ij} \tag{2}$$

(2) Assume that the initial weights of total carbon emissions in each year is $1/n$, then the average value and standard deviation of E_{ij} are as follows:

$$\overline{E} = \frac{1}{n} \sum_{j=1}^{n} \overline{E}_j \tag{3}$$

$$\sigma = \sqrt{\frac{\sum\limits_{j=1}^{n} (\overline{E}_j - \overline{\overline{E}})^2}{n}} \tag{4}$$

(3) Standardize the total carbon emissions based on the above average value and standard deviation, and the calculation equation is as follows:

$$\beta_j = \frac{\overline{E}_j - \overline{\overline{E}}}{\sigma} \tag{5}$$

(4) Using the standard normal distribution density function, the corresponding values of α_j under the specific β_j are as follows:

$$\alpha_j = \varphi(\beta_j) = \frac{1}{\sqrt{2\pi}} e^{-\frac{\beta_j^2}{2}} \tag{6}$$

(5) Normalize the obtained value of α_j to calculate the time weights, and the formula is as follows:

$$\omega_j = \frac{\alpha_j}{\sum\limits_{j=1}^{n} \alpha_j} \tag{7}$$

3.4.3. Geographically- and Temporally-Weighted Regression

When exploring the relationship between agricultural carbon emissions and influencing factors in the past, the ordinary least squares method and the spatial econometric model were usually used. Normal panel models usually only represent the correlation between dependent and independent variables in the mean sense, but cannot effectively reflect the spatial heterogeneity of the regression. Therefore, the model estimates are biased and lack an explanation. Besides, in the study of spatial heterogeneity, the geographically-weighted regression model (GWR) has been widely used because it can describe the variability of different geographic locations [53,54]; however, the GWR model does not consider the influence of the time factor [55]. Therefore, as an extension of the geographically-weighted regression model, the geographically- and temporally-weighted regression incorporates the time dimension in the geographic space, effectively expanding the multiple linear regression model and GWR model [56,57]. In this paper, both temporal and spatial effects are included in the model to analyze the characteristics of the regression relationship changing with space and time. That is, this paper uses the GTWR model to analyze the data of agricultural carbon emission and its influencing factors in Fujian from 2008–2017, as well as to explore the direction and degree of influencing factors on agricultural carbon emission in each region in each year. The specific model of GTWR is as follows [55]:

$$y_i = \beta_0(u_i, v_i, t_i) + \sum_{k=1}^{d} \beta_k(u_i, v_i, t_i) x_{ik} + \varepsilon_i \tag{8}$$

where y_i denotes the observations of agricultural carbon emissions, while x_{ik} represents the influencing factors at the specific point (u_i, v_i, t_i). In addition, β_0 represents the constant coefficients. (u_i, v_i, t_i) denotes the longitude coordinate u_i and the latitude coordinate v_i, and the time point t_i of the specific location. $\beta_k(u_i, v_i, t_i)$ represents the unknown parameter at the specific location (u_i, v_i, t_i), while it is also the arbitrary function of (u_i, v_i, t_i). ε_i denotes an independently and identically distributed (iid) error and is assumed to obey the $N(0, \sigma^2)$ distribution. The cross-validation method is applied in this paper to determine the optimal bandwidth. Ultimately, this paper chooses agricultural carbon emissions y_i as the dependent variable, selects RDI, ALF, AVA, AIS, disposable income of rural residents (DIR), and

ALA as the independent variables, denoted as x_1, x_2, x_3, x_4, x_5, and x_6, then constructs the model as follows:

$$
\begin{aligned}
y_i = \ & \beta_0(u_i, v_i, t_i) + \beta_1(u_i, v_i, t_i)x_{i1} + \beta_2(u_i, v_i, t_i)x_{i2} + \beta_3(u_i, v_i, t_i)x_{i3} + \\
& \beta_4(u_i, v_i, t_i)x_{i4} + \beta_5(u_i, v_i, t_i)x_{i5} + \beta_6(u_i, v_i, t_i)x_{i6} + \varepsilon_i
\end{aligned}
\tag{9}
$$

where i equals the interval of natural numbers from 1–9 and $\beta_1(u_i,v_i,t_i)$ denotes the change range in which agricultural carbon emissions follow RDI. Similarly, $\beta_2(u_i,v_i,t_i)$, $\beta_3(u_i,v_i,t_i)$, $\beta_4(u_i,v_i,t_i)$, $\beta_5(u_i,v_i,t_i)$, and $\beta_6(u_i,v_i,t_i)$ represent the change range in which agricultural carbon emissions follow ALF, AVA, AIS, DIR, and ALA, respectively.

3.4.4. Data Sources

The agricultural carbon emissions from the agricultural land use, rice paddies and crop production, and livestock enteric fermentation and manure storage, as well as their emission coefficients are used in this paper. The applied emission coefficients were mainly released by IPCC in 2006 and WRI in 2015. Besides, the original data of the agricultural carbon emissions from 2008–2017 were from the Fujian Statistical Yearbook and the statistical yearbooks of all prefecture-level cities without any other processing. In the end, the original data covering 9 prefectural-level cities for 10 years were obtained.

Furthermore, this paper adopts the formula of agricultural carbon emissions and the OWA operator to measure the agricultural carbon emissions of 9 prefecture-level cities in Fujian Province from 2008–2017 and analyzes the spatial and temporal characteristics of agricultural carbon emissions in the past 10 years. In order to further clarify the influences of the 6 driving factors selected above on agricultural carbon emissions in Fujian, this paper will also use GTWR to measure the direction and degree of impacts of driving factors in each prefecture-level city. According to the results, this paper puts forward countermeasures and suggestions to promote effectively agricultural carbon emission reduction and the development of low-carbon agriculture in Fujian in the next stages.

4. Results

4.1. Evolution Trends of Agricultural Carbon Emissions

According to the calculation process of Formula (1), agricultural carbon emissions in Fujian from 2008–2017are shown in Table 2. In order to consider fully the dynamic evaluation of the panel data, the weights of each year based on OWA are listed in the last row of Table 2, while the average agricultural carbon emissions of each region calculated based on OWA are listed in the last column. As shown in Table 2, agricultural carbon emissions in Fujian showed a fluctuating downward trend from 2008–2017. That is, agricultural carbon emissions decreased from 708.88 thousand tonnes in 2008 to 615.78 thousand tonnes in 2017. The fluctuating evolution with the basic spatial pattern of "M" can be divided into four stages: fluctuating increase, low speed reduction, rapid increase, and finally, rapid reduction. The result shows that in the process of agricultural development, Fujian has taken some measures to control agricultural carbon emissions and strengthen the awareness of ecological agriculture. Besides, the composition of carbon emissions in the agricultural sector varied from year to year. Moreover, as shown in Table 3, agricultural land use was the main source of agricultural carbon emissions, exceeding 40% in each year. Rice paddies also accounted for more than 30% of carbon emissions in each year. The proportion of crop production and livestock enteric fermentation in agricultural carbon emissions was relatively small. One explanation might be that with the increase of population, Fujian increased the utilization rate of agricultural land and relied heavily on chemical fertilizers and pesticides, so as to ensure food supply, which ultimately contributed to the increase of agricultural carbon emissions.

Table 2. Average agricultural carbon emissions (ACE) in Fujian (units: 10^3 tonnes of carbon).

Cities	2008	2009	2010	2011	2012	2013	2014	2015	2016	2017	ACE
Fuzhou	752.01	749.26	748.13	761.97	754.29	748.75	741.54	728.36	658.57	645.79	740.87
Xiamen	96.19	93.27	87.28	86.33	86.19	82.60	66.00	62.09	61.48	63.72	80.38
Putian	337.20	334.26	328.08	322.95	311.75	299.66	278.62	278.43	267.48	228.45	307.12
Sanming	944.54	925.30	914.94	921.81	907.74	906.69	901.55	903.80	770.21	779.90	904.92
Quanzhou	700.73	698.66	705.85	691.92	677.60	663.43	640.98	633.50	628.09	655.61	672.45
Zhangzhou	1151.33	1158.86	1175.91	1181.57	1181.36	1165.64	1155.07	1143.35	1131.38	1043.93	1160.47
Nanping	1044.41	1047.00	1053.41	1065.75	1065.76	1299.81	1088.80	1086.94	1145.93	933.96	1087.33
Longyan	835.92	838.23	844.88	847.72	844.65	844.59	835.18	769.20	716.58	715.98	822.82
Ningde	517.60	513.59	513.09	515.18	511.42	505.39	497.19	489.83	478.49	474.64	505.35
Average	708.88	706.49	707.95	710.58	704.53	724.06	689.44	677.28	650.91	615.78	n/a
Weights (%)	11.44	11.95	11.65	11.06	12.33	7.59	13.80	12.79	6.50	0.89	n/a

Table 3. The proportion of sources of agricultural carbon emissions in Fujian (units: %).

Sources	2008	2009	2010	2011	2012	2013	2014	2015	2016	2017
agricultural land use	46.65	45.97	44.09	43.08	40.66	41.71	41.25	41.08	41.07	40.92
rice paddies	31.49	30.70	32.77	32.68	34.93	32.88	33.45	34.01	34.09	34.48
crop production	3.92	4.16	4.61	4.40	4.09	4.12	4.05	3.99	3.96	3.90
livestock: manure storage	12.23	12.84	11.65	12.48	13.03	13.83	13.78	13.40	13.38	13.26
livestock: enteric fermentation	5.71	6.33	6.88	7.36	7.29	7.46	7.47	7.52	7.50	7.44
CO_2	46.65	45.97	44.09	43.08	40.66	41.71	41.25	41.08	41.07	40.92
CH_4	43.64	44.21	46.08	47.08	49.72	48.43	48.95	49.37	49.47	49.78
N_2O	9.71	9.82	9.83	9.84	9.62	9.86	9.80	9.55	9.46	9.30

4.2. Regional Differences of Agricultural Carbon Emissions

As can be seen in Table 2, agricultural carbon emissions of prefectural-level cities showed different trends. According to the variation of agricultural carbon emissions, the trend can be roughly divided into four types: (1) a rise-drop feature; (2) a slow decline feature; (3) a drop-rise-drop-rise feature; (4) a drop-rise-drop feature. The representative cities of the first type are Fuzhou, Zhangzhou, Nanping, and Longyan. However, in 2017, Zhangzhou and Nanping were still the cities with the highest carbon emissions in Fujian, with the agricultural carbon emissions of these two cities reaching over 35.6% of the total. In addition, the second type of agricultural carbon emissions showed a downward trend over time. Putian and Xiamen are the representative cities of the second type. Besides, agricultural carbon emissions in Xiamen increased slightly in 2017. The third type of carbon emissions presents a typical "W" trend, mainly represented by Sanming and Quanzhou. The fourth type shows a typical inverted "N" trend, mainly represented by Ningde. It should be noted that there were also significant differences in agricultural carbon emissions among cities. For instance, the average agricultural carbon emission of Zhangzhou based on OWA was about 14.44-times as much as that of Xiamen. Zhangzhou is famous for its flowers and fruits. In 2018, Zhangzhou's total agricultural output value accounted for 20.89% of the whole province, becoming the largest prefectural-level city of agricultural carbon emissions in Fujian. By comparison, Xiamen, which is dominated by services, is in the rapid-developing economic circle on the west side of the Taiwan Strait. Thus, Xiamen's agricultural output value accounted for only 0.47% of the regional gross product and 1.37% of Fujian's agricultural output value in 2018. In addition, the continuous reduction of agricultural land use in Xiamen in recent years has further led to the lowest agricultural carbon emissions.

Although agricultural carbon emissions in Fujian showed a trend of fluctuating downward during the investigation period, there are still some problems to be solved, such as an unreasonable agricultural structure, extensive management, and unreasonable allocation of resources. Accordingly, vigorously developing low-carbon agriculture will be the main measure of agricultural carbon emission reduction in Fujian in the future. This paper analyses the influencing factors of agricultural carbon emissions and clarifies the reasons for the growth of agricultural carbon emissions. Then, according to the direction and force of the influencing factors, this paper puts forward differentiated measures for agricultural emission reduction, which is of great significance to promote the development of low-carbon agriculture.

4.3. Analysis Results of Influencing Factors

The descriptive statistics of all variables used in the GTWR model, including agricultural carbon emissions and the influencing factors, can be seen in Table 4. The standard deviation of some variables reflects the great difference among cities. For instance, the maximum value of AVA was 24.68-times the minimum, while the maximum values of RDI and ALA were 13.52- and 9.00-times the minimum, respectively. In addition, in order to overcome the shortcomings of heteroscedasticity, all the original data used in GTWR were adopted in logarithmic form without changing the nature and relevance. That is, the coefficients calculated by GTWR measure the elasticity of the dependent variable with respect to the independent variable, i.e., the percentage of the dependent variable when the independent variable changes by 1%. In addition, this paper mainly adopts ArcGIS 10.4 to realize the regression coefficient estimation based on the properties of time and space. Moreover, the descriptive statistics and geographical distribution of the regression coefficients calculated by the GTWR model can be seen in Table 5 and Figure 2.

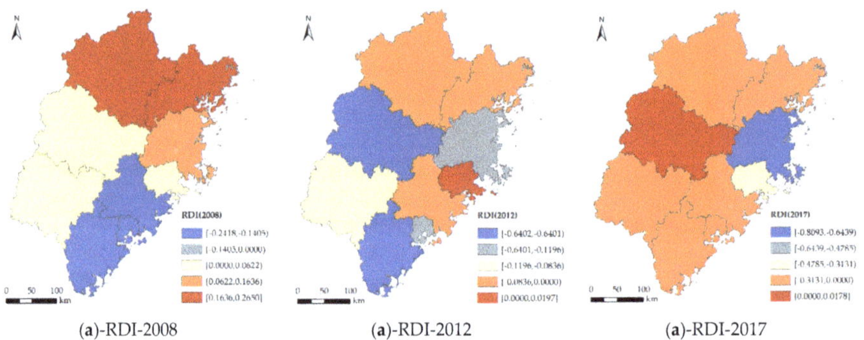

(a)-RDI-2008 (a)-RDI-2012 (a)-RDI-2017

Figure 2. *Cont.*

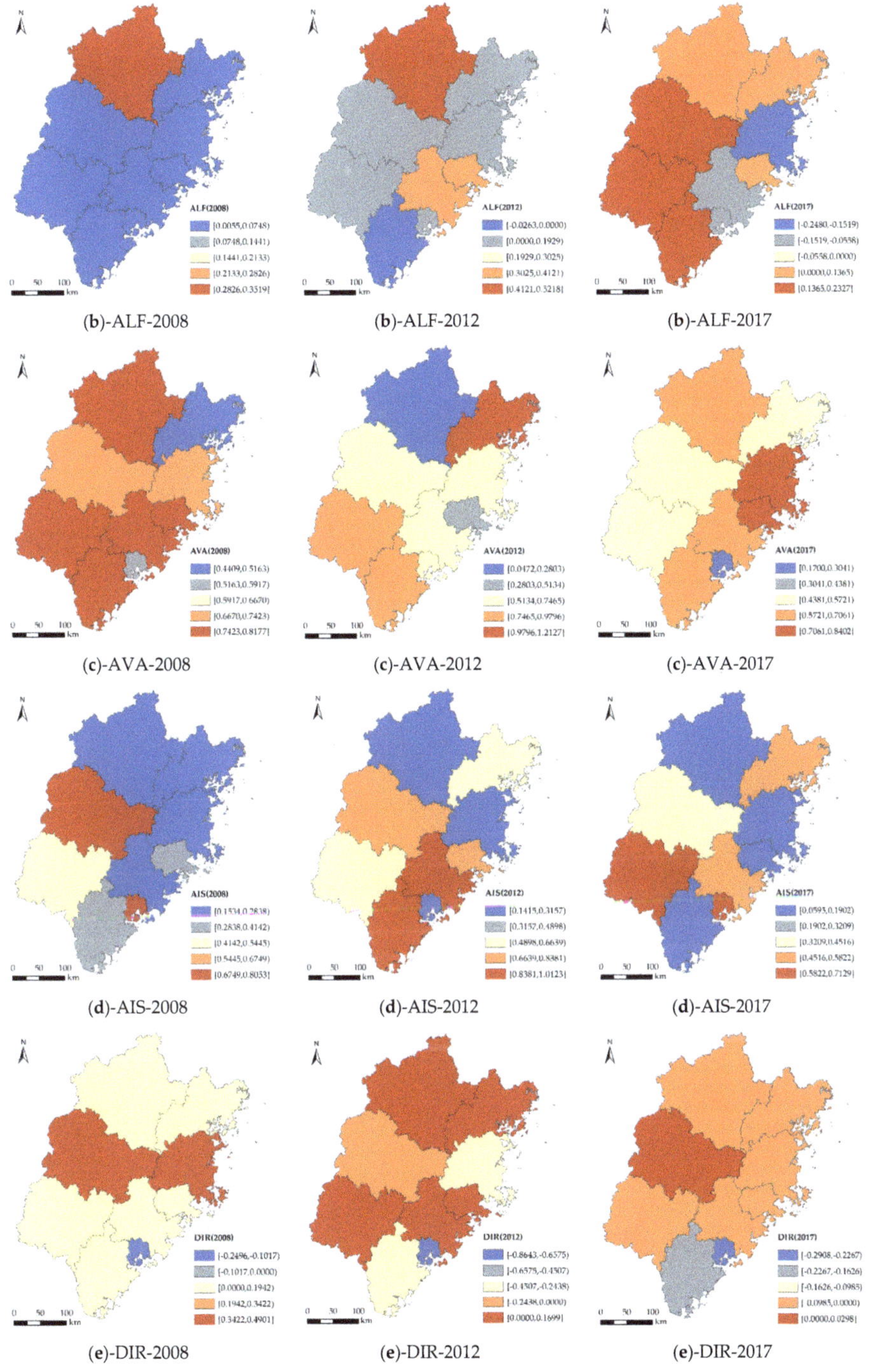

Figure 2. *Cont.*

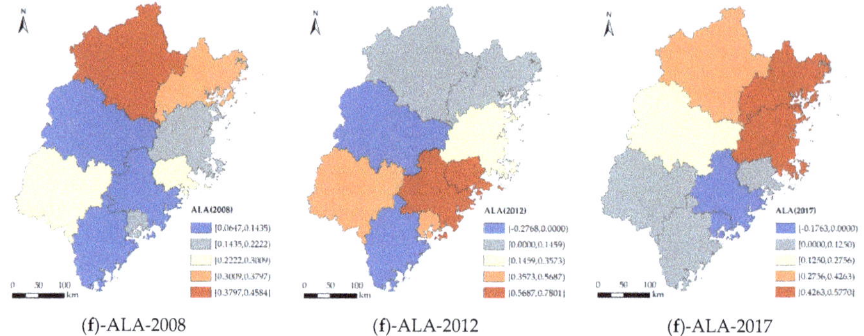

(f)-ALA-2008   (f)-ALA-2012   (f)-ALA-2017

Figure 2. Spatial distribution of regression coefficients calculated by GTWR in 2008, 2012, and 2017.

Table 4. The descriptive statistics of the original data of the variables used in GTWR. RDI, research and development intensity; ALF, agricultural labor force; AVA, added value of agriculture; AIS, agricultural industrial structure; DIR, disposable income of rural residents; ALA, arable land area.

Variables	Units	Mean	SD	Minimum	Q_1	Median	Q_3	Maximum
ACE	10^3 tonnes	689.59	335.51	61.48	486.99	734.95	922.68	1299.81
RDI	%	1.13	0.65	0.23	0.76	1.02	1.27	3.11
ALF	%	28.47	13.62	0.26	17.83	32.57	38.55	49.53
AVA	10^8 CNY	85.87	45.92	7.81	50.27	85.82	119.49	192.74
AIS	%	40.84	8.81	22.72	36.40	43.30	46.36	56.28
DIR	10^3 CNY	11.24	3.68	5.40	7.95	11.28	13.94	20.46
ALA	ha/person	0.05	0.02	0.01	0.03	0.04	0.05	0.09

Table 5. The comparison of the parameter estimation of OLS, GWR, and GTWR.

Variables	OLS			GWR					GTWR				
	Coefficients	t	p	Minimum	Mean	Maximum	t	p	Minimum	Mean	Maximum	t	p
Intercept	4.7102	5.613	0.000	3.2891	7.6363	11.8337	23.154	0.000	−4.0447	6.4822	15.7804	13.157	0.000
LNRDI	−0.8575	−2.020	0.047	−0.4406	−0.0524	0.2281	−2.254	0.027	−1.0764	−0.1829	0.4611	−5.306	0.000
LNALF	0.0682	1.231	0.222	−1.1752	0.1024	0.2054	2.407	0.018	−0.7462	0.0512	0.5217	2.356	0.021
LNAVA	0.8681	12.535	0.000	0.2318	0.6099	1.0202	25.091	0.000	0.0473	0.6955	1.3152	27.777	0.000
LNAIS	0.1534	0.343	0.732	−1.4618	0.1844	1.5394	2.254	0.027	−0.6021	0.4426	1.2361	13.984	0.000
LNDIR	−0.0513	−7.498	0.000	−1.2432	−0.6536	−0.2494	−16.857	0.000	−1.4131	−0.0711	0.5832	−2.187	0.031
LNALA	0.0133	0.151	0.881	−0.3998	0.4033	1.1930	7.369	0.000	−0.4468	0.2873	0.8727	9.183	0.000
R^2	0.9321				0.9950					0.9960			
F	189.598				2752.833					3444.500			
RSS	4.0754				0.289					0.213			
AIC	−9.1264				−130.719					−189.513			

Note: LNRDI, LNALF, LNAVA, LNAIS, LNDIR and LNALA represent the logarithmic form of RDI, ALF, AVA, AIS, DIR and ALA, respectively.

In this paper, the ordinary least squares method (OLS) and the GWR and GTWR models are used for regression analysis and comparison. The comparisons of the regression analyses are shown in Table 5. As shown in Table 5, the goodness of fit of the GTWR model was superior to OLS and GWR. For instance, the adjusted R^2 of the GTWR model was 0.9960, which was larger than that of GWR and OLS, which equaled 0.9950 and 0.9321, respectively. The residual sum of squares (RSS) and F-value of GTWR were also better than those of the other models. Moreover, the smaller the AIC value is, the higher the precision of the model is [58]. Furthermore, if the difference of the AIC values between two models is more than three, this shows that there is a significant difference between the two models. As can be seen in Table 5, the AIC value of GTWR was the smallest, and the difference between the AIC of GTWR and that of GWR or OLS exceeded three, which indicates that the GTWR estimation was much better than the GWR and OLS estimation. Because GWR and GTWR have a set of corresponding coefficients at each point, only the maximum, minimum, and mean coefficients of each independent variable are listed in Table 5. By comparing the coefficients estimated by GWR and GTWR, it was found that the coefficients varied greatly, which indicates that the direction and degree of the impacts of influencing factors on agricultural carbon emissions were different in both the time and space dimensions, which shows significant spatial and temporal non-stationarity characteristics. The regression coefficients estimated by GTWR are analyzed in detail with the spatial and temporal distribution of the coefficients.

4.3.1. The Influence of RDI on Agricultural Carbon Emissions

Considering the time lag effect of RDI on agricultural carbon emissions, this paper measures the influences of the one and two lag stages of RDI on agricultural carbon emissions. When the time lag stage was one year, the average impact of RDI on agricultural carbon emissions was −0.0524%, but the impact did not prove to be statistically significant ($p > 0.05$). When the time lag stage was two years, the average impact of RDI on agricultural carbon emissions was −0.1829%, and it was significantly negative under the 1% significant level. It should be noted that the influence of RDI in inhibiting agricultural carbon emissions requires a process that usually takes two years. Therefore, the RDI data used in this paper refer to the two-year lag stage. As shown in Table 5, RDI significantly inhibited agricultural carbon emissions in Fujian. That is, when RDI increased by 1%, agricultural carbon emissions decreased by an average of 0.1829%. One explanation might be that the inputs of agricultural technical research and development promote the progress of agricultural technology, reduce the dependence on pesticides, chemical fertilizers, and other energy consumptions in agricultural production, and finally, play a role in reducing agricultural carbon emissions. However, the influence of RDI showed strong temporal and spatial differences. For instance, as demonstrated in Figure 2a, RDI in Xiamen, Quanzhou, and Zhangzhou in 2008 inhibited agricultural carbon emissions. Except Sanming, RDI in other cities in both 2012 and 2017 significantly inhibited agricultural carbon emissions. Although the RDI of Sanming was relatively low and the effect on agricultural carbon emission reduction had not been reflected yet, with the continuous enhancement of agricultural technological innovation, the promoting influence of RDI in Sanming on agricultural carbon emissions gradually decreased from 2008–2017. Moreover, from the spatial perspective, the restraint impact of RDI on agricultural carbon emissions in eastern Fujian was significantly stronger than that in western Fujian, because the economic development level in eastern Fujian was much higher than that in the west, which can support the increasing intensity of agricultural R & D investment.

R & D investment in technology can significantly reduce agricultural carbon emissions, so the governments should continue to increase R & D investment in technology in the agricultural field and develop new technologies that are low carbon and have higher efficiency. Moreover, the governments should also promote new production models, i.e., organic agriculture and ecological agriculture, and finally, give full play to the role of agricultural technology in low-carbon agriculture [58]. In addition, there is a time lag effect of R & D investment on agricultural carbon emissions. When the time lag period was two years, the impact of R & D investment reached the maximum. It should be noted that the restraining effect of RDI on agricultural carbon emissions in Sanming was not reflected, so the government should strengthen the education of agricultural technicians and farmers, cultivate their long-term strategic thinking, and avoid short-term behavior.

4.3.2. The Influence of ALF on Agricultural Carbon Emissions

The results listed in Table 5 show that ALF significantly promoted the increase of agricultural carbon emissions. That is, when ALF increased by 1%, agricultural carbon emissions increased by an average of 0.0512%. One possible explanation is that the disorderly increase of ALF enlarged the scale of agricultural production to a certain extent, which was not conducive to reducing the agricultural carbon emissions. As shown in Figure 2b, the increase of ALF in all prefectural-level cities promoted the increase of agricultural carbon emissions, and the positive influence of ALF on agricultural carbon emissions in Nanping was the strongest. Nanping's agricultural labor force accounted for nearly 50% of the total labor force, which had a strong role in promoting agricultural carbon emissions. Moreover, the ALF in Quanzhou in 2012 and in Fuzhou, Quanzhou, and Xiamen in 2017 demonstrated negative correlations with agricultural carbon emissions, which was inconsistent with the significant positive correlation results of other relevant literature. The possible reason is that ALF showed obvious spatial heterogeneity. That is, Fuzhou, Quanzhou, and Xiamen have been facing a rapid development of economic and agricultural modernization in recent years. With the vigorous support of human resources and financial policies, the increase of ALF has brought advanced technology and professionals, improved the level of agricultural technology, and thus, to a certain extent, restrained the large growth of carbon emissions. However, the relationship between ALF and agricultural carbon emissions in other cities is significantly positive, which makes it more important to think about how to train the labor force in other regions and formulate corresponding policies to attract talents in the progress of agricultural technological innovation, so as to restrain agricultural carbon emissions.

The increase of the agricultural labor force in Fujian has promoted agricultural carbon emissions as a whole. However, in Fuzhou, Quanzhou, and Xiamen, the increase of technical personnel and the improvement of the quality of the agricultural labor force can reduce agricultural carbon emissions. Therefore, on the one hand, the government needs to guide actively the transfer of rural surplus labor to manufacturing and service industries, so as to reduce the agricultural labor force. On the other hand, the government needs to improve the education level of rural residents and strengthen their low carbon awareness, so that farmers can rationally use advanced agricultural technology. At the same time, the government needs to establish, improve, and strengthen the training and introduction of high-quality agricultural technological talents and provide human support for agricultural modernization.

4.3.3. The Influence of AVA on Agricultural Carbon Emissions

AVA significantly promoted the increase of agricultural carbon emissions and mainly promoted the average increases of agricultural carbon emissions by 0.6955% when AVA increased by 1%. That is, AVA is the main driving factor for the increase of agricultural carbon emissions. Besides, Fujian is a populous province, but the land resources are limited, so agriculture depends on chemical fertilizer and pesticides to meet people's demand for agricultural products and agricultural by-products, which also leads to the increase of AVA and the sharp increase of agricultural carbon emissions. As shown in Figure 2c, in 2008, 2012, and 2017, the regression coefficients of AVA in all prefectural-level cities in Fujian were all positive, and there was no obvious downward trend. Besides, AVA in Putian, Quanzhou, and Zhangzhou had a stronger impact on agricultural carbon emissions. For instance, in 2017, AVA in Putian, Quanzhou, and Zhangzhou accounted for nearly 2.9322%, 9.9831%, and 19.9504% of Fujian's total AVA, respectively, while chemical fertilizer use accounted for 5.0622%, 12.4713%, and 32.2809% of Fujian's total chemical fertilizer use, respectively. However, it should be noted that there existed a tendency of diminishing marginal utility in the use of chemical fertilizers, pesticides, plastic sheeting, and other products. It is obviously unsustainable for AVA to depend too much on the inputs of pesticides and fertilizers, nor can it enhance the comprehensive production capacity of agriculture.

Although AVA promotes agricultural carbon emissions, it cannot reduce agricultural carbon emissions by directly reducing the added value of agriculture. Thus, the increase of AVA mainly depends on the progress of agricultural technology and the promotion of production efficiency, so as to achieve the dual objectives of increasing AVA and controlling agricultural carbon emissions, and this can also fundamentally reduce the positive impact of AVA on agricultural carbon emissions.

4.3.4. The Influence of AIS on Agricultural Carbon Emissions

According to Table 5, AIS significantly promoted the increase of agricultural carbon emissions. That is, when AIS increased by 1%, agricultural carbon emissions increased by 0.4426% on average in the same direction. This shows that the plant products industry is the main source of agricultural carbon emissions, and the increase of its proportion will correspondingly increase the total carbon emissions. According to Figure 2d, AIS in Fuzhou had less influence on agricultural carbon emissions. This is mainly due to the high proportion of the fishery industry in Fuzhou's agricultural structure. In 2017, the proportion of the fishery industry output in Fuzhou's agricultural sector output reached 58%. Moreover, although the proportion of the plant products industry in Zhangzhou is inferior to Sanming, nearly 50%, the influence of AIS on agricultural carbon emissions was less. The results in Figure 2d also show that the coefficient of AIS in Zhangzhou was 0.3665 in 2008, 0.2055 in 2012, and 0.0903 in 2017, which shows a large downward trend. The possible reason is that the cultivated land in Zhangzhou is relatively concentrated and flat, which is suitable for large-scale mechanized farming. That is, the mechanized level of the plant products industry in Zhangzhou has reached the leading level. Thus, the development of the plant products industry would promote the increase of agricultural carbon emissions, but the influence was less, which is closely related to the production mode of its own planting industry.

AIS was second only to AVA in promoting agricultural carbon emissions. Thus, on the premise of food security, the government should actively optimize the structure of agriculture, guide farmers to reduce the planting of crops with high resource input and energy consumption, then increase the proportion of low-carbon productions such as fruits, flowers, and vegetables. This can not only improve the economic benefits, but also have a certain carbon-sink function. Besides, AIS played the most important role in promoting agricultural carbon emissions in Xiamen and Sanming. Thus, Xiamen and Sanming should further reduce the proportion of the planting industry. For instance, Xiamen can take advantage of its coastal location to develop fisheries and flowers, while Sanming can continue to develop special forestry industries such as *Camellia oleifera*, bamboo shoots, and forest tourism.

4.3.5. The Influence of DIR on Agricultural Carbon Emissions

DIR significantly inhibited the increase of agricultural carbon emissions. When DIR increased by 1%, agricultural carbon emissions decreased by 0.0711% on average. As a whole, the relationship between per capita disposable income of rural residents and environmental pollution in Fujian has jumped over the turning point of the inverted "U" shape and reached the right side. It has entered the stage that the more the agricultural economy develops, the smaller the agricultural carbon emissions. Under these circumstances, on the one hand, the increase of disposable income of rural residents can promote farmers to adopt modern machinery in agricultural production, improve the mechanization level and agricultural production efficiency, and then achieve the effect of increasing production and reducing carbon emissions. On the other hand, the increase of rural residents' disposable income can encourage farmers to choose less polluting products, such as organic fertilizer and bio-fertilizer, so as to reduce agricultural carbon emissions. As shown in Figure 2e, in 2008, only Xiamen's DIR was negatively correlated with agricultural carbon emissions. In 2012, the relationship between DIR and environmental pollution in Fuzhou, Quanzhou, and Zhangzhou also surpassed the turning point of the inverted "U" shape, and DIR was negatively correlated with agricultural carbon emissions.

In 2017, except Sanming, the increase of DIR in the other eight prefectural-level cities all reduced agricultural carbon emissions, indicating that DIR of Sanming had not reached the inflection point of the environmental Kuznets curve, and the increase of DIR at this stage increased agricultural carbon emissions.

The increase of DIR can significantly inhibit agricultural carbon emissions. The governments can adopt diversified policies to increase rural residents' disposable income. For areas with rural economic underdevelopment, such as Longyan, Ningde, Sanming, and Nanping, the governments can reduce the cost of agricultural production through fiscal policies and the promotion of agricultural mechanization. The governments can also help these areas establish and develop leading industries through planning guidance and technical services. For cities with large plains, i.e., Putian, Quanzhou, Zhangzhou, and Fuzhou, the governments can guide the transfer of agricultural labor forces to non-agricultural industries, make the agricultural labor force employ more agricultural production materials, and expand the scale of agricultural operation, increasing the income of agricultural workers. In addition, Xiamen has a relatively high degree of industrialization and urbanization. The government can vigorously develop the processing industry of agricultural products and promote the integration of the industries. In addition, in 2017, except Sanming, the relationship between DIR and agricultural carbon emissions has jumped the turning point of the inverted "U" shape. Therefore, Sanming should transfer its environmental Kuznets curve to the decline stage of the inverted "U" shape, that is to take into account the dual responsibilities of income growth and agricultural carbon emission reduction.

4.3.6. The Influence of ALA on Agricultural Carbon Emissions

According to Table 5, ALA significantly promoted the increase of agricultural carbon emissions. ALA mainly promoted the average increases of agricultural carbon emissions by 0.2873% when ALA increased by 1%. As shown in Figure 2f, the positive impact of ALA on agricultural carbon emissions mainly existed in the northern cities, especially in Ningde and Nanping. In comparison, Xiamen and Quanzhou, which are located in the southern part, became the cities where arable land area had a negative influence on agricultural carbon emissions in 2017. A possible explanation for this is due to the high level of industrialization and urbanization in Quanzhou and Xiamen. In these two cities, agricultural land has been gradually transformed into industrial land, and the arable land area has been seriously insufficient. However, economic development and residents' demand for agricultural products have led to the high investment and intensive use of arable land, which leads to the increase of agricultural carbon emissions beyond the land carrying threshold. Besides, the ALA of Putian and Longyan had little impact on agricultural carbon emissions, which is related to the policy of "returning farmland to forestry" in the two cities. Farmland with serious pollution and declining soil biological activity stopped being tilled and gradually turned into woodland. Meanwhile, reclamation of exploitable barren hills could reduce the promotion of arable land area on agricultural carbon emissions.

There exists a positive correlation between ALA and agricultural carbon emissions, but if the government blindly returns farmland to forestry or converts agricultural land into industrial land, it may also increase agricultural carbon emissions. Thus, the governments should develop and utilize the land rationally and optimize the land use structure. For instance, the governments of Quanzhou and Xiamen should control the total amount of industrial land to avoid the serious shortage of arable land threatening food security and bringing about the large use of energy products. Similarly, Ningde and Fuzhou can implement cultivated land protection measures to improve the ecological carrying capacity of agricultural land, that is to stop cultivating land with serious pollution and the decline of soil biological activity, and gradually turn it into woodland or grassland.

5. Conclusions

This paper measured agricultural carbon emissions in Fujian Province based on carbon emission coefficients released by IPCC and WRI, then applied GTWR to test empirically the influencing factors on agricultural carbon emissions. As a whole, the main conclusions can be listed as follows:

(1) Average agricultural carbon emissions in Fujian decreased from 708.88×10^3 tonnes of carbon in 2008 to 615.78×10^3 tonnes of carbon in 2017, and total agricultural carbon emissions showed a fluctuating downward trend. The fluctuating evolution with the basic spatial pattern of "M" can be divided into four stages: fluctuating increase, low speed reduction, rapid increase, and finally, rapid reduction. In addition, among all kinds of carbon sources, agricultural land use had the largest agricultural carbon emissions, followed by rice paddies and livestock manure storage, accounting for 40.92%, 34.48%, and 13.26% of the total agricultural carbon emissions in Fujian in 2017, respectively.

(2) From the perspective of direction, among factors affecting agricultural carbon emissions in Fujian, RDI and DIR mainly had inhibitory effects, while ALF, AVA, AIS, and ALA had promoting effects. From the perspective of degree, AVA had the greatest influence, followed by AIS and ALA. When these three factors changed by 1%, agricultural carbon emissions would change by 0.6955%, 0.4426%, and 0.2873% on average, respectively.

(3) According to the GTWR regression results, the six factors selected in this paper had different directions and different degrees of effects on the nine prefectural-level cities in Fujian Province. For instance, ALF mainly inhibited agricultural carbon emissions in Fuzhou and Xiamen, but promoted agricultural carbon emissions in other cities. Moreover, the effect of RDI and DIR on agricultural carbon emissions in eastern Fujian was stronger than that in western Fujian. Besides, for Xiamen, the main factor affecting agricultural carbon emissions was AIS, but for other cities of Fujian, the main factor was AVA. Therefore, it is necessary to adopt different means to reduce carbon emissions according to the actual situation.

(4) Combining the total amount of agricultural carbon emissions and the spatial and temporal characteristics of the influencing factors in Fujian, some policy recommendations can be put forward to achieve the ultimate goal of reducing agricultural carbon emissions.

Author Contributions: Conceptualization, Y.C. and M.L.; investigation, Y.C., K.S. and X.L.; data collection, M.L.; methodology and manuscript, Y.C. and M.L. All authors read and approved the final manuscript.

Funding: This research was funded by The Ministry of Agriculture and Rural Affairs: The Construction of Anxi Modern Agricultural Industrial Park (Grant No. KMD18003A) and The Social Science Planning Project of Fujian Province (Grant No. FJ2018C045) in China.

Conflicts of Interest: The authors declare no conflict of interest.

References

1. Lamb, A.; Green, R.; Bateman, I.; Broadmeadow, M.; Bruce, T.; Burney, J.; Carey, P.; Chadwick, D.; Crane, E.; Field, R.; et al. The potential for land sparing to offset greenhouse gas emissions from agriculture. *Nat. Clim. Chang.* **2016**, *6*, 488–492. [CrossRef]

2. Pellerin, S.; Bamiere, L.; Angers, D.; Beline, F.; Benoit, M.; Butault, J.P.; Chenu, C.; Colnenne-David, C.; De Cara, S.; Delame, N.; et al. Identifying cost-competitive greenhouse gas mitigation potential of French agriculture. *Environ. Sci. Policy* **2017**, *77*, 130–139. [CrossRef]

3. Nayak, D.; Saetnan, E.; Cheng, K.; Wang, W.; Koslowski, F.; Cheng, Y.F.; Zhu, W.Y.; Wang, J.K.; Liu, J.X.; Moran, D.; et al. Management opportunities to mitigate greenhouse gas emissions from Chinese agriculture. *Agric. Ecosyst. Environ.* **2015**, *209*, 108–124. [CrossRef]

4. Wang, W.; Koslowski, F.; Nayak, D.R.; Smith, P.; Saetnan, E.; Ju, X.T.; Guo, L.D.; Han, G.; de Perthuis, C.; Lin, E.; et al. Greenhouse gas mitigation in Chinese agriculture: Distinguishing technical and economic potentials. *Glob. Environ. Chang.* **2014**, *26*, 53–62. [CrossRef]

5. ACIL Tasman Pty Ltd. *Agriculture and GHG Mitigation Policy: Options in Addition to the CPRS*; Industry & Investment NSW: Dubbo, Australia, 2009; Available online: http://farminstitute.org.au/LiteratureRetrieve.aspx?ID=53067 (accessed on 13 August 2019).

6. Eggleston, S.; Buendia, L.; Miwa, K.; Ngara, T.; Tanabe, K. Volume 4: Agriculture, forestry and other land use. In *2006 IPCC Guidelines for National Greenhouse Gas Inventories*; Cambridge University Press: Cambridge, UK, 2006.

7. Wang, W.; Guo, L.P.; Li, Y.C.; Su, M.; Lin, Y.B.; de Perthuis, C.; Ju, X.T.; Lin, E.D.; Moran, D. Greenhouse gas intensity of three main crops and implications for low-carbon agriculture in China. *Clim. Chang.* **2015**, *128*, 57–70. [CrossRef]

8. Xiong, C.H.; Yang, D.G.; Xia, F.Q.; Huo, J.W. Changes in agricultural carbon emissions and factors that influence agricultural carbon emissions based on different stages in Xinjiang, China. *Sci. Rep.* **2016**, *6*, 36912. [CrossRef] [PubMed]

9. Tian, J.X.; Yang, H.L.; Xiang, P.A.; Liu, D.W.; Li, L. Drivers of agricultural carbon emissions in Hunan Province, China. *Environ. Earth Sci.* **2016**, *75*, 121. [CrossRef]

10. Han, H.B.; Zhong, Z.Q.; Guo, Y.; Xi, F.; Liu, S.L. Coupling and decoupling effects of agricultural carbon emissions in China and their driving factors. *Environ. Sci. Pollut. Res.* **2018**, *25*, 25280–25293. [CrossRef]

11. Zhang, L.; Pang, J.X.; Chen, X.P.; Lu, Z. Carbon emissions, energy consumption and economic growth: Evidence from the agricultural sector of China's main grain-producing areas. *Sci. Total Environ.* **2019**, *665*, 1017–1025. [CrossRef]

12. Bell, M.J.; Cloy, J.M.; Rees, R.M. The true extent of agriculture's contribution to national greenhouse gas emissions. *Environ. Sci. Policy* **2014**, *39*, 1–12. [CrossRef]

13. Wisniewski, P.; Kistowski, M. Assessment of greenhouse gas emissions from agricultural sources in order to plan for needs of low carbon economy at local level in Poland. *Geogr. Tidsskr.-Den.* **2018**, *118*, 123–136. [CrossRef]

14. Yue, Q.; Xu, X.R.; Hillier, J.; Cheng, K.; Pan, G.X. Mitigating greenhouse gas emissions in agriculture: From farm production to food consumption. *J. Clean. Prod.* **2017**, *149*, 1011–1019. [CrossRef]

15. Ismael, M.; Srouji, F.; Boutabba, M.A. Agricultural technologies and carbon emissions: Evidence from Jordanian economy. *Environ. Sci. Pollut. Res.* **2018**, *25*, 10867–10877. [CrossRef]

16. Gomiero, T.; Paoletti, M.G.; Pimentel, D. Energy and environmental issues in organic and conventional agriculture. *Crit. Rev. Plant Sci.* **2008**, *27*, 239–254. [CrossRef]

17. Cui, H.R.; Zhao, T.; Shi, H.J. STIRPAT-based driving factor decomposition analysis of agricultural carbon emissions in Hebei, China. *Pol. J. Environ. Stud.* **2018**, *27*, 1449–1461. [CrossRef]

18. Gerlagh, R. Measuring the value of induced technological change. *Energy Policy* **2007**, *35*, 5287–5297. [CrossRef]

19. Zhao, R.Q.; Liu, Y.; Tian, M.M.; Ding, M.L.; Cao, L.H.; Zhang, Z.P.; Chuai, X.W.; Xiao, L.G.; Yao, L.G. Impacts of water and land resources exploitation on agricultural carbon emissions: The water-land-energy-carbon nexus. *Land Use Policy* **2018**, *72*, 480–492. [CrossRef]

20. Lu, X.H.; Kuang, B.; Li, J.; Han, J.; Zhang, Z. Dynamic evolution of regional discrepancies in carbon emissions from agricultural land utilization: Evidence from Chinese provincial data. *Sustainability* **2018**, *10*, 552. [CrossRef]

21. Sarauer, J.L.; Coleman, M.D. Converting conventional agriculture to poplar bioenergy crops: Soil greenhouse gas flux. *Scand. J. For. Res.* **2018**, *33*, 781–792. [CrossRef]

22. West, T.O.; Marland, G. Net carbon flux from agriculture: Carbon emissions, carbon sequestration, crop yield, and land-use change. *Biogeochemistry* **2003**, *63*, 73–83. [CrossRef]

23. Zafeiriou, E.; Mallidis, I.; Galanopoulos, K.; Arabatzis, G. Greenhouse gas emissions and economic performance in EU agriculture: An empirical study in a non-linear framework. *Sustainability* **2018**, *10*, 3837. [CrossRef]

24. Owusu, P.A.; Asumadu-Sarkodie, S. Is there a causal effect between agricultural production and carbon dioxide emissions in Ghana? *Environ. Eng. Res.* **2017**, *22*, 40–54. [CrossRef]

25. Khan, M.; Ali, Q.; Ashfaq, M. The nexus between greenhouse gas emission, electricity production, renewable energy and agriculture in Pakistan. *Renew. Energy* **2018**, *118*, 437–451. [CrossRef]

26. Liu, X.; Zhang, S.; Bae, J. The impact of renewable energy and agriculture on carbon dioxide emissions: Investigating the environmental Kuznets curve in four selected ASEAN countries. *J. Clean. Prod.* **2017**, *164*, 1239–1247. [CrossRef]

27. Mourao, P.R.; Martinho, V.D. Portuguese agriculture and the evolution of greenhouse gas emissions-can vegetables control livestock emissions? *Environ. Sci. Pollut. Res.* **2017**, *24*, 16107–16119. [CrossRef]

28. Xiong, C.H.; Yang, D.G.; Huo, J.W. Spatial-temporal characteristics and LMDI-based impact factor decomposition of agricultural carbon emissions in Hotan Prefecture, China. *Sustainability* **2016**, *8*, 262. [CrossRef]

29. Tian, Y.; Zhang, J.B.; He, Y.Y. Research on spatial-temporal characteristics and driving factor of agricultural carbon emissions in China. *J. Integr. Agric.* **2014**, *13*, 1393–1403. [CrossRef]

30. Appiah, K.; Du, J.G.; Poku, J. Causal relationship between agricultural production and carbon dioxide emissions in selected emerging economies. *Environ. Sci. Pollut. Res.* **2018**, *25*, 24764–24777. [CrossRef]

31. Yadav, D.; Wang, J.Y. Modelling carbon dioxide emissions from agricultural soils in Canada. *Environ. Pollut.* **2017**, *230*, 1040–1049. [CrossRef]

32. Ye, R.; Espe, M.B.; Linquist, B.; Parikh, S.J.; Doane, T.A.; Horwath, W.R. A soil carbon proxy to predict CH4 and N2O emissions from rewetted agricultural peatlands. *Agr. Ecosyst. Environ.* **2016**, *220*, 64–75. [CrossRef]

33. Jiang, X.; Fang, W.; Zhuang, G.; Bai, W.; Zhu, S.; Lu, L.; Feng, J. *Greenhouse Gas Accounting Tool for Chinese Cities (Version 2.0)*; World Resources Institute: Washington, DC, USA, 2015.

34. Snyder, C.S.; Davison, E.A.; Smith, P.; Venterea, R.T. Agriculture: Sustainable crop and animal production to help mitigate nitrous oxide emissions. *Curr. Opin. Environ. Sustain.* **2014**, *9*, 46–54. [CrossRef]

35. Sasmal, T.K. Adoption of new agricultural technologies for sustainable agriculture in eastern India: An empirical study. *Indian Res. J. Ext. Educ.* **2015**, *15*, 38–42.

36. Xiong, C.H.; Yang, D.G.; Huo, J.W.; Zhao, Y.N. The relationship between agricultural carbon emissions and agricultural economic growth and policy recommendations of a low-carbon agriculture economy. *Pol. J. Environ. Stud.* **2016**, *25*, 2187–2195. [CrossRef]

37. You, D.; Jiang, K. Research into dynamic lag effect of R&D input on economic growth based on the vector auto-regression model. *J. Comput. Theor. Nanosci.* **2016**, *13*, 6787–6796.

38. Zhang, Y.; Fang, G. Research on spatial-temporal characteristics and affecting factors decomposition of agricultural carbon emission in Suzhou City, Anhui Province, China. *Appl. Mech. Mater.* **2013**, *291*, 1385–1388. [CrossRef]

39. Yao, C.; Qian, S.; Mao, Y.; Li, Z. Decomposition of impacting factors of animal husbandry carbon emissions change and its spatial differences in China. *Trans. Chin. Soc. Agric. Eng.* **2017**, *33*, 10–19. (In Chinese)

40. Satterthwaite, D. The implications of population growth and urbanization for climate change. *Environ. Urban.* **2009**, *21*, 545–567. [CrossRef]

41. Al-Mulali, U.; Sab, C.N.B.S.; Fereidouni, H.G. Exploring the bi-directional long run relationship between urbanization, energy consumption, and carbon dioxide emission. *Energy* **2012**, *46*, 156–167. [CrossRef]

42. Murad, W.; Ratnatunga, J. Carbonomics of the Bangladesh agricultural output: Causality and long-run equilibrium. *Manag. Environ. Qual. An Int. J.* **2013**, *24*, 256–271. [CrossRef]

43. Jebli, M.B.; Youssef, S.B. The role of renewable energy and agriculture in reducing CO_2, emissions: Evidence for North Africa countries. *Ecol. Indic.* **2015**, *74*, 295–301. [CrossRef]

44. Rafiq, S.; Salim, R.; Apergis, N. Agriculture, trade openness and emissions: An empirical analysis and policy options. *Aust. J. Agric. Resour. Econ.* **2016**, *60*, 348–365. [CrossRef]

45. Alamdarlo, N.H. Water consumption, agriculture value added and carbon dioxide emission in Iran, environmental Kuznets curve hypothesis. *Int. J. Environ. Sci. Technol.* **2016**, *13*, 2079–2090. [CrossRef]

46. Panayotou, T. *Empirical Tests and Policy Analysis of Environmental Degradation at Different Stages of Economic Development*; World Employment Research Programme, Working Paper; International Labour Office: Geneva, Switzerland, 1993.

47. Liu, L.H.; Xin, H.P. Research on spatial-temporal characteristics of agricultural carbon emissions in Guangdong Province and the relationship with economic growth. *Adv. Mater. Res.* **2014**, *1010*, 2072–2079. [CrossRef]

48. Zhang, J.; Yuan, C.; Zhang, L.; Ding, L. Do technological innovations promote urban green development?—A spatial econometric analysis of 105 cities in China. *J. Clean. Prod.* **2018**, *182*, 395–403. [CrossRef]

49. Li, X.; Jiang, D.; Bian, Z.; Yan, J.; Qu, F. China cultivated land change and its carbon budget measurement based on the system dynamics. *World Agric.* **2015**, *7*, 19–24.

50. Smith, P.; Martino, D.; Cai, Z.; Gwary, D.; Janzen, H.; Kumar, P.; McCarl, B.; Ogle, S.; O'Mara, F.; Rice, C.; et al. Greenhouse gas mitigation in agriculture. *Philos. Trans. R. Soc. B Biol. Sci.* **2008**, *363*, 789–813. [CrossRef]

51. Yager, R.R. On ordered weighted averaging aggregation operators in multicriteria decisionmaking. *Read. Fuzzy Sets Intell. Syst.* **1988**, *18*, 80–87. [CrossRef]

52. Xu, Z. An overview of methods for determining OWA weights. *Int. J. Intell. Syst.* **2005**, *20*, 843–865. [CrossRef]

53. Brunsdon, C.; Fotheringham, A.S.; Charlton, M.E. Geographically weighted regression: A method for exploring spatial nonstationarity. *Geogr. Anal.* **1996**, *28*, 281–298. [CrossRef]

54. Brunsdon, C.; Fotheringham, A.S.; Charlton, M. Some notes on parametric significance tests for geographically weighted regression. *J. Reg. Sci.* **1999**, *39*, 497–524. [CrossRef]

55. Li, M.; Wang, J.; Chen, Y. Evaluation and influencing factors of sustainable development capability of agriculture in countries along the Belt and Road route. *Sustainability* **2019**, *11*, 2004. [CrossRef]

56. Huang, B.; Wu, B.; Barry, M. Geographically and temporally weighted regression for modeling spatio-temporal variation in house prices. *Int. J. Geogr. Inf. Sci.* **2010**, *24*, 383–401. [CrossRef]

57. He, Q.; Bo, H. Satellite-based high-resolution PM 2.5 estimation over the Beijing-Tianjin-Hebei region of China using an improved geographically and temporally weighted regression model. *Environ. Pollut.* **2018**, *236*, 1027–1037. [CrossRef]

58. Mohamad, R.S.; Verrastro, V.; Al Bitar, L.; Roma, R.; Moretti, M.; Al Chami, Z. Effect of different agricultural practices on carbon emission and carbon stock in organic and conventional olive systems. *Soil Res.* **2016**, *54*, 173–181. [CrossRef]

Article

Analysis of Influencing Factors and Trend Forecast of Carbon Emission from Energy Consumption in China Based on Expanded STIRPAT Model

Zhen Li *, Yanbin Li and Shuangshuang Shao

School of Economics and Management, North China Electric Power University, Beijing 102206, China
* Correspondence: lizhen2020@ncepu.edu.cn; Tel.: +18-810-668-262

Received: 1 July 2019; Accepted: 5 August 2019; Published: 8 August 2019

Abstract: With the convening of the annual global climate conference, the issue of global climate change has gradually become the focus of attention of the international community. As the largest carbon emitter in the world, China is facing a serious situation of carbon emission reduction. This paper uses the IPCC (The Intergovernmental Panel on Climate Change) method to calculate the carbon emissions of energy consumption in China from 1996 to 2016, and uses it as a dependent variable to analyze the influencing factors. In this paper, five factors, total population, per capita GDP (Gross Domestic Product), urbanization level, primary energy consumption structure, technology level, and industrial structure are selected as the influencing factors of carbon emissions. Based on the expanded STIRPAT (Stochastic Impacts by Regression on Population, Affluence, and Technology) model, the influencing degree of different factors on carbon emissions of energy consumption is analyzed. The results show that the order of impact on carbon emissions from high to low is total population, per capita GDP, technology level, industrial structure, primary energy consumption structure, and urbanization level. On the basis of the above research, the carbon emissions of China's energy consumption in the future are predicted under eight different scenarios. The results show that, when the population and economy keep a low growth rate, while improving the technology level can effectively control carbon emissions from energy consumption, China's carbon emissions from energy consumption will reach 302.82 million tons in 2020.

Keywords: energy consumption; carbon emission; IPCC method; factor analysis; trend forecast

---

1. Introduction

According to the fourth IPCC assessment report, the average surface temperature has risen by about 0.74 °C in the past 100 years [1]. Global climate change is mainly caused by greenhouse gases such as carbon dioxide and methane emitted by human activities. The warming effect of carbon dioxide is the most clear. In order to reduce carbon dioxide emissions and the negative impact of human activities on the environment, the international community has made many efforts in this century. China is the largest developing country whose science and technology are inferior to developed countries. Additionally, its economic development largely relies on traditional energy, so the environmental problems and energy crisis need to be solved urgently in China [2]. China is also a big energy-consuming country. Under the background of global low carbon emission reduction, China has an obligation to contribute to the development of environmental friendliness.

As for the influencing factors of carbon emissions, scholars at home and abroad have made fruitful research. The representative research results are driving force analysis based on the IPAT equation and driving factor analysis based on the Kaya model. Ehrlich et al. put forward the IPAT equation, believing that the driving force of carbon emissions is the comprehensive effect of population size, the economic development level, and scientific and technological progress [3]. Dietz et al. combined stochastic theory with the IPAT model and established the STIRPAT model. He introduced the index into the model so that the model could be used to analyze the non-proportional impact of human factors on the environment [4]. Kaya, who is a Japanese scholar, established a mathematical model to reflect the quantitative relationship between population, economy, energy, and carbon dioxide produced by human activities. He believed that the total amount of carbon dioxide produced by social and economic activities in a region was equal to the product of factors such as total population, per capita GDP, energy intensity, and carbon emissions per unit energy consumption [5]. Wang et al. decomposed the influencing factors of carbon emissions by logarithmic mean Dirichlet decomposition (LMDI). The results showed that energy intensity was the most important factor to reduce carbon emission [6]. Wang Feng and others used the logarithmic average Divisia index decomposition method to study the growth rate of carbon dioxide emissions from China's energy consumption. He thought that per capita GDP growth was the greatest factor affecting the increase of carbon emissions, and that the decrease of energy intensity in the production sector was the most important factor to restrain the increase of carbon emissions [7].

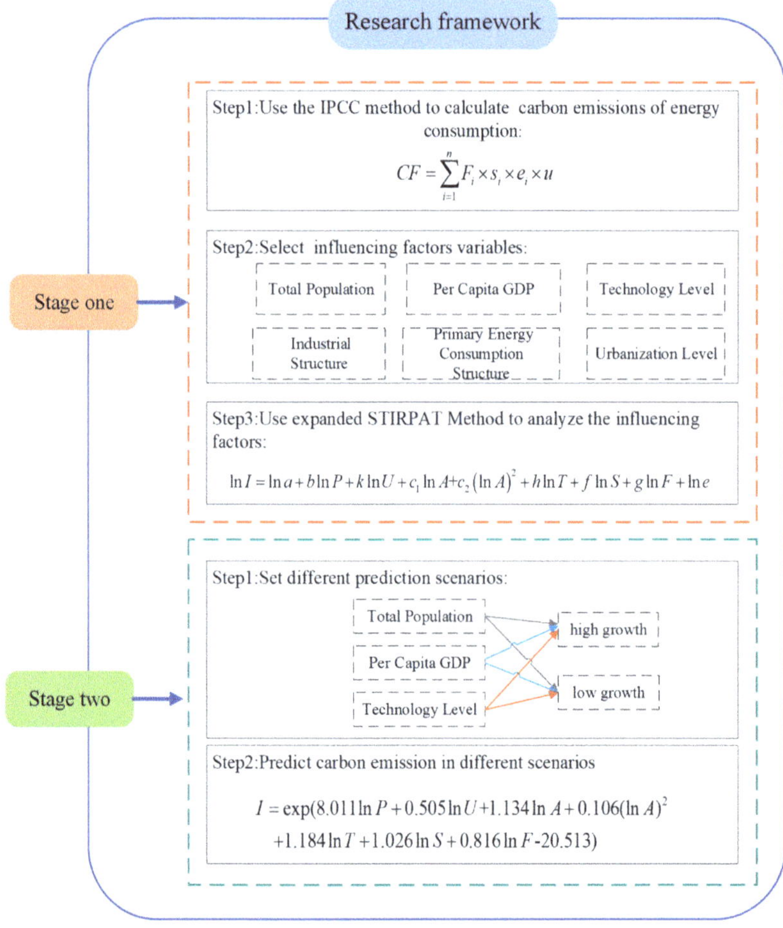

$$CF = \sum_{i=1}^{n} F_i \times s_i \times e_i \times u$$

$$\ln I = \ln a + b\ln P + k\ln U + c_1 \ln A + c_2 (\ln A)^2 + h\ln T + f\ln S + g\ln F + \ln e$$

$$I = \exp(8.011\ln P + 0.505\ln U + 1.134\ln A + 0.106(\ln A)^2$$
$$+ 1.184\ln T + 1.026\ln S + 0.816\ln F - 20.513)$$

Figure 1. The research structure of the article.

Research shows that different factors have different effects on carbon emissions, so it is necessary to further explore the factors affecting carbon emissions. In view of the shortcomings of previous studies, this paper uses the expanded STIRPAT model to analyze the impact of population size, affluence, primary energy consumption structure, technology level, industrial structure, and urbanization level on total carbon emissions. On the basis of factor analysis, the three most influential factors are taken as variables to set different scenarios. In addition, the carbon emissions of China's future energy consumption under different scenarios are predicted and analyzed. The research structure of the article is shown in Figure 1. Within the figure, b, k, c, h, f, and g are elastic coefficients. Descriptions of other symbols are shown in following article.

2. Establishment of Influencing Factors Regression Model

2.1. Sources of Data

Energy consumption data are from the China Energy Statistics Yearbook (1996–2016) [8]. Population, GDP, urban population, and output value of secondary industry all come from the "China Statistical Yearbook" (1996–2016) [9].

2.2. Introduction of the Modeling Method

2.2.1. Introduction of the Carbon Emission Calculating Method

This paper analyzes carbon emissions from energy consumption in China. Considering the availability of data and the purpose of research, the IPCC method is adopted in this paper. The selected energy consumption categories are raw coal, washed coal, coke, coke oven gas, other gas, crude oil, gasoline, kerosene, diesel oil, fuel oil, liquefied petroleum gas, refinery dry gas, and natural gas. The calculation formula is shown below.

$$CF = \sum_{i=1}^{n} F_i \times s_i \times e_i \times u \tag{1}$$

F_i is the total energy consumption per ton. s_i is the standard coal coefficient for energy conversion, as shown in Table 1. e_i is the carbon emission factor of each energy source. The energy carbon emission coefficients used in this paper refer to the various energy carbon emission coefficients in the 2006 IPCC National Greenhouse Gas Emission Inventory Guidelines, as shown in Table 2. Since the statistical unit of each energy consumption is not meaningful, u is the unit conversion coefficient.

Table 1. Standard coal coefficient for 13 energy conversions.

	Standard Coal Coefficient	Energy Types	Standard Coal Coefficient
raw coal	0.7143	kerosene	1.4714
washed coal	0.9000	diesel oil	1.4571
coke	0.9714	fuel oil	1.4286
coke oven gas	0.5926	liquefied petroleum gas	1.7143
other gas	0.3214	refinery dry gas	1.5714
crude oil	1.4286	natural gas	1.2128
gasoline	1.4714		

Table 2. Carbon emission coefficient for 13 energy types.

Energy Types	Carbon Emission Coefficient	Energy Types	Carbon Emission Coefficient
raw coal	0.7559	kerosene	0.5714
washed coal	0.7559	diesel oil	0.5921
coke	0.855	fuel oil	0.6185
coke oven gas	0.3548	liquefied petroleum gas	0. 5042
other gas	0.3548	refinery dry gas	0.4602
crude oil	0.5857	natural gas	0.4483
gasoline	0.5538		

2.2.2. Introduction of Influencing Factors Analysis Method

The IPAT (I = Human Impact, P = Population, A = Affluence, T = Technology) model is widely used in the research of carbon emission related issues. The IPAT model was first proposed by American ecologists Ehrlich and Commoner to study the relationship between human activities and the natural environment. The major variables are population size (P), affluence (A), technology level (T), and environment (I) [10]. It has been widely used by scholars to analyze the influencing factors of environmental change since it is simple and easy to understand. However, the factors leading to environmental problems are complex. The IPAT model only involves three influencing factors, which cannot fully reflect the actual problems. The model can only analyze the problem by changing one factor while keeping other factors fixed, so that the influence of independent variables on dependent variables is proportional. However, this is not in line with the actual situation. In order to make up for the deficiency of the IPAT model, York and other scholars put forward the STIRPAT model on the basis of this model [11], which is expressed as follows.

$$I = aP^b \times A^c \times T^d e \tag{2}$$

In the formula, a is the coefficient of the model, b, c, and d are the index of population size, affluence degree, and technology level, respectively, and e is the random error term. In practical application, according to the STIRPAT model, the elasticity of influence factors on the environment is obtained by taking a natural logarithm on both sides of the equation. The logarithmic form is as follows.

$$\ln I = \ln a + b \ln P + c \ln A + d \ln T + e \tag{3}$$

Among them, b, c, and d are the elasticity coefficients of the population, affluence, and the level of technology, and $\ln a$ is a constant term.

In order to analyze the influencing factors of carbon emission intensity in China, this paper introduces six indicators: total population, urbanization level, per capita GDP, technology level, industrial structure, and primary energy consumption structure to extend the original STIRPAT model. The expanded STIRPAT model is expressed below.

$$\ln I = \ln a + b \ln P + k \ln U + c \ln A + h \ln T + f \ln S + g \ln F + \ln e \tag{4}$$

Among them, I represents the total carbon emissions from China's energy consumption, P represents the total population, U represents the urbanization level, A represents the per capita GDP, T represents the level of technology, S represents the industrial structure, and F represents the primary energy consumption structure. b, k, c, h, f, and g are elastic coefficients, which indicate that, when P, U, A, T, S and F change by 1%, the carbon emissions will change b%, k%, c%, h%, f%, and g%, respectively.

The explanations of each variable are given in Table 3.

Table 3. The explanations of variables.

Variable	Symbol	Variable Description	Unit
carbon emission	I	total carbon dioxide emissions	ten thousand tons
population size	P	total population	ten thousand
affluence	A	per capita GDP	yuan
technology level	T	energy intensity	–
urbanization level	U	urbanization rate	%
industrial structure	S	the proportion of the output value of the secondary industry	%
primary energy consumption structure	F	the proportion of coal consumption in primary energy consumption	%

Technology level refer to energy intensity. Energy intensity is energy consumption per unit of GDP, which reflects the input-output characteristics of the energy system, and reflects the overall efficiency of energy economic activities.

The urbanization level is expressed by the urbanization rate, that is, the proportion of urban population to the permanent population. The urbanization level is one of the important factors affecting carbon emissions. Cities are the concentration areas of population, transportation, industry, and other resources, as well as energy consumption and carbon emissions [12].

The industrial structure is explained by the proportion of the output value of the second industry in the total output value of that year. Among the three major industries, the secondary industry consumes the largest energy, especially the heavy industry.

The structure of the primary energy consumption is the proportion of coal consumption in primary energy consumption in that year. China's energy consumption structure is still dominated by coal, and coal and other fossil energy consumption is the main reason for carbon dioxide production.

In order to test whether there is an inverted U-shaped curve between economic growth and carbon emissions, the $\ln A$ in model (4) is decomposed into $\ln A$ and $(\ln A)^2$ [13]. The model is adjusted below.

$$\ln I = \ln a + b \ln P + k \ln U + c_1 \ln A + c_2 (\ln A)^2 + h \ln T + f \ln S + g \ln F + \ln e \tag{5}$$

c_1 and c_2 are coefficients of the logarithm of per capita GDP and logarithm quadratic of per capita GDP, respectively.

From Equation (5), the elasticity coefficient EE_{IA} of per capita GDP to carbon emissions from energy consumption can be obtained as follows.

$$EE_{IA} = c_1 + 2c_2 \ln A \tag{6}$$

If c_2 is negative, there is an inverted U-shaped curve between per capita GDP and carbon emissions.

In data regression, because the nature and unit of each variable are different, if the original data is directly used for regression, it will result in unfair regression. Therefore, before using principal component analysis, the data should be standardized. The standard processing method adopted in this paper is the Z-score processing method [14].

2.3. Regression Model Results

2.3.1. Results of China's Energy Consumption Carbon Emissions

According to the China Energy Statistics Yearbook (1996–2016) and Formula (1), China's energy consumption and carbon emissions from 1996 to 2016 are calculated as shown in Figure 2.

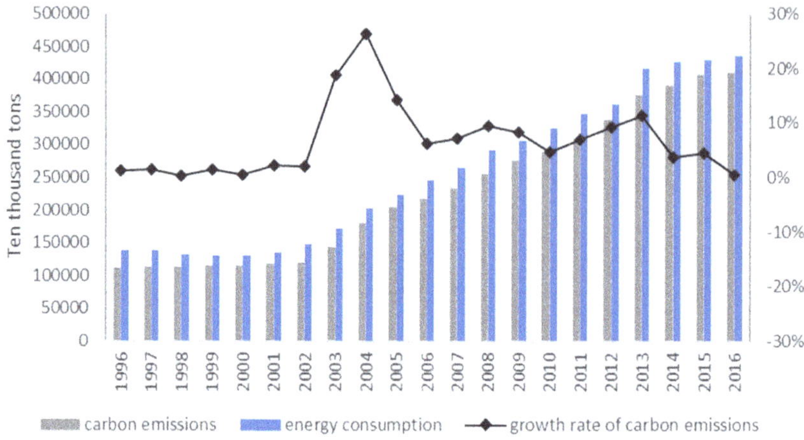

Figure 2. China's energy consumption and carbon emissions from 1996 to 2016.

As can be seen from the figure, since 1996, China's total energy consumption has maintained a growing trend. China's total energy consumption in 1996 was 138.948 million tons, while, in 2016, it was 435.819 million tons, which reached an increase of 213.66%. Similarly, carbon emissions from China's energy consumption were increasing. Before 2003, the growth rate was low and stable at 2%. However, the growth rate in the three years from 2003 to 2005 exceeded 15%, of which the growth rate in 2004 reached 26.25%. After 2005, the growth rate has decreased slightly, but the carbon emissions are still increasing. From 111.263 million tons in 1996 to 410.052 million tons in 2016, it increased by 268.54% in 20 years.

2.3.2. Results of the Regression Model

The least squares regression is performed on the variables $\ln P$, $\ln U$, $\ln A$, $(\ln A)^2$, $\ln T$, $\ln S$, and $\ln F$. The model test results include the adjustable coefficient $R^2 = 0.71505$, the F value is 43.31292, and the p value is $0.00000 < 0.05$. The multivariate regression parameters are estimated as shown in Table 4, and the value p of the variables $\ln A$, $\ln T$, and $\ln S$ are all greater than the significance level of 0.05. The Variance Inflation Factor (VIF) is calculated on the basis of multiple explanatory variables to assist the regression equation, and reflects the severity of multicollinearity between explanatory variables. It is generally believed that there is a serious multiple collinearity between explanatory variables and residual explanatory variables when VIF is greater than 10. Additionally, the multiple collinearity between explanatory variables will affect the results of least squares regression [15]. In this study, the variable VIF values for each indicator are shown in Table 4. There are multiple collinearity between variables. The main methods to eliminate multicollinearity are partial least squares [16], principal component regression [17], and ridge regression [18]. In this study, principal component regression is used.

Before eliminating multiple collinearity by principal component regression, the independent variables $\ln P$, $\ln U$, $\ln A$, $(\ln A)^2$, $\ln T$, $\ln S$, and $\ln F$ should be tested by KMO (Kaiser-Meyer-Olkin) test and Bartlett sphericity test. The two tests can confirm whether the above variables are suitable for principal component regression [19]. This can be seen from Table 5. The result shows that KMO = 0.747. The Bartlett sphericity test has a significant value less than 0.05, so we should reject the zero hypothesis. Additionally, consider that the correlation coefficient matrix cannot be a unit matrix. Therefore, there is correlation between the original variables. This indicates that these seven independent variables are suitable for principal component analysis.

Table 4. Estimation of multivariate regression parameters.

Variable	Coefficient	Standard Error	t-Statistic	Probability	VIF
C	8.07×10^{-8}	0.0131	6.17×10^{-6}	1.0000	–
$\ln P$	1.8444	0.7040	2.6199	0.0212	2759.3270
$\ln U$	−1.8825	0.7706	−2.4430	0.0296	2951.6632
$\ln A$	−0.7634	3.5876	−0.2128	0.8348	1754.8335
$(\ln A)^2$	2.5871	3.3737	0.7669	0.4569	109.1082
$\ln T$	−0.7992	0.1163	−6.8732	0.0000	75.0465
$\ln S$	0.0170	0.0668	0.2549	0.8028	24.1146
$\ln F$	0.0163	0.0841	0.1943	0.8489	19.7655

Table 5. KMO test and Bartlett sphericity test.

Kaiser-Meyer-Olkin Measure of Sampling Adequacy		0.747
Bartlett's Test of Sphericity	Approx. Chi-Square	502.132
	df	21
	sig	0.000

This paper uses SPSS22.0 software to carry out principal component analysis. The eigenvalues of each factor must be greater than 1. The contribution rate of cumulative variance is shown in Table 6 and the principal component load matrix is shown in Table 7. The contribution rate of variance of the first principal component extracted is 75.336% and, of the first two, is 97.787%, which means it meets the requirement of principal component extraction. Therefore, this study extracts two principal components Z1 and Z2 after seven independent variables of principal component analysis.

Table 6. The cumulative variance contribution rate.

Component	Eigenvalue	Percentage of Variance of Initial Eigenvalue	Cumulative Percentage	Total	Percentage of Square Sum Loading Variance Extracted	Cumulative Percentage
1	5.273	75.336	75.336	5.273	75.336	75.336
2	1.572	22.451	97.787	1.572	22.451	97.787
3	0.137	1.958	99.745			
4	0.011	0.152	99.897			
5	0.007	0.100	99.997			
6	0.000	0.003	100			
7	7.378×10^{-6}	0.000	100			

Extraction method: Principal component analysis.

Table 7. Principal component load matrix.

Variable	1	2
$\ln P$	0.989	0.124
$\ln U$	0.987	0.146
$\ln A$	0.989	0.096
$(\ln A)^2$	0.990	0.072
$\ln T$	0.991	−0.057
$\ln S$	−0.611	0.750
$\ln F$	0.084	0.977

According to Table 7, we can get the principal component coefficient, as shown in Table 8. With the result of principal component analysis, the regression equations of Z1 and Z2 can be established.

$$Z1 = 0.187 \ln P^* + 0.187 \ln U^* + 0.188 \ln A^* + 0.188(\ln A^*)^2 + 0.188 \ln T^* - 0.188 \ln S^* + 0.016 \ln F^*$$
$$Z1 = 0.079 \ln P^* + 0.093 \ln U^* + 0.061 \ln A^* + 0.046(\ln A^*)^2 - 0.036 \ln T^* + 0.477 \ln S^* + 0.622 \ln F^*$$

Establishing regression equation as follows,

$$\ln I^* = \beta_0 + \beta_1 Z1 + \beta_2 Z2 \tag{7}$$

β_0 is a constant term. β_1 and β_2 are coefficients.

Table 8. Principal component coefficient.

Variable	1	2
$\ln P$	0.187	0.079
$\ln U$	0.187	0.093
$\ln A$	0.188	0.061
$(\ln A)^2$	0.188	0.046
$\ln T$	0.188	−0.036
$\ln S$	−0.116	0.477
$\ln F$	0.016	0.622

The regression result is shown in Table 9. From Table 9, we can see that VIF is 1.000, so there is no multiple collinearity between the extracted two principal components. After calculating the score, we make a regression analysis with a standardized dependent variable ln I^* and the adjusted R-squared is 0.823. The P value of the T-test of the constant term (β_0) is not significant, but the coefficient is very small and can be neglected. Therefore, it has no influence on this study. The T-test of Z1 and Z2 showed a significant difference.

Table 9. Component coefficient.

Coefficient	Non-Standardized Coefficient	Standard Error	Standardized Coefficient	t	Significant	VIF
β_0	–	0.046	–	0	1.000	–
Z1	0.769	0.047	0.793	20.450	0.000	1.000
Z2	0.316	0.047	0.325	2.961	0.008	1.000

* means at 0.1 level. Therefore, the regression result is as follows.

$$\begin{aligned} \ln I^* &= 0.793 Z1 + 0.325 Z2 \\ &= 0.412 \ln P^* + 0.711 \ln U^* + 1.656 \ln A^* + 1.826 (\ln A^*)^2 \\ &\quad + 2.266 \ln T^* + 0.752 \ln S^* - 0.205 \ln F^* \end{aligned}$$

The above formula is a regression equation for standardized variables. According to the principle of standardization, the final regression equation can be obtained by restoring the data.

$$\begin{aligned} \ln I &= 8.011 \ln P + 0.505 \ln U + 1.314 \ln A + 0.106 (\ln A)^2 \\ &\quad + 1.184 \ln T + 1.026 \ln S + 0.816 \ln F - 20.513 \end{aligned}$$

The regression results show that the order of impact on carbon emissions from high to low is population, per capita GDP, level of science and technology, proportion of secondary industry, primary energy consumption structure, and urbanization level. Their elasticity coefficients are 8. 011, (1.314 + 0.212ln A), 1.184, 1.026, 0.816, and 0.505, respectively. Among them, A is per capita GDP.

The coefficient of $(\ln A)^2$ is positive, which indicates that there is no inverted U-shaped relationship between China's economic growth and carbon emissions. With the economic growth, environmental pressures are increasing day by day, and there is no equilibrium inflection point yet.

3. Prediction of Carbon Emission Trend under Different Scenarios

3.1. Introduction of Prediction Model

Based on the simulation results of the STIRPAT model, the future carbon emissions of China are predicted. The prediction formulas are shown below.

$$I = \exp(8.011\ln P + 0.505\ln U + 1.134\ln A + 0.106(\ln A)^2 + 1.184\ln T \\ + 1.026\ln S + 0.816\ln F - 20.513) \tag{8}$$

According to the historical population, per capita GDP, primary energy consumption, energy intensity, the proportion of secondary industry, and urbanization rate, this paper uses the above model to simulate the carbon emissions of China's historical energy consumption, and makes regression between the simulated and historical values. The comparison chart is shown in Figure 3. The results show that the simulated R^2 reaches 0.9984. Therefore, it is feasible to use the above model to forecast the carbon emissions of energy consumption in China in the future.

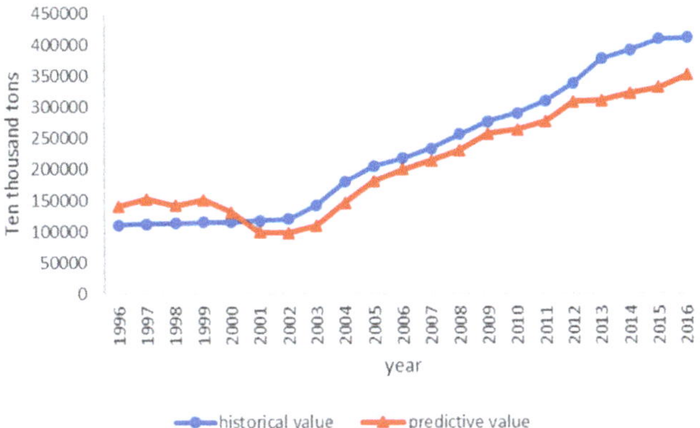

Figure 3. Historical and forecast values of China's energy consumption carbon emissions.

3.2. Scenarios Setting

In the factor analysis above, population, per capita GDP, and the technological level are the three factors that have the greatest impact, respectively. Therefore, taking these three indicators as variables in the scenario and growth rate as scenario conditions, the specific scenario settings are shown in Table 10.

Table 10. Scenario settings.

Scenario	Population	Per Capita GDP	Technological Level
scenario 1	low growth	low growth	high growth
scenario 2	low growth	low growth	low growth
scenario 3	high growth	low growth	high growth
scenario 4	high growth	low growth	low growth
scenario 5	low growth	high growth	low growth
scenario 6	low growth	high growth	high growth
scenario 7	high growth	high growth	low growth
scenario 8	high growth	high growth	high growth

3.3. Prediction of Carbon Emission Trend

According to the Development Strategy of the Development Research Center of the State Council and the Report of the Ministry of Regional Economic Research on the Forecast and Analysis of the Speed of Urbanization in China [No. 99 of 2017], the growth rate of urbanization in China slowed down from 2016 to 2050. The urbanization rates in 2020, 2030, 2040, and 2050 were 60.34%, 68.38%, 75.37%, and 81.63%, respectively [20]. In this way, we can deduce that the growth rate of the urbanization rate in each stage is 0.793%, and then calculate the urbanization rate of China in the coming years.

According to the national energy strategic action plan and related research, the slowdown of coal demand growth will become a new normal for the development of the coal industry. By 2020, 2030, 2040, and 2050, the proportion of coal in China's primary energy structure will remain 62%, 55%, 53%, and 50% [21]. Similarly, it can deduce the primary energy consumption structure in different stages, and calculate the primary energy consumption structure in future years in China.

According to the Long-term Forecast of China's Economy in the 21st Century, the proportion of the secondary industry output value to GDP in 2020, 2030, 2040, and 2050 is 50.2%, 48%, 45.3%, and 42.1%, respectively [22]. According to the same method mentioned above, we can predict the proportion of secondary industry output value in China in the coming years.

Based on the historical data of China from 1996 to 2016, the population growth rate and per capita GDP growth rate of each year are calculated. The maximum and minimum growth rates of population growth rate (1.13% vs 0.39%)and per capita GDP growth rate (13.64% vs. 6.12%) are used as their respective high and low growth rates, respectively, to estimate China's future population and per capita GDP. The maximum and minimum of the energy intensity decline rate (14.26% vs. 1.75%) are selected as the high and low growth rates of technology, respectively. Additionally, the future technological level of China is estimated.

Based on the above scenarios, the carbon emissions from China's energy consumption are predicted by using Formula (8). The calculation results are shown in Figure 4. Table 11 provides a forecast of carbon emissions from China's energy consumption in 2020, 2030, 2040, and 2050.

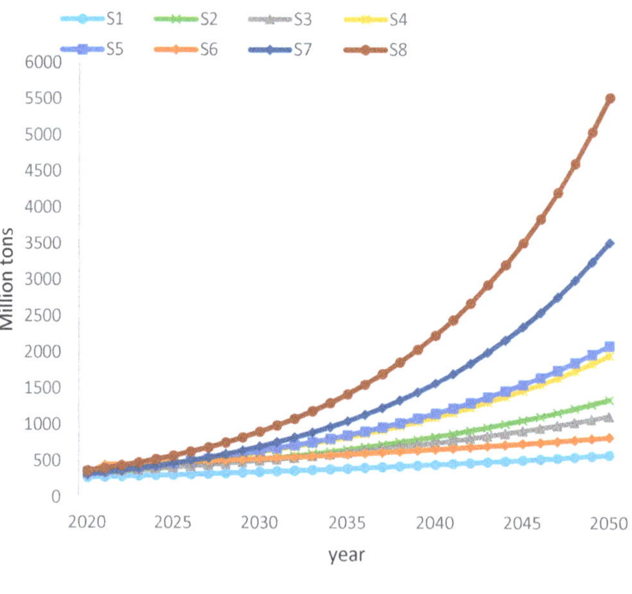

S means scenario

Figure 4. Prediction trends of carbon emissions in different scenarios.

Table 11. Prediction of China's energy consumption carbon emissions in different scenarios.

Scenario	2020	2030	2040	2050
S1	302.81	359.96	362.27	692.05
S2	337.72	552.06	704.77	835.34
S3	341.82	429.13	663.62	1123.48
S4	354.29	579.03	1285.48	1963.68
S5	355.61	562.59	1188.6	2103.72
S6	327.33	437.52	770.38	1351.39
S7	316.47	603.37	1547.23	3527.45
S8	368.43	733.56	2003.69	5532.06

Unit: million tons.

4. Discussion

4.1. Discussion on Influencing Factors

Population size has the highest impact on carbon emissions, and the elasticity coefficient is as high as 8.011. It shows that if the population increases by 1%, the total carbon emissions will increase by 8.011%. China has a large population base, and its lifestyle and production activities depend on traditional energy. Population size has the most direct impact on carbon emissions. China is the largest manufacturing concentration country in the world. Developed countries have set up high-pollution and high-energy-consuming manufacturing links in China. Labor-intensive and high-carbon industries make the environmental pressure worse.

Per capita GDP is another important factor affecting China's energy consumption carbon emissions. For every 1% increase in per capita GDP, energy consumption carbon emissions will increase $(1.314 + 0.2121 \text{n} A)$. It is the second important factor affecting China's energy consumption carbon emissions. This shows that China's economic development and social life are highly dependent on energy consumption.

The elasticity coefficient of the impact of technology level on carbon emissions from energy consumption is 1.184. For every 1% increase in energy intensity, carbon emissions from China's energy consumption will grow by 1.184%. China's energy consumption structure is dominated by coal, which leads to a large consumption of fossil energy and a large amount of carbon emissions when GDP increases.

Similarly, the industrial structure also plays an important role in China's energy consumption carbon emissions. Regression results show that every 1% increase in the proportion of secondary industry output value will generate 1.026% carbon emissions. In China's industrial structure, heavy industry and manufacturing industry occupy the main position. This kind of industrial structure with high energy consumption and emission has a negative impact on reducing carbon emissions from energy consumption in China.

The impact coefficient of primary energy consumption structure, especially coal consumption, on China's carbon emissions is 0.816%. The current situation of energy consumption in China is that coal accounts for 60% of primary energy. China is one of the few countries in the world where coal is the most important energy resource. It is also one of the most polluted areas in the world because of coal combustion. The long-term high proportion of coal resources in energy consumption is also one of the main reasons for China's high carbon emissions. Therefore, China needs to optimize its energy structure and vigorously develop low-carbon energy.

The level of urbanization is also a factor contributing to the increase of carbon emissions from energy consumption in China. The regression results show that an increase of 1% in the urbanization level will result in an increase of 0.505% in carbon emissions. From 29.37% urbanization in 1996 to 57. 37% in 2016, China's urbanization level has developed rapidly in the past 20 years. Yet, with the advancement of urbanization, the demand for energy in urban buildings, transportation, and residential buildings is also increasing, which increases China's carbon emissions.

4.2. Discussion on Trend Prediction

As can be seen from Figure 4, in scenario 1, when China maintains low population growth rate, low per capita GDP growth rate, and high-tech growth rate in the future, the growth rate of carbon emissions from energy consumption in China is the slowest. By 2050, the carbon emissions will be 692.05 million tons. In scenario 8, when the population and per capita GDP keep a high growth, and the technological progress rate keeps low growth, the growth of carbon emissions from energy consumption will be the fastest. There will be 5532.06 million tons of carbon emissions by 2050. It is also eight times higher than the result in scenario 1. By comparing Scenario 2, Scenario 3, and Scenario 6 with Scenario 1, we can find that, when the population keeps low growth, the per capita technological level keeps high growth, and the per capita GDP keeps high growth, while the growth rate of carbon emissions from energy consumption is the fastest. While, when the population and per capita GDP keep a low growth, the growth rate of energy consumption is the slowest when the technological level keeps a low growth. Therefore, reducing carbon emissions from China's energy consumption should enhance the technological level and reduce energy intensity.

Comparing Scenario 4 with Scenario 8, when the population keeps a high growth and the technological level keeps a low growth rate, the carbon emissions of energy consumption can be effectively reduced by reducing the growth rate of per capita GDP. Therefore, in order to control China's future carbon emissions, it is necessary not only to improve energy utilization technology and control the population, but also to reduce the growth rate of per capita GDP. This means that China should slow down its economic development in the future, change its mode of economic growth, and make its economic growth tend to a new normal.

5. Conclusions

First, this paper calculates the carbon emissions of energy consumption in China from 1996 to 2016, and then uses the STIRPAT model to decompose the influencing factors of carbon emissions of energy consumption and analyze the influence degree of different factors. This combines different scenarios to predict the future trend of carbon emissions from energy consumption in China.

(1) During the two decades from 1996 to 2016, China's energy consumption and carbon emissions of energy consumption showed an increasing trend. Among them, energy consumption increased by 213.66% and carbon emissions of energy consumption increased by 268.54%.

(2) Among the factors affecting carbon emissions from energy consumption in China, population factors have the highest impact on carbon emissions, with an elasticity coefficient of 8.011. The impact of per capita GDP on carbon emissions is second only to that of the population. The high demand for energy in China's economic development has greatly increased China's carbon emissions from energy consumption. The order of impact degree is population quantity > per capita GDP > technology level > industrial structure > primary energy consumption structure > urbanization level.

(3) There is no inverted U-shaped relationship between China's economic growth and carbon emissions. Therefore, with economic growth, environmental pressures are increasing, and there is no equilibrium inflection point.

(4) By forecasting the carbon emissions of China's future energy consumption in different scenarios, we can see that the growth rate of China's energy consumption carbon emissions is the slowest while maintaining low population growth rate, low per capita GDP growth rate, and a high-tech growth rate. When in the scenarios of high population growth rate, high per capita GDP growth rate, and a low-tech growth rate, the carbon emissions are the highest. Therefore, in order to reduce China's energy consumption carbon emissions, we should not only control the population size and control the speed of economic development, but also improve energy utilization technology and reduce the dependence of economic development on energy.

Author Contributions: Conceptualization and Methodology: Y.L. Formal Analysis, Investigation, Writing—Original Draft Preparation and Writing—Review & Editing: Z.L. Resources, Data Duration, Validation and Supervision: S.S.

Funding: The Beijing social science foundation research base project (Grant No. 17JDGLA009), *"Natural Science Foundation of China Project" (Grant No. 71471058),* the Fundamental Research Funds for the Central Universities under No. 2019QN075, supported this paper.

Acknowledgments: First of all, I would like to extend my sincere gratitude to my supervisor, Professor Li, for her instructive advice and useful suggestions on my thesis. I am deeply grateful of his help in the completion of this thesis. Second, I would like to express my heartfelt gratitude to the anonymous reviewers and referees, which gave me many inspirations and suggestions for this paper. Last but not least, I owe much to my friends especially Shuangshuang Shao for their valuable suggestions and critiques, which are helpful and important for making this thesis a reality.

Conflicts of Interest: The authors declare that they have no competing interest.

References

1. The Fourth IPCC Assessment Report. Available online: https://www.ipcc.ch/reports/ (accessed on 1 March 2019).
2. Bei, X. Research on Temporal and Spatial Migration of Energy Consumption Carbon Footprint and its Influence Factors in China. Master's Thesis, Anhui University of Finance and Economics, Bengbu, China, December 2016.
3. Ehrlich, P.R.; Ehrlich, A.H. *Population, Resources, Environment. Issues in Human Ecology*; Freeman: San Francisco, CA, USA, 1970; pp. 89–157.
4. Dietz TRosa, E.A. Rethinking the environmental impacts of population affluence and technology. *Hum. Ecol. Rev.* **1994**, *1*, 277–300.
5. Kaya, Y. *Impact of Carbon Dioxide Emission on GNP Growth: Interpretation of Proposed Scenarios*; Presentation to the Energy and Industry Subgroup; Response Strategies Working Group, IPCC: Paris, France, 1989; pp. 1–25.
6. Can, W.; Jining, C.; Ji, Z. Decomposition of energy-related CO_2 emission in China: 1957–2000. *Energy* **2005**, *30*, 73–83.
7. Feng, W.; Lihua, W.; Chao, Y. Study on the Driving Factors of Carbon Emission Growth in China's Economic Development. *Econ. Res.* **2010**, *2*, 0629.
8. *China Energy Statistics Yearbook (1996–2016)*; China Statistics Press: Beijing, China, 1996–2016.
9. China Statistical Yearbook (1996–2016). Available online: http://www.stats.gov.cn/tjsj/ndsj/ (accessed on 1 December 2017).
10. Shuang, L.; Tao, D.; Xia, Q. Study on the Influencing Factors of Carbon Emission in China's Construction Industry Based on Extended STIRPAT Model. *Mod. Manag.* **2017**, *37*, 96–98.
11. Jingshui, S.; Zhirui, C.; Zhijian, L. Research on the Influencing Factors of China's Low Carbon Economy Development—Based on the Expanded STIRPAT Model Analysis. *Audit Econ. Res.* **2011**, *26*, 85–93.
12. Bangli, C.; Meiping, X. Analysis of the Influencing Factors of Carbon Emissions in China: An Empirical Study of STIRPAT-Alasso Model Based on Panel Data. *Ecoeconomy* **2018**, *1*, 20–24.
13. Yue, P. Analysis of Driving Factors of Carbon Emission in Jiangsu Province Based on STIRPAT Model. *Environ. Pollut. Prev.* **2014**, *36*, 104–109.
14. Gao, T. An Empirical Study on the Influencing Factors of Housing Prices in Nanjing. *Econ. Res. Guide.* **2019**, *12*, 125–127.
15. Vu, D.H.; Muttaqi, K.M.; Agalgaonkar, A.P. A variance inflation factor and backward elimination based robust regression model for forecasting monthly electricity demand using climatic variables. *Appl. Energy* **2015**, *140*, 385–394. [CrossRef]
16. Xu, O.; Fu, Y.; Su, H. A Selective Moving Window Partial Least Squares Method and Its Application in Process Modeling. *Chin. J. Chem. Eng.* **2014**, *22*, 799–804. [CrossRef]
17. LV, F.; Liang, B.; Sun, W.J.; Wang, Y. Gas emission quantity prediction of working face based on principal component regression analysis method. *J. China Coal Soc.* **2012**, *37*, 113–116.
18. Ngo, S.H.; Kemény, S.; Deák, A. Performance of the ridge regression method as applied to complex linear and nonlinear models. *Chemom. Intell. Lab. Syst.* **2003**, *67*, 69–78. [CrossRef]

19. Wu, H.C.; Ai, C.H.; Yang, L.J.; Li, T. A Study of Revisit Intentions, Customer Satisfaction, Corporate Image, Emotions and Service Quality in the Hot Spring Industry. *J. China Tour. Res.* **2015**, *11*, 371–401. [CrossRef]
20. Shantong, L. The Prediction and Analysis of the Speed of Urbanization in China. Development Strategy and Regional Economic Research Department of Development Research Center of the State Council. Available online: http://www.drc.gov.cn/n/20170824/1-224-2894327.html (accessed on 1 December 2017).
21. Xianzheng, W. Wang Xianzheng reports on the development of coal industry in China University of Mining and Technology. *Coal Eng.* **2014**, *7*, 120.
22. Jingwen, L. Long-term Forecast of China's Economy in the 21st Century. *Metall. Econ. Manag.* **2000**, *3*, 4–7.

Article

Analysis of the Nexus of CO_2 Emissions, Economic Growth, Land under Cereal Crops and Agriculture Value-Added in Pakistan Using an ARDL Approach

Sajjad Ali [1], Liu Ying [1,2,*], Tariq Shah [3], Azam Tariq [4], Abbas Ali Chandio [5] and Ihsan Ali [6]

[1] College of Economics and Management, Huazhong Agricultural University, Wuhan 430070, China; sajjad@webmail.hzau.edu.cn
[2] Hubei Collaborative Innovation Center for Grain Industry, Yangtze University, Jingzhou 434025, China
[3] Department of Economics & Development Studies, University of Swat, Khyber Pakhtunkhwa 19130, Pakistan; shah6833@yahoo.com
[4] College of Humanities and Social Sciences, Huazhong Agricultural University, Wuhan 430070, China; azo.durrani@gmail.com
[5] College of Economics, Sichuan Agricultural University, Chengdu 625014, China; alichandio@sicau.edu.cn
[6] College of Plant Science and Technology, Huazhong Agricultural University, Wuhan 430070, China; immok9@gmail.com
*   Correspondence: liuyhzau@gmail.com; Tel.: +86-136-6720-1370 or +86-27-5982-4673

Received: 30 October 2019; Accepted: 27 November 2019; Published: 2 December 2019

Abstract: The present study attempts to explore the correlation between carbon dioxide emissions (CO_2 e), gross domestic product (GDP), land under cereal crops (LCC) and agriculture value-added (AVA) in Pakistan. The study exploits time-series data from 1961 to 2014 and further applies descriptive statistical analysis, unit root test, Johansen co-integration test, autoregressive distributed lag (ARDL) model and pairwise Granger causality test. The study employes augmented Dickey–Fuller (ADF) and Phillips–Perron (PP) tests to check the stationarity of the variables. The results of the analysis reveal that there is both short- and long-run association between agricultural production, economic growth and carbon dioxide emissions in the country. The long-run results estimate that there is a positive and insignificant association between carbon dioxide emissions, land under cereal crops, and agriculture value-added. The results of the short-run analysis point out that there is a negative and statistically insignificant association between carbon dioxide emissions and gross domestic product. It is very important for the Government of Pakistan's policymakers to build up agricultural policies, strategies and planning in order to reduce carbon dioxide emissions. Consequently, the country should promote environmentally friendly agricultural practices in order to strengthen its efforts to achieve sustainable agriculture.

Keywords: carbon dioxide emissions; cereal crops; gross domestic product; ARDL model; granger causality; Pakistan

---

1. Introduction

The changes in climate affect the productivity of the agriculture sector through a variation in global temperatures, the variability of precipitation and other related factors. It is estimated that about 15%–30% of the output of agriculture would be affected negatively by 2080–2100 [1]. A further decline in crop yield may occur in Africa, Latin America and Asia because adaptive measures are overlooked. The Fourth Assessment Report of the Intergovernmental Panel on Climate Change stated that it would cost about 5%–10% of GDP for Africa to take adaptation measures to combat climate change [2]. Moreover, they predicted that about a 50% drop in agricultural crops would be observed by 2020 and the crop revenue may further decrease even up to 90% by 2100. The variation in the pattern of rainfall has also affected more than one billion people in South Asia [3]. Researchers including [4–15] and many others have shown that climate change poses threats to agriculture, food and water supplies, especially in the developing economies. Most of the models indicate that climate variation would adversely affect the yield of wheat in South Asia. The Intergovernmental Panel on Climate Change (IPCC) 4th Assessment Report put forth that in South Asia the crop yield would reduce proportionately from 1820 m^3 to 1140 m^3 from 2001 to 2050.

The increasing population of Pakistan and non-assurance of food security for its society is a challenge, since the residents are expected to double by 2050 [16]. Climate change and adaptation strategies are increasingly becoming the main focus of scientific research these days, for instance, the effect on the production of crops such as wheat, rice and maize [17]. The vulnerability index of the fluctuation of climate in Pakistan is remarkably rising in comparison to numerous countries around the globe, due to variable climatic conditions. Of late, Pakistan has been confronted with a lot of climatic variations, for instance; a rise in temperature, changes in the pattern of precipitation, floods, earthquakes and weather shifts. The development of the agriculture sector in developing countries is hampered by increasing climatic risk and projected changes in climate over the 21st century [18]. Pakistan is affected the most by climate change owing to inadequate and substandard infrastructure and limited adaptive capacity [19]. It is projected that by 2050, there would be a 2%–3% rise in temperature causing a significant variation in the pattern of rainfall [20]. The country is ranked 8 among the most negatively affected countries by adverse weather conditions and climate change over the period 1995–2014 as reported by the Global Climate Risk Index (GCRI) [21]. The productivity of the main crops including wheat, rice, cotton and sugarcane and rural livelihoods has been affected significantly due to climate variability and extreme events over the last two decades [22]. The vulnerability of rural livelihood to climate change can be seen from the historic floods during 2010–2014 and severe droughts from 1999 to 2003 [22]. Greenhouse gas emissions may cause an unproductive effect on the environment up to a great extent and this issue becomes substantially critical for all countries in the world. Recently, many researchers have been paying attention to the carbon dioxide emissions as one of the essential causes of global warming [23–26]. There has been an unprecedented increase in population, agricultural production, energy demand and economic growth to achieve food security, and carbon dioxide emissions have also increased over the decades [27–30].

In this study, we conducted an in-depth investigation of the entire country (Pakistan) which explores the variety of responses of the carbon dioxide emissions (CO$_2$ e), gross domestic product (GDP), land under cereal crops (LCC) and agriculture value-added (AVA) based on historical data during 1961 to 2014. The autoregressive distributed lag (ARDL) model is employed simultaneously to observe the effect of the CO$_2$ e, GDP, LCC and AVA in order to identify a certain correlation between them. This enabled us to determine the long-run relationships among several variables [31]. Johansen and Juseliu's estimation to carefully investigate this subject in-depth. In addition, we also conducted generalized impulse response functions and variance decomposition methods to find out the effects of shocks on the adjustment path of the variables.

The rest of the study is structured as follows: the second section entails a brief part of the literature review. The third section is about the research methodology refers to the processing for the data collection. The fourth section is the results and discussion part and the final section is the conclusion and policy recommendations of the study in hand.

2. Literature Review

A wide range of literature is accessible on determining the factor of economic growth, agricultural production and the emissions of carbon dioxide. The long-run equilibrium relationship between carbon dioxide emissions, income growth, energy consumption and agriculture for Pakistan from 1971 to 2014 have been verified and tested. The results confirmed that there were bidirectional causalities between GDP, agriculture, energy use and CO$_2$ emissions. They also found that AVA had a positive inelastic effect on CO$_2$ emissions and that GDP had a positive elastic impact on CO$_2$ emissions [32].

The previous study investigated the impact of AVA and per capita renewable energy consumption on carbon dioxide emissions in Asian countries. They found that agricultural and renewable energy had negative impacts on CO_2 emissions [33]. Evidence from the study revealed long-run equilibrium association flowing from consumption of electricity industrialization, gross domestic product and carbon dioxide emissions [34]. The study employed the vector error correction model (VECM) and ordinary least squares (OLS) regression revealed the effect of population progression, energy intensity and GDP on carbon dioxide emissions in Ghana. The study provided evidence of the existence of long-run equilibrium association flowing from population growth, energy intensity and gross domestic product to carbon dioxide emissions. The study also revealed that there was a bi-directional causality among energy consumption and carbon dioxide emissions [35]. Another study in Ghana investigated the association between population growth, use of energy, gross domestic product and carbon dioxide emissions by using both ARDL regression analysis and VECM. The study found that there will be fluctuation in carbon dioxide emissions due to the use of energy in the future. Evidence from the study showed a unidirectional causality running from carbon dioxide emissions to the use of energy and population [29]. Another study in China employed the ARDL model, the Granger causality test based on VECM, and impulse response and variance decomposition to test the relationship between CO_2 emissions, energy consumption and economic growth in the agricultural sector. The estimated results illustrated that there is bidirectional causality between agricultural carbon emissions and agricultural economic growth in both the short run and long run and there exist unidirectional causality from agricultural energy consumption to agricultural carbon emissions and agricultural economic growth [36]. The empirical results derived from the study confirmed the validty of the environmental Kuznets curve (EKC) hypothesis for three countries namely France, Portugal and Spain during the period under the study in the long-run as well as in short-run with exception the case of Portugal [37].

It is evident that a rise in temperature can have a devastating effect on the productivity of the agriculture sector, food security and farmers' incomes. This phenomenon varies in tropical and temperate zones. In the middle- and the high-latitude zones, the output of crops is anticipated to increase and spread northwards and vice versa for several other countries in tropical regions [38]. It has been found that high latitudes can cause an expansion in the production by nearly 10% due to a 2 °C rise in temperature, whereas it reduced production just by the same percent in the low latitude. Considering the inevitable effect of contemporary technology, it is projected that an increase in temperature would increase the productivity of yield by 37% and 101% by 2050s for the Russian Federation [39].

As compared to other developing countries, the effects of escalating temperature on agriculture are harsher in Sub-Saharan Africa [40]. It has been observed that some important climatic conditions such as temperatures and rainfall had persisted at their pre-1960 status, then the gap of agricultural production between different developing countries and Sub- Saharan Africa at the end of the 20th century would have remained only 32% of the existing shortfall. A study for the period of 1980–2005 in Nigeria indicated that temperature exerts a negative effect while rainfall has a positive effect on agricultural production [41].

Another study developed a two-chain logarithmic mean divisia index (LMDI) decomposition method and derived the results that technology, distribution and population effects could not suppress China's agricultural carbon dioxide emissions simultaneously in most years [42]. Developed countries have the ability to maintain a minimum level of technology for the improvement of living standards and increasing agricultural productivity [43]. Generally, developed countries are capable of counterbalancing the negative consequences of climate change. Developed states usually have a low level of susceptibility but a high level of adaptive ability, which itself has a role of technological expertise, dissemination and supply of assets, and human social and political capital [44]. The developed world has very standard levels of water filtration and sanitation; on the other hand, developing countries have insecure and unreliable water supplies and often the sanitation system is non-satisfactory and below the margin. The concept of crop insurance is utterly missing in developing countries to protect their farmers from the negative consequences of climate change which may destroy their livelihoods.

Since the last decade, the country's (Pakistan) per capita GDP has observed a diverse or unlike trend and lack of equilibrium. During the period from 2005 to 2014 the per capita GDP increased from 974.5\$ to 1111.2\$ respectively. In 2011, the government gave great importance ton upgrading the country's economy and can be witnessed that per capita GDP has consistently increased during the period 2011 to 2014. During the period of 2011 to 2014, even though there were several types of socio-economic challenges such as energy crises, a war against terrorism and poverty, still there was a rise of 64.71\$ in per capita GDP (Pakistan Economic Survey 2017). In consequence, it is evident that the Government of Pakistan has taken actions to raise economic growth and enriched living conditions.

3. Methodology and Data Collection

3.1. Data Sources and Description

The fundamental purpose of the aforementioned study is to find out the relationship between CO_2 e, GDP, LCC and AVA in Pakistan. The study adopted the time series data spanning from 1961 to 2014 using the ARDL method to test the relationship between study variables. To fulfill the study objectives, the data sets of the selected variable in the study were procured from the Food and Agriculture Organization Corporate Statistical Database FAOSTATS (www.fao.org) and World Development Indicators (http://data.worldbank.org).

Four variables were considered throughout the analysis where carbon dioxide emissions CO_2 e (kt) was taken as a dependent variable and explanatory variables include GDP (current US\$), LCC (hectares) and AVA (percentage of GDP). This study employed the actual CO_2 emissions instead of potential CO_2 (i.e., CO_2 eq.). Previous studies [29,45,46] put into practice the actual CO_2 emissions which show that the use of actual CO_2 emissions improves the efficiency of the model. Table 1 shows the source of data and variable description. The trend analysis of the study variables are given Figure 1.

Table 1. Detail of variables.

Variable Name	Abbreviation	Unit of Measurement	Source
Carbon dioxide emission	CO_2 e	Kilotons (kt)	FAOSTAT (2018)
Gross domestic product	GDP	Current US \$	WDI (2018)
Land under cereal crop	LCC	Hectares '	WDI (2018)
Agriculture value added	AVA	Percentage of GDP	WDI (2018)

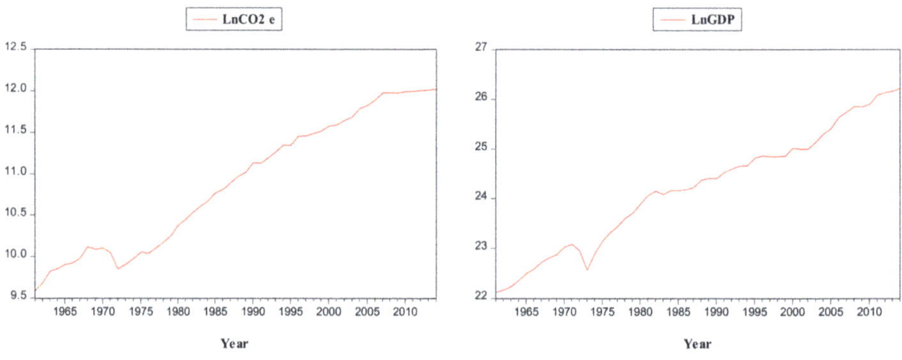

Figure 1. *Cont.*

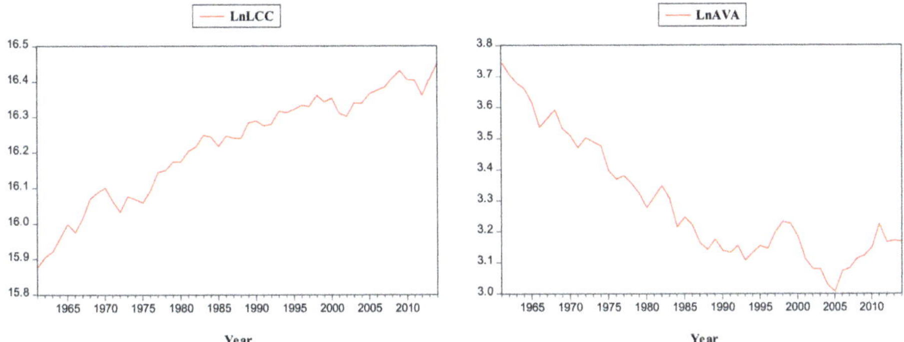

Figure 1. Trend of the study variables.

3.2. Econometric Model

The current study entails the co-integration and autoregressive distributed lag model to find out the association between carbon dioxide emissions, gross domestic product, land under cereal crop and agriculture value-added in Pakistan. The following steps show our study analysis. In the first step, we have to find out the stationarity in the time series data. For this objective, we conducted the augmented Dickey-Fuller (ADF) test [47] and Phillips and Perron (PP) unit root tests [48]. The step second was to find out the optimal lag length of the study variables. To determine the lag lengths we used the Akaike information criterion (AIC) [49] or Schwarz information criterion (SIC) [50]. In the third step, we estimated the Johansen co-integration test to seek the long-run relationship between the study variables. Were there a co-integration, then we moved to the next step. In the last step, were a co-integration to exist then we estimated an ARDL model. Furthermore, we also estimated the pairwise Granger causality test to establish causal links between variables. The econometric model used in this study is given as:

$$CO_{2\,et} = f(GDP_t, LCC_t, AVA_t) \tag{1}$$

where in the above equation (1), $CO_2\,e$ is the carbon dioxide emissions, GDP is the gross domestic product, LCC is the land under cereal crop, AVA is the agriculture value-added and t is the time period. We then applied the Cobb Douglas production function in its stochastic form as:

$$CO_{2\,et} = \alpha_0\alpha_1 GDP_t, \alpha_2 LCC_t, \alpha_3 AVA_t \tag{2}$$

Then we employed the log-linear model, for this purpose, we log-transform the above model to get the linear regression model which is given as:

$$log_e(CO_{2\,et}) = \alpha_0 + \sum log_e(\alpha_1 GDP_t, \alpha_2 LCC_t, \alpha_3 AVA_t) \tag{3}$$

Then we transformed the variable's value into their natural logarithm form to find out the long-run association between the study variables. This transformation of the data into their natural logarithm is to ensure the results were efficient, reliable and consistent. Equation (4) shows the logarithm form for the study variables.

$$lnCO_{2\,et} = \alpha_0 + \alpha_1 lnGDP_t + \alpha_2 lnLCC_t + \alpha_3 lnAVA_t + \varepsilon_t \tag{4}$$

where $lnCO_{2\,et}$ $lnGDP_t$, $lnLCC_t$ and $lnAVA_t$ expressed the natural logarithm of carbon dioxide emissions, gross domestic product, crop production index, land under cereal crop and agriculture value-added, respectively. In the above equation (4), $t =1, \ldots \ldots .N$ represents the time period and ε_t is the error term. The parameters $\alpha_0, \alpha_1,$ α_2, and α_3 measure the long-run elasticity of carbon dioxide emissions with respect to the real GDP, land under cereal crop and agriculture value-added respectively.

4. Results and Discussions

4.1. Descriptive Analysis and Correlation Matrix

The descriptive analysis shows mean, coefficient of variation, skewness, kurtosis and normality of distribution over the study variables. Table 2 provides the descriptive analysis and the kurtosis results display that all the variables exhibit platykurtic distribution. The results of the skewness indicate that both carbon dioxide emissions and agriculture value-added have long right-tail distribution while the remaining variables indicate long left-tail distribution. The outcome from the Jarque–Bera test shows that we accept the null hypothesis of normal distribution at the 5% level of significance for all variables except agriculture value-added. The mean results show that the gross domestic product generate a high value of 24.21. The standard deviation analysis show that the gross domestic product is also the most explosive variable with the highest deviation of 1.19 followed by carbon dioxide emissions.

Table 2. Descriptive statistics and correlation matrix of all the variables.

Variables	$LnCO_2$ e	LnGDP	LnLCC	LnAVA
Mean	10.88533	24.21358	16.22058	3.290421
Median	10.92998	24.30179	16.24839	3.222102
Maximum	12.02154	26.22191	16.45170	3.746831
EMinimum	9.592673	22.12312	15.87711	3.006656
Std. Dev.	0.801811	1.193248	0.152553	0.196416
Skewness	0.002686	−0.087221	−0.548516	0.734181
Kurtosis	1.510778	1.953711	2.226820	2.385217
Jarque-Bera	4.990073	2.531587	4.052896	5.701605
Probability	0.082493	0.282015	0.131803	0.057798
Correlation				
$LnCO_2$ e	1.000000			
LnGDP	0.978197	1.000000		
LnLCC	0.956098	0.973567	1.000000	
LnAVA	−0.884455	−0.897413	−0.930411	1.000000

4.2. Lag Selection for Vector Error Correction Model

After the unit root test, in the next step we need to find out the optimum lag length for co-integration analysis by using the AIC criteria [49] or SIC [50] criteria. The AIC results in Table 3 indicate that the most suitable lag value is lag 2 for the model.

Table 3. Selection of lag length.

Lag	LogL	LR	FPE	AIC	SC	HQ
0	112.9309	NA	1.51e−07	−4.357238	−4.204276	−4.298989
1	346.9699	421.2700	2.46e−11	−13.07879	−12.31398*	−12.78755*
2	366.4259	31.90793*	2.17e−11*	−13.21704*	−11.84038	−12.69280
3	380.8168	21.29858	2.39e−11	−13.15267	−11.16417	−12.39544
4	395.6585	19.59104	2.67e−11	−13.10634	−10.50599	−12.11611

* indicates lag order selected by the criterion; e: stands for exponential constant; LR: sequential modified LR test statistic (each test at 5% level); FPE: Final prediction error; AIC: Akaike information criterion; SC: Schwarz information criterion; HQ: Hannan-Quinn information criterion.

It is important to find out how many lags to be used in ARDL model. Therefore, to figure out the optimal number of lags for the model, the unrestricted vector autoregression (VAR) lag selection criteria is tested. Table 3 formulates the lag selection criteria for the model but the most commonly employed criteria are AIC and SIC. The previous study used AIC for a small sample size [51].

4.3. Unit Root Test

Before estimating the co-integration analysis, it is important to determine where the study variables are stationary at first difference i.e., $I(1)$. The stationarity of the variables is tested using the ADF test [47] and PP test [48] in order to have a robust result and avoid spurious regression results. Table 4 shows the unit root test results. Our findings in Table 4 indicate that all the study variables are non-stationary at a level. However, the variables became stationary at their first difference and rejected the null hypothesis that unit root exists at first difference. The results show that all the study variables are stationary at first difference which means that variables are integrated at $I(1)$. Since the variables entailed in the study are $I(1)$, so this indicates the spurious regression problem occurs here. Hence it is important to find out the co-integration test among the time series variables.

Table 4. Unit root test (Augmented Dickey–Fuller).

Variables	Akaike Info Criterion				Philips–Perron			
	LEVEL		1ST DIFFERENCE		LEVEL		1ST DIFFERENCE	
	Intercept	Trend and Intercept	Intercept	Trend and Intercept	Intercept	Trend and Intercept	Intercept	Trend and Intercept
$LnCO_2$ e	−0.63771	−2.107644	−5.915923	−2.908476	−0.809440	−1.554595	−5.928838	−5.897317
	0.8528	0.5292	0.0000	0.1689	0.8082	0.7974	0.0000	0.0001
LnGDP	−0.512237	−3.102790	−6.128411	−6.074545	−0.501008	−2.682416	−6.117041	−6.043380
	0.8803	0.1165	0.0000	0.0000	0.8825	0.2478	0.0000	0.0000
LnLCC	−1.845078	−3.097552	−7.310103	−5.882637	−2.177064	−3.058810	−7.399540	−7.703130
	0.3552	0.1175	0.0000	0.0001	0.2168	0.1268	0.0000	0.0000
LnAVA	−2.617304	−1.487037	−6.708506	−4.529780	−2.720270	−1.506937	−6.708506	−7.242419
	0.0959	0.8218	0.0000	0.0039	0.0773	0.8148	0.0000	0.0000
Conclusion	Non-stationary		Stationary		Non-stationary		Stationary	

4.4. Johansen Co-Integration Test

A summary of the Johansen co-integration [52] test is presented in Table 5. The purpose of the Johansen co-integration test is to find out the long-run relationship between the study variables in the model. Maximum eigenvalue and trace statistic tests [53] were conducted to determine the co-integration among the study variables. The results of the maximum eigenvalue and trace statistic showed 4 co-integrating equations at the 5 percent level. Here, the results of co-integration would determine whether we have to apply a VAR model or VECM model.

Table 5. Results of Johansen co-integration test.

Hypothesized No. of CE(s)	Rank Test (Trace)				Rank Test (Maximum Eigenvalue)			
	Eigenvalue	Trace Statistic	0.05 Critical Value	Prob.	Max-Eigen Statistic	0.05 Critical Value	Prob.	
None	0.423351	52.89693	47.85613	0.0156	28.07661	27.58434	0.0432	
At most 1	0.300607	24.82032	29.79707	0.1679	18.23469	21.13162	0.1213	
At most 2	0.121100	6.585634	15.49471	0.6263	6.583293	14.26460	0.5395	
At most 3	0.0000459	0.002341	3.841466	0.9593	0.002341	3.841466	0.9593	

4.5. Autoregressive Distributed Lag (ARDL) Bound Testing of Co-Integration

The current study uses an ARDL bound testing approach suggested by [54] to find out both short-run and long-run association of the CO_2 e, GDP, LCC and AVA. The ARDL bound testing method is appropriate for those models in which there is a mixture of $I(0)$ and $I(1)$ variables. Another characteristic of this model is that it is appropriate for a small sample size as our sample size is only 52 [54].

After the estimation of unit root testing which shows that all variables are integrated at $I(1)$, now we carried out the ARDL method of co-integration (bounds testing) to estimate the relationship between the selected variables in this study. The results of the ARDL bound testing are reported in Table 6. The results indicate that the f-statistic value (5.805114) is greater than the 10% and 5% upper critical values of $I(0)$ bound. The results of the bounds testing validate significant long-run relationships among variables and showing the rejection of null hypothesis of no co-integration association among $LnCO_2$ e, LnGDP, LnLCC and LnAVA.

Furthermore, the study estimates the AIC to prefer the optimal model by employing long-run and short-run association among variables. Employing the Akaike information criterion shows the top 20 possible ARDL models in Figure 2. Based on the model specification in equation (4), the short-run and long-run equilibrium relation $LnCO_2$ e, LnGDP, LnLCC and LnAVA is estimated using the ARDL regression analysis shown in equation (5) where

$$\alpha_0 = 19.2356, \ \alpha_1 = 0.3246, \ \alpha_2 = -0.2867 \text{ and } \alpha_3 = -3.3902. \tag{5}$$

Table 6. ARDL bound testing.

Test Statistic	Value	k
F-statistic	5.805114	3
	Critical Value Bounds	
Significance	$I(0)$ Bound	$I(1)$ Bound
10%	2.37	3.2
5%	2.79	3.67
2.5%	3.15	4,08
1%	3.65	4.66

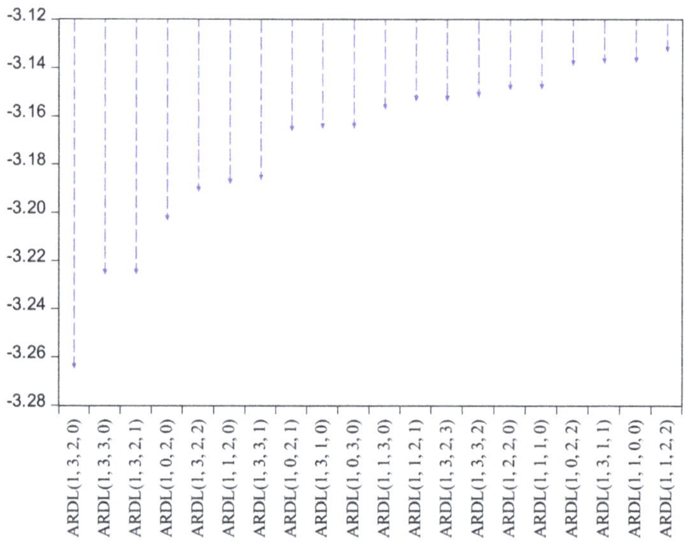

Figure 2. ARDL model selection criterion. **Source.** Authors' calculation.

4.6. Short-Run and Long-Run Equation Models

Table 7 summarizes the results of short-run equation of the ARDL model. The results show that the speed of adjustment (error correction term ECT(−1)) value is –0.077780 which shows that there are a long run and short-run equilibrium relationships running from LnGDP, LnLCC and LnAVA to $LnCO_2$ e. The speed of adjustment is approximately 7.7 % in one period of long-run equilibrium.

Table 7. Short-run and long-run relationship estimates selected model for autoregressive distributed lag (ARDL) (1,3,2,0).

Short Run Coefficients				
Variable	**Coefficient**	**Std. Error**	**t-Statistic**	**Prob.**
D(LnGDP)	0.033643	0.056658	0.593784	0.5559
D(LnGDP(−1))	0.014904	0.055969	0.266287	0.7914
D(LnGDP(−2))	−0.173688	0.058569	−2.965542	0.0050
D(LnLCC)	0.863260	0.241211	3.578864	0.0009
D(LnLCC(−1))	0.716364	0.243073	2.947112	0.0053
ECT(−1)	−0.077780	0.013781	−5.644230	0.0000
Long Run Coefficients				
Variable	**Coefficient**	**Std. Error**	**t-Statistic**	**Prob.**
C	1.496155	4.774608	0.313357	0.7556
$Ln\,CO_2$ e(−1)	−0.077780	0.038989	−1.994952	0.0527
LnGDP(−1)	0.025249	0.042403	0.595470	0.5548
LnLCC(−1)	−0.022297	0.326479	−0.068295	0.9459
LnAVA	−0.263688	0.097057	−2.716831	0.0096
D(LnGDP)	0.033643	0.063982	0.525816	0.6018
D(LnGDP(−1))	0.014904	0.065642	0.227049	0.8215
D(LnGDP(−2))	−0.173688	0.066271	−2.620874	0.0122
D(LnLCC)	0.863260	0.298943	2.887709	0.0062
D(LnLCC(−1))	0.716364	0.289813	2.471814	0.0177

EC = $LnCO_2$ e − (0.3246(LnGDP) − 0.2867(LnLCC) − 3.3902(LnAVA) + 19.2356))

Table 5 also shows the results of long-run equation results of the ARDL approach. The results of long-run equilibrium relationship show that a 1% increase in LnGDP will increase $LnCO_2$ e by 2%, a 1% increase in LnLCC will decrease $LnCO_2$ e by 0.02% and a 1% increase in LnAVA will decrease $LnCO_2$ e by 26% in long-run estimates.

The evidence of the following studies reveals that carbon dioxide emissions increase in the early phases of economic growth and then decline after a threshold point. The findings of these studies such as [10,55–62] examined the relationship between carbon dioxide emissions and GDP growth.

The findings of previous studies such as [63] for China, [59] for Tunisia, [64] for Iran, [65] for Pakistan, [66] for Malaysia, [57] for Turkey and [55] for India examined a unidirectional causality running from GDP income to carbon dioxide emissions without response which suggests that emission reduction plans will not restrain trade and industry growth and which seems to be a feasible policy instrument in the aforementioned studied countries to accomplish its long-run sustainable growth.

Furthermore, we applied generalized impulse response functions for the verification of the results. The generalized impulse response results show an in-depth understanding of shocks to gross domestic product, land under cereal crop, agriculture value-added affected carbon dioxide emissions. The results of generalized impulse responses for carbon dioxide emissions, gross domestic product, land under cereal crop and agriculture value-added are provided in Figure 3.

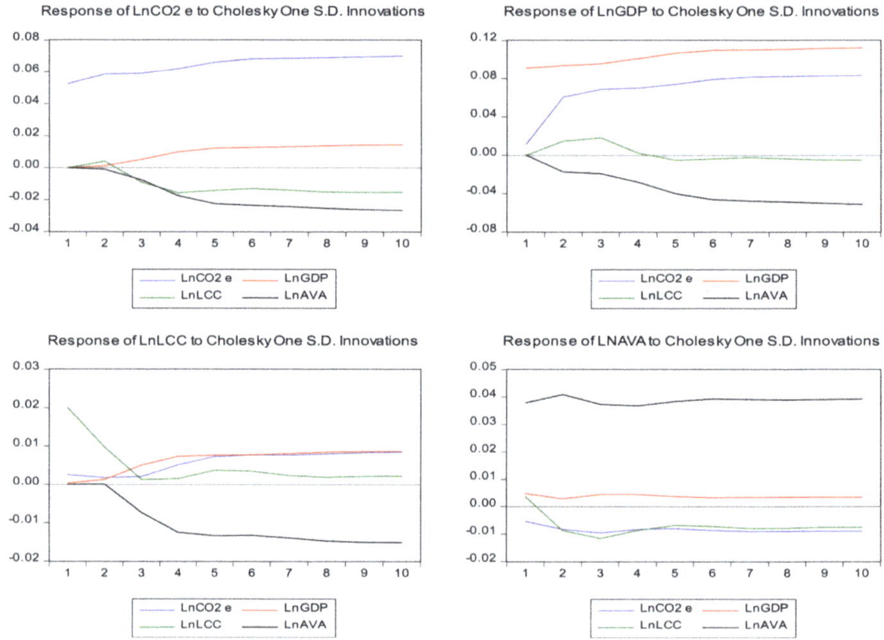

Figure 3. Results of generalized impulse response functions.

4.7. Diagnostic Test

As suggested by [67], both the cumulative sum of the recursive residuals (CUSUM) and the cumulative sum of the square of the recursive residuals (CUSUMsq) tests were implemented to run the ARDL model in a befitting manner. Figure 4 reveals that both the graphs of CUSUM and CUSUMsq tests lie between the critical bounds indicated with red colored lines at a 5% confidence interval. The blue color lines in the middle represent the measurements for the cumulative sum of the recursive residuals and the cumulative sum of the squares of the recursive residuals. Both CUSUM and CUSUMsq graphs show that the model of our study is well stable.

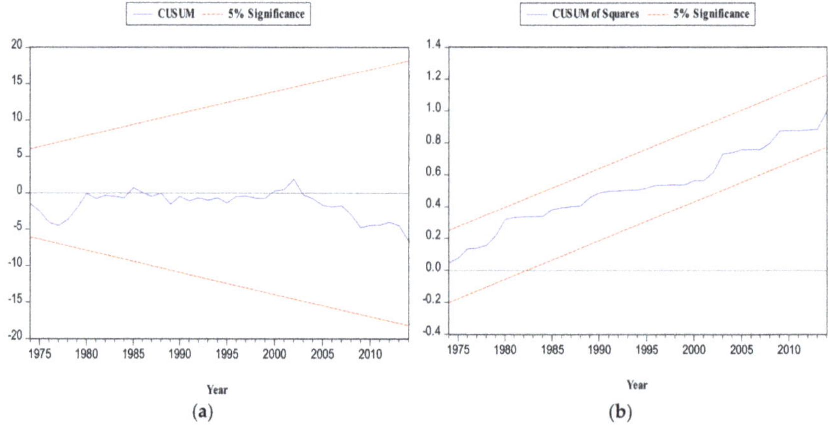

Figure 4. Stability test based on (**a**) cumulative sum of the recursive residuals (CUSUM) and (**b**) CUSUM of squares. **Source.** Authors' calculation.

Several diagnostic tests were operated to check the good fit of the ARDL model. Table 8 shows that estimation is fine regarding the serial correlation Lagrange Multiplier (LM) test, where the F-statistics (0.237056) have insignificant probability. The heteroskedasticity test under Breusch–Pagan–Godfrey also signifies that there is no sign of serial correlation. The value of F-statistics (1.190498) shows an insignificant probability, which means there is no heteroskedasticity issue in the model estimation.

Table 8. Diagnostic tests results.

Breusch-Godfrey Serial Correlation LM Test:	
F-statistic	0.237056
Obs R-squared	0.936929
Prob. F(3,38)	0.8700
Prob. Chi-Square(3)	0.8165
Heteroskedasticity Test: Breusch–Pagan–Godfrey	
F-statistic	1.190498
Obs R-squared	10.56646
Scaled explained SS	10.87174
Prob. F(9,41)	0.3268

Furthermore, the inverse root of AR polynomial graph displaying the stability of the model where are blue dots is within the circle. Figure 5 shows the inverse root of AR polynomial estimation.

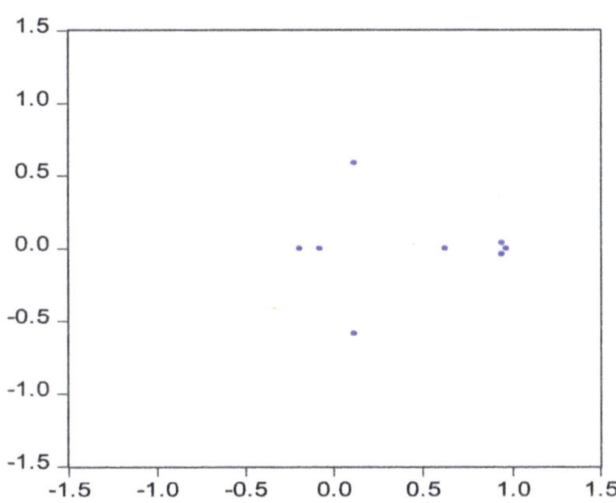

Figure 5. Checking the stability of vector autoregression (VAR).

4.8. Pairwise Granger Causality Tests

The pairwise Granger causality test is estimated to find out the robustness of the model, which elaborates the directional linkages between the two variables at a time. The results of the pairwise Granger causality is exhibited in Table 9. The estimations of the pairwise Granger causality shows unidirectional causality between LnGDP to $LnCO_2$ e, LnLCC to LnGDP and LnAVA to LnLCC.

Table 9. Pairwise Granger causality test.

Null Hypothesis:	Obs	F-Statistic	Prob.
LnGDP does not Granger Cause $LnCO_2$ e	52	0.34510	0.7099
$LnCO_2$ e does not Granger Cause LnGDP		8.51829	0.0007
LnLCC does not Granger Cause $LnCO_2$ e	52	1.91090	0.1593
$LnCO_2$ e does not Granger Cause LnLCC		1.81672	0.1738
LnAVA does not Granger Cause $LnCO_2$ e	52	2.63228	0.0825
$LnCO_2$ e does not Granger Cause LnAVA		0.42783	0.6544
LnLCC does not Granger Cause LnGDP	52	1.78823	0.1784
LnGDP does not Granger Cause LnLCC		6.22181	0.0040
LnAVA does not Granger Cause LnGDP	52	1.03562	0.3630
LnGDP does not Granger Cause LnAVA		0.13864	0.8709
LnAVA does not Granger Cause LnLCC	52	3.95660	0.0258
LnLCC does not Granger Cause LnAVA		1.43426	0.2485

4.9. Impulse Response and Variance Decomposition Analysis

Finally, the study employed impulse response analysis between $LnCO_2$ e, LnGDP, LnLCC and LnAVA to describe random innovations among them. As the pairwise Granger causality test does not indicate any random response, so in this case, we have to run the impulse response analysis. Figure 6 displays that the response of carbon dioxide emissions to a gross domestic product, land under cereal crops and agriculture value-added are insignificant within 10-period horizons. On the other hand, the initial response of carbon dioxide emissions to land under cereal crop is significant in the beginning. A one standard deviation shock to land under cereal crop first increases carbon dioxide emissions to 1-period horizon and then starts decreasing to the 10-periods horizon. Figure 7 illustrates the response of gross domestic product, land under cereal crop and agricultural value-added to carbon dioxide emissions.

Table 10 estimates Cholesky's method of variance decomposition to random innovation affecting the variables in the VAR [68]. The results indicate that almost 4.3% of the future fluctuations in $LnCO_2$ e is due to shocks in the LnGDP, 0.27% of future fluctuations in the $LnCO_2$ e is due to shocks in LnLCC and 0.27% of future fluctuations in the $LnCO_2$ e is due to shocks in LnAVA, respectively. Evidence from the table shows that almost 25% of future fluctuations in LnGDP is due to shocks in $LnCO_2$ e, 10% of future fluctuations in LnGDP is due to shocks in LnAVA and 2.9% of future fluctuations in LnGDP is due to shocks in LnLCC. Moreover, evidence from the results shows that almost 37% of future fluctuations in LnLCC is due to shocks in $LnCO_2$ e, 24% of future fluctuations in LnLCC is due to shocks in LnAVA and 10% of future fluctuations in LnLCC is due to shocks in LnGDP. Finally, the evidence from Table 9 shows that almost 6.2% of the future fluctuations in LnAVA is due to shocks in LnLCC, 1.9% of future fluctuations in LnAVA is due to shocks in LnGDP and 0.4% of future fluctuations in LnAVA is due to shocks in $LnCO_2$ e, respectively.

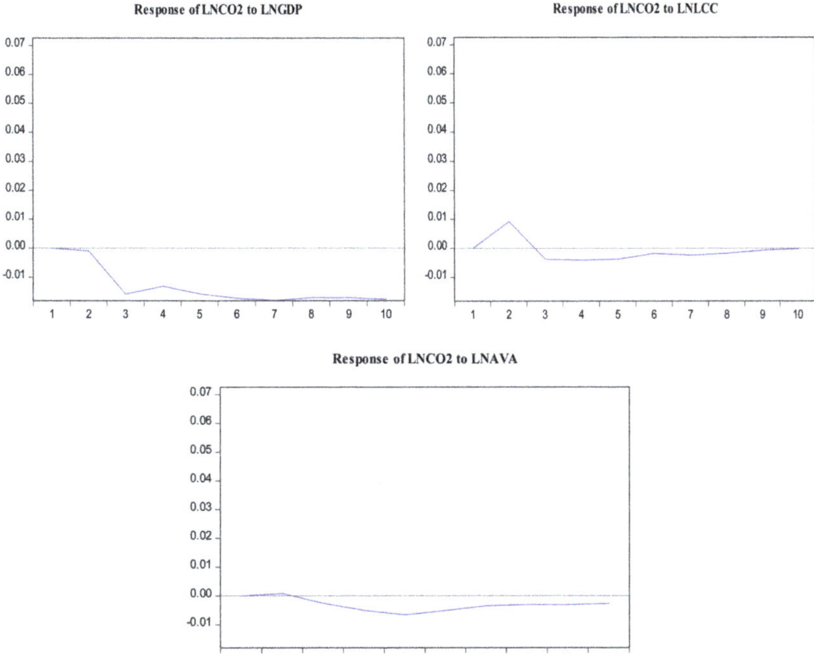

Figure 6. Impulse response of $LnCO_2$ e to Cholesky one standard deviation (S.D.).

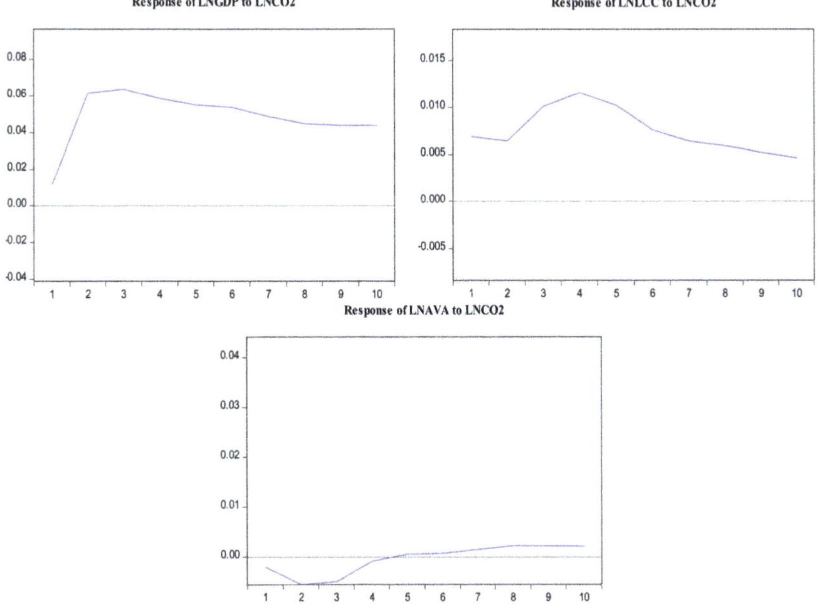

Figure 7. Impulse-response of other variables to Cholesky one S.D. Innovations in $LnCO_2$ e.

Table 10. Variance decomposition Cholesky ordering; $LnCO_2$ e_EMISSIONS LnGDP LnLCC LnAVA.

		Variance Decomposition of $LnCO_2$ e:			
Period	S.E.	$LNCO_2$ e	LnGDP	LnLCC	LnAVA
1	0.052387	100.0000	0.000000	0.000000	0.000000
2	0.082091	98.71976	0.013097	1.256378	0.010761
3	0.110017	97.05350	2.074796	0.812947	0.058761
4	0.131880	96.72981	2.427748	0.658399	0.184048
5	0.151464	96.20375	2.913165	0.557815	0.325269
6	0.168350	95.78747	3.397230	0.461823	0.353472
7	0.183532	95.44742	3.816423	0.404535	0.331620
8	0.197549	95.30147	4.032265	0.356521	0.309743
9	0.210862	95.20442	4.188896	0.313796	0.292891
10	0.223536	95.10645	4.340165	0.279285	0.274097

		Variance Decomposition of LnGDP:			
Period	S.E.	$LnCO_2$ e	LnGDP	LnLCC	LnAVA
1	0.090540	1.714846	98.28515	0.000000	0.000000
2	0.147440	18.08957	79.94849	0.319798	1.642142
3	0.180273	24.58352	71.58001	0.872488	2.963985
4	0.202225	27.94786	67.03118	1.197606	3.823346
5	0.227024	28.07898	64.92696	1.894400	5.099661
6	0.250104	27.76311	63.32763	2.031381	6.877877
7	0.268642	27.36151	62.39127	2.146152	8.101068
8	0.285511	26.69581	62.00309	2.495791	8.805307
9	0.302657	25.87071	61.89515	2.792302	9.441847
10	0.319132	25.16367	61.79875	2.942672	10.09490

		Variance Decomposition of LnLCC:			
Period	S.E.	$LnCO_2$ e	LnGDP	LnLCC	LnAVA
1	0.019616	12.45762	0.679255	86.86313	0.000000
2	0.022818	17.21273	0.890578	78.32556	3.571125
3	0.025397	29.76056	3.976938	63.25735	3.005150
4	0.029421	37.65852	8.366014	47.75381	6.221656
5	0.032195	41.56618	7.848681	40.22046	10.36468
6	0.033756	42.90460	7.661578	36.78005	12.65377
7	0.035466	42.15718	8.493156	34.06056	15.28910
8	0.037344	40.54667	9.617245	31.05127	18.78481
9	0.038945	39.10538	10.27839	28.82381	21.79242
10	0.040393	37.68193	10.95538	27.30356	24.05913

		Variance Decomposition of LnAVA:			
Period	S.E.	$LnCO_2$ e	LnGDP	LnLCC	LnAVA
1	0.040325	0.255478	3.255885	4.319678	92.16896
2	0.060142	0.952231	1.551848	2.327764	95.16816
3	0.071323	1.153774	2.165245	2.590248	94.09073
4	0.081077	0.902385	2.715711	4.100540	92.28136
5	0.091163	0.718266	2.498629	5.149917	91.63319
6	0.100451	0.597671	2.267443	5.312943	91.82194
7	0.108566	0.532349	2.227840	5.480631	91.75918
8	0.116045	0.503706	2.175351	5.836543	91.48440
9	0.123284	0.479624	2.061993	6.111924	91.34646
10	0.130224	0.457998	1.967759	6.249748	91.32449

5. Conclusions and Policy Recommendations

The purpose of the study was to determine the relationships between CO_2 e as a dependent variable and GDP, LCC and AVA as independent variables in Pakistan. These independent variables have been tested to determine their effect on Pakistan's carbon dioxide emissions. Therefore, an empirical study was necessary to notify the policymakers and place Pakistan properly in efforts directed to mitigate the consequences of global warming. The study uses time-series data from 1961 to 2014. In the study, we run a descriptive analysis, Johansen co-integration test, pairwise Granger causality test and autoregressive distributed lag model.

The ARDL bounds test co-integration analysis displayed evidence of both short-run and long-run equilibrium relationship between the study variables. The speed of adjustment (ECT) is approximately 7.7 percent in one period of long-run equilibrium. Furthermore, the outcome of CUSUM and CUSUMsq showed that the model used in the study is stable. The pairwise Granger causality test was applied to find out the robustness of the model.

Our study findings have few policy implications for promoting agricultural development. To maintain economic growth and to reduce carbon dioxide emissions, it is very important to adjust and optimize the industrial structure. Pakistan's industrial sectors are generating heavy and high emissions of carbon dioxide. Therefore, the policymakers need to promote zero and light emissions industries for the development of the country. The results of the study show unidirectional Granger causality from gross domestic product to carbon dioxide emissions which indicates that ensuring a continuous increase in economic growth is a necessary condition for achieving high carbon dioxide emissions. Therefore, the government of Pakistan should take necessary actions to achieve high economic growth with less carbon dioxide emissions. As Pakistan predominantly is an agricultural country, thus, it is summarized that variations in climate change might have negative consequences for agricultural production and industrial growth, poverty reduction and job creation. As a South Asian country, Pakistan is not an exception, and the vulnerability index of climate change in the country is quite high. The country is listed among the countries severely affected by climate change [69] despite being a low producer of CO_2 gases [70] because of its increasing dependence on agriculture for food and fiber needs [71]. In addition, the agriculture sector of Pakistan consists of a majority of small resource, poor farmers with less adaption capacity. For the major crop production of mainly cereals, fruits and vegetables in Pakistan, the policymakers or government need to develop new crop farming methods, introducing new crop varieties, and an extension services role is also very important for spreading the updated science-based information.

Author Contributions: Conceptualization, S.A.; Data curation, T.S. and Abbas A.A.C.; Formal analysis, S.A. and I.A.; Methodology, S.A., L.Y. and A.T.; Supervision, L.Y.; Writing-original draft, S.A.; Writing-review and editing, L.Y.

Funding: This research received no external funding.

Acknowledgments: The authors are thankful to the Chinese Scholarship Council (CSC) for providing financial assistance to carry out this research as part of his PhD studies in China. In addition, the authors would also like to extend gratitude to anonymous reviewers for providing helpful suggestions on an earlier draft of this paper.

Conflicts of Interest: The authors declare no conflict of interest in this manuscript.

References

1. FAO. *The State of Food Insecurity in the World the Multiple Dimensions of Food Security*; Food and Agriculture Organization of the United Nations: Rome, Italy, 2013.
2. Boko, M.; Niang, I.; Nyong, A.; Vogel, A.; Githeko, A.; Medany, M.; Osman-Elasha, B.; Tabo, R.; Yanda Pius, Z. *Africa, Climate Change 2007: Impacts, Adaptation and Vulnerability: Contribution of Working Group II to the Fourth Assessment Report of the Intergovernmental Panel on Climate Change*; Cambridge University Press: Cambridge, UK, 2007; pp. 433–467.
3. Turner, A.G.; Annamalai, H. Climate change and the South Asian summer monsoon. *Nat. Clim. Chang.* **2012**, *2*, 587–595. [CrossRef]
4. Deressa, T.T.; Hassan, R.M.; Ringler, C. Perception of and adaptation to climate change by farmers in the Nile basin of Ethiopia. *J. Agric. Sci.* **2011**, *149*, 23–31. [CrossRef]
5. Fatuase, A.; Ajibefun, I. Perception and Adaptation to Climate Change among Farmers in Selected Communities of Ekiti State, Nigeria. *Gaziosmanpasa Univ. Ziraat Fak. Derg.* **2014**, *31*, 101–114.

6. Gömann, H. How Much did Extreme Weather Events Impact Wheat Yields in Germany?—A Regionally Differentiated Analysis on the Farm Level. *Procedia Environ. Sci.* **2015**, *29*, 119–120. [CrossRef]

7. Kurukulasuriya, P.; Mendelsohn, R.; Hassan, R.; Benhin, J.; Deressa, T.; Diop, M.; Eid, H.M.; Fosu, K.Y.; Gbetibouo, G.; Jain, S.; et al. Will African agriculture survive climate change? *World Bank Econ. Rev.* **2006**, *20*, 367–388. [CrossRef]

8. Özdoğan, M. Modeling the impacts of climate change on wheat yields in Northwestern Turkey. *Agric. Ecosyst. Environ.* **2011**, *141*, 1–12. [CrossRef]

9. Potgieter, A.; Meinke, H.; Doherty, A.; Sadras, V.O.; Hammer, G.; Crimp, S.; Rodriguez, D. Spatial impact of projected changes in rainfall and temperature on wheat yields in Australia. *Clim. Chang.* **2013**, *117*, 163–179. [CrossRef]

10. Song, T.; Zheng, T.; Tong, L. An empirical test of the environmental Kuznets curve in China: A panel cointegration approach. *China Econ. Rev.* **2008**, *19*, 381–392. [CrossRef]

11. Tao, F.; Yokozawa, M.; Xu, Y.; Hayashi, Y.; Zhang, Z. Climate changes and trends in phenology and yields of field crops in China, 1981–2000. *Agric. For. Meteorol.* **2006**, *138*, 82–92. [CrossRef]

12. Tao, F.; Zhang, S.; Zhang, Z. Spatiotemporal changes of wheat phenology in China under the effects of temperature, day length and cultivar thermal characteristics. *Eur. J. Agron.* **2012**, *43*, 201–212. [CrossRef]

13. Tao, F.; Zhang, Z.; Xiao, D.P.; Zhang, S.; Rötter, R.P.; Shi, W.J.; Liu, Y.J.; Wang, M.; Liu, F.S.; Zhang, H. Responses of wheat growth and yield to climate change in different climate zones of China, 1981–2009. *Agric. For. Meteorol.* **2014**, *189–190*, 91–104. [CrossRef]

14. Webb, L.B.; Whetton, P.H.; Barlow, E.W.R. Climate change and winegrape quality in Australia. *Clim. Res.* **2008**, *36*, 99–111. [CrossRef]

15. You, L.; Rosegrant, M.W.; Wood, S.; Sun, D. Impact of growing season temperature on wheat productivity in China. *Agric. For. Meteorol.* **2009**, *149*, 1009–1014. [CrossRef]

16. Ahmad, M.; Farooq, U. The state of food security in Pakistan: Future challenges and coping strategies. *Pak. Dev. Rev.* **2010**, *49*, 903–923. [CrossRef]

17. Kang, Y.; Khan, S.; Ma, X. Climate change impacts on crop yield, crop water productivity and food security—A review. *Prog. Nat. Sci.* **2009**, *19*, 1665–1674. [CrossRef]

18. IPCC. *Climate Change 2014: Synthesis Report. Contribution of Working Groups I, II and III to the Fifth Assessment Report of the Intergovernmental Panel on Climate Change*; IPCC: Geneva, Switzerland, 2014.

19. Stocker, T.F.; Qin, D.; Plattner, G.K.; Tignor, M.M.B.; Allen, S.K.; Boschung, J.; Nauels, A.; Xia, Y.; Bex, V.; Midgley, P.M. *IPCC, 2013: Climate Change 2013: The Physical Science Basis. Contribution of Working Group I to the Fifth Assessment Report of the Intergovernmental Panel on Climate Change*; IPCC: Geneva, Switzerland, 2013; p. 1535.

20. Gorst, A.; Groom, B.; Dehlavi, A. Crop Productivity and Adaptation to Climate Change in Pakistan Centre for Climate Change. *Environ. Dev. Econom.* **2018**, *23*, 1–23. [CrossRef]

21. Sönke, K.; Eckstein, D.; Dorsch, L.; Fischer, L. *Global Climate Risk Index 2016: Who Suffers most from Extreme Weather Events? Weather-Related Loss Events in 2014 and 1995 to 2014*; Report from Germanwatch; Germanwatch: Bonn, Germany, 2015; pp. 1–28. Available online: https://germanwatch.org/sites/germanwatch.org/files/publication/16411.pdf (accessed on 30 October 2019).

22. Abid, M.; Scheffran, J.; Schneider, U.A.; Ashfaq, M. Farmers' perceptions of and adaptation strategies to climate change and their determinants: The case of Punjab province, Pakistan. *Earth Syst. Dyn.* **2015**, *6*, 225–243. [CrossRef]

23. Tunç, G.I.; Türüt-Aşık, S.; Akbostancı, E. A decomposition analysis of CO_2 emissions from energy use: Turkish case. *Energy Policy* **2009**, *37*, 4689–4699. [CrossRef]

24. Sarkar, M.S.K.; Sadeka, S.; Sikdar, M.M.H.; Zaman, B. Energy Consumption And CO_2 Emission In Bangladesh: Trends And Policy. *Asia Pac. J. Energy Environ.* **2015**, *2*, 175–182. [CrossRef]

25. Appiah, K.; Dua, J.; Boamah, K.B. The Effect of Environmental Performance on Firm's Performance—Evidence from Ghana. *Br. J. Interdiscip. Res.* **2017**, *8*, 340–348.

26. Appiah, K.; Du, J.; Musah, A.A.I.; Afriyie, S. Investigation of the Relationship between Economic Growth and Carbon Dioxide (CO_2) Emissions as Economic Structure Changes: Evidence from Ghana. *Resour. Environ.* **2017**, *7*, 160–167.

27. McAusland, C. Globalisation's Direct and Indirect Effects on the Environment. *Glob. Transp. Environ.* **2010**, 31–53. [CrossRef]

28. Adom, P.K.; Bekoe, W.; Amuakwa-mensah, F.; Mensah, J.T.; Botchway, E. Carbon dioxide emissions, economic growth, industrial structure, and technical efficiency: Empirical evidence from Ghana, Senegal, and Morocco on the causal dynamics. *Energy* **2012**, *47*, 314–325. [CrossRef]

29. Asumadu-Sarkodie, S.; Owusu, P.A. The relationship between carbon dioxide and agriculture in Ghana: A comparison of VECM and ARDL model. *Environ. Sci. Pollut. Res.* **2016**, *23*, 10968–10982. [CrossRef] [PubMed]

30. Shahbaz, M.; Khan, S.; Ali, A.; Bhattacharya, M. The impact of globalization on CO_2 emissions in china. *Singap. Econ. Rev.* **2017**, *62*, 929–957. [CrossRef]

31. Pesaran, M.H.; Shin, Y. An autoregressive distributed-lag modelling approach to cointegration analysis. *Econom. Soc. Monogr.* **1998**, *31*, 371–413.

32. Gokmenoglu, K.K.; Taspinar, N. Testing the agriculture-induced EKC hypothesis: The case of Pakistan. *Environ. Sci. Pollut. Res.* **2018**, *25*, 22829–22841. [CrossRef]

33. Liu, X.; Zhang, S.; Bae, J. The impact of renewable energy and agriculture on carbon dioxide emissions: Investigating the environmental Kuznets curve in four selected ASEAN countries. *J. Clean. Prod.* **2017**, *164*, 1239–1247. [CrossRef]

34. Asumadu-sarkodie, S.; Owusu, P.A.; Asumadu-sarkodie, S.; Owusu, P.A. Carbon dioxide emission, electricity consumption, industrialization, and economic growth nexus: The Beninese case. *Energy Sources Part B Econ. Plan. Policy* **2016**, *11*, 1089–1096. [CrossRef]

35. Asumadu-Sarkodie, S. Carbon dioxide emissions, GDP, energy use, and population growth: A multivariate and causality analysis for Ghana, 1971–2013. *Environ. Sci. Pollut. Res.* **2016**, *23*, 13508–13520. [CrossRef]

36. Zhang, L.; Pang, J.; Chen, X.; Lu, Z. Carbon emissions, energy consumption and economic growth: Evidence from the agricultural sector of China's main grain-producing areas. *Sci. Total Environ.* **2019**, *665*, 1017–1025. [CrossRef] [PubMed]

37. Zafeiriou, E.; Azam, M. CO_2 emissions and economic performance in EU agriculture: Some evidence from Mediterranean countries. *Ecol. Indic.* **2017**, *81*, 104–114. [CrossRef]

38. Gornall, J.; Betts, R.; Burke, E.; Clark, R.; Camp, J.; Willett, K.; Wiltshire, A. Implications of climate change for agricultural productivity in the early twenty-first century. *Philos. Trans. R. Soc. B. Biol. Sci.* **2010**, *365*, 2973–2989. [CrossRef] [PubMed]

39. Victor, P.A. Growth, degrowth and climate change: A scenario analysis. *Ecol. Econ.* **2012**, *84*, 206–212. [CrossRef]

40. Barrios, S.; Ouattara, B.; Strobl, E. The impact of climatic change on agricultural production: Is it different for Africa? *Food Policy* **2008**, *33*, 287–298. [CrossRef]

41. Ayinde, O.E.; Muchie, M.; Olatunji, G.B. Effect of Climate Change on Agricultural Productivity in Nigeria: A Co-integration Model Approach. *J. Hum. Ecol.* **2011**, *35*, 189–194. [CrossRef]

42. Chen, J.; Cheng, S.; Song, M. Changes in energy-related carbon dioxide emissions of the agricultural sector in China from 2005 to 2013. *Renew. Sustain. Energy Rev.* **2018**, *94*, 748–761. [CrossRef]

43. Goklany, I.M. Integrated strategies to reduce vulnerability and advance adaptation, mitigation, and sustainable development. *Mitig. Adapt. Strateg. Glob. Chang.* **2007**, *12*, 755–786. [CrossRef]

44. Tol, R.S.J.; Downing, T.E.; Kuik, O.J.; Smith, J.B. Distributional aspects of climate change impacts. *Glob. Environ. Chang.* **2004**, *14*, 259–272. [CrossRef]

45. Li, W.; Ou, Q.; Chen, Y. Decomposition of China's CO_2 emissions from agriculture utilizing an improved Kaya identity. *Environ. Sci. Pollut. Res.* **2014**, *21*, 13000–13006. [CrossRef]

46. Ismael, M.; Srouji, F.; Boutabba, M.A. Agricultural technologies and carbon emissions: Evidence from Jordanian economy. *Environ. Sci. Pollut. Res.* **2018**, *25*, 10867–10877. [CrossRef] [PubMed]

47. Dickey, D.A.; Fuller, W.A. Likelihood Ratio Statistics For Autoregressive Time Series With A Unit. *Econometrica* **1981**, *49*, 1057–1072. [CrossRef]

48. Phillips, P.; Perron, P. Testing for a unit root in time series regression. *Biometrika* **1988**, *75*, 335–346. [CrossRef]

49. Akalke, H. A New Look at the Statistical Model Identification. *IEEE Trans. Autom. Control* **1974**, *19*, 716–723.

50. Schwarz, G. Institute of Mathematical Statistics is collaborating with JSTOR to digitize, preserve, and extend access to The Annals of Statistics. *Ann. Stat.* **1978**, *6*, 461–464. [CrossRef]

51. Khim, V.; Liew, S. Which Lag Length Selection Criteria Should We Employ? *Econ. Bull.* **2004**, *3*, 1–9.

52. Johansen, S. Statistical Analysis of Cointegration Vectors. *J. Econ. Dyn. Control* **1988**, *12*, 231–254. [CrossRef]

53. Johansen, S.; Juselius, K. Maximum Likelihood Estimation and Inference on Cointegration—With applications to the demand for money. *Oxf. Bull. Econ. Stat.* **1990**, *52*, 169–210. [CrossRef]

54. Pesaran, M.H.; Shin, Y.; Smith, R.J. Bounds testing approaches to the analysis of level relationships. *J. Appl. Econom.* **2001**, *16*, 289–326. [CrossRef]

55. Boutabba, M.A. The impact of fi nancial development, income, energy and trade on carbon emissions: Evidence from the Indian economy. *Econ. Model.* **2014**, *40*, 33–41. [CrossRef]

56. Shahbaz, M.; Hooi, H.; Shahbaz, M. Environmental Kuznets Curve hypothesis in Pakistan: Cointegration and Granger causality. *Renew. Sustain. Energy Rev.* **2012**, *16*, 2947–2953. [CrossRef]

57. Ozturk, I.; Acaravci, A. The long-run and causal analysis of energy, growth, openness and fi nancial development on carbon emissions in Turkey. *Energy Econ.* **2013**, *36*, 262–267. [CrossRef]

58. Hooi, H.; Smyth, R. CO_2 emissions, electricity consumption and output in ASEAN. *Appl. Energy* **2010**, *87*, 1858–1864.

59. Fodha, M.; Zaghdoud, O. Economic growth and pollutant emissions in Tunisia: An empirical analysis of the environmental Kuznets curve. *Energy Policy* **2010**, *38*, 1150–1156. [CrossRef]

60. Ferda, H. An econometric study of CO_2 emissions, energy consumption, income and foreign trade in Turkey. *Energy Policy* **2009**, *37*, 1156–1164.

61. Kanjilal, K.; Ghosh, S. Environmental Kuznet's curve for India: Evidence from tests for cointegration with unknown structural breaks. *Energy Policy* **2013**, *56*, 509–515. [CrossRef]

62. Jayanthakumaran, K.; Verma, R.; Liu, Y. CO_2 emissions, energy consumption, trade and income: A comparative analysis of China and India. *Energy Policy* **2012**, *42*, 450–460. [CrossRef]

63. Jalil, A.; Mahmud, S.F. Environment Kuznets curve for CO_2 emissions: A cointegration analysis for China. *Energy Policy* **2009**, *37*, 5167–5172. [CrossRef]

64. Lotfalipour, M.R.; Falahi, M.A.; Ashena, M. Economic growth, CO_2 emissions, and fossil fuels consumption in Iran. *Energy* **2010**, *35*, 5115–5120. [CrossRef]

65. Nasir, M.; Rehman, F.U. Environmental Kuznets Curve for carbon emissions in Pakistan: An empirical investigation. *Energy Policy* **2011**, *39*, 1857–1864. [CrossRef]

66. Saboori, B.; Sulaiman, J.; Mohd, S. Economic growth and CO_2 emissions in Malaysia: A cointegration analysis of the Environmental Kuznets Curve. *Energy Policy* **2012**, *51*, 184–191. [CrossRef]

67. Brown, R.L.; Durbin, J.; Evans, J.M. Techniques for testing the constancy of regression relationships over time. *J. R. Stat. Soc. Ser. B* **1975**, *37*, 149–192. [CrossRef]

68. Payne, J.E. Inflationary dynamics of a transition economy: The Croatian experience. *J. Policy Model.* **2002**, *24*, 219–230. [CrossRef]

69. Smadja, J.; Aubriot, O.; Puschiasis, O.; Duplan, T.; Hugonnet, M. Climate change and water resources in the Himalayas: Field study in four geographic units of the Koshi. *J. Alp. Res.* **2015**, *103*, 2–22. [CrossRef]

70. Yousuf, I.; Ghumman, A.R.; Hashmi, H.N.; Kamal, M.A. Carbon emissions from power sector in Pakistan and opportunities to mitigate those. *Renew. Sustain. Energy Rev.* **2014**, *34*, 71–77. [CrossRef]

71. Ahmad, M.; Mustafa, G.; Iqbal, M. Impact of Farm Households' Adaptations to Climate Change on Food Security: Evidence from Different Agro-Ecologies of Pakistan. *Pakistan development review.* **2016**, *55*, 561–588.

Article

The Relationship between Carbon Dioxide Emissions, Economic Growth and Agricultural Production in Pakistan: An Autoregressive Distributed Lag Analysis

Sajjad Ali [1], Li Gucheng [1,*], Liu Ying [1,2], Muhammad Ishaq [3] and Tariq Shah [4]

[1] College of Economics and Management, Huazhong Agricultural University, Wuhan 430070, China; sajjad@webmail.hzau.edu.cn (S.A.); liuyhzau@gmail.com (L.Y.)
[2] Hubei Collaborative Innovation Center for Grain Industry, Yangtze University, Jingzhou 434025, China
[3] Agricultural Pricing and Trade Policy, Social Sciences Division, Pakistan Agricultural Research Council, Islamabad 44000, Pakistan; ishaqecon@gmail.com
[4] Department of Economics & Development Studies, University of Swat, Khyber Pakhtunkhwa 19130, Pakistan; shah6833@yahoo.com
* Correspondence: lgcabc@mail.hzau.edu.cn; Tel.: +86-027-87286896

Received: 26 October 2019; Accepted: 29 November 2019; Published: 6 December 2019

Abstract: This study aims to explore the casual relationship between agricultural production, economic growth and carbon dioxide emissions in Pakistan. An autoregressive distributed lag (ARDL) model is applied to examine the relationship between agricultural production, economic growth and carbon dioxide emissions using time series data from 1960 to 2014. The Augmented Dickey–Fuller (ADF), Phillips–Perron (PP) and Kwiatkowski–Phillips–Schmidt–Shin (KPSS) tests are used to check the stationarity of variables. The results show both short-run and long-run relationships between agricultural production, gross domestic product (GDP) and carbon dioxide emissions in Pakistan. From the short-run estimates, it is found that a 1% increase in barley and sorghum production will decrease carbon dioxide emissions by 3% and 4%, respectively. The pairwise Granger causality test shows unidirectional causality of cotton, milled rice, and sorghum production with carbon dioxide emissions. Due to the aforementioned cause, it is essential to manage the effects of carbon dioxide emissions on agricultural production. Appropriate steps are needed to develop agricultural adaptation policies, improve irrigation facilities and introduce high-yielding and disease-resistant varieties of crops to ensure food security in the country.

Keywords: carbon dioxide emissions; agricultural production; GDP; ARDL bounds test; Granger causality; Pakistan

---

1. Introduction

The issue of climate change is now a global challenge and has attracted attention of world leaders for proactive and expedited planning for low carbon industrial growth, clean and renewable energy sources, agricultural sustainability and low-level energy-intensive economic growth [1–7]. To ensure food safety and food security, dedicated actions are needed on climate change and its impacts on food production [3,8,9].

Climate change can affect agriculture productivity through a change in global temperatures, variability in precipitation and other related factors. It is estimated that about 15–30% of the output of agriculture would be affected globally by 2080–2100 [10]. If timely and adequate adaptive measures are not taken, a decline in crop yield may occur in Africa, Latin America and Asia. Further, it would cost about 5–10% of gross domestic product (GDP) for Africa to take adaptation measures to combat climate change. Moreover, the results of the study predicted that about 50% of the decline in agricultural crops would be observed by 2020 and the crop revenue may further decrease, even up to 90% by 2100 [11]. Changes in the pattern of rainfall may also affect more than one billion people in South Asia [12]. Most of the studies envisage that climate variation would adversely affect the yield of wheat crops in South Asia. According to the Intergovernmental Panel on Climate Change (IPCC) 4th Assessment Report, crop yield in South Asia would reduce proportionately from 1820 m^3 to 1140 m^3 from 2001 to 2050.

For estimating the effects of climate change on yield and growth, there are usually two approaches that are being followed: (1) discovering the effects of long-term variation via crop simulation models [13,14] and (2) implementation of experiments related to artificial climate change [15–17]. Crop simulation modeling in combination with simulation models and climate change scenarios is the most frequently used approach. Modeling depends upon several factors, such as nutrition, soil, evapotranspiration, rainfall, temperature, carbon circulation, economic environment and atmospheric circulation. Climate change and adaptation strategies are increasingly becoming the main focus of current scientific research; for instance, the effect on the production of crops such as wheat, rice and maize [18]. The vulnerability index of the changing climate in Pakistan is relatively high in comparison to numerous countries around the globe, due to variable climatic conditions. Recently, Pakistan has faced numerous climatic variations, for instance, increased temperature, changes in the pattern of precipitation, floods, earthquakes and weather shifts. The development of the agriculture sector in developing countries is hampered by increasing climatic risk and projected changes in climate over the 21st century [19]. Pakistan is affected the most by climate change due to poor infrastructure and limited adaptive capacity [20]. It is projected that by 2050, there would be a 2–3% increase in temperature causing a significant variation in the pattern of rainfall [21]. Pakistan is ranked eighth among the countries most negatively affected by adverse weather conditions and climate change over the period 1995–2014 as reported by the Global Climate Risk Index (GCRI) [22]. The productivity of major crops, including wheat, rice, cotton and sugarcane, and rural livelihoods, has been affected greatly due to climate variability and extreme events over the last two decades [23]. The vulnerability of rural livelihoods to climate change can be seen from the historic floods during 2010–2014 and severe droughts from 1999 to 2003 [23].

Several conceptual works of literature have been established which show different ways in which climate change affects economic growth. The negative consequences of climate change are proved both theoretically and empirically. First, the devastation of the ecosystem by numerous intensive weather conditions, such as flood, drought, erosion, leading to the extinction of endangered species, has resulted in perpetual harm to economic growth. Secondly, the necessary resources to oppose the warming impact reduce investment in the economy, as well as the physical framework, research and development, and human capital, thus minimizing growth [24,25].

Climate change has resulted in crop reduction in many regions; for example, it was estimated that global maize production reduced by 12 Mt from 1981 to 2002 [26]. Recently, this methodology has been used in various regions, such as Europe [27], Pakistan [28], India [29] and Ghana [30], for the identification of the relationship between climate change and various factors on agriculture. Even the effect of a single weather variable can harm the long-term benefits of economic development [31]. In South Asia, the production of cereal crops has been already under heat stress. Consequently, in Central and South Asia, the crop yields will decline by up to 30% by 2050 [32]. The production of these crops is an important factor in food security around the Asian region.

For decades, researchers globally have struggled to address the problem of endogeneity. A researcher briefly stated that there is no way to empirically test whether a variable is correlated with the regression error terms because the error term is unobservable [33]. This is why exogenous latent variables, and the disturbance term, in particular, as the most common case, is the cause of so much difficulty for empirical researchers. Because many key exogenous variables of concern are not measured, "there is no way to statistically ensure that an endogeneity problem has been solved" [33]. This means that the problem of endogeneity is not so much a problem as it is a dilemma, hence, the title of this paper. Dilemmas do not call for solutions, they call for choices. In the statistical sense, the dilemma boils down to a trading one set of untestable assumptions for another. There are no direct tests of endogeneity, and the consequences of this must be understood. However, there are many indirect tests that give the researcher useful information to guide their decisions and conclusions. Therefore, this paper echoes the call for reasonable endogeneity standards found in the recent method literature [34–36].

Many environmental factors, such as floods, wind speed, sunshine, monsoon patterns and relative humidity, can affect agriculture production. We only include CO_2 emissions in our model, so the endogeneity problem arises here, which can affect the results. Not only environmental variables but also other factors, such as agricultural land use, fertilizer used, agriculture inputs and population, are included as control variables. Endogeneity is a problematic situation in which explanatory variables correlate with the error term. In this case, when there is an endogeneity problem in our model or variables, we need to remove it with the help of an included instrumental variable. Technically, a Two-Stage Least Square (2SLS) model is applied when there is endogeneity in time series data. Ideally, it is only applied to cross-sectional data, as if you apply 2SLS to time series, it will not be able to ensure co-integration, and results may be spurious. Secondly, if we apply 2SLS to panel data, it might not incorporate the cross-sectional heterogeneity. Thus, in the case of panel data, most researchers have used a Generalized Method of Moments (GMM) model as an advanced version of 2SLS. It is very rare to see endogeneity in time series data because co-integration solves that issue. Some previous researchers have used the 2SLS model in their studies [37,38].

This study explores the responses of carbon dioxide emissions to gross domestic product (GDP) and agricultural production based on historical data in Pakistan. An autoregressive distributed lag (ARDL) model is employed to examine the effect of agricultural production, gross domestic product and carbon dioxide emissions to determine the long-run relationships among several variables [39]. The remainder of this study is structured as follows: Section 2 consists of a literature review. Section 3 briefly describes the materials and methods, including the study area, data sources and description, model specification, and econometric model. Section 4 describes the results and discussions, which consist of descriptive statistics, unit root tests, lag order selection criteria, ARDL bounds tests, analysis of long-run and short-run estimates, and ARDL diagnostic tests and normality plots. Section 5 contains the conclusion and policy implications of the study.

2. Literature Review

Many previous studies have employed modern econometric techniques to determine the association between environmental greenhouse gasses, energy consumption and socio-economic variables in various nations globally [5,40–47]. A previous study investigated the relationship between the consumption of electricity, industrialization, GDP and carbon dioxide emissions in Benin using an autoregressive distributed lag (ARDL) model [42]. Evidence from the study revealed a long-run equilibrium association flowing from consumption of electricity industrialization, GDP and carbon dioxide emissions [42]. Another study employed the vector error correction model (VECM) and ordinary least squares (OLS) regression to reveal the impact of population progression, energy intensity and GDP on carbon dioxide emissions in Ghana [48]. The study found evidence of the existence of a long-run equilibrium association flowing from population growth, energy intensity and GDP to carbon dioxide emissions. The study also revealed that there was a bi-directional causality among energy consumption and carbon dioxide emissions [48]. Another study in Ghana investigated the association between population growth, use of energy, GDP and carbon dioxide emissions using both autoregressive distributed lag (ARDL) regression analysis and a vector error correction model (VECM). The study found that there will be fluctuation in carbon dioxide emissions due to the use of energy in the future. Furthermore, evidence from the study showed a unidirectional causality running from carbon dioxide emissions to use the energy and population [49].

Theoretically, an association could be established through microeconomic and macroeconomic dimensions. From the view of the macroeconomic dimension, the two important areas which are stressed include the impact on the output level, such as yields and the ability of the economy to grow [50]. On the microeconomic side, we have factors such as physical productivity of labor, health and conflict. These factors have economy-wide implications [51–53]. Moreover, climate change can have such effects as political inconstancy, which may obstruct factor accumulation and growth in productivity [54].

It has been reported that a rise in temperature can have a profound influence on the productivity of the agriculture sector, food security and farmer's income. This effect varies in tropical and temperate areas. In middle and high latitudes, the aptness and output of crops are anticipated to increase and spread northwards, and vice versa is true for several countries in tropical regions [55]. It is found that in high latitudes, production can be increased by nearly 10% due to a 2 °C rise in temperature, while it reduces production by the same percentage in low latitudes. By taking into account the effect of technology, it is projected that an increase in temperature would increase the productivity of yields by between 37% and 101% by the 2050s in the Russian Federation [54].

In comparison to other developing countries, the effects of escalating temperature on agriculture are harsher in sub-Saharan Africa [56]. It has been observed that if some important climatic conditions, such as temperature and rainfall, had persisted at their pre-1960 status, then the gap of agricultural production between different developing countries and sub-Saharan Africa at the end of the 20th century would have remained only 32% of the existing shortfall. A study of the period of 1980–2005 in Nigeria indicated that temperature exerts a negative influence while rainfall has a positive effect on agricultural production [57].

Some illumination of the effects of climate change on African development was provided in the 4th assessment report of the IPCC. For instance, it was estimated that yield could be reduced by 50% by 2020 in some countries, and the revenue generated from crops could fall nearly 90% by 2100. Smallholder farmers would be affected the most. This will also provoke water problems, as almost 25% of the population in Africa has recently encountered high water stress. Because of increasing water stress in Africa, the population at risk is projected to be between 350 and 600 million by 2050 and about 25–40% of mammals may become endangered in national parks in sub-Saharan Africa [11].

Developed countries have the ability to maintain a minimum level of technology for the improvement of living standards and increasing agricultural productivity [58]. These countries are generally also capable of offsetting the negative consequences of climate change. Developed states usually have a low level of susceptibility but a high level of adaptive ability, which itself is a function of technological expertise, dissemination and supply of assets, and human social and political capital [59]. The developed world has good levels of water filtration and sanitation. On the other hand, developing countries have insecure and unreliable water supplies, and often sanitation systems are non-satisfactory. The notion of crop insurance to protect farmers from the negative consequences of climate change, which may destroy their livelihoods, is missing in developing countries.

During the past decade, Pakistan's per capita gross domestic product (GDP) has experienced a diverse trend. During the period from 2005 to 2014, per capita GDP increased from USD 974.5 to USD 1111.2. In 2011, the government placed significant emphasis on upgrading the country's economy, resulting in a consistent increase of per capita GDP during the period 2011 to 2014. During this period, despite several types of socio-economic challenges, such as energy crises, a war against terrorism, and poverty, per capita GDP (Pakistan Economic Survey 2017) increased by USD 64.71, providing evidence that the Government of Pakistan has taken actions to raise economic growth and enriched the living conditions of the hinterlands.

3. Materials and Methods

3.1. Data Sources and Description

The key purpose of this study is to answer the question: is there any causal effect between carbon dioxide emissions, gross domestic product and agricultural production in Pakistan? The study used time series data from 1960 to 2014. The data for different variables of this study was acquired from Index Mundi and World Development Indicators of the World Bank. Based on the review of literature, the current study uses nine variables: carbon dioxide emissions CO_2 (kt), gross domestic product (GDP) (constant 2010 US$), barley production (1000 Mt), corn production (1000 Mt), cotton production (1000 Mt), milled rice production (1000 Mt), millet production (1000 Mt), sorghum production (1000 Mt) and wheat production (1000 Mt). The trends of the study variables are given in Figure 1.

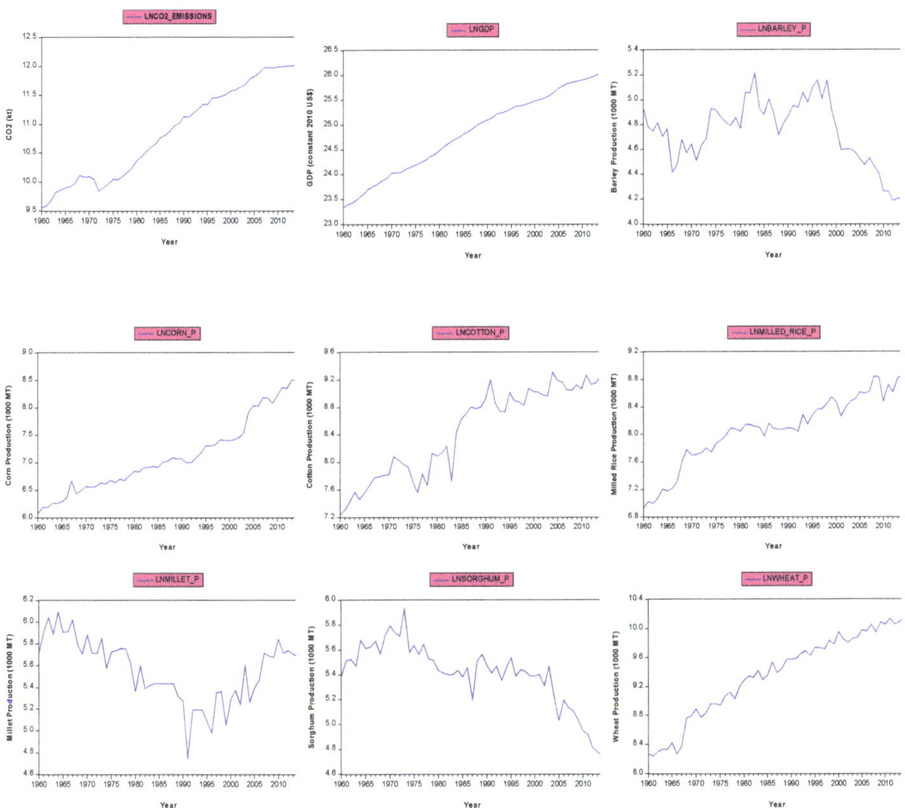

Figure 1. Trend of study variables.

3.2. Econometric Model

Descriptive statistics are estimated to determine the features of the study variables. To find out the integration order of the study variables, in the first step, we have to identify stationarity in the time series data. For this purpose, we employed the Augmented Dickey–Fuller (ADF) [60], Kwiatkowski–Phillips–Schmidt–Shin (KPSS) and Phillips–Perron (PP) unit root tests [61], and the ARDL bounds test was then estimated. Furthermore, the pairwise Granger causality test and variance decomposition analysis were carried out to examine the direction of causality and improve the study variables in the future. Figure 2 presents the schematic diagram of the study.

The econometric specification of the study variables can be written as:

$$CO_{2t} = f(GDP_t, \; BARLEY_t, \; CORN_t, COTTON_t, MILLED \; RICE_t, MILLET_t, SORGHUM_t, WHEAT_t) \tag{1}$$

The empirical specification of the proposed model is written as:

$$
\begin{aligned}
LnCO_{2t} &= \alpha_0 + \alpha_1 LnGDP_t + \alpha_2 LnBARLEY_t + \alpha_3 LnCORN_t + \alpha_4 LnCOTTON_t + \alpha_5 LnMILLED\ RICE_t \\
&\quad + \alpha_6 LnMILLET_t + \alpha_7 LnSORGHUM_t + \alpha_8 LnWHEAT_t \\
&\quad + \varepsilon_t)
\end{aligned}
\tag{2}
$$

In Equation (2), $LnCO_{2t}$ is the logarithmic form of carbon dioxide emissions, $LnGDP_t$ is the gross domestic product (GDP), $LnBARLEY_t$ is the barley production, $LnCORN_t$ is the corn production, $LnCOTTON_t$ is the cotton production, $LnMILLED\ RICE_t$ is the milled rice production, $LnMILLET_t$ is the millet production, $LnSORGHUM_t$ is the sorghum production and $LnWHEAT_t$ is the wheat production in year t, ε_t is the error term, and $\alpha_0, \alpha_1, \alpha_2, \alpha_3, \alpha_4, \alpha_5, \alpha_6, \alpha_7\ and\ \alpha_8$ are the elasticities to be estimated in Equation (2).

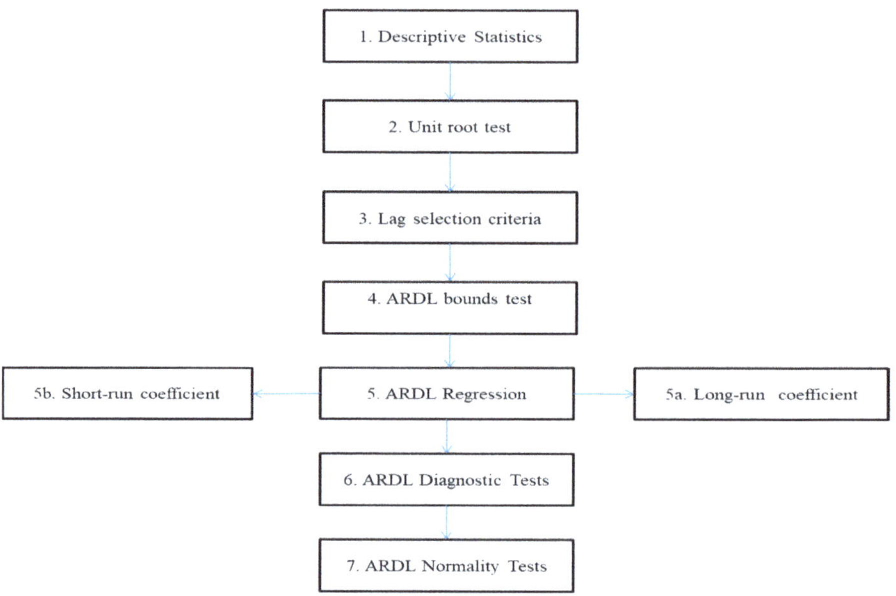

Figure 2. A schematic presentation of the study.

4. Results and Discussion

4.1. Descriptive Analysis

The descriptive analysis shows the mean, coefficient of variation, skewness, kurtosis and normality of distribution of the study variables. The results of descriptive statistics of the study variables are estimated in Table 1. Evidence shows that CO_2, gross domestic product (GDP), barley, corn, cotton, milled rice, millet and wheat exhibit positive skewness, while sorghum exhibits a negative skewness. The result of the kurtosis test shows that the CO_2, gross domestic product (GDP), barley, cotton, milled rice, millet and wheat exhibit a platykurtic distribution, while corn and sorghum exhibit a leptokurtic distribution. The outcome from the Jarque–Bera test shows that we accept the null hypothesis of normal distribution at the 5% level of significance for barley, milled rice, millet, sorghum and wheat crops.

Table 1. Descriptive statistics analysis.

Statistic	CO_2 Emissions (kt)	GDP (M USD)	Barley (1000 Mt)	Corn (1000 Mt)	Cotton (1000 Mt)	Milled-Rice (1000 Mt)	Millet (1000 Mt)	Sorghum (1000 Mt)	Wheat (1000 Mt)
Mean	70,590.69	8,140,000	118.4182	1567.873	5618.964	3575.291	269.3455	230.3636	13,477.24
Median	53,535.00	6,770,000	118.0000	1100.000	6250.000	3272.000	274.0000	231.0000	12,675.00
Maximum	166,299.0	2,060,000	185.0000	4944.000	11,138.00	7003.000	446.0000	378.0000	25,979.00
Minimum	14,155.00	1,370,000	66.00000	439.0000	1398.000	1030.000	115.0000	115.0000	3814.000
Std. Dev.	52,092.32	5,740,000	29.42013	1213.157	3070.710	1601.680	74.69896	53.63780	6683.929
Skewness	0.627348	0.616588	0.116649	1.464809	0.106986	0.411594	0.226418	−0.102608	0.169114
Kurtosis	1.960562	2.149885	2.372942	3.998844	1.527590	2.529724	2.451225	3.302701	1.841895
Jarque–Bera	6.083672	5.141172	1.025816	21.95496	5.073237	2.059748	1.160074	0.306490	3.335760
Probability	0.047747	0.076491	0.598752	0.000017	0.079134	0.357052	0.559878	0.857919	0.188647

Source "Authors' calculation".

4.2. Unit Root Tests

Before estimating the ARDL bounds test co-integration, it is necessary to determine the stationarity of the variables. To meet the stationarity requirement, the study estimates the unit root using the Augmented Dickey–Fuller (ADF) [62], Phillips–Perron (PP) [61] and Kwiatkowski–Phillips–Schmidt–Shin (KPSS) tests in order to have a robust result. The results of the unit root tests are reported in Table 2. The result of the ADF test shows that the null hypothesis of the unit root cannot be rejected at a 5% significance level. The results of the KPSS test show the null hypothesis of stationarity is rejected at a 5% significance level. Evidence from the results of ADF, PP and KPSS unit root tests shows that the series are integrated at $I(1)$.

Table 2. Unit root test.

Model	ADF Level t-Stat (p-Vale)	ADF 1st Diff t-Stat (p-Vale)	KPSS Level t-Stat (5% Critical Level)	KPSS 1st Diff t-Stat (5% Critical Level)	PP Level t-stat (p-Vale)	PP 1st Diff t-Stat (p-Vale)
			Intercept			
$LnCO_2$	−0.806182 (0.8092)	−5.953051 (0.0000)	0.882144 (0.463000)	0.121843 (0.463000)	−0.761270 (0.8218)	−5.991025 (0.0000)
$LnGDP$	−3.144898 (0.0291)	−5.525176 (0.0000)	0.893568 (0.463000)	0.482889 (0.463000)	−2.886320 (0.0536)	−5.623884 (0.0000)
$LnBarley$	−1.278657 (0.6332)	−8.855400 (0.0000)	0.304102 (0.463000)	0.177851 (0.463000)	−1.278657 (0.6332)	−8.825782 (0.0000)
$LnCorn$	0.467842 (0.9840)	−8.517140 (0.0000)	0.861313 (0.463000)	0.147426 (0.463000)	0.631484 (0.9894)	−8.526955 (0.0000)
$LnCotton$	−1.423831 (0.5638)	−9.945326 (0.0000)	0.853528 (0.463000)	0.170778 (0.463000)	−1.450853 (0.5506)	−11.39586 (0.0000)
$LnMilled\ rice$	−1.631603 (0.4597)	−9.582429 (0.0000)	0.954980 (0.463000)	0.204843 (0.463000)	−1.988090 (0.2911)	−10.16939 (0.0000)
$LnMillet$	−1.647754 (0.4515)	−11.71139 (0.0000)	0.453031 (0.463000)	0.056429 (0.463000)	−2.143656 (0.2290)	−13.01371 (0.0000)
$LnSorghum$	0.550452 (0.9869)	−11.35154 (0.0000)	0.775917 (0.463000)	0.187032 (0.463000)	0.052072 (0.9589)	−11.65376 (0.0000)
$LnWheat$	−2.155233 (0.2248)	−7.468655 (0.0000)	0.867938 (0.463000)	0.286169 (0.463000)	−1.991421 (0.2897)	−11.88184 (0.0000)

Table 2. *Cont.*

Model	ADF Level	ADF 1st Diff	KPSS Level	KPSS 1st Diff	PP Level	PP 1st Diff
	t-Stat (*p*-Vale)	*t*-Stat (*p*-Vale)	*t*-Stat (5% Critical Level)	*t*-Stat (5% Critical Level)	*t*-stat (*p*-Vale)	*t*-Stat (*p*-Vale)
			Intercept and Trend			
$LnCO_2$	−1.146817 (0.9110)	−5.943842 (0.0000)	0.109365 (0.146000)	0.109589 (0.146000)	−1.573362 (0.7905)	−6.031511 (0.0000)
$LnGDP$	−0.822407 (0.9569)	−6.292270 (0.0000)	0.229191 (0.146000)	0.056925 (0.146000)	−0.994228 (0.9362)	−6.301469 (0.0000)
$LnBarley$	−1.549158 (0.7997)	−8.936464 (0.0000)	0.211863 (0.146000)	0.101501 (0.146000)	−1.549158 (0.7997)	−9.066157 (0.0000)
$LnCorn$	−1.541615 (0.8025)	−8.669357 (0.0000)	0.207887 (0.146000)	0.062252 (0.146000)	−1.328646 (0.8700)	−8.690926 (0.0000)
$LnCotton$	−3.418161 (0.0596)	−9.913682 (0.0000)	0.129370 (0.146000)	0.108013 (0.146000)	−3.338073 (0.0711)	−12.56710 (0.0000)
$LnMilled\ rice$	−3.199369 (0.0954)	−9.593674 (0.0000)	0.156707 (0.146000)	0.124665 (0.146000)	−3.079200 (0.1216)	−10.85739 (0.0000)
$LnMillet$	−1.342745 (0.8660)	−8.664871 (0.0000)	0.217252 (0.146000)	0.032753 (0.146000)	−2.391047 (0.3799)	−14.16708 (0.0000)
$LnSorghum$	−1.812902 (0.6845)	−8.306334 (0.0000)	0.154413 (0.146000)	0.093016 (0.146000)	−2.956875 (0.1538)	−13.87512 (0.0000)
$LnWheat$	−1.569448 (0.7912)	−7.764154 (0.0000)	0.239256 (0.146000)	0.126430 (0.146000)	−2.593371 (0.2849)	−23.15530 (0.0001)

4.3. ARDL Bounds Testing of Co-Integration and Regression Analysis

The current study uses an autoregressive distributed lag (ARDL) bounds testing approach suggested by [63] to determine both short-run and long-run associations of carbon dioxide emissions, gross domestic product and agricultural production. The ARDL bounds testing method is appropriate for those models in which there is a mixture of $I(0)$ and $I(1)$ variables. Another characteristic of this model is that it is appropriate for small sample size, as our sample size is only 54 [63].

It is important to determine how many lags are to be used in an ARDL model. Therefore, to find the optimal number of lags for the model, the unrestricted vector autoregression (VAR) lag selection criteria are tested. Table 3 formulates the lag selection criteria for the model, but the most commonly employed criteria are the Akaike information criterion (AIC) and the Schwarz information criterion (SIC). A previous study used AIC for small sample size [64]. In this study, we employed the Akaike information criterion, which revealed that the most suitable lag value for the model is lag 3.

Table 3. Optimal lags selection.

Lag	LogL	LR	FPE	AIC	SC	HQ
0	223.6244	NA	2.10×10^{-15}	−8.254784	−7.917069	−8.125312
1	599.5208	607.2173	2.61×10^{-20} *	−19.59696	−16.21980 *	−18.30223 *
2	660.1237	76.91906	7.52×10^{-20}	−18.81245	−12.39586	−16.35248
3	772.8762	104.0792 *	5.06×10^{-20}	−20.03370 *	−10.57767	−16.40848

* indicates lag order selected by the criterion; Likelihood Ratio LR: sequential modified LR test statistic (each test at 5% level); FPE: Final prediction error; AIC: Akaike information criterion; SC: Schwarz information criterion; HQ: Hannan–Quinn information criterion. **Source**" Authors' calculation".

After unit root testing, which showed all variables are integrated at $I(1)$, we carried out the ARDL method of co-integration (bounds testing) to estimate the relationship between the selected variables in this study. The results of the ARDL bounds testing are reported in Table 4. The results indicate that the f-statistic value (4.954551) is greater than the 10% and 5% upper critical values of $I(0)$ bound. The results of the bounds testing validate significant long-run relationships among variables and show the rejection of the null hypothesis of no co-integration association among *LnCO2*, *LnGDP*, *Lnbarley*, *Lncorn*, *Lncotton*, *Lnmilled rice*, *Lnmillet*, *Lnsorghum* and *Lnwheat*.

Table 4. ARDL Bound Test.

Test Statistic	Value	k
F-statistic	4.954551	8
	Critical value bounds	
Significance	$I(0)$ Bound	$I(1)$ Bound
10%	1.85	2.85
5%	2.11	3.15
2.5%	2.33	3.42
1%	2.62	3.77

Furthermore, the study uses the Akaike information criterion (AIC) to select the optimal model by employing long-run and short-run associations among variables. Employing the Akaike information criterion shows the top twenty possible ARDL models in Figure 3. Based on the model specification in Equation (2), the short-run and long-run equilibrium relationships of *LnCO2*, *LnGDP*, *Lnbarley*, *Lncorn*, *Lncotton*, *Lnmilledrice*, *Lnmillet*, *Lnsorghum* and *Lnwheat* are estimated using the ARDL regression analysis shown in Equation (3).

$$Cointeq = LnCO_2\_EMISSIONS - (2.0507 \times LnGDP + 0.3425 \times LnBARLEY\_P + 0.2393$$
$$\times LnCORN\_P - 0.3300 \times LnCOTTON\_P - 0.4678 \times LnMILLED\_RICE\_P - 0.2392 \times \qquad (3)$$
$$LnMILLET\_P - 0.0549 \times LnSORGHUM\_P - 0.9790 \times LnWHEAT\_P - 26.2134)$$

where $\alpha_0 = -26.2134$, $\alpha_1 = 2.0507$, $\alpha_2 = 0.3425$, $\alpha_3 = 0.2393$, $\alpha_4 = -0.3300$, $\alpha_5 = -0.4678$, $\alpha_6 = -0.2393$, $\alpha_7 = -0.0549$ and $\alpha_8 = -0.9790$.

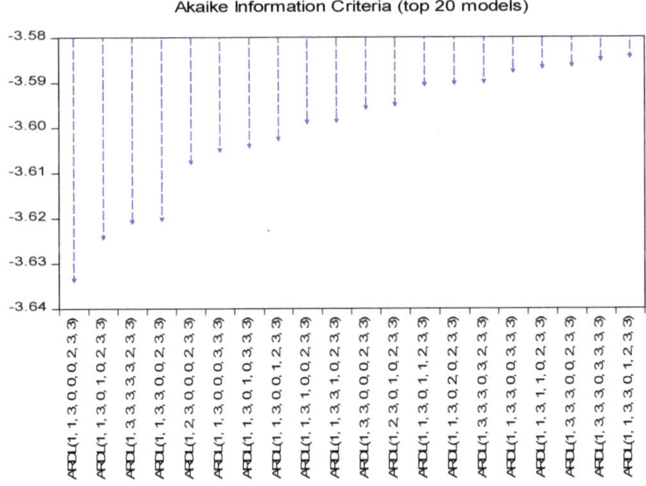

Figure 3. ARDL model selection criterion. **Source** "Authors' calculation".

4.4. Short-Run and Long-Run Equation Model

Table 5 summarizes the results of the short-run equation of the ARDL model. The results show that the speed of adjustment Error Correction Term ECT(−1) value is −0.30225 which shows that there is a long-run and short-run equilibrium relationship running from *LnGDP*, *LnBARLEY*, *LnCORN*,

LnCOTTON, LnMILLED RICE, LnMILLET, LnSORGHUM and *LnWHEAT* to *LnCO$_2$*. The speed of adjustment is approximately 30.2% in one period of the long-run equilibrium.

Table 5. Short-run and long-run relationship estimates for the selected model ARDL(1,1,3,0,0,0,2,3,3).

Short Run Coefficients				
Variable	Coefficient	Std. Error	t-Statistic	Prob.
D(LnGDP)	1.603111	0.134533	11.91612	0.0000
D(LnBARLEY_P)	−0.033465	0.044624	−0.749932	0.4591
D(LnBARLEY_P(−1))	−0.028035	0.045818	−0.611888	0.5452
D(LnBARLEY_P(−2))	−0.182478	0.047232	−3.863424	0.0006
D(LnMILLET_P)	−0.003291	0.030180	−0.109053	0.9139
D(LnMILLET_P(−1))	0.120510	0.030014	4.015104	0.0004
D(LnSORGHUM_P)	−0.044541	0.041888	−1.063337	0.2961
D(LnSORGHUM_P(−1))	0.031559	0.047437	0.665288	0.5109
D(LnSORGHUM_P(−2))	0.156119	0.046536	3.354824	0.0022
D(LnWHEAT_P)	0.160592	0.066146	2.427827	0.0214
D(LnWHEAT_P(−1))	0.254506	0.068919	3.692827	0.0009
D(LnWHEAT_P(−2))	0.198675	0.070262	2.827628	0.0083
ECT(−1)	−0.302533	0.037696	−8.025532	0.0000
Long Run Coefficients				
Variable	Coefficient	Std. Error	t-Statistic	Prob.
LnGDP	2.050716	0.354124	5.790953	0.0000
LnBARLEY_P	0.342550	0.211505	1.619578	0.1158
LnCORN_P	0.239295	0.236151	1.013316	0.3190
LnCOTTON_P	−0.330019	0.148359	−2.224459	0.0338
LnMILLED_RICE_P	−0.467837	0.214080	−2.185334	0.0368
LnMILLET_P	−0.239205	0.298329	−0.801816	0.4290
LnSORGHAM_P	−0.054855	0.227681	−0.240930	0.8112
LnWHEAT_P	−0.978995	0.346072	−2.828881	0.0082
C	−26.21344	7.839747	−3.343659	0.0022

EC = LnCO$_2$_EMISSIONS − (2.0507 × LnGDP + 0.3425 × LnBARLEY_P + 0.2393 × LnCORN_P − 0.3300 × LnCOTTON_P − 0.4678 × LnMILLED_RICE_P − 0.2392 × LnMILLET_P − 0.0549 × LnSORGHUM_P − 0.9790 × LnWHEAT_P − 26.2134)

Source "Authors' calculation".

Table 5 also shows the results of long-run equation results of the ARDL approach. The results of the long-run equilibrium relationship show that a 1% increase in *LnBARLEY* will decrease LnCO$_2$ by 3%, a 1% increase in *LnMILLET* will decrease *LnCO$_2$* by 0.03%, and a 1% increase in *LnSORGHUM* will decrease *LnCO$_2$* by 3% in short-run estimates. The evidence of the following studies reveals that carbon dioxide emissions increase in the early phases of economic growth and then decline after a threshold point. The findings of these studies (such as [10,48–55]) show the relationship between carbon dioxide emissions and GDP growth. The findings of previous studies, such as [65] for China, [66] for Tunisia, [67] for Iran, [68] for Pakistan, [69] for Malaysia, [70] for Turkey and [71] for India, indicate that there is a unidirectional causality running from GDP income to carbon dioxide emissions without response, suggesting that emission reduction plans will not restrain trade and industry growth and that the implementation of such plans seems to be a feasible policy strategy in the aforementioned studied countries to accomplish their long-run sustainable growth.

4.5. Diagnostic Test

Once the cointegration relationship was confirmed for the different variables, the cumulative sum (CUSUM) and the cumulative sum of the square of the recursive residuals (CUSUM^2) were implemented to run the ARDL model in a befitting manner. The CUSUM and CUSUM^2 tests were employed based on the recursive regression residuals as suggested by [72]. Evidence from the cumulative sum (CUSUM) and cumulative sum of squares (CUSUM^2) tests show that the plots lie within the 5% significance level. The two straight lines (red color) show the critical bounds at the 5% significant level. The lines (blue color) in the middle represent the measurements for the cumulative sum of the recursive residuals and the cumulative sum of the square of the recursive residuals. The above statements mean that the ARDL model is constant and stable for estimation of the parameters of the ARDL co-integration bounds test, and the long-run and short-run causality relationship. Figure 4 presents the diagnostic and stability tests for the ARDL model and validates the model.

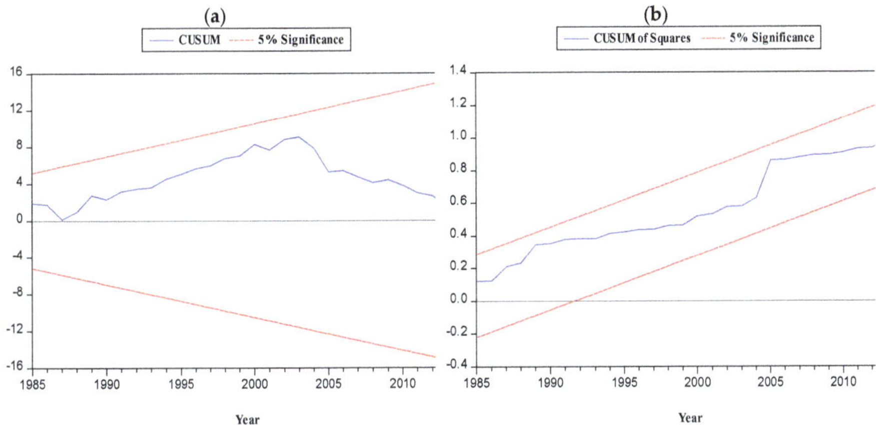

Figure 4. Stability test based on (**a**) CUSUM and (**b**) CUSUM of squares. **Source** "Authors' calculation".

Several diagnostic tests were undertaken to check for a good fit of the ARDL model. Table 6 shows that the estimation was suitable with regard to serial correlation and heteroskedasticity, and the inverse root of the AR graph shows the stability of the model.

Table 6. Diagnostic test results.

Breusch–Godfrey Serial Correlation Lagrange Multiplier LM Test:	
F-statistic	2.958497
Obs R-squared	12.86465
Prob. F(3,27)	0.0501
Prob. Chi-Square(3)	0.0049
Heteroskedasticity Test: Breusch–Pagan–Godfrey	
F-statistic	1.858453
Obs R-squared	29.40032
Scaled explained sum of square SS	9.473970
Prob. F(21,30)	0.0588

Inverse Roots of AR Characteristic Polynomial

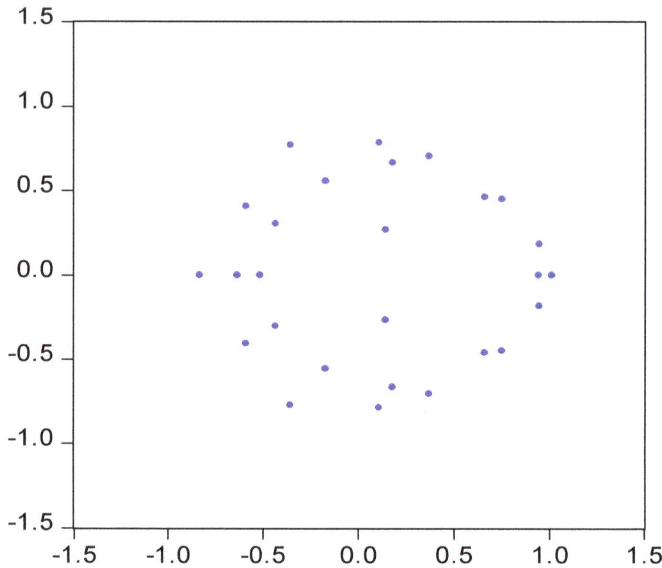

4.6. Pairwise Granger-Causality Tests

In this study, we applied an ARDL testing model to determine the short-run and long-run relationship between variables. To find out the causality between $LnCO_2$, *LnGDP, LnBARLEY, LnCORN, LnCOTTON, LnMILLEDRICE, LnMILLET, LnSORGHUM and LnWHEAT*, we used pairwise Granger causality [73] estimations. The results of the pairwise Granger causality test are presented in Table 7. The null hypothesis that $LnCO_2\_EMISSIONS$ does not Granger cause *LnCOTTON_P*, $LnCO_2\_EMISSIONS$ does not Granger cause *LnMILLED_RICE_P*, $LnCO_2\_EMISSIONS$ does not Granger cause *LnSORGHUM_P, LnGDP* does not Granger cause *LnCOTTON_P, LnGDP* does not Granger cause *LnMILLED RICE_P, LnGDP* does not Granger cause *LnSORGHUM_P, LnGDP* does not Granger cause *LnWHEAT_P, LnSORGHUM_P* does not Granger cause *LnBARLEY_P, LnCORN_P* does not Granger cause *LnMILLED_RICE_P, LnCOTTON_P* does not Granger cause *LnSORGHUM_P, LnWHEAT_P* does not Granger cause *LnCOTTON_P, LnCOTTON_P* does not Granger cause *LnWHEAT_P, LnMILLED_RICE_P* does not Granger cause *LnSORGHUM_P, LnWHEAT_P* does not Granger cause *LnMILLED_RICE_P, LnMILLED_RICE_P* does not Granger cause *LnWHEAT_P, and LnWHEAT_P* does not Granger cause *LnSORGHUM_P* is rejected at the 5% significance level. The results of Granger causality shows unidirectional causality between: *LnCOTTON_P* $\rightarrow LnCO_2$, *LnMILLED RICE_P* $\rightarrow LnCO_2$, *LnSORGHUM_P* $\rightarrow LnCO_2$, *LnCOTTON_P* $\rightarrow$ *LnGDP, LnMILLED RICE_P* $\rightarrow$ *LnGDP, LnSORGHUM_P* $\rightarrow$ *LnGDP, LnWHEAT_P* $\rightarrow$ *LnGDP, LnSORGHUM_P* $\rightarrow$ *LnBARLEY_P, LnMILLED_RICE_P* $\rightarrow$ *LnCORN_P, LnSORGHUM_P* $\rightarrow$ *LnCOTTON_P, LnSORGHUM_P* $\rightarrow$ *LnMILLED_RICE_P, and LnWHEAT_P* $\rightarrow$ *LnSORGHUM_P*, and bidirectional causality between: *LnWHEAT_P* $\leftrightarrow$ *LnCOTTON_P and LnWHEAT_P* $\leftrightarrow$ *LnMILLED_RICE_P*.

Table 7. **Pairwise** Granger causality test.

Pairwise Granger Causality Tests			
Null Hypothesis:	**Obs**	**F-Statistic**	**Prob.**
$LnGDP$ does not Granger cause $LnCO_2\_EMISSIONS$	54	1.65000	0.2048
$LnCO_2\_EMISSIONS$ does not Granger cause $LnGDP$		0.00268	0.9589
$LnBARLEY\_P$ does not Granger cause $LnCO_2\_EMISSIONS$	54	3.66357	0.0612
$LnCO_2\_EMISSIONS$ does not Granger cause $LnBARLEY\_P$		1.63838	0.2063
$LnCORN\_P$ does not Granger cause $LnCO_2\_EMISSIONS$	54	0.41262	0.5235
$LnCO_2\_EMISSIONS$ does not Granger cause $LnCORN\_P$		2.10540	0.1529
$LnCOTTON\_P$ does not Granger cause $LnCO_2\_EMISSIONS$	54	0.57836	0.4505
$LnCO_2\_EMISSIONS$ does not Granger cause $LnCOTTON\_P$		13.3746	0.0006
$LnMILLED\_RICE\_P$ does not Granger cause $LnCO_2\_EMISSIONS$	54	0.00024	0.9877
$LnCO_2\_EMISSIONS$ does not Granger cause $LnMILLED\_RICE\_P$		4.64682	0.0359
$LnMILLET\_P$ does not Granger cause $LnCO_2\_EMISSIONS$	54	1.58702	0.2135
$LnCO_2\_EMISSIONS$ does not Granger cause $LnMILLET\_P$		0.85293	0.3601
$LnSORGHUM\_P$ does not Granger cause $LnCO_2\_EMISSIONS$	54	0.07604	0.7839
$LnCO_2\_EMISSIONS$ does not Granger cause $LnSORGHUM\_P$		8.42960	0.0054
$LnWHEAT\_P$ does not Granger cause $LnCO_2\_EMISSIONS$	54	0.66557	0.4184
$LnCO_2\_EMISSIONS$ does not Granger cause $LnWHEAT\_P$		2.66479	0.1088
$LnBARLEY\_P$ does not Granger cause $LnGDP$	54	1.4×10^{-05}	0.9970
$LnGDP$ does not Granger cause $LnBARLEY\_P$		0.82411	0.3683
$LnCORN\_P$ does not Granger cause $LnGDP$	54	0.08792	0.7680
$LnGDP$ does not Granger cause $LnCORN\_P$		1.06664	0.3066
$LnCOTTON\_P$ does not Granger cause $LnGDP$	54	1.78409	0.1876
$LnGDP$ does not Granger cause $LnCOTTON\_P$		13.9597	0.0005
$LnMILLED\_RICE\_P$ does not Granger cause $LnGDP$	54	0.17989	0.6733
$LnGDP$ does not Granger cause $LnMILLED\_RICE\_P$		7.41034	0.0089
$LnMILLET\_P$ does not Granger cause $LnGDP$	54	0.39985	0.5300
$LnGDP$ does not Granger cause $LnMILLET\_P$		1.45823	0.2328
$LnSORGHUM\_P$ does not Granger cause $LnGDP$	54	1.45676	0.2330
$LnGDP$ does not Granger cause $LnSORGHUM\_P$		9.16817	0.0039
$LnWHEAT\_P$ does not Granger cause $LnGDP$	54	4.3×10^{-07}	0.9995
$LnGDP$ does not Granger cause $LnWHEAT\_P$		11.3023	0.0015
$LnCORN\_P$ does not Granger cause $LnBARLEY\_P$	54	1.82014	0.1833
$LnBARLEY\_P$ does not Granger cause $LnCORN\_P$		0.83645	0.3647
$LnCOTTON\_P$ does not Granger cause $LnBARLEY\_P$	54	0.16781	0.6838
$LnBARLEY\_P$ does not Granger cause $LnCOTTON\_P$		0.01421	0.9056
$LnMILLED\_RICE\_P$ does not Granger cause $LnBARLEY\_P$	54	0.72632	0.3981
$LnBARLEY\_P$ does not Granger cause $LnMILLED\_RICE\_P$		1.33097	0.2540
$LnMILLET\_P$ does not Granger cause $LnBARLEY\_P$	54	0.19762	0.6585
$LnBARLEY\_P$ does not Granger cause $LnMILLET\_P$		1.73499	0.1937
$LnSORGHUM\_P$ does not Granger cause $LnBARLEY\_P$	54	6.36879	0.0148
$LnBARLEY\_P$ does not Granger cause $LnSORGHUM\_P$		1.78782	0.1871
$LnWHEAT\_P$ does not Granger cause $LnBARLEY\_P$	54	0.76246	0.3867
$LnBARLEY\_P$ does not Granger cause $LnWHEAT\_P$		2.48626	0.1210
$LnCOTTON\_P$ does not Granger cause $LnCORN\_P$	54	0.02533	0.8742
$LnCORN\_P$ does not Granger cause $LnCOTTON\_P$		2.74956	0.1034
$LnMILLED\_RICE\_P$ does not Granger cause $LnCORN\_P$	54	0.03956	0.8431
$LnCORN\_P$ does not Granger cause $LnMILLED\_RICE\_P$		7.44094	0.0087
$LnMILLET\_P$ does not Granger cause $LnCORN\_P$	54	0.11889	0.7317
$LnCORN\_P$ does not Granger cause $LnMILLET\_P$		0.19930	0.6572
$LnSORGHUM\_P$ does not Granger cause $LnCORN\_P$	54	1.09323	0.3007
$LnCORN\_P$ does not Granger cause $LnSORGHUM\_P$		21.4589	3×10^{-05}
$LnWHEAT\_P$ does not Granger cause $LnCORN\_P$	54	0.13426	0.7156
$LnCORN\_P$ does not Granger cause $LnWHEAT\_P$		3.26729	0.0766
$LnMILLED\_RICE\_P$ does not Granger cause $LnCOTTON\_P$	54	3.47557	0.0680
$LnCOTTON\_P$ does not Granger cause $LnMILLED\_RICE\_P$		1.58936	0.2132
$LnMILLET\_P$ does not Granger cause $LnCOTTON\_P$	54	0.48885	0.4876

Table 7. *Cont.*

Pairwise Granger Causality Tests			
Null Hypothesis:	**Obs**	**F-Statistic**	**Prob.**
LnCOTTON_P does not Granger cause *LnMILLET_P*		1.61906	0.2090
LnSORGHUM_P does not Granger cause *LnCOTTON_P*	54	0.78492	0.3798
LnCOTTON_P does not Granger cause *LSORGHUM_P*		5.29439	0.0255
LnWHEAT_P does not Granger cause *LnCOTTON_P*	54	8.20250	0.0061
LnCOTTON_P does not Granger cause *LnWHEAT_P*		5.52918	0.0226
LnMILLET_P does not Granger cause *LnMILLED_RICE_P*	54	0.17433	0.6780
LnMILLED_RICE_P does not Granger cause *LnMILLET_P*		1.31917	0.2561
LnSORGHUM_P does not Granger cause *LnMILLED_RICE_P*	54	1.37066	0.2471
LnMILLED_RICE_P does not Granger cause *LnSORGHUM_P*		8.25064	0.0059
LnWHEAT_P does not Granger cause *LnMILLED_RICE_P*	54	4.53110	0.0381
LnMILLED_RICE_P does not Granger cause *LnWHEAT_P*		5.60364	0.0218
LnSORGHUM_P does not Granger cause *LnMILLET_P*	54	0.99540	0.3231
LnMILLET_P does not Granger cause *LnSORGHUM_P*		0.19501	0.6606
LnWHEAT_P does not Granger cause *LnMILLET_P*	54	2.35497	0.1311
LnMILLET_P does not Granger cause *LnWHEAT_P*		0.09127	0.7638
LnWHEAT_P does not Granger cause *LnSORGHAM_P*	54	8.43335	0.0054
LnSORGHUM_P does not Granger cause *LnWHEAT_P*		0.17899	0.6740

Source "Authors' calculation".

4.7. Two-Stage Least Square (2SLS) Method for Endogeneity Problem

Endogeneity is a problem when the explanatory variables correlate with the error term. When an endogeneity problem is found in a model or variables, it is resolved by including an instrumental variable. To identify if an endogeneity problem exists, we applied the 2SLS method to the time series data. In the case of endogeneity in the model, there is a need for instrumental variables. We added agriculture value-added (AVA) as an instrumental variable in our model. Table 8 shows the two-stage least square method for the study variables. The model also shows the Durbin–Watson, J-statistic and second-stage results (SSR) for the study variables.

Table 8. Two-stage least square (2SLS) method.

Dependent Variable: LNCO2_EMISSIONS Method: Two-Stage Least Squares				
Instrument specification: *LnBARLEY LnCORN LnCOTTON LnMILLED_RICE* *LnMILLET LnSORGHUM LnWHEAT LnAVA C*				
Variable	**Coefficient**	**Std. Error**	***t*-Statistic**	**Prob.**
LnGDP	0.057291	0.635302	0.090178	0.9285
LnBARLEY	0.167707	0.171722	0.976616	0.3340
LnCORN	0.738820	0.355009	2.081133	0.0431
LnCOTTON	0.359947	0.199564	1.803670	0.0780
LnMILLED_RICE	−0.518871	0.215862	−2.403720	0.0204
LnMILLET	−0.059969	0.192319	−0.311823	0.7566
LnSORGHUM	0.028461	0.182736	0.155750	0.8769
LnWHEAT	0.559824	0.525935	1.064437	0.2928
C	−0.535170	9.044380	−0.059172	0.9531
R-squared	0.972394	Mean dependent var		10.88533
Adjusted R-squared	0.967486	S.D. dependent var		0.801814
S.E. of regression	0.144579	Sum squared resid		0.940644
F-statistic	197.8743	Durbin-Watson stat		0.873537
Prob(F-statistic)	0.000000	Second-Stage SSR		0.984332
J-statistic	2.41×10^{-32}	Instrument rank		9

4.8. Impulse Response and Variance Decomposition Analysis

Finally, we employed impulse response analysis in which we employ the response of *LnCO₂*, *LnGDP, LnBARLEY, LnCORN, LnCOTTON, LnMILLED RICE, LnMILLET, LnSORGHUM,* and *LnWHEAT* to explain random innovations among them. The random response is not described by the pairwise Granger causality test. The impulse-response of carbon dioxide emissions to Cholesky One S.D. innovations in other variables are displayed in Figure 5.

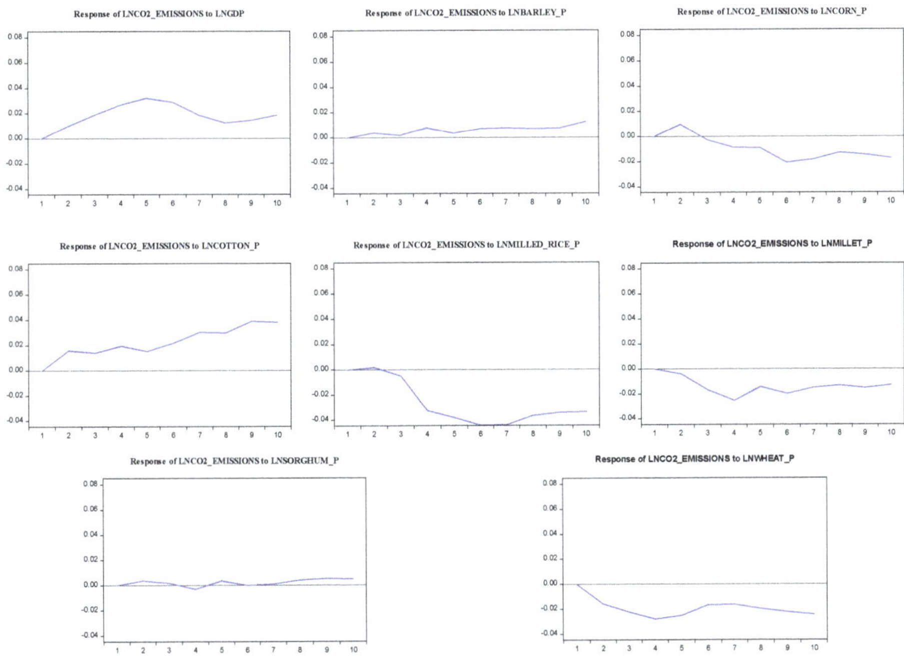

Figure 5. Impulse response of LCO_2 to Cholesky One S.D.

This study employed the variance decomposition method, which estimates the percentage of influence of each independent variable on the error variance of the dependent variable [39]. Figure 5 shows that the response of carbon dioxide emissions to corn production, millet production, milled rice production, sorghum production, and wheat production are insignificant within 10-period horizons. On the other hand, the initial response of carbon dioxide emissions to all other variables, for example, GDP, barley production and cotton production, is significant. On the other hand, a one standard deviation shock to GDP causes carbon dioxide emissions to steadily increase within a 10-period horizon. Similarly, a one standard deviation shock to barley production causes carbon dioxide emissions to gradually increase within a 10-period horizon, while corn production first increases carbon dioxide emissions over a 2-period horizon, and then starts decreasing over a 10-period horizon. A one standard deviation shock to cotton production causes carbon dioxide emissions to exhibit and up-and-down motion within a 10-period horizon.

Figure 6 shows the response of GDP, barley production, corn production, cotton production, milled rice production, millet production, sorghum production and wheat production to carbon dioxide emissions.

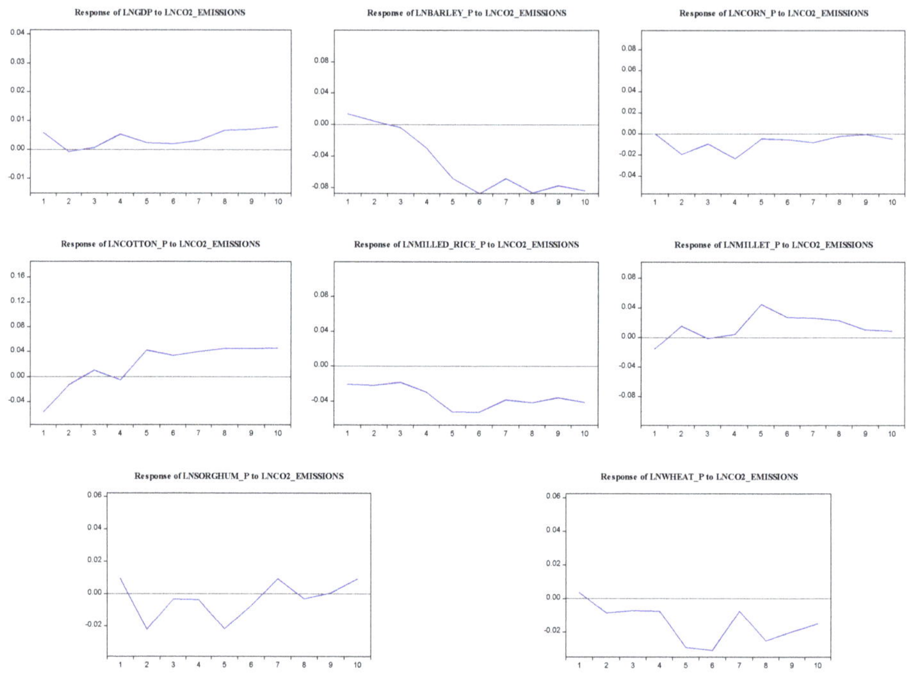

Figure 6. Impulse response of other variables to Cholesky One S.D. Innovations in LCO$_2$.

Table 9 shows the variance decomposition of *LCO$_2$*, *LnGDP*, *LnBARLEY*, *LnCORN*, *LnCOTTON*, *LnMILLED RICE*, *LnMILLET*, *LnSORGHUM* and *LnWHEAT* within a 10-period horizon. The variance decomposition provides evidence of the relative importance of each random innovation in affecting *LnCO$_2$*, *LnGDP*, *LnBARLEY*, *LnCORN*, *LnCOTTON*, *LnMILLED RICE*, *LnMILLET*, *LnSORGHUM* and *LnWHEAT* in the VAR model.

Table 9. Variance decomposition Cholesky ordering; LnCO$_2$_EMISSIONS LnGDP LnBARLEY_P LnCORN_P LnCOTTON_P LnMILLED_RICE_P LnMILLET_P LnSORGHUM_P LnWHEAT_P.

				Variance Decomposition of LnCO$_2$_EMISSIONS:					
Period	*S.E.*	*LnCO$_2$_EMISSIONS*	*LnGDP*	*LnBARLEY_P*	*LnCORN_P*	*LnCOTTON_P*	*LnMILLED_RICE_P*	*LnMILLET_P*	*LnSORGHUM_P*	*LnWHEAT_P*
1	0.06	100.00	0.00	0.00	0.00	0.00	0.00	0.00	0.00	0.00
2	0.09	93.65	0.95	0.03	1.56	0.81	0.01	0.05	0.08	2.85
3	0.13	89.12	0.68	0.29	1.47	0.71	0.41	1.57	0.08	5.68
4	0.16	87.56	0.86	0.19	1.39	0.74	0.47	2.56	0.22	6.01
5	0.18	86.15	1.31	0.17	2.01	0.66	0.40	2.72	0.16	6.42

				Variance Decomposition of LnGDP:					
Period	*S.E.*	*LnCO$_2$_EMISSIONS*	*LnGDP*	*LnBARLEY_P*	*LnCORN_P*	*LnCOTTON_P*	*LnMILLED_RICE_P*	*LnMILLET_P*	*LnSORGHUM_P*	*LnWHEAT_P*
1	0.02	29.19	70.81	0.00	0.00	0.00	0.00	0.00	0.00	0.00
2	0.04	16.85	70.69	0.01	3.98	6.36	0.82	0.43	0.34	0.51
3	0.05	20.05	64.50	0.08	4.31	6.83	2.48	0.41	0.20	1.13
4	0.07	22.61	57.10	0.13	6.66	6.85	3.32	0.89	0.21	2.23
5	0.08	24.28	52.49	0.16	7.48	6.97	3.44	1.63	0.15	3.40

				Variance Decomposition of LnBARLEY_P:					
Period	*S.E.*	*LnCO$_2$_EMISSIONS*	*LnGDP*	*LnBARLEY_P*	*LnCORN_P*	*LnCOTTON_P*	*LnMILLED_RICE_P*	*LnMILLET_P*	*LnSORGHUM_P*	*LnWHEAT_P*
1	0.09	0.49	0.88	98.62	0.00	0.00	0.00	0.00	0.00	0.00
2	0.14	3.07	0.58	64.29	5.27	1.04	0.35	0.02	11.15	14.25
3	0.19	3.82	0.71	52.53	3.69	11.85	0.63	2.66	6.58	17.52
4	0.24	4.58	3.52	45.68	3.67	11.87	0.67	4.05	5.38	20.58
5	0.28	6.31	5.76	44.51	3.84	12.56	0.96	3.99	4.99	17.08

				Variance Decomposition of LnCORN_P:					
Period	*S.E.*	*LnCO$_2$_EMISSIONS*	*LnGDP*	*LnBARLEY_P*	*LnCORN_P*	*LnCOTTON_P*	*LnMILLED_RICE_P*	*LnMILLET_P*	*LnSORGHUM_P*	*LnWHEAT_P*
1	0.10	5.67	0.47	1.08	92.79	0.00	0.00	0.00	0.00	0.00
2	0.14	5.49	0.37	0.74	88.99	0.56	1.11	1.49	1.04	0.19
3	0.16	4.96	0.39	0.68	85.45	1.69	0.92	2.78	2.06	1.07
4	0.19	4.83	0.29	0.75	83.99	2.34	0.74	3.95	1.71	1.40
5	0.21	5.10	0.33	0.61	83.96	2.21	0.60	4.21	1.53	1.44

Table 9. *Cont.*

Variance Decomposition of LnCOTTON_P:									
Period	*S.E.*	*LnCO$_2$_EMISSIONS*	*LnGDP*	*LBARLEY_P*	*LnCORN_P*	*LnCOTTON_P*	*LnMILLED_RICE_P*	*LnMILLET_P*	*LnSORGHUM_P*	*LnWHEAT_P*
1	0.20	2.69	5.47	11.66	4.67	75.51	0.00	0.00	0.00	0.00
2	0.24	1.82	7.70	16.16	4.15	67.59	0.22	0.93	1.09	0.34
3	0.28	1.97	8.72	20.77	3.69	59.26	0.74	1.12	3.40	0.33
4	0.31	3.21	8.15	23.22	3.11	57.21	0.63	1.04	2.84	0.59
5	0.33	4.52	7.76	26.19	3.54	52.69	0.87	1.08	2.86	0.50

Variance Decomposition of LnMILLED_RICE_P:									
Period	*S.E.*	*LnCO$_2$_EMISSIONS*	*LnGDP*	*LnBARLEY_P*	*LnCORN_P*	*LnCOTTON_P*	*LnMILLED_RICE_P*	*LnMILLET_P*	*LnSORGHUM_P*	*LnWHEAT_P*
1	0.11	6.53	10.45	0.02	3.20	9.10	70.70	0.00	0.00	0.00
2	0.15	4.14	7.31	0.52	14.61	11.64	57.57	0.13	0.39	3.69
3	0.18	3.27	6.36	0.40	21.71	8.52	54.20	0.35	0.31	4.88
4	0.20	3.17	7.79	0.39	21.43	7.22	51.84	0.67	0.63	6.86
5	0.23	3.30	9.31	0.65	23.90	6.42	49.11	0.59	0.59	6.12

Variance Decomposition ofLnMILLET_P:									
Period	*S.E.*	*LnCO$_2$_EMISSIONS*	*LnGDP*	*LnBARLEY_P*	*LnCORN_P*	*LnCOTTON_P*	*LnMILLED_RICE_P*	*LnMILLET_P*	*LnSORGHUM_P*	*LnWHEAT_P*
1	0.17	0.51	7.17	1.96	1.86	34.61	4.02	49.87	0.00	0.00
2	0.20	0.38	7.13	3.84	1.74	39.32	3.11	38.61	5.54	0.32
3	0.23	0.54	8.54	3.57	1.41	36.34	2.79	37.53	7.39	1.90
4	0.26	0.50	7.79	3.24	1.12	38.74	2.58	38.06	6.36	1.60
5	0.28	0.44	9.07	4.00	0.99	38.82	2.18	36.75	6.20	1.54

Variance Decomposition of LnSORGHUM_P:									
Period	*S.E.*	*LnCO$_2$_EMISSIONS*	*LnGDP*	*LnBARLEY_P*	*LnCORN_P*	*LnCOTTON_P*	*LnMILLED_RICE_P*	*LnMILLET_P*	*LnSORGHUM_P*	*LnWHEAT_P*
1	0.11	1.85	8.84	2.74	0.54	5.02	0.02	1.20	79.79	0.00
2	0.14	2.80	10.33	3.04	13.22	6.00	0.40	1.16	61.66	1.39
3	0.15	2.68	12.01	4.12	12.92	5.66	1.40	0.98	59.11	1.12
4	0.17	2.26	14.32	3.70	11.91	5.74	1.13	0.85	58.97	1.12
5	0.19	1.98	14.27	5.10	11.16	5.13	1.80	0.85	58.45	1.27

Variance Decomposition of LnWHEAT_P:									
Period	*S.E.*	*LnCO$_2$_EMISSIONS*	*LnGDP*	*LnBARLEY_P*	*LnCORN_P*	*LnCOTTON_P*	*LnMILLED_RICE_P*	*LnMILLET_P*	*LnSORGHUM_P*	*LnWHEAT_P*
1	0.06	2.17	2.71	4.38	1.57	0.24	11.54	6.11	0.04	71.25
2	0.09	5.14	3.74	4.92	6.53	6.39	23.22	14.23	1.65	34.18
3	0.12	3.20	2.78	3.54	8.84	9.89	30.56	15.71	4.11	21.36
4	0.13	2.50	5.68	2.93	14.12	7.92	28.58	18.44	3.19	16.64
5	0.15	3.94	7.63	2.56	15.30	8.74	25.99	19.42	2.53	13.90

5. Conclusions and Policy Implications

This study explored the causal relationship between carbon dioxide emissions, economic growth and agricultural production in Pakistan for the time period from 1960 to 2014. By employing the ARDL optimal model, there was evidence of short-run and long-run associations between gross domestic product, barley, corn, cotton, milled rice, millet, sorghum and wheat to carbon dioxide emissions. The evidence from the unit root tests (ADF, PP and KPSS) showed that all study variables are integrated at $I(1)$. The results of the ARDL bounds test showed that there is a co-integration relationship between all the study variables.

The results of the Granger causality test indicated that there is both unidirectional and bidirectional causality between the study variables. The study also applied the two-stage least square method to describe the endogeneity problem in our variables or model. The paper aimed to employ variance decomposition and Cholesky ordering to investigate the future effect of variables on carbon dioxide emissions in the VAR model.

Agriculture plays a very important role and is considered a backbone in a nation's growth. The government of Pakistan is trying to achieve a healthy living style and increase its economic growth. There is a need to improve agricultural productivity through advanced agriculture production techniques. The country is listed among the countries severely affected by climate change [74] despite being a low producer of CO_2 gasses [75] because of its increasing dependence on agriculture for food and fiber needs [76]. The role of extension services is also very important for spreading updated scientific information to farmers.

Author Contributions: Conceptualization, L.Y.; Data curation, S.A.; Formal analysis, S.A. and M.I.; Methodology, S.A. and T.S.; Research funding—L.G.; Writing—original draft, S.A.; Writing—review & editing, L.Y. and M.I.

Funding: The research is financially supported by the National Natural Sciences Foundation of China (NSFC No. 71873050).

Acknowledgments: The authors are thankful to the Chinese Scholarship Council (CSC) for providing financial assistance to carry-out this research as part of his Ph.D. studies in China. In addition, the authors would also like to extend gratitude to anonymous reviewers for providing helpful suggestions on an earlier draft of this paper.

Conflicts of Interest: The authors declare no conflict of interest.

References

1. Owusu, P.A.; Asumadu-sarkodie, S. CIVIL & ENVIRONMENTAL ENGINEERING | REVIEW ARTICLE A review of renewable energy sources, sustainability issues and climate change mitigation. *Cogent Eng.* **2016**, *15*, 1–14.
2. Asumadu-sarkodie, S.; Owusu, P.A. CIVIL & ENVIRONMENTAL ENGINEERING | REVIEW ARTICLE A review of Ghana's energy sector national energy statistics and policy framework. *Cogent Eng.* **2016**, *3*, 1155274.
3. Asumadu-sarkodie, S.; Owusu, P.A. Feasibility of biomass heating system in Middle East Technical University, Northern Cyprus Campus Feasibility of biomass heating system in Middle East. *Cogent Eng.* **2016**, *3*, 1134304. [CrossRef]
4. Owusu, P.A.; Asumadu-sarkodie, S.; Ameyo, P. CIVIL & ENVIRONMENTAL ENGINEERING | REVIEW ARTICLE A review of Ghana's water resource management and the future prospect CIVIL & ENVIRONMENTAL ENGINEERING | REVIEW ARTICLE A review of Ghana 's water resource management and the future prospect. *Cogent Eng.* **2016**, *3*, 1164275.
5. Mohiuddin, O.; Asumadu-sarkodie, S.; Obaidullah, M. The relationship between carbon dioxide emissions, energy consumption, and GDP: A recent evidence from Pakistan energy consumption, and GDP: A recent evidence. *Cogent Eng.* **2016**, *3*, 1210491. [CrossRef]
6. Asumadu-sarkodie, S.; Owusu, P.A.; Asumadu-sarkodie, S.; Owusu, P.A. Recent evidence of the relationship between carbon dioxide emissions, energy use, GDP, and population in Ghana: A linear regression approach regression approach. *Energy Sour. Part B Econ. Plan. Policy* **2017**, *12*, 495–503. [CrossRef]

7. Asumadu-sarkodie, S.; Owusu, P.A. A review of Ghana ' s solar energy potential. *AIMS Energy* **2016**, *4*, 675–696. [CrossRef]

8. Asumadu-sarkodie, S.; Owusu, P.A.; Asumadu-sarkodie, S.; Owusu, P.A. The potential and economic viability of wind farms in Ghana The potential and economic viability of wind farms in Ghana. *Energy Sour. Part A Recovery Util. Environ. Eff.* **2016**, *38*, 695–701. [CrossRef]

9. Asumadu-sarkodie, S.; Owusu, P.A.; Asumadu-sarkodie, S.; Owusu, P.A. The potential and economic viability of solar photovoltaic power in Ghana in Ghana. *Energy Sour. Part A Recovery Util. Environ. Eff.* **2016**, *38*, 709–716. [CrossRef]

10. FAO. *The stAte of Food Insecurity in the World the Multiple Dimensions of Food Security*; Food and Agriculture Organization of the United Nations: Rome, Italy, 2013.

11. Boko, M.; Niang, I.; Nyong, A.; Vogel, A.; Githeko, A.; Medany, M.; Osman-Elasha, B.; Tabo, R.; Yanda, P.Z. *Africa, Climate change 2007: Impacts, adaptation and vulnerability. Contribution of Working Group II to the Fourth Assessment Report of the Intergovernmental Panel on Climate Change*; Intergovernmental Panel on Climate Change: Geneva, Switzerland, 2018; pp. 433–467.

12. Turner, A.G.; Annamalai, H. Climate change and the South Asian summer monsoon. *Nat. Clim. Chang.* **2012**, *2*, 587–595. [CrossRef]

13. Li, Z.; Jin, X.; Zhao, C.; Wang, J.; Xu, X.; Yang, G.; Li, C.; Shen, J. Estimating wheat yield and quality by coupling the DSSAT-CERES model and proximal remote sensing. *Eur. J. Agron.* **2015**, *71*, 53–62. [CrossRef]

14. el Chami, D.; Daccache, A. Assessing sustainability of winter wheat production under climate change scenarios in a humid climate-An integrated modelling framework. *Agric. Syst.* **2015**, *140*, 19–25. [CrossRef]

15. Jalota, S.K.; Vashisht, B.B.; Kaur, H.; Kaur, S.; Kaur, P. Location specific climate change scenario and its impact on rice and wheat in Central Indian Punjab. *Agric. Syst.* **2014**, *131*, 77–86. [CrossRef]

16. Wilcox, J.; Makowski, D. A meta-analysis of the predicted effects of climate change on wheat yields using simulation studies. *Field Crop. Res.* **2014**, *156*, 180–190. [CrossRef]

17. Özdoğan, M. Modeling the impacts of climate change on wheat yields in Northwestern Turkey. *Agric. Ecosyst. Environ.* **2011**, *141*, 1–12. [CrossRef]

18. Kang, Y.; Khan, S.; Ma, X. Climate change impacts on crop yield, crop water productivity and food security-A review. *Prog. Nat. Sci.* **2009**, *19*, 1665–1674. [CrossRef]

19. IPCC. Climate Change 2014: Synthesis Report. In *Contribution of Working Groups I, II and III to the Fifth Assessment Report of the Intergovernmental Panel on Climate Change*; IPCC: Geneva, Switzerland, 2014.

20. IPCC. Climate Change 2013: The Physical Science Basis. In *Contribution of Working Group I to the Fifth Assessment Report of the Intergovernmental Panel on Climate Change*; Stocker, F.T., Ed.; Cambridge University Press: Cambridge, UK, 2014; p. 1535.

21. Gorst, A.; Groom, B.; Dehlavi, A. Crop productivity and adaptation to climate change in Pakistan Centre for Climate Change Economics and Policy the Environment. *Environ. Dev. Econ.* **2015**, *23*, 679–701. [CrossRef]

22. Kreft, S.; Eckstein, D.; Dorsch, L.; Fischer, L. *Global Climate Risk Index 2016: Who Suffers Most from Extreme Weather Events? Weather-Related Loss Events in 2014 and 1995 to 2014*; Germanwatch Nord-Süd Initiative: Bonn, Germany, 2015.

23. Abid, M.; Scheffran, J.; Schneider, U.A.; Ashfaq, M. Farmers' perceptions of and adaptation strategies to climate change and their determinants: The case of Punjab province, Pakistan. *Earth Syst. Dyn.* **2015**, *6*, 225–243. [CrossRef]

24. Pindyck, R.S. Fat tails, thin tails, and climate change policy. *Rev. Environ. Econ. Policy* **2011**, *5*, 258–274. [CrossRef]

25. Ali, S. Climate Change and Economic Growth in a Rain-Fed Economy: How Much Does Rainfall Variability Cost Ethiopia? Available online: https://ssrn.com/abstract=2018233 (accessed on 8 February 2012).

26. Lobell, D.B.; Field, C.B. Global scale climate-crop yield relationships and the impacts of recent warming. *Environ. Res. Lett.* **2007**, *2*, 014002. [CrossRef]

27. Acaravci, A.; Ozturk, I. On the relationship between energy consumption, CO_2 emissions and economic growth in Europe. *Energy* **2010**, *35*, 5412–5420. [CrossRef]

28. Janjua, P.Z.; Samad, G.; Khan, N. Climate change and wheat production in Pakistan: An autoregressive distributed lag approach. *NJAS Wagening. J. Life Sci.* **2014**, *68*, 13–19. [CrossRef]

29. Asumadu-Sarkodie, S.; Owusu, P.A. The relationship between carbon dioxide and agriculture in Ghana: A comparison of VECM and ARDL model. *Environ. Sci. Pollut. Res.* **2016**, *23*, 10968–10982. [CrossRef] [PubMed]

30. Arshed, N.; Abduqayumov, S. Economic Impact of Climate Change on Wheat and Cotton in Major Districts of Punjab. *Int. J. Econ. Financ. Res.* **2016**, *2*, 183–191.

31. FAO. *Food and Agriculture Organization*; FAO: Roma, Italy, 2014.

32. IPCC. Summary for Policymakers: C. Current knowledge about future impacts. In *Climate Change 2007: Impacts, Adaptation and Vulnerability. Contribution of Working Group II to the Fourth Assessment Report of the Intergovernmental Panel on Climate Change*; Parry, M.L., Ed.; IPCC: Geneva, Switzerland, 2007.

33. Robertsa, M.R.; Whited, T.M. Endogeneity in Empirical Corporate Finance. In *Handbpook of the Econonics of Finance*; Constantinides, G.M., Harris, M., Stulz, R.M., Eds.; Elsevier: Amsterdam, The Netherlands, 2013.

34. Conley, T.G.; Hansen, C.B.; Rossi, P.E. Plausibly exogenous. *Rev. Econ. Stat.* **2012**, *94*, 260–272. [CrossRef]

35. Antonakis, J.; Bendahan, S.; Jacquart, P.; Lalive, R. On making causal claims: A review and recommendations. *Leadersh. Q.* **2010**, *21*, 1086–1120. [CrossRef]

36. Ashley, R.A.; Parmeter, C.F. When is it justifiable to ignore explanatory variable endogeneity in a regression model? *Econ. Lett.* **2015**, *137*, 70–74. [CrossRef]

37. Samuel, O.O.; Sylvia, T.S. Establishing the nexus between climate change adaptation strategy and smallholder farmers' food security status in South Africa: A bi-casual effect using instrumental variable approach. *Cogent Soc. Sci.* **2019**, *5*, 1–12. [CrossRef]

38. HusnaiN, A.S.M.I.; Jan, I.; Mahmood, T. Does Endogeneity Undermine Temperature Impact on Agriculture in South Asia? *Sarhad J. Agric.* **2018**, *34*, 334–341. [CrossRef]

39. Pesaran, M.H.; Shin, Y. An autoregressive distributed-lag modelling approach to cointegration analysis. *Econom. Soc. Monogr.* **1998**, *31*, 371–413.

40. Asumadu-sarkodie, S.; Owusu, P.A.; Asumadu-sarkodie, S.; Owusu, P.A. Forecasting Nigeria's energy use by 2030, an econometric approach approach. *Energy Sour. Part B Econ. Plan. Policy* **2016**, *11*, 990–997. [CrossRef]

41. Asumadu-sarkodie, S.; Owusu, P.A.; Asumadu-sarkodie, S.; Owusu, P.A. Energy use, carbon dioxide emissions, GDP, industrialization, financial development, and population, a causal nexus in Sri Lanka: With a subsequent prediction of energy use using neural network network. *Energy Sour. Part B Econ. Plan. Policy* **2016**, *11*, 889–899. [CrossRef]

42. Asumadu-sarkodie, S.; Owusu, P.A.; Asumadu-sarkodie, S.; Owusu, P.A. Carbon dioxide emission, electricity consumption, industrialization, and economic growth nexus: The Beninese case. *Energy Sour. Part B Econ. Plan. Policy* **2016**, *11*, 1089–1096. [CrossRef]

43. Asumadu-sarkodie, S.; Owusu, P.A.; Asumadu-sarkodie, S.; Owusu, P.A. The causal nexus between energy use, carbon dioxide emissions, and macroeconomic variables in Ghana and macroeconomic variables in Ghana. *Energy Sour. Part B Econ. Plan. Policy* **2017**, *12*, 533–546. [CrossRef]

44. Asumadu-sarkodie, S.; Owusu, P.A.; Asumadu-sarkodie, S.; Owusu, P.A. The causal effect of carbon dioxide emissions, electricity consumption, economic growth, and industrialization in Sierra Leone Sierra Leone. *Energy Sour. Part B Econ. Plan. Policy* **2017**, *12*, 32–39. [CrossRef]

45. Ko, P.; Bekoe, W.; Amuakwa-mensah, F.; Mensah, J.T.; Botchway, E. Carbon dioxide emissions, economic growth, industrial structure, and technical ef fi ciency: Empirical evidence from Ghana, Senegal, and Morocco on the causal dynamics. *Energy* **2012**, *47*, 314–325.

46. Asumadu-sarkodie, S.; Owusu, P.A.; Asumadu-sarkodie, S.; Owusu, P.A. The relationship between carbon dioxide emissions electricity production and consumption in Ghana production and consumption in Ghana. *Energy Sour. Part B Econ. Plan. Policy* **2017**, *6*, 547–558. [CrossRef]

47. Asumadu-sarkodie, S.; Owusu, P.A.; Asumadu-sarkodie, S.; Owusu, P.A. A multivariate analysis of carbon dioxide emissions, electricity consumption, economic growth, financial development, industrialization, and urbanization in Senegal. *Energy Sour. Part B Econ. Plan. Policy* **2017**, *12*, 77–84. [CrossRef]

48. Asumadu-sarkodie, S.; Owusu, P.A. Multivariate co-integration analysis of the Kaya factors in Ghana. *Environ. Sci. Pollut. Res.* **2016**, *23*, 9934–9943. [CrossRef]

49. Asumadu-sarkodie, S. Carbon dioxide emissions, GDP, energy use, and population growth: A multivariate and causality analysis for Ghana, 1971–2013. *Environ. Sci. Pollut. Res.* **2016**, *23*, 13508–13520. [CrossRef]

50. Dell, M.; Jones, B.F.; Olken, B.A. Temperature shocks and economic growth: Evidence from the last half century. *Am. Econ. J. Macroecon.* **2012**, *4*, 66–95. [CrossRef]

51. Pachauri, R.K.; Reisinger, A. *Contribution of Working Groups I, II and III to the Fourth Assessment Report of the Intergovernmental Panel on Climate Change*; IPCC: Geneva, Switzerland, 2014.

52. Parry, M.L.; Canziani, O.F.; Palutikof, J.P.; van der Linden, P.J.; Hanson, C.E. *Climate Change 2007: Impacts, Adaptation and Vulnerability: Contribution of Working Group II to the Fourth Assessment Report of the Intergovernmental Panel*; Intergovernmental Panel on Climate Change: Geneva, Switzerland, 2007.

53. Gallup, J.L.; Sachs, J.D.; Mellinger, A.D. Geography and Economic Development. *Int. Reg. Sci. Rev.* **1999**, *22*, 179–232. [CrossRef]

54. Victor, P.A. Growth, degrowth and climate change: A scenario analysis. *Ecol. Econ.* **2012**, *84*, 206–212. [CrossRef]

55. Gornall, J.; Betts, R.; Burke, E.; Clark, R.; Camp, J.; Willett, K.; Wiltshire, A. Implications of climate change for agricultural productivity in the early twenty-first century. *Philos. Trans. R. Soc. Lond. B. Biol. Sci.* **2010**, *365*, 2973–2989. [CrossRef]

56. Barrios, S.; Ouattara, B.; Strobl, E. The impact of climatic change on agricultural production: Is it different for Africa? *Food Policy* **2008**, *33*, 287–298. [CrossRef]

57. Ayinde, O.E.; Muchie, M.; Olatunji, G.B. Effect of Climate Change on Agricultural Productivity in Nigeria: A Co-integration Model Approach. *J. Hum. Ecol.* **2011**, *35*, 189–194. [CrossRef]

58. Goklany, I.M. Integrated strategies to reduce vulnerability and advance adaptation, mitigation, and sustainable development. *Mitig. Adapt. Strateg. Glob. Chang.* **2007**, *12*, 755–786. [CrossRef]

59. Tol, R.S.J.; Downing, T.E.; Kuik, O.J.; Smith, J.B. Distributional aspects of climate change impacts. *Glob. Environ. Chang.* **2004**, *14*, 259–272. [CrossRef]

60. Dickey, D.A.; Fuller, W.A. Likelihood ratio statistics for autoregressive time series with a unit. *Econometrica* **1981**, *49*, 1057. [CrossRef]

61. Phillips, P.; Perron, P. Testing for a unit root in time series regression. *Biometrika* **1988**, *75*, 335–346. [CrossRef]

62. Taylor, P.; Dickey, D.A.; Fuller, W.A.; Dickey, D.A.; Fuller, W.A. Journal of the American Statistical Association Distribution of the Estimators for Autoregressive Time Series with a Unit Root Distribution of the Estimators for Autoregressive Time Series With a Unit Root. *J. Am. Stat. Assoc.* **1979**, *74*, 427–431.

63. Hashem, Y.P.; Shin, Y.; Smith, R.J. Bounds testing approaches to the analysis. *J. Appl. Econom.* **2001**, *16*, 289–326.

64. Khim, V.; Liew, S. Which Lag Length Selection Criteria Should We Employ? *Econ. Bull.* **2004**, *3*, 1–9.

65. Jalil, A.; Mahmud, S.F. Environment Kuznets curve for CO_2 emissions: A cointegration analysis for China. *Energy Policy* **2009**, *37*, 5167–5172. [CrossRef]

66. Fodha, M.; Zaghdoud, O. Economic growth and pollutant emissions in Tunisia: An empirical analysis of the environmental Kuznets curve. *Energy Policy* **2010**, *38*, 1150–1156. [CrossRef]

67. Lotfalipour, M.R.; Falahi, M.A.; Ashena, M. Economic growth CO_2 emissions and fossil fuels consumption in Iran. *Energy* **2010**, *35*, 5115–5120. [CrossRef]

68. Nasir, M.; Rehman, F.U. Environmental Kuznets Curve for carbon emissions in Pakistan: An empirical investigation. *Energy Policy* **2011**, *39*, 1857–1864. [CrossRef]

69. Saboori, B.; Sulaiman, J.; Mohd, S. Economic growth and CO_2 emissions in Malaysia: A cointegration analysis of the Environmental Kuznets Curve. *Energy Policy* **2012**, *51*, 184–191. [CrossRef]

70. Ozturk, I.; Acaravci, A. The long-run and causal analysis of energy, growth, openness and fi nancial development on carbon emissions in Turkey. *Energy Econ.* **2013**, *36*, 262–267. [CrossRef]

71. Boutabba, M.A. The impact of fi nancial development, income, energy and trade on carbon emissions: Evidence from the Indian economy Mohamed Amine Boutabba. *Econ. Model.* **2014**, *40*, 33–41. [CrossRef]

72. Brown, R.L.; Durbin, J.; Evans, J.M. Techniques for testing the constancy of regression relationships over time. *J. R. Stat. Soc. Ser. B* **1975**, *37*, 149–192. [CrossRef]

73. Granger, C.W.J. Concept of causality. *J. Econom.* **1988**, *39*, 199–211. [CrossRef]

74. Smadja, J.; Aubriot, O.; Puschiasis, O.; Duplan, T.; Hugonnet, M. Climate change and water resources in the Himalayas: Field study in four geographic units of the Koshi. *J. Alp. Res.* **2015**, *103*, 2–22. [CrossRef]

75. Yousuf, I.; Ghumman, A.R.; Hashmi, H.N.; Kamal, M.A. Carbon emissions from power sector in Pakistan and opportunities to mitigate those. *Renew. Sustain. Energy Rev.* **2014**, *34*, 71–77. [CrossRef]
76. Ahmad, M.; Mustafa, G.; Iqbal, M. *Impact of Farm. Households' Adaptations to Climate Change on Food Security: Evidence from Different Agro-Ecologies of Pakistan*; Munich Personal RePEc Archive: Islamabad, Pakistan, 2015; Volume 6.

Article

Emission-Intensity-Based Carbon Tax and Its Impact on Generation Self-Scheduling

Ping Che [1],*, Yanyan Zhang [2] and Jin Lang [2]

[1] Department of Mathematics, College of Sciences, Northeastern University, Shenyang 110819, China
[2] Key Laboratory of Data Analytics and Optimization for Smart Industry (Northeastern University), Ministry of Education, Shenyang 110819, China; zhangyanyan@ise.neu.edu.cn (Y.Z.); langjin@ise.neu.edu.cn (J.L.)
* Correspondence: cheping@mail.neu.edu.cn; Tel.: +86-024-83686202

Received: 26 January 2019; Accepted: 21 February 2019; Published: 26 February 2019

Abstract: We propose an emission-intensity-based carbon-tax policy for the electric-power industry and investigate the impact of the policy on thermal generation self-scheduling in a deregulated electricity market. The carbon-tax policy is designed to take a variable tax rate that increases stepwise with the increase of generation emission intensity. By introducing a step function to express the variable tax rate, we formulate the generation self-scheduling problem under the proposed carbon-tax policy as a mixed integer nonlinear programming model. The objective function is to maximize total generation profits, which are determined by generation revenue and the levied carbon tax over the scheduling horizon. To solve the problem, a decomposition algorithm is developed where the variable tax rate is transformed into a pure integer linear formulation and the resulting problem is decomposed into multiple generation self-scheduling problems with a constant tax rate and emission-intensity constraints. Numerical results demonstrate that the proposed decomposition algorithm can solve the considered problem in a reasonable time and indicate that the proposed carbon-tax policy can enhance the incentive for generation companies to invest in low-carbon generation capacity.

Keywords: generation self-scheduling; emission intensity; carbon tax; mixed integer linear programming

1. Introduction

Thermal-power generation is one of the main sources of carbon dioxide emissions. About 41 percent of global carbon dioxide emissions come from thermal-power generation. Since excessive emission of carbon dioxide leads to global warming and environmental deterioration, it is important to set up a reasonable policy to reduce carbon dioxide emissions from thermal-power generation for the sustainable development of the electric-power industry.

1.1. Literature Review

Three policies are generally implemented to reduce emissions from the electric power industry. The first policy is the emissions-cap policy that specifies a limit or cap on the total quantity of emissions over a certain time period. Related studies include an emission-constrained typical unit commitment problem [1], a tabu search and Benders decomposition based short term unit commitment solution approach [2], emission-constrained generation self-scheduling problems [3,4], the emission-constrained robust self-scheduling of a hydro-thermal generation company [5], a stochastic long-term security-constrained unit commitment formulation [6], and the stochastic self-scheduling of thermal units with emission constraints [7]. The second policy is the emissions-trading scheme (ETS) that allocates each emission entity a specified emission allowance, and allows emission entities to purchase or sell allocated allowances in the emissions-trading market. The ETS was first introduced in European Union [8] and then developed in many other countries. The effects of the ETS on generation scheduling were studied in [9]. Emissions-trading planning for a combined heat and power producer was investigated in [10]. The impacts of the ETS on the Nordic electricity market and electricity consumers were assessed in [11]. An agent-based model was established in [12] to study the potential impact of carbon-emissions trading on the power sector in China. Fuel switching in electricity production under the ETS was discussed in [13]. A computable general equilibrium model was established in [14] to analyze the impact of carbon-allowance allocation on the electric-power industry in China. A review of the carbon-trading market in China was presented in [15]. The third policy is the carbon-tax policy, under which generation companies are obliged to pay for their carbon dioxide emissions according to the carbon-tax rate and the quantity of their emissions. The carbon-tax policy was first implemented in Finland and gradually spread in many other countries, such as Sweden, Denmark, and Germany. A revenue and distributionally neutral approach was described in [16] to design a carbon tax to reduce greenhouse-gas emissions. The mitigation effects of carbon tax on carbon dioxide emissions are comprehensively estimated in [17]. A carbon tax generation self-scheduling model was presented and the effects of generation profits and emissions profiles under different carbon-tax scenarios are analyzed in [18]. Different evolutions of carbon dioxide taxes that might be applied to the national electricity sector in Portugal were studied in [19]. A choice experiment study of Chinese companies was summarized in [20] to measure the choice preferences of Chinese companies to carbon-tax policy. A bibliometric analysis of the carbon-tax literature from 1989 to 2014 was provided in [21].

The emissions-cap policy has the advantage of regulating the total quantity of emissions, but lacks impetus from market forces. The ETS has the advantages of broad participation, international equity, and controlling the total quantity of emissions [22], but has to face the impact of large uncertainties from emissions-trading platforms on policy efficiency [16]. Compared with the emissions-cap policy and the ETS, the carbon-tax policy not only provides a price signal to impel emitters to develop low-carbon technologies or cleaner energy sources [23], but can also be implemented without trading platforms; consequently, it is highly recommended by economists and organizations [17].

1.2. Work and Contribution of the Paper

A challenge of carbon-tax policy is how to determine the tax rate, which is the levied charge per unit quantity of emissions. If the tax rate is low, emissions cannot be effectively reduced. If the tax rate is high, both production and economic profits are badly hurt. In the previous literature, the tax rate is usually a designated constant [18,24,25]. However, the use of a constant tax rate does not take into account the difference in pollution levels between generation companies, and lacks flexibility in incentives to reduce emissions. It was indicated in [26] that changing the tax rate based on energy efficiency is more effective than applying the same tax rate to all manufacturers. It was suggested in [27] to allow the tax rate to vary among generation units.

Inspired by the idea in [26,27], we propose in this paper an emission intensity-based carbon tax policy in the context of the electric-power industry. Emission intensity is the quantity of emissions per unit power generation. It is an index to measure the pollution level of power generation. The higher emission intensity is, the more serious the pollution in power generation is. In order to improve controlling power-generation pollution, and impel generation companies to invest in low-carbon power generation, we provide different carbon-price signals for different power-generation levels. That is, we adjust the tax rate according to the pollution level of power generation. The more serious the pollution in power generation is, the higher the assigned tax rate is. Therefore, the tax rate is designed to stepwise increase as emission intensity of power generation increases. By introducing emission intensity, the proposed carbon-tax policy increases taxation on high pollution production while balancing generation contribution and environmental pollution in determining the tax rate.

To analyze the effect of the proposed carbon-tax policy, we investigate the impact of the policy on generation self-scheduling decisions. The generation self-scheduling problem plays an important role in the daily-operation planning of generation companies in the deregulated electricity market.

A typical generation self-scheduling problem is deciding the operation of generation units according to electricity prices and the physical characteristics of generation units, with the objective of maximizing total profits [28,29]. Effective generation self-scheduling can promote the economic operation of generation companies. Under the proposed carbon-tax policy, generation companies need to not only adjust power output according to price signal, but also reduce emissions according to the carbon-tax policy, which brings a new challenge for generation self-scheduling decisions.

Our work is different from [26,27] in the following two aspects. First, the mechanisms for designing the tax rate are different. The tax rate was designed based on game theory in both [26] and [27], while it is designed by explicitly formulating the relation function between tax rate and emission intensity in our work. Second, the involved models are different. A simple numerical example was considered in [26], a simplified economic dispatch model omitting time-coupled unit-operation constraints was considered in [27], while an explicit generation self-scheduling model that includes conventional unit-operation constraints and decides not only economic dispatch but also unit commitment is considered in our work. A tax rate that varies based on the quantity of emissions was presented in [30]. Our work is different from [30] in that tax-rate variation in our work is emission intensity-based. Compared with [30], our work in determining tax rate takes into account not only generation emissions but also generation contribution. The major contributions of the paper can be summarized as follows:

(1) A carbon-tax policy with variable tax rate is proposed with regard to the electric-power industry. Tax rate is designed to increase stepwise with the increase of power-generation emission intensity, which can strengthen tax collection from high-pollution generation companies and balance electricity supply and environmental pollution in determining tax rate.
(2) The impact of the proposed carbon-tax policy is investigated by formulating the generation self-scheduling problem under the proposed carbon-tax policy as a mixed integer nonlinear programming (MINLP) model and developing a decomposition algorithm to solve the problem.
(3) Numerical experiments are carried out to test the performance of the proposed algorithm and discuss the effectiveness of the proposed carbon tax policy.

The remainder of this paper is organized as follows: Formulation of the generation self-scheduling problem under the proposed carbon-tax policy is described in Section 2. The solution approach to solving this problem is developed in Section 3. Numerical experiments and results are presented in Section 4. The conclusions of the paper are provided in Section 5.

2. Problem Description and Formulation

2.1. Parameteres and Decision Variables

The parameters and decision variables used in the formulation are defined as follows:

Parameters

n	Number of generation units.
T	Scheduling horizon.
M	Number of carbon-tax rate values.
τ_i^D / τ_i^U	Minimum down-time/up-time of generation unit i.
R_i^D / R_i^U	Ramp-down/ramp-up rate of generation unit i.
S_i^D / S_i^U	Shutdown/startup ramp rate of generation unit i.
P_i^L / P_i^U	Minimum/maximum power output of generation unit i.
α_{ih} / β_{ih}	Slope/intercept of the h-th segment line for the fuel-cost curve of generation unit i.
B_i	Startup cost of generation unit i.
r_i	Emission coefficient of generation unit i.
μ	Emission intensity.
ρ	Carbon-tax rate.
w_m	The m-th carbon-tax rate value.

Q_m	Upper bound of emission intensity corresponding to the m-th carbon-tax rate value.
λ_t	Expected electricity price in hour t.
u_{i0}	Binary parameter to indicate the initial on/off status of generation unit i.
p_{i0}	Generation output of generation unit i in hour 0.
σ_i	Last hour of the periods during which the on/off status of generation unit i is determined by the initial status of the generation unit.

Decision Variables

u_{it}	Binary decision variable to indicate on/off status of generation unit i in hour t.
v_{it}	Binary decision variable to indicate if generation unit i is started up in hour t.
p_{it}	Generation output of generation unit i in hour t.
ϕ_{it}	Fuel cost of generation unit i in hour t.
δ_m	Binary decision variable to indicate whether carbon-tax rate equals m-th carbon-tax rate value.

2.2. Description and Formulation of Carbon-Tax Policy

For the carbon tax policy, we consider M levels for the tax rate as illustrated in Figure 1. Each level corresponds to a tax-rate value and a range of emission intensity. Each tax rate value provides a carbon-tax signal for the corresponding emission-intensity range.

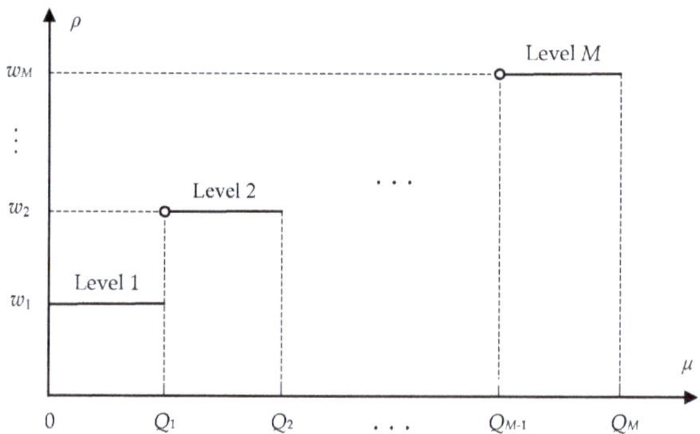

Figure 1. Tax rate with M levels.

Tax rate is variable and equals a tax-rate value according to the range to which emission intensity belongs. The relationship between tax rate and emission intensity can be formulated by a step-up function as follows:

$$\rho = \begin{cases} w_1, & \mu \leq Q_1 \\ w_m, & Q_{m-1} < \mu \leq Q_m, \quad m = 2, \ldots, M-1 \\ w_M, & \mu > Q_{M-1} \end{cases} \tag{1}$$

where $0 \leq w_1 \leq w_2 \leq \cdots \leq w_M$, and emission intensity μ over the scheduling horizon is determined by the ratio of the quantity of total emissions [1,9] to the total generation level as follows:

$$\mu = \begin{cases} \sum_{i=1}^{n} \sum_{t=1}^{T} r_i \phi_{it} / \sum_{i=1}^{n} \sum_{t=1}^{T} p_{it}, & \text{if } \sum_{i=1}^{n} \sum_{t=1}^{T} p_{it} > 0 \\ 0, & \text{otherwise} \end{cases} \tag{2}$$

$0 < Q_1 < Q_2 < \cdots < Q_{M-1} < Q_M$, and Q_M is the upper bound of possible emission intensity. Expression (1) shows that tax rate increases stepwise as emission intensity increases. Particularly, when M = 1, the tax rate is constant. Therefore, the proposed carbon tax policy is a generalization of the typical carbon tax policy with a constant tax rate.

2.3. Generation Self-Scheduling Model under the Proposed Carbon-Tax Policy

To investigate the effect of the proposed carbon tax policy, we consider the generation self-scheduling problem under the proposed carbon-tax policy, which is described as follows. A generation company schedules a certain number of thermal generation units and sells the produced electricity at the market price. The objective is to maximize generation profits. The generation company is obliged to pay for generation emissions in accordance with the proposed carbon-tax policy. The scheduling of generation units needs to satisfy operating constraints such as minimum up-time and down-time constraints, generation-capacity constraints, and ramp-rate constraints.

Based on the above description, the generation self-scheduling problem under consideration can be formulated as follows:

(P)

$$\max \sum_{t=1}^{T} \lambda_t \sum_{i=1}^{n} p_{it} - \sum_{i=1}^{n} \sum_{t=1}^{T} (\phi_{it} + B_i v_{it}) - \rho \sum_{i=1}^{n} \sum_{t=1}^{T} r_i \phi_{it} \tag{3}$$

subject to (1), (2), and:

$$u_{it} = u_{i0}, \quad \forall i, \quad t = 1, \ldots, \sigma_i \tag{4}$$

$$v_{it} = 0, \quad \forall i, \quad t = 1, \ldots, \sigma_i \tag{5}$$

$$u_{it} - u_{i,t-1} \le u_{ik}, \quad \forall i, \quad k = t+1, \ldots, \min\left\{T, \tau_i^U + t - 1\right\}, \quad t = \sigma_i + 1, \ldots, T \tag{6}$$

$$u_{i,t-1} - u_{it} \le 1 - u_{ik}, \quad \forall i, \quad k = t+1, \ldots, \min\left\{T, \tau_i^D + t - 1\right\}, \quad t = \sigma_i + 1, \ldots, T \tag{7}$$

$$u_{it} - u_{i,t-1} \le v_{it}, \quad \forall i, \quad t = \sigma_i + 1, \ldots, T \tag{8}$$

$$u_{it} P_i^L \le p_{it} \le u_{it} P_i^U, \quad \forall i, \quad \forall t \tag{9}$$

$$-R_i^D u_{i,t+1} - S_i^D (1 - u_{i,t+1}) \le p_{i,t+1} - p_{it} \le R_i^U u_{it} + S_i^U (1 - u_{it}), \quad \forall i, \quad t = 0, \ldots, T - 1 \tag{10}$$

$$\phi_{it} \ge \alpha_{ih} p_{it} + \beta_{ih} u_{it}, \quad \forall h, \quad \forall i, \quad \forall t \tag{11}$$

$$\phi_{it}, p_{it} \ge 0, \quad \forall i, \quad \forall t \tag{12}$$

$$u_{it}, v_{it} \in \{0, 1\}, \quad \forall i, \quad \forall t. \tag{13}$$

In formulation (P), function (3) represents generation profits, which consist of generation revenue, generation cost, and paid tax for generation emissions. Constraints (4) and (5) represent the impact of the initial statuses of generation units on the on/off statuses, and the start-up actions of generation units over the scheduling horizon, respectively [31,32]. Constraints (6) and (7) represent the minimum up-time and down-time requirements of generation units, respectively. Constraints (8) represent the relationship between the on/off statuses and the start-up actions of generation units. Constraints (9) represent the range of output power for the committed generation units. Constraints (10) define the ramp-rate limits to represent the relationship between output levels in adjacent hours. Constraints (11) represent the piecewise linear approximation of the quadratic fuel cost function. Constraints (12) and (13) provide the value fields of the decision variables.

3. Solution Methodology

Formulation (P) includes continuous variables, integer variables, a piecewise expression in the function (1), and a bilinear term in the objective function (3), which corresponds to an MINLP model and makes it impossible to directly solve the problem by using a commercial solver such as CPLEX. Therefore, it is necessary to develop an efficient algorithm to solve the problem. According to the characteristics of (P), we develop the following decomposition algorithm to solve the problem.

3.1. Reformulation of the Problem

First, by introducing binary variables δ_m, $m = 1, \ldots, M$, to indicate whether the tax rate equals the m-th value or not, we can obtain the integer linear formulation of the tax rate as follows:

$$\rho = \sum_{m=1}^{M} \delta_m w_m \tag{14}$$

where:

$$\sum_{m=1}^{M} \delta_m = 1 \tag{15}$$

$$\delta_m \sum_{i=1}^{n} \sum_{t=1}^{T} r_i \phi_{it} \leq \delta_m Q_m \sum_{i=1}^{n} \sum_{t=1}^{T} p_{it}, \quad m = 1, \ldots, M-1 \tag{16a}$$

$$\delta_m \sum_{i=1}^{n} \sum_{t=1}^{T} r_i \phi_{it} > \delta_m Q_{m-1} \sum_{i=1}^{n} \sum_{t=1}^{T} p_{it}, \quad m = 2, \ldots, M \tag{16b}$$

$$\delta_m \in \{0, 1\}, \quad \forall m. \tag{17}$$

Then, based on the above expressions, we can reformulate the considered generation self-scheduling problem as:

(P1)

$$\max \sum_{t=1}^{T} \lambda_t \sum_{i=1}^{n} p_{it} - \sum_{i=1}^{n} \sum_{t=1}^{T} (\phi_{it} + B_i v_{it}) - \sum_{m=1}^{M} \delta_m w_m \sum_{i=1}^{n} \sum_{t=1}^{T} r_i \phi_{it} \tag{18}$$

subject to (4)–(13), and (15)–(17).

Finally, according to the formulation of the objective function (18), we replace constraints (16b) with:

$$\delta_m \sum_{i=1}^{n} \sum_{t=1}^{T} r_i \phi_{it} \geq \delta_m Q_{m-1} \sum_{i=1}^{n} \sum_{t=1}^{T} p_{it}, \quad m = 2, \ldots, M \tag{19}$$

without changing the optimal solution of (P1).

3.2. Decomposition Algorithm

For an arbitrary m, we can obtain the following formulation by fixing $\delta_m = 1$ and $\delta_{m'} = 0$, $m' \neq m$, in (P1):

(P$_m$)

$$\max \sum_{t=1}^{T} \lambda_t \sum_{i=1}^{n} p_{it} - \sum_{i=1}^{n} \sum_{t=1}^{T} (\phi_{it} + B_i v_{it}) - w_m \sum_{i=1}^{n} \sum_{t=1}^{T} r_i \phi_{it} \tag{20}$$

subject to constraints (4)–(13), and:

$$\sum_{i=1}^{n} \sum_{t=1}^{T} r_i \phi_{it} \leq Q_m \sum_{i=1}^{n} \sum_{t=1}^{T} p_{it}, \text{ if } 1 \leq m \leq M-1 \tag{21a}$$

$$Q_{m-1} \sum_{i=1}^{n} \sum_{t=1}^{T} p_{it} \leq \sum_{i=1}^{n} \sum_{t=1}^{T} r_i \phi_{it}, \text{ if } 2 \leq m \leq M. \tag{21b}$$

Formulation (P$_m$) corresponds to a generation self-scheduling problem with a constant tax rate w_m and emission-intensity constraints (21a) and (21b). It is a mixed integer linear programming (MILP) model and can be solved optimally by a commercial solver if the feasible domain determined by (4)–(13), (21a), and (21b) is not empty.

The above observation motivates us to enumerate all feasible δ_m to decompose (P1) into M subproblems (P$_m$) and develop the following algorithm:

Step 0   Initialization: set $u\_best_{it} = p\_best_{it} = 0$ for all i and all t, $m = 1$, $m\_best = 1$, and $f\_best = M_f$ where M_f is a large-enough positive number.

Step 1   Solve (P_m) by calling an MILP solver. If (P_m) can be optimally solved, denote the optimally objective function value as f_m. If $f_m < f\_best$, update f_best, m_best, and $\{u\_best_{it}, p\_best_{it}, i = 1, \ldots, n, t = 1, \ldots, T\}$ with f_m, m, and the optimal unit commitment and economic dispatch decisions of (P_m), respectively.

Step 2   If $m = M$, set

$$\delta_m = \begin{cases} 1, & \text{if } m = m\_best \\ 0, & \text{otherwise} \end{cases}$$

and stop. Otherwise, set $m = m + 1$ and go to Step 1.

Using the above algorithm, we can obtain optimal objective function value f_best, and optimal unit commitment and economic dispatch decisions $\{u\_best_{it}, p\_best_{it}, i = 1, \ldots, n, t = 1, \ldots, T\}$ for (P1).

4. Numerical Results

In this section, we present the numerical experiments in four parts. In the first part, we use test cases of different sizes to test the performance of the proposed algorithm. In the second part, we test the impact of tax-rate values on generation self-scheduling and provide a method to choose appropriate tax-rate values. In the third part, we test the effect of the proposed carbon-tax policy on emission reduction. In the fourth part, we compare the proposed carbon-tax policy with carbon-tax policy with a constant tax rate. The test cases are described as follows:

(1) In the first part of the experiments, the scheduling horizon is set to 24, 72, 120, and 168 h, respectively, and the number of generation units is set to 10, 40, 70, and 100, respectively. The combination of the two configurations forms 16 problem sizes. Under each problem size, ten test cases are randomly generated. For each test case, parameters for startup costs and emission coefficients are uniformly generated from the ranges provided in Table 1, in which ranges for startup costs are based on those for cool-start fuel costs in [33] and ranges for emission coefficients are based on data in [9]. Other unit parameters are uniformly generated from ranges provided in [33].

Table 1. Value ranges for startup costs and emission coefficients.

Parameter (Unit)	Value Range
B_i (\$)	
If $P_i^U < 600$	$(4P_i^U, 6P_i^U)$
If $P_i^U \geq 600$	$(2P_i^U, 4P_i^U)$
r_i (t/\$)	
If $P_i^U < 600$	$(0.05, 0.09)$
If $P_i^U \geq 600$	$(0.005, 0.04)$

(2) In the remaining three parts of the experiments, the test is performed by using the ten sets of generation-unit data corresponding to the problem size of 10 generation units and 24 h in the first part of the experiments. To clearly show the impact of the proposed carbon-tax policy on generation self-scheduling, the initial status of each generation unit is modified so that the on/off statuses of the generation unit during the scheduling horizon are not affected by the initial status of the generation unit. For each set of generation-unit data, the scheduling horizon is set to 24, 72, 120, and 168 h, respectively. The combination of the generation-unit data and the scheduling horizon forms 40 test cases.

(3) The number of tax rate values can be set according to the actual requirement. If the number of tax rate values is too small, pollution levels cannot be well-distinguished. If the number of tax rate values is too large, it is inconvenient to implement carbon-tax policy. As a tradeoff between the two aspects, the number of tax rate values is set as five in the experiments. Based on the range of

emission intensity, Q_m is set to $0.1m$ and w_m is set to $m\Delta$, in which tax-rate increment Δ is set to 15 without loss of generality in the first part of the experiments; its value is discussed in the second part of the experiments. In practice, tax-rate value w_m can be set in other form according to the requirements. We use Pennsylvania–New Jersey–Maryland (PJM) Interconnection Real Time data from 2005 to 2006 [34] to estimate electricity prices. The estimated price data for 168 h are presented in Figure 2. A ten-piece piecewise linear function is used to approximate each fuel cost function.

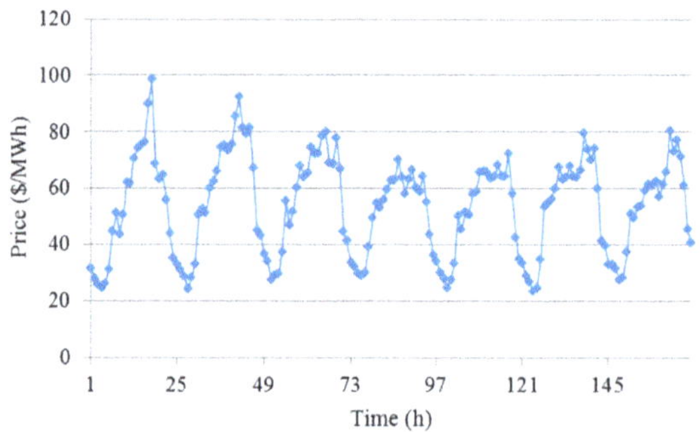

Figure 2. Estimated electricity price data for 168 hours ($/MWh).

The solution algorithm is tested under the computing environment of 3.40 GHz Intel Core i7 CPU and 16.0 GB memory. The MILP involved is solved by calling CPLEX 12.5.

4.1. Performance of Proposed Algorithm

To test the performance of the proposed algorithm, the MILP approach in which (P1) is transformed into an MILP model by using the method in [35] is also used to solve the test cases. The computation time and the number of cases that can be solved optimally are reported in Table 2, in which each computation time is the average of solvable test cases in the same problem size. If no test cases in a problem size can be solved optimally, the average computation time is not reported for that problem size. The number of cases that can be solved optimally in the same problem size is denoted by N. From Table 2, we can obtain the following observations:

(1) For the MILP approach, the computation time increases exponentially as the problem size increases. All cases can be solved optimally for the problem in small sizes corresponding to 10×24, 40×24, 10×72, 10×120 and 10×168. Only partial cases can be solved optimally for the problem in medium sizes corresponding to 70×24, 100×24, 40×72, 70×72, 100×72, 40×120, 70×120, and 40×168. CPLEX is out of memory for all cases for the problem in large sizes corresponding to 100×120, 70×168, and 100×168. The average number of cases that can be solved optimally in the same problem size is 5.

(2) Compared with the MILP approach, the proposed algorithm requires a shorter computation time. All the cases in all problem sizes can be solved optimally by the proposed algorithm. The average computation time is 55.77 s, and the longest computation time is 288.06 s.

The above observations indicate that the proposed algorithm is more effective than the MILP approach for solving the considered problem and can solve the considered test cases in a reasonable time.

Table 2. The computation time and the number of cases that can be solved optimally for the MILP approach and the proposed algorithm.

No.	Size ($n \times T$)	MILP		Proposed	
		Time (s)	N [a]	Time (s)	N [a]
1	10×24	1.40	10	1.04	10
2	40×24	23.56	10	4.42	10
3	70×24	131.35	8	8.76	10
4	100×24	136.17	7	14.07	10
5	10×72	8.79	10	3.19	10
6	40×72	101.60	4	18.03	10
7	70×72	149.65	1	38.45	10
8	100×72	468.42	3	68.39	10
9	10×120	30.47	10	5.97	10
10	40×120	365.90	2	42.42	10
11	70×120	284.56	2	77.52	10
12	100×120	-	0	131.72	10
13	10×168	116.63	10	9.74	10
14	40×168	363.93	6	64.14	10
15	70×168	-	0	116.33	10
16	100×168	-	0	288.06	10
	Average	-	5	55.77	10

[a] The number of cases that can be solved optimally.

4.2. Impact of Tax-Rate Values on Generation Self-Scheduling

To analyze the impact of the tax-rate values, w_m, $m = 1, \ldots, 5$, on generation self-scheduling, we test the model under different w_m settings by allowing tax-rate increment Δ to vary within $\{0, 5, 10, 15, \ldots, 75\}$. For each tax-rate increment, test cases are solved and four indices are obtained in the optimal solution of each test case, including total profits, total generation level, the quantity of total emissions, and emission intensity over the scheduling horizon. The variations of the four indices, along with the increase of the tax-rate increment, are shown in Figure 3, in which each point depicts the average of the ten test cases in the same problem size. From Figure 3, it can be observed that the four indices show different sensitivity to the change of tax-rate increment as follows:

(1) As tax-rate increment increases, total profits decrease and the sensitivity of the total profits to the tax-rate-increment change gradually decreases.

(2) As tax-rate increment increases, total generation level, the quantity of total emissions, and emission intensity decrease. The decrease rates of the three indices present oscillating behavior when tax-rate increment increases from 0 to 25, and relatively stable behavior when the tax rate increment increases from 25 to 75.

(3) Because emission intensity is the ratio of total emissions quantity to total generation level, the decrease of emission intensity with the increase of tax-rate increment indicates that the quantity of total emissions is more sensitive to the increase of tax-rate increment than total generation level.

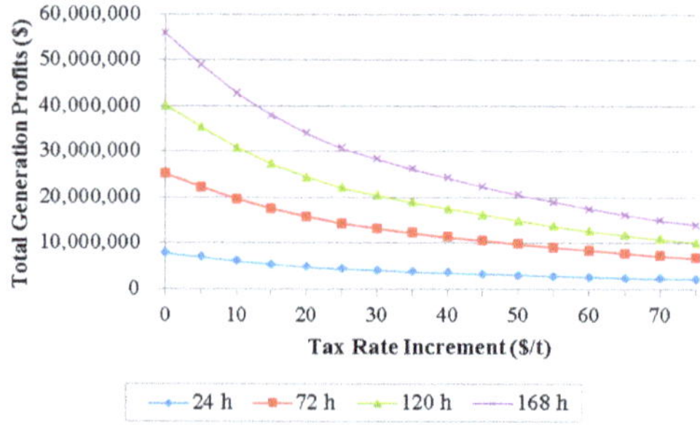

(**a**) Variation of total generation profits with the increase of tax-rate increment.

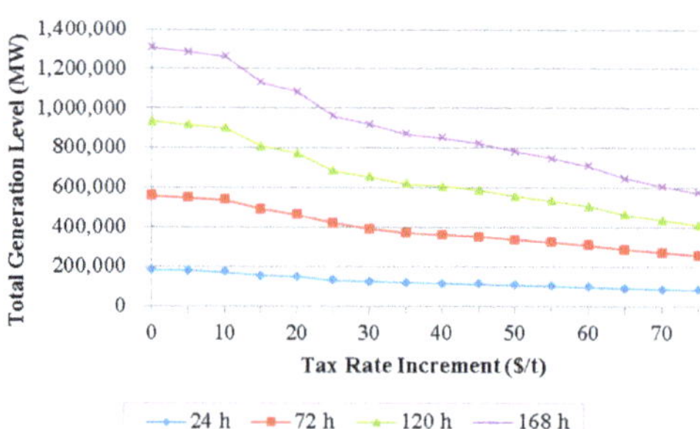

(**b**) Variation of total generation level with the increase of tax-rate increment.

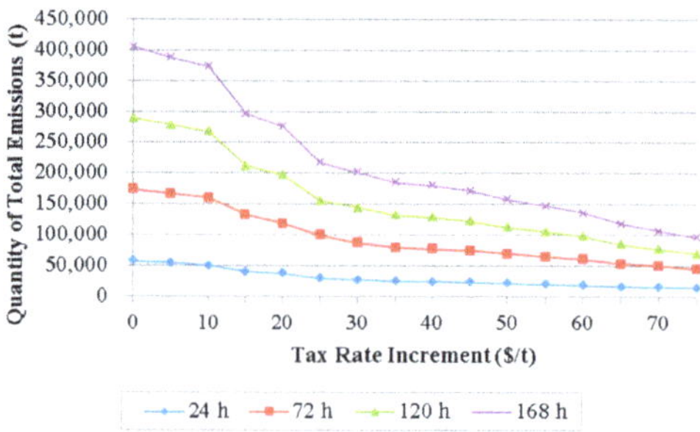

(**c**) Variation of the quantity of total emissions with the increase of tax-rate increment.

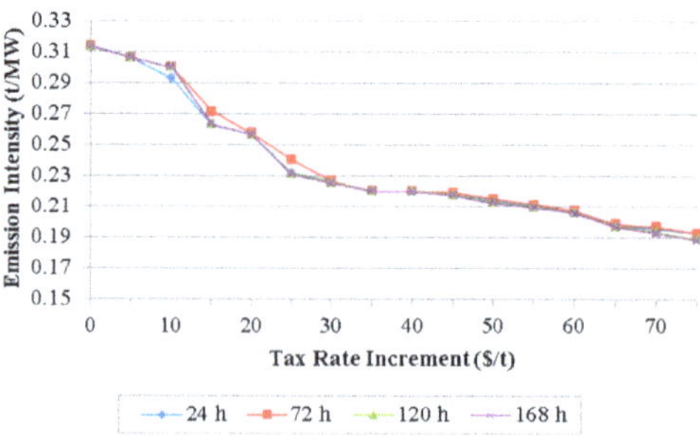

(**d**) Variation of emission intensity with the increase of tax-rate increment.

Figure 3. Variations of total generation profits, total generation level, the quantity of total emissions, and emission intensity with the increase of tax-rate increment.

Based on the above observations, we design the following method to choose an appropriate tax-rate increment. Note that a reduction in emission intensity is at the expense of generation profits. To evaluate profit loss in reducing unit emission intensity, we define the ratio of profits-reduction percentage to emission intensity reduction percentage as follows:

$$\theta_\Delta = \frac{(Prof_0 - Prof_\Delta)/Prof_0}{(\mu_0 - \mu_\Delta)/\mu_0},$$

in which $Prof_0$ and μ_0 are total profits and emission intensity under no carbon-tax policy (corresponding to $\Delta = 0$), respectively, and $Prof_\Delta$ and μ_Δ are total profits and emission intensity corresponding to a nonzero tax-rate increment Δ, respectively, in which $\Delta \in \{5, 10, 15, \dots, 75\}$. The variation of θ_Δ with the increase of tax-rate increment Δ is presented in Figure 4, in which each point depicts the average of all test cases. From Figure 4, it can be observed that θ_Δ presents obvious decline behavior when tax-rate increment increases from 5 to 15, reaches the first local minimum value at $\Delta = 15$, presents oscillating behavior when tax-rate increment is larger than 15, and reaches the minimum at $\Delta = 25$. The difference between θ_{15} and θ_{25} is small. To balance between emission intensity and generation profits, tax-rate increment is set to 15 for the test cases. Correspondingly, tax-rate value w_m is set to $15m$, $m = 1, \dots, 5$, which is used in the following experiments.

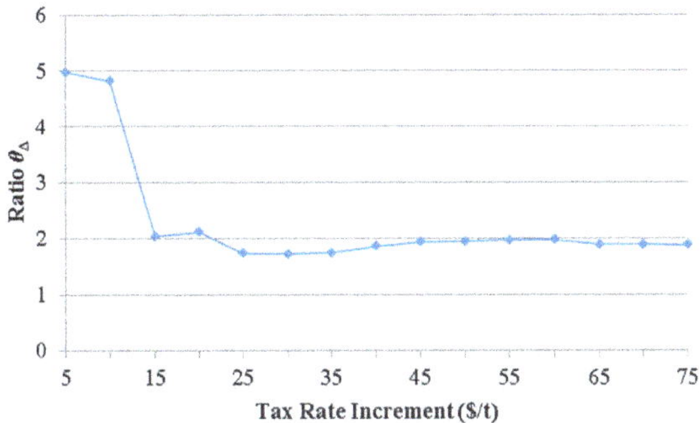

Figure 4. Variation of θ_Δ with the increase of tax-rate increment.

4.3. Effect of Proposed Carbon Tax Policy on Emission Reduction

To test the effect of the proposed carbon-tax policy on emission reduction, we compare the quantity of total emissions and emission intensity under the proposed carbon-tax policy with those under no carbon-tax policy, respectively. The quantity of total emissions and emission intensity under the proposed carbon-tax policy are reported in columns 4 and 5 of Table 3, respectively, and those under no carbon-tax policy are reported in Table 4. From the results, we can make the following observations:

(1) For the quantity of total emissions, the average value under the proposed carbon-tax policy is 170,037.51 t, whereas the average value under no carbon-tax policy is 230,946.30 t. The ratio between the two quantities is 0.74. The results show that the quantity of total emissions is reduced by 26% by adopting the proposed carbon-tax policy.

(2) For emission intensity, the average value under the proposed carbon-tax policy is 0.26, whereas the average value under no carbon-tax policy is 0.31. The ratio between the two emission intensities is 0.84. The results show that the quantity of emissions per unit power generation is reduced by 16% by adopting the proposed carbon-tax policy.

The above observations indicate the effectiveness of the proposed carbon-tax policy in reducing total generation emissions and the pollution level of power generation.

Table 3. Total generation profits, total generation level, the quantity of total emissions, and emission intensity under the proposed carbon-tax policy.

T (h)	Generation Profits ($)	Generation Level (MW)	Quantity of Emissions (t)	Emission Intensity (t/MW)
24	5,423,040.63	156,285.49	40,663.25	0.26
72	17,369,095.36	491,025.77	131,917.18	0.27
120	27,175,633.46	807,196.34	211,287.38	0.26
168	37,811,819.12	1,131,651.47	296,282.21	0.26
Average	21,944,897.14	646,539.77	170,037.51	0.26

Table 4. The quantity of total emissions and emission intensity under no carbon-tax policy.

T (h)	Quantity of Emissions (t)	Emission Intensity (t/MW)
24	56,966.69	0.31
72	172,953.10	0.31
120	288,939.50	0.31
168	404,925.90	0.31
Average	230,946.30	0.31

4.4. Comparison between Proposed Carbon-Tax Policy and Carbon-Tax Policy with a Constant Tax Rate

To show the difference between the proposed carbon-tax policy and carbon-tax policy with a constant tax rate, we make the following comparisons. First, we compare the optimal solutions under the two carbon-tax policies without changing the test-case parameter setting. Second, we adjust value ranges for emission coefficients and compare the sensitivities of the optimal solutions to the change of emission coefficients under the two carbon-tax policies. For the carbon-tax policy with a constant tax rate, five constant tax rates are considered, with values set from w_1 to w_5, respectively.

4.4.1. Comparison between Two Carbon-Tax Policies without Changing Parameter Setting

The optimal solution under each carbon-tax policy is presented by the same indices used in Section 4.2. The results under the proposed carbon-tax policy are reported in Table 3, and those under the carbon-tax policy with a constant tax rate corresponding to different settings of the constant tax rate are reported in Table 5.

From Tables 3 and 5, we can make the following observations:

(1) Average emission intensity under the proposed carbon-tax policy is 0.26, which indicates that the average tax rate corresponding to the proposed carbon-tax policy is equal to w_3. Compared with the carbon-tax policy with a constant tax rate of no more than w_3 (w_1, w_2, or w_3), the proposed carbon-tax policy corresponds to lower generation profits and generation level, but much-reduced emission quantity and intensity.

(2) Compared with the carbon-tax policy with relatively large constant tax rate w_4, the proposed carbon-tax policy corresponds to a lower generation level, and emission quantity and intensity, but higher generation profits.

(3) Compared with the carbon-tax policy with the largest constant tax rate w_5, the proposed carbon-tax policy is corresponding to higher generation level and quantity of emissions, but an equal emission intensity and much-increased generation profits.

The above comparison indicates the superiority of the proposed carbon-tax policy over the carbon-tax policy with a constant tax rate in comprehensive consideration of generation profits, generation level, emission quantity, and emission intensity.

Table 5. Total generation profits, total generation level, the quantity of total emissions, and emission intensity under the carbon-tax policy with a constant tax rate.

T (h)	Tax Rate ($/t)	Generation Profits ($)	Generation Level (MW)	Quantity of Emissions (t)	Emission Intensity (t/MW)
24	w_1	7,042,972.81	183,806.99	56,954.39	0.31
	w_2	6,212,003.63	178,441.98	53,104.73	0.30
	w_3	5,453,691.95	168,627.56	47,995.38	0.29
	w_4	4,776,296.57	156,731.18	42,564.42	0.28
	w_5	4,181,646.67	144,142.39	36,981.80	0.26
72	w_1	22,473,333.53	558,384.99	172,915.76	0.31
	w_2	19,923,699.57	547,249.87	164,666.40	0.30
	w_3	17,556,184.34	522,160.40	150,658.46	0.29
	w_4	15,414,841.59	490,051.75	135,105.06	0.28
	w_5	13,513,722.19	453,420.58	117,811.99	0.26
120	w_1	35,631,936.10	932,979.61	288,892.02	0.31
	w_2	31,363,452.90	915,292.42	275,542.84	0.30
	w_3	27,399,348.57	873,602.74	251,806.51	0.29
	w_4	23,827,289.19	816,426.34	222,986.41	0.28
	w_5	20,710,184.75	751,676.87	191,643.08	0.26
168	w_1	49,660,909.40	1,307,523.04	404,822.36	0.31
	w_2	43,680,701.04	1,282,775.65	386,336.73	0.30
	w_3	38,123,525.06	1,224,525.39	353,144.53	0.29
	w_4	33,109,787.72	1,145,776.37	313,255.36	0.28
	w_5	28,739,390.04	1,053,084.53	268,178.16	0.26
Average	w_1	28,702,287.96	745,673.66	230,896.13	0.31
	w_2	25,294,964.28	730,939.98	219,912.68	0.30
	w_3	22,133,187.48	697,229.02	200,901.22	0.29
	w_4	19,282,053.77	652,246.41	178,477.81	0.28
	w_5	16,786,235.91	600,581.09	153,653.76	0.26

4.4.2. Comparison between Two Carbon-Tax Policies under Different Emission-Coefficient Settings

Emission coefficients are important parameters that determine the pollution level of power generation. To show how significantly the optimal solution changes as emission coefficients increase, we test the generation self-scheduling model corresponding to each carbon-tax policy under different value range settings for emission coefficients. In the test, eight sets of value ranges for emission coefficients are considered and provided in Table 6 in increasing order, in which each set of value ranges is denoted by $R_j, j = 1, \ldots, 8$. Results for tax-rate value, emission intensity, and total generation profits in the optimal solution corresponding to different carbon-tax policies under different value-range settings for emission coefficients are depicted in Figure 5, in which each point depicts the average over all test cases. From Figure 5, we can make the following observations:

(1) Compared with the carbon-tax policy with a constant tax rate, the proposed carbon-tax policy provides a tax rate that can be adaptively adjusted according to the change of emission coefficients and increases in a stepwise manner as emission coefficients increase. The result indicates that the proposed carbon-tax policy provides more flexibility than the carbon-tax policy with a constant tax rate.

(2) For each carbon-tax policy, emission intensity increases as emission coefficients increase. Compared with each carbon-tax policy with a constant tax rate, the proposed carbon-tax policy corresponds to the smallest average increase of emission intensity. The results indicate that the proposed carbon-tax policy can slow down the rise of emission intensity caused by the increase of emission coefficients.

(3) For each carbon-tax policy, total generation profits increase with the decrease of emission coefficients. As emission coefficients decrease, the proposed carbon-tax policy corresponds to a larger increase in total generation profits compared with the carbon-tax policy with a constant tax rate that is equal to the tax rate value of the proposed carbon-tax policy. The results indicate that the proposed carbon-tax policy can provide more economic incentives for generation companies to develop emission-reduction technologies or cleaner energy sources, compared with a carbon-tax policy with a constant tax rate.

Table 6. Value ranges for emission coefficients.

Situation	R_1	R_2	R_3	R_4
If $P_i^U < 600$	(0.05, 0.055)	(0.055, 0.06)	(0.06, 0.065)	(0.065, 0.07)
If $P_i^U \geq 600$	(0, 0.005)	(0.005, 0.01)	(0.01, 0.015)	(0.015, 0.02)

Situation	R_5	R_6	R_7	R_8
If $P_i^U < 600$	(0.07, 0.075)	(0.075, 0.08)	(0.08, 0.085)	(0.085, 0.09)
If $P_i^U \geq 600$	(0.02, 0.025)	(0.025, 0.03)	(0.03, 0.035)	(0.035, 0.04)

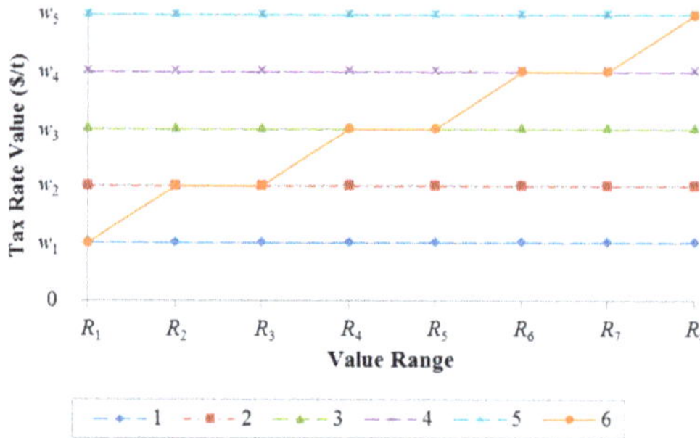

(**a**) Variation of tax-rate value with the increase of emission coefficients.

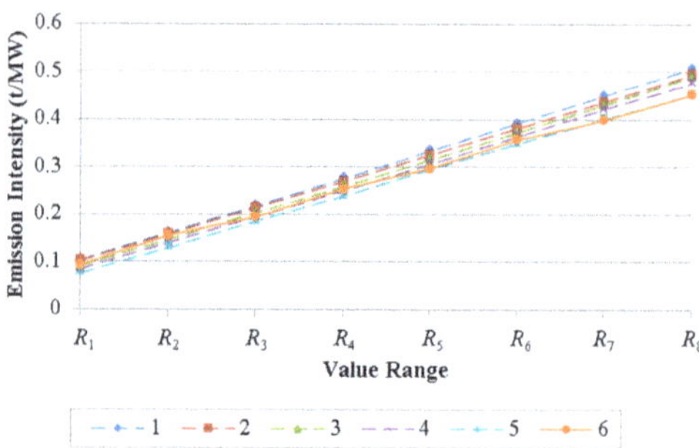

(**b**) Variation of emission intensity with the increase of emission coefficients.

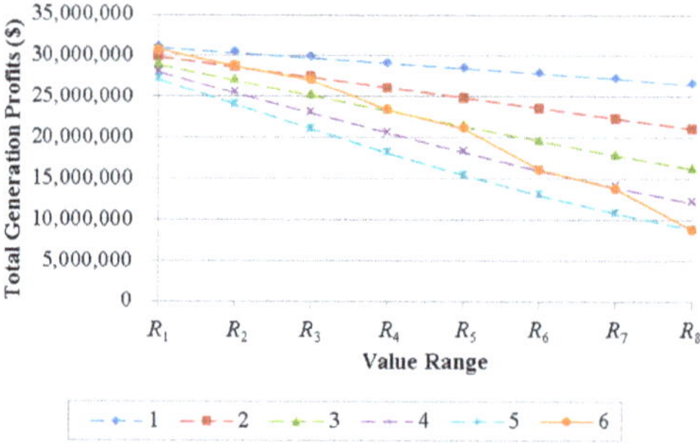

(**c**) Variation of total generation profits with the increase of emission coefficients.

Figure 5. Variations of tax-rate value, emission intensity, and total generation profits with the increase of emission coefficients. Curves 1, 2, 3, 4, and 5 depict the results corresponding to the carbon-tax policy with constant tax rate w_1, w_2, w_3, w_4, and w_5, respectively; Curve 6 depicts the results corresponding to the proposed carbon-tax policy.

5. Conclusions

In this paper, we propose a rate-variable carbon-tax policy and analyze the impact of the policy on generation self-scheduling of a generation company. The tax rate is designed to change along with emission-intensity variation, which is more flexible for emission reduction and distinguishes the proposed carbon-tax policy from existing carbon-tax policies. The variable tax rate is expressed by a step function, and the generation self-scheduling problem under the proposed carbon-tax policy is formulated as an MINLP model. A decomposition algorithm where multiple MILP procedures are implemented is developed to solve the problem. Computational results indicate that the decomposition algorithm is more effective than the MILP approach. It is also observed from numerical results that the proposed carbon-tax policy is effective in emission reduction and has more advantages than a carbon-tax policy with a constant tax rate in: (1) comprehensive consideration of generation profits, generation output, emission quantity s, and emission intensity; (2) slowing down the rise of emission intensity; and (3) stimulating generation companies to invest in low-carbon electricity-generation methods.

Our research provides new advice for policy makers to establish an effective emission-reduction mechanism for the electric-power industry. Modeling the generation self-scheduling problem under the proposed carbon-tax policy provides a theoretical framework for the implementation of carbon-tax policy. However, no uncertainties on electricity market are considered in the model, which is still a practical limitation of our research.

Besides addressing this limitation, future research can focus on: (1) analyzing the proposed policy under a generation-expansion planning framework to assess the impact of the proposed policy on investment in renewable energies and clean-generation techniques, and (2) studying environmental policies in other energy-intensive industries.

Author Contributions: P.C. contributed in developing the idea of the research. P.C. and Y.Z. contributed in writing the original draft of the manuscript; P.C., Y.Z., and J.L. contributed in reviewing and editing the manuscript.

Funding: This research is supported by the National Key Research and Development Program of China (2016YFB0901900), the National Natural Science Foundation of China (71402021, 61573086), the Major International Joint Research Project of the National Natural Science Foundation of China (71520107004), and the 111 Project (B16009).

Acknowledgments: The authors would like to thank the editor and reviewers for their valuable suggestions on improving the quality of this paper.

References

1. Gjengedal, T. Emission constrained unit-commitment (ECUC). *IEEE Trans. Energy Convers.* **1996**, *11*, 132–138. [CrossRef]
2. Bai, X.; Shahidehpour, S.M. Extended neighborhood search algorithm for constrained unit commitment. *Int. J. Electr. Power Energy Syst.* **1997**, *19*, 349–356. [CrossRef]
3. Ghadi, M.J.; Karin, A.I.; Baghramian, A.; Imani, M.H. Optimal power scheduling of thermal units considering emission constraint for gencos' profit maximization. *Int. J. Electr. Power Energy Syst.* **2016**, *82*, 124–135. [CrossRef]
4. Yamin, H.Y.; El-Dwairi, Q.; Shahidehpour, S.M. A new approach for GenCos profit based unit commitment in day-ahead competitive electricity markets considering reserve uncertainty. *Int. J. Electr. Power Energy Syst.* **2007**, *29*, 609–616. [CrossRef]
5. Soroudi, A. Robust optimization based self scheduling of hydro-thermal Genco in smart grids. *Energy* **2013**, *61*, 262–271. [CrossRef]
6. Wu, L.; Shahidehpour, M.; Li, T. Stochastic security-constrained unit commitment. *IEEE Trans. Power Syst.* **2007**, *22*, 800–811. [CrossRef]
7. Laia, R.; Pousinho, H.M.I.; Melíco, R.; Mendes, V.M.F. Self-scheduling and bidding strategies of thermal units with stochastic emission constraints. *Energy Convers. Manag.* **2015**, *89*, 975–984. [CrossRef]

8.   Ellerman, A.D.; Buchner, B.K. The European Union Emissions Trading Scheme: Origins, Allocation, and Early Results. *Rev. Environ. Econ. Policy* **2007**, *1*, 66–87. [CrossRef]

9.   Kockar, I.; Conejo, A.J.; McDonald, J.R. Influence of the emissions trading scheme on generation scheduling. *Int. J. Electr. Power Energy Syst.* **2009**, *31*, 465–473. [CrossRef]

10.   Rong, A.; Lahdelma, R. CO_2 emissions trading planning in combined heat and power production via multi-period stochastic optimization. *Eur. J. Oper. Res.* **2007**, *176*, 1874–1895. [CrossRef]

11.   Kara, M.; Syri, S.; Lehtilä, A.; Helynen, S.; Kekkonen, V.; Ruska, M.; Forsström, J. The impacts of EU CO_2 emissions trading on electricity markets and electricity consumers in Finland. *Energy Econ.* **2008**, *30*, 193–211. [CrossRef]

12.   Cong, R.; Wei, Y. Potential impact of (CET) carbon emissions trading on China's power sector: A perspective from different allowance allocation options. *Energy* **2010**, *35*, 3921–3931. [CrossRef]

13.   Considine, T.; Larson, D.F. Short term electric production technology switching under carbon cap and trade. *Energies* **2012**, *5*, 4165–4185. [CrossRef]

14.   Zhang, L.; Li, Y.; Jia, Z. Impact of carbon allowance allocation on power industry in China's carbon trading market: Computable general equilibrium based analysis. *Appl. Energy* **2018**, *229*, 814–827. [CrossRef]

15.   Weng, Q.; Xu, H. A review of China's carbon trading market. *Renew. Sustain. Energy Rev.* **2018**, *91*, 613–619. [CrossRef]

16.   Metcalf, G. Designing a carbon tax to reduce US greenhouse gas emissions. *Rev. Environ. Econ. Policy* **2009**, *3*, 63–83. [CrossRef]

17.   Lin, B.; Li, X. The effect of carbon tax on per capita CO_2 emissions. *Energy Policy* **2011**, *39*, 5137–5146. [CrossRef]

18.   Liu, S.; Chevallier, J.; Zhu, B. Self-scheduling of a power generating company: Carbon tax considerations. *Comput. Oper. Res.* **2016**, *66*, 384–392. [CrossRef]

19.   Gerbelová, H.; Amorim, F.; Pina, A.; Melo, M.; Ioakimidis, C.; Ferrão, P. Potential of CO_2 (carbon dioxide) taxes as a policy measure towards low-carbon Portuguese electricity sector by 2050. *Energy* **2014**, *69*, 113–119. [CrossRef]

20.   Liu, X.; Wang, C.; Niu, D.; Suk, S.; Bao, C. An analysis of company choice preference to carbon tax policy in China. *J. Clean. Prod.* **2015**, *103*, 393–400. [CrossRef]

21.   Zhang, K.; Wang, Q.; Liang, Q.-M.; Chen, H. A bibliometric analysis of research on carbon tax from 1989 to 2014. *Renew. Sustain. Energy Rev.* **2016**, *58*, 297–310. [CrossRef]

22.   Keohane, N. Cap and trade, rehabilitated: Using tradable permits to control U.S. greenhouse gases. *Rev. Environ. Econ. Policy* **2009**, *3*, 42–62. [CrossRef]

23.   Johnson, K.C. A decarbonization strategy for the electricity sector: New-source subsidies. *Energy Policy* **2010**, *38*, 2499–2507. [CrossRef]

24.   Alton, T.; Arndt, C.; Davies, R.; Hartley, F.; Makrelov, K.; Thurlow, J.; Ubogu, D. Introducing carbon taxes in South Africa. *Appl. Energy* **2014**, *116*, 344–354. [CrossRef]

25.   Palanichamy, C.; Sundar Babu, N. Analytical solution for combined economic and emission dispatch. *Electr. Power Syst. Res.* **2008**, *78*, 1129–1137. [CrossRef]

26.   Li, Y.; Cui, J. A method of designing energy tax rate based on game theory. In Proceedings of the 7th International Symposium on Operations Research and Its Applications (ISORA'08), Lijiang, China, 31 October–3 November 2008.

27.   Wei, W.; Liang, Y.; Liu, F.; Mei, S.; Tian, F. Taxing Strategies for carbon emissions: A bilevel optimization approach. *Energies* **2014**, *7*, 2228–2245. [CrossRef]

28.   Li, T.; Shahidehpour, M. Price-based unit commitment: A case of Lagrangian relaxation versus mixed integer programming. *IEEE Trans. Power Syst.* **2005**, *20*, 2015–2025. [CrossRef]

29.   Che, P.; Tang, Z.; Gong, H.; Zhao, X. An improved Lagrangian relaxation algorithm for the robust generation self-scheduling problem. *Math. Probl. Eng.* **2018**, *2018*, 1–12. [CrossRef]

30.   Tang, L.; Che, P. Generation scheduling under a CO_2 emission reduction policy in the deregulated market. *IEEE Trans. Eng. Manag.* **2013**, *60*, 386–397. [CrossRef]

31.   Che, P.; Tang, L.; Wang, J. Two-stage minimax stochastic unit commitment. *IET Gener. Transm. Distrib.* **2018**, *12*, 947–956. [CrossRef]

32.   Tang, L.; Che, P.; Wang, J. Corrective unit commitment to an unforeseen unit breakdown. *IEEE Trans. Power Syst.* **2012**, *27*, 1729–1740. [CrossRef]

33. Bard, J.F. Short-term scheduling of thermal-electric generators using Lagrangian relaxation. *Oper. Res.* **1988**, *36*, 756–766. [CrossRef]

34. PJM Interconnected Hourly Real-Time LMP. Available online: https://www.pjm.com/markets-and-operations/energy/real-time/monthlylmp.aspx (accessed on 21 September 2016).

35. Jiang, R.; Wang, J.; Guan, Y. Robust unit commitment with wind power and pumped storage hydro. *IEEE Trans. Power Syst.* **2012**, *27*, 800–810. [CrossRef]

Article

Green Activity-Based Costing Production Decision Model for Recycled Paper

Chu-Lun Hsieh, Wen-Hsien Tsai * and Yao-Chung Chang

Department of Business Administration, National Central University, Jhongli, Taoyuan 32001, Taiwan;
hsiehcl@gmail.com (C.-L.H.); ko902014@yahoo.com.tw (Y.-C.C.)
* Correspondence: whtsai@mgt.ncu.edu.tw; Tel.: +886-3-4267247; Fax: +886-3-4222891

Received: 22 March 2020; Accepted: 5 May 2020; Published: 12 May 2020

Abstract: Using mathematical programming with activity-based costing (ABC) and based on the theory of constraints (TOC), this study proposed a green production model for the traditional paper industry to achieve the purpose of energy saving and carbon emission reduction. The mathematical programming model presented in this paper considers (1) revenue of main products and byproducts, (2) unit-level, batch-level, and product-level activity costs in ABC, (3) labor cost with overtime available, (4) machine cost with capacity expansion, (5) saved electric power and steam costs by using the coal as the main fuel in conjunction with Refuse Derived Fuel (RDF). This model also considers the constraint of the quantity of carbon equivalent of various gases that are allowed to be emitted from the mill paper-making process to conform to the environmental protection policy. A numerical example is used to demonstrate how to apply the model presented in this paper. In addition, sensitivity analysis on the key parameters of the model are used to provide further insights for this research.

Keywords: green production; Activity-Based Costing (ABC); Theory of Constraints (TOC); green supply chain; energy saving; carbon emission reduction

---

1. Introduction

A total of 195 nations signed the Paris Agreement in December 2015 in order to solve the problem of environmental climate change [1]. Green issues have received considerable attentions in many industries worldwide in recent decades [2]. Enterprises have also tried to recover renewable raw materials to achieve profit, as well as to protect the environment [3,4], to actively reduce energy and resource consumption, waste output and virgin material consumption [5].

When implementing environmental strategies, the government should formulate and promote laws and regulations that all industries should comply with, and draw up plans to promote the entire industry, and provide sufficient funds and resources for enterprises [6]. However, when the government faces pressure on limited natural resources and waste disposal, they formulate resource recovery policies [3], and as the public continue to increase pressure on the government in terms of environmental pollution, it forces the government to develop stricter environment regulations and fines for environmental pollution, aiming to reduce the pollution caused by enterprises by adopting environmental management regulations [7]. The purpose of environmental management is to solve the problem of ecological environment pollution during the growth of enterprises [8], and to promote the production efficiency and effective use of raw materials of enterprises, including the use of alternative raw materials, and recycling and reusing raw materials, in order to effectively use raw materials and reduce resource waste [9]. Environmental problems can be considered and solved in the engineering or product development stage, especially through the life cycle assessment (LCA) method [10], where product designers can make a more environmentally friendly design [11]. The solution to environmental pollution largely depends on a combination of pollution prevention technologies and environmental management, which are keys to achieve the goals of environmental protection and pollution reduction. Through process innovations, enterprises can achieve both cost reduction and environment protection [7].

Early paper mills used wood, straw, sugarcane bagasse, and waste paper as raw materials to make pulp, which was used to make paper products. Concern for the environment has forced many enterprises to develop local policies for environment protection. Regarding the paper industry, practicing paper recycling is a cost-reducing choice, as compared to using wood as the raw material [12]. Some paper mills use a variety of waste paper to produce recycled paper, purchase pulp and paper equipment to improve efficiency and reduce the consumption of water and coal by using cogeneration equipment (also known as combined heat and power) in order to efficiently use energy. Then, paper mills can sell excess heat and electricity to recover costs. Coal, given its advantage of low cost, is the main fuel used to power the generator units of the paper industry; however, it greatly influences the environment, for example, the emitted pollutants of Carbon Dioxide (CO_2), Sulphur Oxides (SO_x), Nitrogen oxides (NO_x), and suspended matter, which brings severe harm to the environment. The paper-making process consists of pulping, papermaking, coating, and packing. The waste paper recycling rate in high-income countries is higher than that in less developed countries, meaning that countries with higher economic development need more attention on the problems of waste management and environmental protection [13].

This paper describes the acts related to the environmental management of enterprises, takes the paper industry as an example, and provides feasible pollution prevention and control technologies for the production processes of traditional paper mills. This paper is organized as follows: Section 1 introduces the research background and purposes; Section 2 explores the sustainable management under green paper industry; Section 3 describes the problem statement and model formulation; Section 4 presents a numerical example to demonstrate how to apply the model explored in this paper; Section 5 explains the managerial implications and limitations; and Section 6 offers the conclusions. This study found that enterprises can make full use of production capacity and waste through precise environmentally-friendly production processes to increase profit.

2. Sustainable Management under Green Paper Industry

The consumers and the government have requested companies to achieve a balance between profitability and environmental protection. The demand for integrating environmental awareness and product recycling into supply chain management has become a hot topic [14–16]. Sustainable management under green paper industry is described as follows.

2.1. Green Innovation in Paper Industry

Green innovation brings competitive advantages to the paper industry; since it helps enterprises reduce costs and increase income, some analysts have put forward that improving a company's environmental performance can result in better economic and financial results, instead of increasing costs [17–19]. The green innovation modes are described as green manufacturing and contamination control. Green manufacturing helps to reduce waste, control pollution [20], and improve energy efficiency and production processes through the integration of the production value chain, in order to elevate the efficiency of greenhouse gas reduction. Pollution prevention helps enterprises to reduce, transform, and prevent pollutants and wastewater by improving their internal processes, including changing production modes and means of transport [9], redesigning products and manufacturing modes, as well as recycling to prevent enterprise pollution [21]. More and more manufacturing companies are taking the environment into consideration before conducting their activities, a trend that is gaining support from the management level [22]. In fact, enterprise efforts in green manufacturing and pollution prevention will pave their way to a greener economy [23]. As the digital wave has swept across the world in recent years, it has indirectly impacted the traditional papermaking industry. This study analyzes successful cases where green manufacturing and pollution prevention are adopted prior to providing methods for increasing the utilization level of raw materials and effective waste recycling for the traditional paper industry, where coal-fired boilers still prevail.

2.2. MIP Model for Green Paper Industry

A Mix-Integer Programming (MIP) model is used to solve problems in the allocation and disposition of limited resource [24], which can effectively handle the multi-item inventory problem in the periodic replenishment plan (replenishment cycles are scheduled) [25], in order to optimize inventory distribution and production plans [26], solve the multilevel capacitated lot-sizing and scheduling problem [27], and provide enterprises with simultaneous multiple decision-making schemes [28]. Therefore, this model can be used for decision-making evaluations for enterprises to estimate operating costs and maximize profit [29,30]. In order to save costs, while the traditional paper industry still widely uses coal to power its boilers, coal is a major source of environmental pollution. The ideal mode of production for this industry is to make full use of raw materials in the production process and to recycle the waste. Using mixed integer linear programming models to optimize the production plan of the paper industry and reduce inventory is an issue recently discussed by the paper industry; it concludes that the application of activity-based costing (ABC) and theory of constraints (TOC) can solve the above problems [31,32]. The paper industry has used the ABC model for cost modeling, as well as analysis of the production flow [33]. The TOC could reduce the inventory and costs of mills [34], and overcome the bottleneck of resource limitation in the production process, thereby enhancing production efficiency [35]. Enterprises can use the ABC and TOC in their decision-making [36], as well as the linear programming (LP) model, to solve product mix decision problems [37,38]. As their complementary nature has gradually formed a trend, this model is expected to become a future trend; thus, this study adopts ABC and TOC to build an environmentally-friendly process model for the paper industry, and for discussion.

2.3. ABC and TOC

ABC is based on two stages, and is an extension of modern cost accounting in order to increase the accuracy of cost calculations. The common expenditures (e.g., management fees) of a business are assigned to activities, and then activity costs are traced to products [39]. On the other hand, TOC enhances scientific decision-making in production plans. According to the principles of TOC, enterprises can analyze manufacturing obstacles, increase their upper limit profits [40], and deal with the interactions of supply constraints through an improved product-mix [41]. Therefore, prominent business corporations will apply ABC and TOC to production planning in order to increase their total profits [42] and enhance the operational efficiency and immediateness of distribution centers [43]. Hence, ABC, TOC, and MIP can be adopted by enterprises during their decision-making processes on product portfolios [2,37,44,45].

Based on ABC and TOC, a model presented in this paper can be applied to the green supply chain [46] by using the Mathematical Programming Approach. In terms of the green nature of resources, it can also be applied to building municipal waste recycling systems, regional sewage streams [47], as well as solid waste collection and transportation systems [48]. The optimal process of waste collection and transportation to an incinerator can thus be planned [49]. In this study, the raw materials for green recycled paper are pulp substitutes, clean waste paper, and ordinary waste paper. The products of recycling paper mills are discussed according to the direct combinations of the aforesaid three materials, while ABC, TOC, and the Mathematical Programming Approach are used to plan new types of energy saving and carbon emission reduction activities for enterprises, in order to maximize profit.

3. Problem Statement and Model Formulation

This section takes the paper industry as an example. The green production flow for recycled paper is shown in Figure 1.

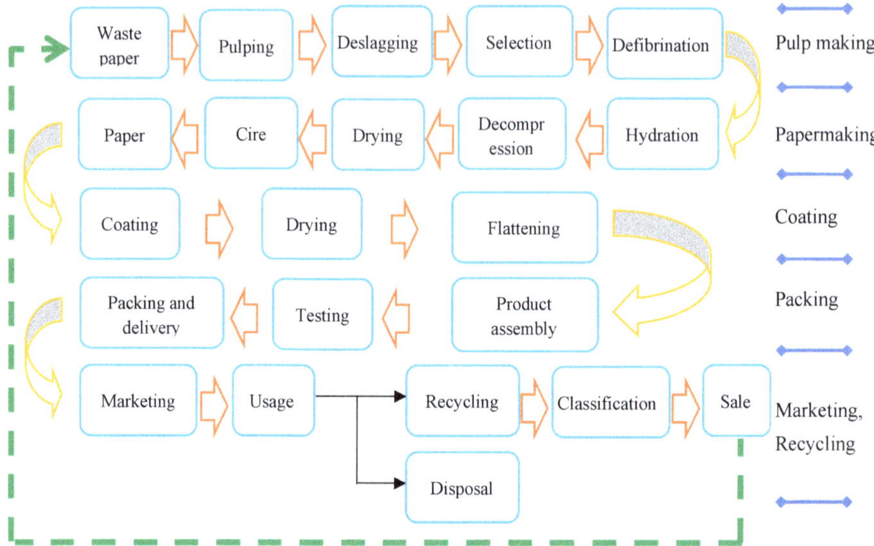

Figure 1. The papermaking process.

The recycled paper company of this case uses purchased pulp and waste paper as raw materials to produce paper products. The manufacturing process consists of pulping, papermaking, coating, and packing. In consideration of costs, in addition to saving water, coal and RDF-5 are used as the main fuels, and the cogeneration equipment integrates steam and electric energy for efficient energy use. The efficient application of waste could increase the profit for enterprises. The optimum mathematical programming decision model, as employed by the recycled paper mill, combines its processes with ABC to meet environmental protection requirements, while considering the maximization of corporate profit.

3.1. The Objective Function

The objective of this model is to maximize the total profit; the total corporate profit function is as follows:

$$
\begin{aligned}
Z = \quad & [(\text{main paper products revenue}) + (\text{relevant byproducts revenue}) + (\text{saved electric power} \\
& \text{cost}) + (\text{saved steam cost})] - [(\text{unit–level activity cost}: \text{total material cost} + \text{total expense} \\
& +\text{total direct labor cost} + \text{total machine cost}) + (\text{batch–level activity cost}) + (\text{product–level} \\
& \text{activity cost}) + (\text{total facility–level activity cost}) + \text{environmental management cost}] \\
= \quad & \left(\sum_{i=1}^{n} p_i X_i + \sum_{i=1}^{n} \sum_{s=1}^{t} c_s b_{is} X_i + \sum_{p=0}^{m} SSC_p \phi_p + \sum_{q=0}^{n} SEC_q \varphi_q \right) - \left[\sum_{i=1}^{n} \sum_{m=1}^{s} c_m a_{im} X_i + \sum_{i=1}^{n} \sum_{j \in U} d_j \lambda_{ij} X_i \right. \\
& \left. + (LC_1 \theta_1 + LC_2 \theta_2) + \sum_{i=1}^{n} \sum_{j \in B} d_j \alpha_{ij} B_{ij} + \sum_{i=1}^{n} \sum_{j \in P} d_j \rho_{ij} R_i + \sum_{k=0}^{r} MC_k \sigma_k + CE \right]
\end{aligned}
\tag{1}
$$

In Equation (1), the total corporate profit is calculated by subtracting various costs/expenses from main products revenue, relevant byproducts revenue, and electric power and steam costs saved. The revenues and costs/expenses and their associated constraints will be described in the following subsections.

3.2. Revenue of Main Products and Byproducts

In the product manufacturing process, the mill uses an Electrostatic Precipitator (ESP) and a Flue-Gas Desulfurization (FGD) to reduce the environmental pollution of the production processes, and to recover any waste byproducts for resale. If the price of the main products and relevant byproducts are represented by p_i and c_s, respectively, and the quantity of byproduct s of product i is $b_{is}X_i$, then the revenue of main products and relevant byproducts are expressed as $\sum_{i=1}^{n} p_i X_i$ and $\sum_{i=1}^{n}\sum_{s=1}^{t} c_s b_{is} X_i$, respectively.

3.3. Direct Material and Expense

In the preparation phase of papermaking, the purchase unit preliminarily sets the prices of various raw materials (a_{im}) for each product (X_i) after investigating the market prices. There are one to n products and one to s raw materials. The total direct material cost is the fifth term of Equation (1), $\sum_{i=1}^{n}\sum_{m=1}^{s} c_m a_{im} X_i$. The decision-maker decides the maximum resources (Q_m) available for each raw material according to the actual cost information previously provided by the accounting division, expressed as Equation (2).

$$\sum_{i=1}^{n} a_{im} X_i \leq Q_m, \quad m = 1, 2, \ldots, s, \tag{2}$$

3.4. Unit-Level Activity Cost

Unit-level activity is executed one time for each unit of a product. Thus, the total unit-level activity cost is the sixth term of Equation (1), $\sum_{i=1}^{n}\sum_{j \in U} d_j \lambda_{ij} X_i$, where λ_{ij} is the activity driver demand of unit-level activity j ($j \in U$) of one unit product i and d_j is the running activity cost per activity driver for activity j.

3.5. Direct Labor Cost

In mill operation, employee operation modes are divided into normal work and overtime work, as shown in Figure 2. The labor hours of normal work are in the range of [0, LH_1], and the cost is LC_1 at LH_1. When the mill receives a large order, and must complete the work quickly, the employees may work overtime. At this point, labor hours are in the range of [LH_1, LH_2], and the cost increases from LC_1 to LC_2. The total labor cost is the seventh term of Equation (1), ($LC_1\theta_1 + LC_2\theta_2$), and the associated constraints are expressed as Equations (3) to (8), where TL in Equation (3) is the total labor hours needed for the company.

$$TL = LH_1\theta_1 + LH_2\theta_2 \tag{3}$$

$$\theta_0 - \gamma_1 \leq 0 \tag{4}$$

$$\theta_1 - \gamma_1 - \gamma_2 \leq 0 \tag{5}$$

$$\theta_2 - \gamma_2 \leq 0 \tag{6}$$

$$\theta_0 + \theta_1 + \theta_2 = 1 \tag{7}$$

$$\gamma_1 + \gamma_2 = 1 \tag{8}$$

(γ_1, γ_2) is a SOS1 set of 0–1 variables, within which exactly one variable must be non-zero; ($\theta_0, \theta_1, \theta_2$) is a SOS2 set of non-negative variables, within which at most two adjacent variables, in the order given to the set, can be non-zero. When the papermaking activities are completed by normal work, then $\gamma_1 = 1$ and $\gamma_2 = 0$ (from Equation (9), and it is in the first segment of Figure 2; $0 \leq \theta_0, \theta_1 \leq 1$ (from Equations (4) and (5)) and $\theta_2 = 0$ (from Equation (6)); and $\theta_0 + \theta_1 = 1$ (from Equation (7)). Thus, the total direct labor hours required is $TL = LH_1\theta_1$ (from Equation (3)), and TL is the linear combination of 0 and LH_1. Similarly, when the papermaking activities need overtime work, then $\gamma_2 = 1$, $\gamma_1 = 0$, $0 \leq \theta_1, \theta_2 \leq 1$, $\theta_0 = 0$, $\theta_1 + \theta_2 = 1$, and the total direct labor hours (TL) required will be the linear combination of LH_1 and LH_2.

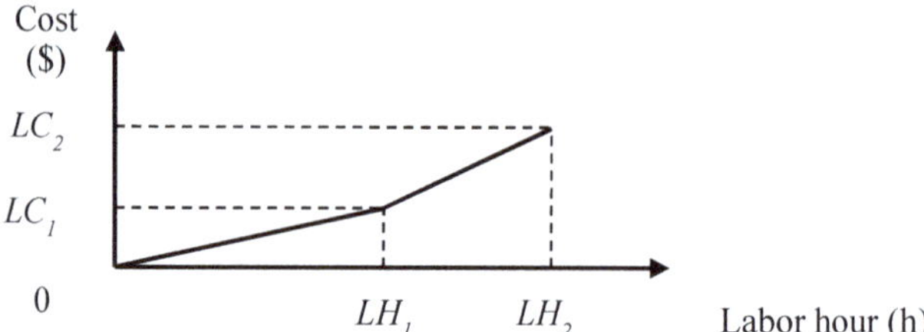

Figure 2. Piecewise direct labor cost.

3.6. Batch-Level Activity Cost

In ABC, batch-level activity is executed one time for each batch of a product. The total batch-level activity cost is the eighth term of Equation (1), $\sum_{i=1}^{n} \sum_{j\in B} d_j \mu_{ij} B_{ij}$, and the associated constraints are Equations (9) and (10):

$$X_i \leq \mu_{ij} B_{ij}, \; i = 1, 2, \ldots, n; \; j \in B \tag{9}$$

$$\sum_{i=1}^{n} \alpha_{ij} B_{ij} \leq T_j, \; B_{ij} \geq 0, \; j \in B \tag{10}$$

Equation (9) is the constraint of the quantity of product i, and Equation (10) is the constraint of resource available for batch-level activity j. If the batch-level activity is the activity "Setup", then Equation (10) may mean that the setup hour available for the batch-level activity "Setup" is T_j.

3.7. Product-Level Activity Cost

In ABC, the product-level activity is the activity consumed by a specific product. The total product-level activity cost is the ninth term of Equation (1), $\sum_{i=1}^{n} \sum_{j\in P} d_j \rho_{ij} R_i$, and the associated constraints are expressed in Equations (11) and (12).

$$X_i \leq V_i R_i, \quad i = 1, 2, \ldots, n \tag{11}$$

$$\sum_{i=1}^{n} \rho_{ij} R_i \leq D_j X_i \geq 0, \quad i = 1, 2, \ldots, n \tag{12}$$

Equation (11) is the constraint of the quantity of product i, and Equation (12) is the constraint of resource available for product-level activity j. If the product-level activity is the activity "Product Design", then Equation (12) may mean that the Computer-Aided Design (CAD) hour available for the product-level activity "Product Design" is D_j.

3.8. Machine Cost

In this paper, it is assumed that machine capacity can be expanded to various levels, as shown in Figure 3 [50]. Assume that the current machine hours available for use are MH_0, and the machine cost is MC_0, i.e., the depreciation of machines. If the machine hours increase to MH_1, the machine cost will be MC_1 after buying or renting additional machines. When the machine hours exceed the upper limit of MH_1 and reach the range of MH_1, MH_2, the machine cost will increase to MC_2.

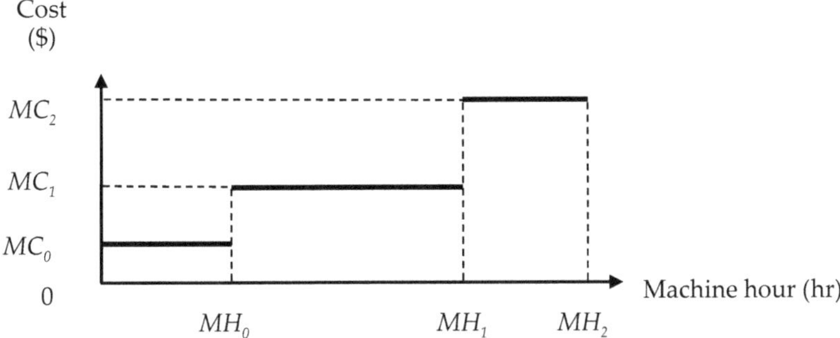

Figure 3. Stepwise machine cost.

Total machine cost is the tenth term of Equation (1), $\sum_{k=0}^{r} MC_k \sigma_k$, and the associated constraints are expressed as Equations (13) and (14).

$$\sum_{i=1}^{q} \delta_i X_i \leq \sum_{k=0}^{r} MH_k \sigma_k \tag{13}$$

$$\sum_{k=0}^{r} \sigma_k = 1 \tag{14}$$

$\sum_{i=1}^{n} \delta_i X_i$ is the total machine hours needed for all products; $(\sigma_0, \sigma_1, \sigma_2, \ldots, \sigma_k)$ is a special ordered set of type 1 (SOS1) of 0–1 variables, within which exactly one variable must be non-zero. If $\sigma_0 = 1$, then it is within the current level of machine capacity, i.e., MH_0. Then, $\sum_{i=1}^{n} \delta_i X_i \leq MH_0$ (from Equation (13)) and total machine cost is MC_0 (from Equation (1)). If $\sigma_1 = 1$, then the machine hours increase to the interval MH_0, MH_1; it is in the first expansion level of machine capacity, i.e., MH_1. Then, $\sum_{i=1}^{n} \delta_i X_i \leq MH_1$ (from Equation (13)) and total machine cost is MC_1 (from Equation (1)). Similarly, if $\sigma_k = 1$, then the machine hours increase to the interval MH_{k-1}, MH_k, within the kth expansion level of machine capacity, i.e., MH_k. Then, $\sum_{i=1}^{n} \delta_i X_i \leq MH_k$ (from Equation (13)) and total machine cost is MC_k (from Equation (1)).

3.9. Benefit of Using RDF-5

The mill uses cogeneration equipment to take advantage of steam and electric energies; therefore, when the consumption of coal and RDF-5 reaches a certain level, per unit steam energy and electric energy costs decrease. Given the power supply, steam supply, and heat supply benefits, RDF-5 can be used in mechanical bed boilers and fluidized bed furnaces as the main or auxiliary fuel for multifuel combustion. Therefore, the electric power and steam costs saved by using RDF-5 are considered in this paper.

First, the saved electric power cost is the sum of saved steam costs (SSC_p) at m power rates, i.e., the third term of Equation (1), $\sum_{p=0}^{m} SSC_p \phi_p$. In the paper-making process, the steam cost saved by using RDF has different conversion benefits due to different degrees of saving electric costs, expressed as Equation (15). The mill adopts the appropriate type of the aforesaid m rates according to the actual state of saving, or allocates according to appropriate proportions in two adjacent schemes, where the sum of the allocation proportions is 1, expressed as Equation (16).

Constraints:

$$\sum_{i=1}^{n} \beta_{ij} X_i \le \sum_{p=0}^{m} SSH_p \phi_p \tag{15}$$

$$\sum_{p=0}^{m} \phi_p = 1 \tag{16}$$

The steam cost saved is the sum of the saved electric costs (SEC_q) at t power rates, i.e., the fourth term of Equation (1), $\sum_{q=0}^{t} SEC_q \varphi_q$. In the paper-making process, the steam cost saved by using RDF has different conversion benefits due to different degrees of electric costs savings, expressed as Equation (17). The mill adopts an appropriate type of the aforesaid n rates according to the actual state of saving, or allocates according to appropriate proportions in two adjacent schemes, where the sum of allocation proportions is 1, expressed as Equation (18).

$$\sum_{i=1}^{n} \gamma_{ij} X_i \le \sum_{q=0}^{t} SGH_q \varphi_q \tag{17}$$

$$\sum_{q=0}^{t} \varphi_q = 1 \tag{18}$$

The quantity of carbon equivalent of various gases allowed to be emitted from the mill paper-making process is G in order to conform to the environmental protection policy, as shown in Equation (19).

$$\sum_{i=1}^{n} \sum_{p=1}^{q} c_p g_{ip} X_i \le G \tag{19}$$

4. Numerical Example

This section provides an example that describes how to apply ABC and TOC to the Mathematical Programming Approach, in order to determine the optimal product mix.

4.1. Description of the Case Problem

In response to environmental protection, Company A of the paper industry has recently used recycled pulp substitutes, clean waste paper, and ordinary waste paper, as the raw materials for paper products. The mixture of the three raw materials can be used to make three kinds of products. Considering the cost, coal remains in use for power generation, and the cogeneration coal-fired machine is used to take full advantage of electric energy and heat energy. The sewage and waste in the production process are properly treated, and a part of the waste is made into RDF-5, which is used together with the coal for cogeneration in the coal-fired machine. In order to take full advantage of resources and exploit financial resources, Company A sells the surplus electric energy of the process to the power company, while the heat energy is sold to nearby residential buildings. In response to environmental protection, the sludge treated from sewage is made into organic compost, while the ash from the incineration treatment is made into cement products. The residual fly ash from the power generation of the coal-fired machine is made into construction materials, the bottom ash is made into structural building materials, and the FGD gypsum is made into fire plate materials. In order to simplify the computing model, the machine costs and labor costs in this section are equally allocated to the unit-level activity of the main products. The costs in the production process include: (1) unit-level activity: including the costs of the three direct materials, direct expenses, required machine hours, and labor hours in the production process; (2) batch-level activity: including pulping costs, papermaking costs, coating costs, and packing costs of the general paper making process; (3) product-level activity:

i.e., product design cost; (4) facility-level cost: the environmental management cost refers to the costs related to routine inspections, effluence, and ensuring the process conforms to the environmental standard assessment specifications as regulated by local government; and the benefit of using RDF-5 is provided.

This paper uses Company A to describe that the present productive capacity determines the maximum profit of products. The aforesaid data are listed in Table 1, and the manufacturing process of Company A is shown in Figure 4. Company A sells three kinds of paper products. In terms of selling price, (Product1, Product2, Product3) = (320, 280, 250), where each product has its allocated cost during the production run. Under ABC, reasonable operating activity analysis and cost driver allocation can increase the correctness of cost information; and the electric energy and heat energy costs, as saved by using RDF-5, are included in the analysis. In order to meet the practical situation, the overtime problem in the production process is considered. The machine hours have three stepwise costs: mill works on demand, general activity, and frequent overtime. The electric and heat energy benefits saved by using RDF-5 are processed piecewise, and the maximum demand for the product is shown at the bottom of Table 1.

Company A uses three raw materials effectively: recycled pulp substitute (X_1), clean waste paper (X_2) and ordinary waste paper (X_3), and generates revenue from seven byproducts of the production process: electricity, steam heat, sludge organic compost, ash cement products, fly ash building materials, bottom ash reinforced structural building materials, and FGD gypsum fire plates. After the optimum production (B_{i1}), preparation (B_{i2}), treatment (B_{i3}), and cutting (B_{i4}) of the batch-level activities, the emissions meet the environmental policy, that CO_2 should not exceed 80,000 units, and the numerical values of NO_x, CO_2, SO_2, CO, COD, BOD, SS, AOX, and product-level constraints (R_i) are obtained. The most important cost is calculated using LINGO software, based on the machine, labor, electricity, and steam costs. The left part of Figure 4 shows the process of paper-making; the middle part shows the required water, coal, electricity, and steam for the paper-making process, as well as the sewage and waste remaining from the production process; the right part shows the utilization of waste from the paper-making process, which can be used as byproducts. The excess electric and heat energies are sold, thereby turning waste into resources, and creating extra profit for the company. Based on Equations (1) to (19), the aforesaid green product mix decision model is described in Appendix A.

Table 1. (**a**) Example data; (**b**) Example data—facility level cost.

(a) Example Data								
Data Category	**Index & Activity-Driver**		**Activity**	**Parameter**	**Product 1**	**Product 2**	**Product 3**	**Available Capacity**
Selling price		Unit cost/price	j	p_i	320	280	250	
Material (Unit-level)	m = 1(pulp substitute)	$20.00/unit		a_{i1}	3	2	1	$Q_1 = 11{,}310$
	m = 2(cleanable waste paper)	$15.00/unit		a_{i2}	4	3	2	$Q_2 = 17{,}020$
	m = 3(ordinary waste paper)	$5.00/unit		a_{i3}	2	4	7	$Q_3 = 24{,}750$
Expense	m = 4(water)	$1.00/unit		a_{i4}	10	12	15	$Q_4 = 70{,}400$
	m = 5(coal)	$2.00/unit		a_{i5}	23	25	30	$Q_5 = 147{,}960$
Selling byproduct	s = 1(electricity)	$0.80/unit		b_{i1}	5	6	7	
	s = 2(steam)	$0.50/unit		b_{i2}	1	2	3	
	s = 3(organic compost)	$0.035/unit		b_{i3}	10	12	15	
	s = 4(cement products)	$0.02/unit		b_{i4}	10	12	15	
	s = 5(building materials)	$0.20/unit		b_{i5}	23	25	30	
	s = 6(reinforced structural building materials)	$0.05/unit		b_{i6}	23	25	30	
	s = 7(fireplates)	$0.70/unit		b_{i7}	23	25	30	
Unit-level activity	Machine hours		1	λ_{i1}	6	7	8	
	Labor hours		2	λ_{i2}	4	5	6	
RDF5 saved cost	Electric energy		3	λ_{i3}	5	6	7	
	Steam (heat) energy		4	λ_{i4}	1	2	3	
Batch-level activity	Activity-driver	Cost per activity driver						
Pulping	Blending	$6	5	α_{i5}	1	2	3	$T_5 = 7750$
				μ_{i5}	3	2	1	
Papermaking	Papermaking	$14	6	α_{i6}	2	3	4	$T_6 = 12{,}470$
				μ_{i6}	8	6	3	
Coating	Treatment	$2	7	α_{i7}	3	4	5	$T_7 = 6100$
				μ_{i7}	5	4	3	
Packing	Packing	$1	8	α_{i8}	2	2	3	$T_8 = 5400$
				μ_{i8}	3	2	1	
Product-level activity-Design Maximum demand	Drawings	$100	9	ρ_{i9}	15	12	18	$D_6 = 50$
				V_i	2000	2500	3000	

Table 1. *Cont.*

(b) Example data—facility level cost		
Environmental Management Cost	**Total Cost**	**$20,000**
Machine hours constraint-Cost	$MC_0 = \$85{,}600$	$MC_1 = \$98{,}975$	$MC_2 = \$115{,}025$
Machine hours	$MH_0 = 42{,}800$	$MH_1 = 48{,}150$	$MH_2 = 53{,}500$
Machine cost rate	$mr_0 = \$2/h$	$mr_1 = \$2.5/h$	$mr_2 = \$3/h$
Direct labor constraint-Cost	$LC_1 = \$38{,}204$		$LC_2 = \$62{,}082$
Labor hours	$LH_1 = 19{,}102$		$LH_2 = 28{,}653$
Wage rate	$wr_1 = \$2/h$		$wr_2 = \$2.5/h$
Power constraint			
Cost	$SEC_1 = \$92{,}000$	$SEC_2 = \$101{,}200$	$SEC_3 = \$108{,}100$
Hour	$SGH_1 = 36{,}800$	$SGH_2 = 41{,}400$	$SGH_3 = 46{,}000$
Rate	$PT_1 = \$2.5/10$ thousand kWh	$PT_2 = \$2/10$ thousand kWh	$PT_3 = \$1.5/10$ thousand kWh
Steam(heat) constraint			
Cost	$SSC_1 = \$32{,}000$	$SSC_2 = \$35{,}200$	$SSC_3 = \$37{,}600$
Hour	$SSH_1 = 12{,}800$	$SSH_2 = 14{,}400$	$SSH_3 = 16{,}000$
Rate	$ST_1 = \$2.5/10$ thousand kWh	$ST_2 = \$2/10$ thousand kWh	$ST_3 = \$1.5/10$ thousand kWh

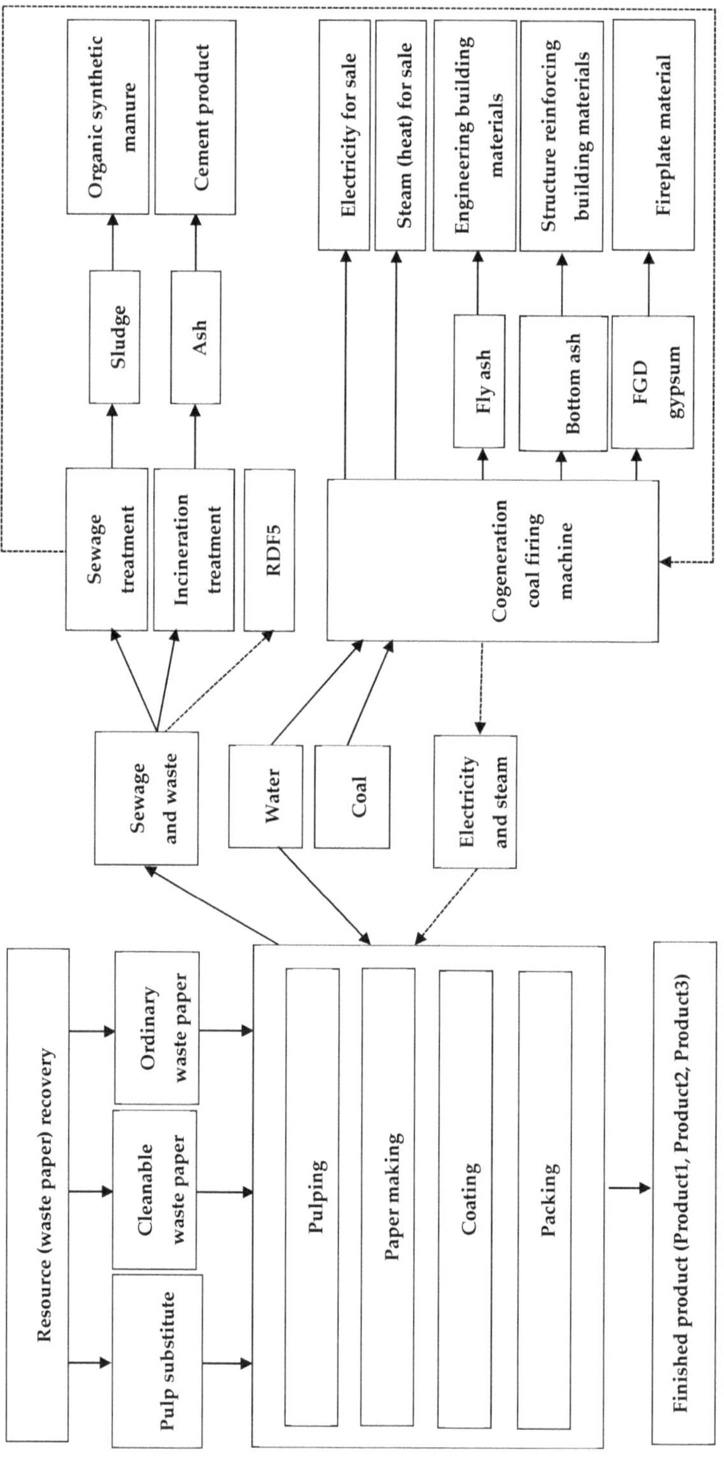

Figure 4. Manufacturing process of Company A.

4.2. Analysis

This MIP model uses LINGO 15.0 [51] software to obtain the optimal solution to the variables of the proposed decision-making model, as shown in Table 2. While decision-makers intend to determine the optimal solutions with limited resources, such optimal solutions are conditional and depend on the preset target structure. This paper suggests using the LINGO model to determine the final target structure and solutions, as LINGO is fast, easy to use, and more effective in establishing linear, nonlinear, quadratic, quadratic constraint, second-order cone, half-definite, stochastic, and integral optimization models.

Table 2. The optimal solution of numerical example *.

Symbol	Value	Symbol	Value	Symbol	Value	Symbol	Value
Z	$643,193.80	B_{21}	1250 batches	CO_2	65,620.8 t	σ_2	0
X_1	1940 t	B_{31}	480 batches	SO_2	0.291648 t	θ_0	0
X_2	2500 t	B_{12}	243 batches	CO	14.5824 t	θ_1	0.577217
X_3	480 t	B_{22}	417 batches	COD	109.368 t	θ_2	0.422783
P_1	$22,448.00	B_{32}	160 batches	BOD	7.2912 t	γ_1	0
P_2	$4190.00	B_{13}	388 batches	SS	9.114 t	γ_2	1
P_3	$1981.00	B_{23}	625 batches	AOX	0.109368 t	ϕ_0	0
P_4	$1132.00	B_{33}	160 batches	R_1	1	ϕ_1	0
P_5	$24,304.00	B_{14}	647 batches	R_2	1	ϕ_2	1
P_6	$6076.00	B_{24}	1250 batches	R_3	1	φ_0	0
P_7	$85,064.00	B_{34}	480 batches	σ_0	1	φ_1	0
B_{11}	647 batches	NOx	72.912 t	σ_1	0	φ_2	1

* Z: Total profit (thousand dollars). P_s: Revenue of byproduct s (thousand dollars).

According to the results, the optimal profit of the product mix of the recycled paper mill is (X_1, X_2, X_3) = (1940, 2500, 480), which requires 11,300 units of the first raw material (=3 × 1940 + 2 × 2500 + 1 × 480), 16,220 units of the second raw material (=4 × 1940 + 3 × 2500 + 2 × 480), 17,240 units of the third raw material (=2 × 1940 + 4 × 2500 + 7 × 480), 56,600 units of water (=10 × 1940 + 12 × 2500 + 15 × 480), 121,520 units of coal (=23 × 1940 + 25 × 2500 + 305 × 480), 32,980 machine hours (=6 × 1940 + 7 × 2500 + 8 × 480), and 23,140 direct labor hours (=4 × 1940 + 5 × 2500 + 6 × 480). The total profit Z is $643,193.80. In terms of byproducts: (1) revenue from selling electricity is $22,448; (2) revenue from selling steam (heat) is $4190; (3) revenue from organic compost is $1981; (4) revenue from cement products is $1132; (5) revenue from building materials is $24,304; (6) revenue from reinforced structural building materials is $6076; (7) revenue from fire plates is $85,064. The total revenue from the aforesaid byproducts is $145,195.

As seen from the above table, through the combination of mathematical programming and accurate cost analysis, cost apportionments of the terminal products, as well as the analytic results of basic operation system, business managers can reference the data for optimal operational decision-making, leading the enterprise towards maximized operating profit before conducting production and business activities. Through the application of this model, enterprises can make decisions regarding whether to continue processing. Moreover, in the decision-making process, enterprises can find the best way to obtain the optimal solution, even under the unfavorable situation of limited resources. Due to its convenience and understandability, calculation tools are based on ABC and TOC, provide a practical method for rendering decisions of a product portfolio which can assist enterprises to use raw materials efficiently in the product mix decisions of GMTs, thereby meeting environmental goals and increasing corporate profit.

4.3. Sensitivity Analysis

Sensitivity analysis on the key parameters of the model provide further insights for this study. This study conducts sensitivity analysis on the cost and available capacity of 3 raw materials: pulp substitute, clean waste paper, and ordinary waste paper, with the unit being 5%. Regarding the unit costs of raw materials (pulp substitute, clean waste paper, and ordinary waste paper = $20, $15, $5 respectively); when all the purchase costs of the three are items decreased by 5%, the increase in the company profit will be (pulp substitute, cleanable waste paper, ordinary waste paper = 1.76%, 1.89%, 0.67%, respectively). When the costs are further decreased by 5%, namely, the decreasing rate changes from 5% to 10% (5% + 5% = 10%), the increase in the company profit will be (pulp substitute, cleanable waste paper, and ordinary waste paper = 3.51%, 3.78%, 1.34%, respectively). As the test results show, when all the purchase costs of the three items are decreased by 5%, the increase in the company profit will be (pulp substitute, cleanable waste paper, and ordinary waste paper = 1.76%, 1.89%, 0.67%, respectively).

Additionally, this study conducts an in-depth exploration into the influence of cost increase on enterprise profit. In terms of the unit cost of raw material, when the three purchase costs are increased by 5%, the reduction in the profit of enterprises will be (pulp substitute, cleanable waste paper, and ordinary waste paper = −1.76%, −1.89%, −0.67%, respectively); when the three purchase costs are increased by another 5% or 10%, the reduction in the profit of enterprises will be (pulp substitute, cleanable waste paper, and ordinary waste paper = −3.51%, −3.78%, −1.34%, respectively). As the test shows, when all the purchase costs of the three items are decreased by 5%, the increase in the profit of enterprises will be (pulp substitute, cleanable waste paper, and ordinary waste paper = 1.76%, 1.89%, 0.67%, respectively). Tables 3–5 show the relationship between the purchase costs and profits of pulp substitute, cleanable waste paper, and ordinary waste paper. The contents of Tables 3–5 are illustrated in Figure 5, which displays the cost decrease and capacity increase over profit increase. As shown in Figure 5, as the cost of the raw materials procured by enterprises gradually decreases, the quantity of the raw materials that can be procured will gradually increase, and enterprises will make more profit. Specifically, cleanable waste paper and ordinary waste paper have a great influence on enterprises, with the former having the greater influence; while the pulp substitute has less influence on the revenue of enterprises.

In terms of available capacity, (pulp substitute, cleanable waste paper, and ordinary waste paper = 11,310, 17,020, 24,750, respectively). The sensitivity analysis showed that when the available capacities of pulp substitute and clean waste paper increased by 5%, respectively, the profit increased by 1.76%. However, when the available capacities further increased by 5%, the profit ceased to increase beyond the 1.76%. Tables 6 and 7 show the relationships between pulp substitute, clean waste paper, and the company profit. When the available capacity of ordinary waste paper increased by 5%, the profit of the company did not increase. If it further increased by 5%, the profit still did not change. Table 8 shows the relationship between cleanable waste paper and the company profit. The sensitivity analysis indicated that the company can further increase its current 5% available capacity of cleanable waste paper to gain more profit.

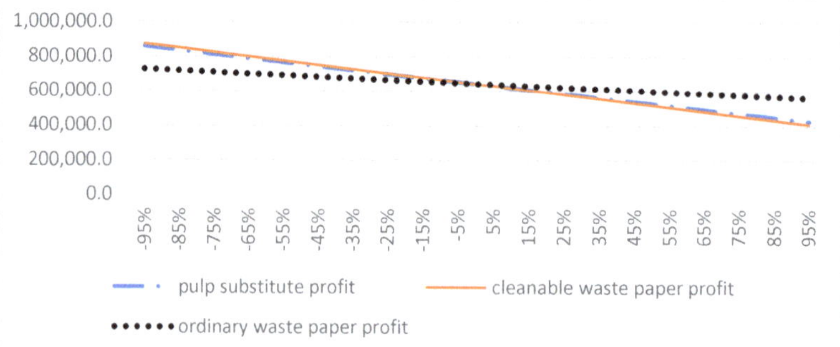

Figure 5. The impacts of cost decrease and capacity increase on profit (unit: thousand dollars).

Table 3. Sensitivity analysis on the cost of pulp substitute.

	Cost Decrease/Increase Ratio (%)	Profit	Increase/Decrease (Compared with the Initial Value)	Increase Profit
pulp substitute	−30%	710,993.8	10.54%	67,800.0
pulp substitute	−25%	699,693.8	8.78%	56,500.0
pulp substitute	−20%	688,393.8	7.03%	45,200.0
pulp substitute	−15%	677,093.8	5.27%	33,900.0
pulp substitute	−10%	665,793.8	3.51%	22,600.0
pulp substitute	−5%	654,493.8	1.76%	11,300.0
pulp substitute	0%	643,193.8	0.00%	0.0
pulp substitute	5%	631,893.8	−1.76%	−11,300.0
pulp substitute	10%	620,593.8	−3.51%	−22,600.0
pulp substitute	15%	609,293.8	−5.27%	−33,900.0
pulp substitute	20%	597,993.8	−7.03%	−45,200.0
pulp substitute	25%	586,693.8	−8.78%	−56,500.0
pulp substitute	30%	575,393.8	−10.54%	−67,800.0

Table 4. Sensitivity analysis on the cost of clean waste paper.

	Cost Decrease/Increase Ratio (%)	Profit	Increase/Decrease (Compared with the Initial Value)	Increase Profit
Cleanable waste paper	−30%	716,183.8	11.35%	72,990.0
Cleanable waste paper	−25%	704,018.8	9.46%	60,825.0
Cleanable waste paper	−20%	691,853.8	7.57%	48,660.0
Cleanable waste paper	−15%	679,688.8	5.67%	36,495.0
Cleanable waste paper	−10%	667,523.8	3.78%	24,330.0
Cleanable waste paper	−5%	655,358.8	1.89%	12,165.0
Cleanable waste paper	0%	643,193.8	0.00%	0.0
Cleanable waste paper	5%	631,028.8	−1.89%	−12,165.0
Cleanable waste paper	10%	618,863.8	−3.78%	−24,330.0
Cleanable waste paper	15%	606,698.8	−5.67%	−36,495.0
Cleanable waste paper	20%	594,533.8	−7.57%	−48,660.0
Cleanable waste paper	25%	582,368.8	−9.46%	−60,825.0
Cleanable waste paper	30%	570,203.8	−11.35%	−72,990.0

Table 5. Sensitivity analysis on the cost of ordinary waste paper.

	Cost Decrease/Increase Ratio (%)	Profit	Increase/Decrease (Compared with the Initial Value)	Increase Profit
Ordinary waste paper	−30%	669,252.8	4.05%	26,059.0
Ordinary waste paper	−25%	664,882.8	3.37%	21,689.0
Ordinary waste paper	−20%	660,512.8	2.69%	17,319.0
Ordinary waste paper	−15%	656,142.8	2.01%	12,949.0
Ordinary waste paper	−10%	651,813.8	1.34%	8620.0
Ordinary waste paper	−5%	647,503.8	0.67%	4310.0
Ordinary waste paper	0%	643,193.8	0.00%	0.0
Ordinary waste paper	5%	638,883.8	−0.67%	−4310.0
Ordinary waste paper	10%	634,573.8	−1.34%	−8620.0
Ordinary waste paper	15%	630,263.8	−2.01%	−12,930.0
Ordinary waste paper	20%	625,953.8	−2.68%	−17,240.0
Ordinary waste paper	25%	621,643.8	−3.35%	−21,550.0
Ordinary waste paper	30%	617,333.8	−4.02%	−25,860.0

Table 6. Sensitivity analysis on the available capacity of pulp substitute.

	Increasing Ratio	Profit	Increase/Decrease (Compared with the Initial Value)	Increase Profit
Pulp substitute	0%	643,193.8	0.00%	0.0
Pulp substitute	5%	654,512.8	1.76%	11,319.0
Pulp substitute	10%	654,512.8	1.76%	11,319.0
Pulp substitute	15%	654,512.8	1.76%	11,319.0
Pulp substitute	20%	654,512.8	1.76%	11,319.0
Pulp substitute	25%	654,512.8	1.76%	11,319.0
Pulp substitute	30%	654,512.8	1.76%	11,319.0

Table 7. Sensitivity analysis on the available capacity of clean waste paper.

	Increasing Ratio	Profit	Increase/Decrease (Compared with the Initial Value)	Increase Profit
Clean waste paper	0%	643,193.8	0.00%	0.0
Clean waste paper	5%	654,512.8	1.76%	11,319.0
Clean waste paper	10%	654,512.8	1.76%	11,319.0
Clean waste paper	15%	654,512.8	1.76%	11,319.0
Clean waste paper	20%	654,512.8	1.76%	11,319.0
Clean waste paper	25%	654,512.8	1.76%	11,319.0
Clean waste paper	30%	654,512.8	1.76%	11,319.0

Table 8. Sensitivity analysis on the available capacity of ordinary waste paper.

	Increasing Ratio	Profit	Increase/Decrease (Compared with the Initial Value)	Increase Profit
Ordinary waste paper	0%	643,193.8	0.00%	0.0
Ordinary waste paper	5%	643,193.8	0.00%	0.0
Ordinary waste paper	10%	643,193.8	0.00%	0.0
Ordinary waste paper	15%	643,193.8	0.00%	0.0
Ordinary waste paper	20%	643,193.8	0.00%	0.0
Ordinary waste paper	25%	643,193.8	0.00%	0.0
Ordinary waste paper	30%	643,193.8	0.00%	0.0

5. Discussion

Recent literature regarding the costs and processes of manufacturing applied ABC and TOC to the allocation of resources [17], and combined ABC and sensitivity analysis [52]; however, few studies have combined all three approaches; thus, this study can serve as reference for future studies regarding costs and resource allocation. The sensitivity analysis of this study implies that although the purchase cost of cleanable waste paper is ranked second among the three raw materials, it has the highest influence on company profit as it accounts for the highest percentage of production costs. The test on available capacity indicates that if the current 5% available capacity of cleanable waste paper can be further increased, the factory will produce more products, thereby obtaining more profit. After providing the findings to the case factory, the Purchasing Department of the case company discovered that the purchase cost of cleanable waste paper has the greatest influence on company profit. As the international price of paper is reduced in the future, the case company can negotiate with their suppliers to reduce their price. Previously, the case company thought that it had already taken full advantage of the available capacity of the raw material; however, after being informed of the findings, it plans to increase its application of cleanable waste paper.

5.1. Managerial Implications

Preliminary research has found that the adoption of ABC can help management to identify purchase behaviors [53] while greatly improving organizational performance, productivity, and profitability; hence, this technique has been widely promoted to enhance enterprise profitability [54,55]. However, as many industries continue to use old production technologies, which consume resources and energy at a rate of more than triple that of new environmental technologies, enterprise pollution cannot be effectively controlled, quality of life is affected, and environmental damage is continuously aggravated [6]. Therefore, when internal environmental technologies of enterprises are enhanced by internal environmental management, pollution is effectively controlled. When new training schemes can be imported into environmental management systems to strengthen employees' environmental awareness and environmental problem-handling modes, the performance of environmental management systems will be significantly influenced [56].

An environmental management system also uses the internal innovations of process innovation, technology innovation, and product innovation for improvement [9]. With the rise of environmental awareness, enterprises actively conduct internal green innovation, which has a significantly positive influence on increasing enterprises' competitive advantages. When the innovation ability of competitors is low, applied innovation can double the competitive advantage of an enterprise [16]. With increased consumer environmental considerations, retailers and manufacturers will benefit by providing extraordinary green business practices [57]. Numerical application and sensitivity analysis demonstrate the applicability of the proposed model and emphasize management insights [58]. The main purpose of business operations is to obtain profit for a company; decreased production costs and increased available capacity are the two most common approaches. Sensitivity analysis shows the proportion of each raw material against production costs. On the basis of changing the procurement environment, a company can negotiate with raw material suppliers to reduce the price of raw materials in order to save purchase costs. During reproduction, the available capacity of raw materials with production potential can be increased to facilitate the company's production of more products, and thereby, create more profits.

5.2. Limitations

This study provides a reference model for energy conservation and carbon emission reduction through a case study, based on the operational procedures of the traditional paper-making industry. The values in the case are all represented by virtual numbers. Paper mills can apply this model to make effective use of their wastes, as based on a power generation method that uses its own fuel and raw materials after obtaining the data of discharged wastes through the testing process. Generalizability to other industries is problematic.

6. Conclusions

Using sustainable operations research to enhance the innovation of environmental technology can help enterprises to improve their environmental problems by reducing environmental pollution [6] within a short time by classifying pollutants and waste; thus, the performance of enterprises in solving environmental problems can be effectively improved [59]. Corporate competitiveness and the usability of managerial accounting information have significant correlation [60]. Therefore, this study combines ABC with TOC, and uses the Mathematical Programming Approach to provide paper-making enterprises with a production mode that includes environmental protection, while taking full advantage of the byproducts of process wastes. This study explores the cost of the production process with ABC.

The TOC is adopted to help managers note the restricted resources of production processing and remove any bottlenecks.

In order to be practical, regarding the situation of occasional overtime work hours, this study handles it with stepwise machine costs and piecewise direct labor costs. The former considers the different rates of plants according to the different electricity consumption rates during production; while the latter considers overtime work in production. In addition to the saved costs of electric and thermal energies of several plants, through the implementation of energy-saving operations, the company expressed their saved costs due to the energy conservation nature of the model. When simultaneously considering a winning corporate image and profit, if recycled paper is used as a raw material for production, products with higher value are created according to the characteristics of the different recycled papers, which may be the future direction for enterprises that have not used recycled resources as raw materials. While coal firing pollutes ambient air, many mills continue to use it to generate power in order to save costs; water, as the industrial blood, is usually wasted without proper recycling. Therefore, this study especially introduces the application of coal and wastewater, and suggests the combination of a boiler and cogeneration for the construction of most resources. Subsequent mills can effectively take full advantage of the waste from boilers, meaning enterprises can become excellent benchmarking enterprises, and find a balance between profitability and energy efficiency through their gradual growing awareness of environmental protection.

Author Contributions: Conceptualization, W.-H.T.; methodology, C.-L.H.; software, Y.-C.C.; formal analysis, C.-L.H. and Y.-C.C.; investigation, C.-L.H. and Y.-C.C.; resources, W.-H.T.; writing—original draft preparation, C.-L.H., W.-H.T., and Y.-C.C.; writing—review and editing, C.-L.H., W.-H.T., and Y.-C.C.; project administration, W.-H.T.; funding acquisition, W.-H.T. All authors have read and agreed to the published version of the manuscript.

Funding: This research was funded by the Ministry of Science and Technology of Taiwan under Grant No. MOST106-2410-H-008-020-MY3.

Acknowledgments: The author is extremely grateful to the energies journal editorial team and reviewers who provided valuable comments for improving the quality of this article. The author also would like to thank the Ministry of Science and Technology of Taiwan for financial support of this research under Grant No. MOST106-2410-H-008-020-MY3.

Conflicts of Interest: The authors declare no conflict of interest.

Notations

The main symbols of variables and parameters used in this model are defined as follows:

Z	Corporate profit;
X_i	Quantity of product i;
p_i	Unit price of product i;
c_s	Unit price of byproduct s
b_{is}	The quantity of byproduct s of one unit of product i
SSC_p	Saved pth steam costs, when p = 1, the preferential rate is applicable; when p = 2, the basic preferential rate is applicable; when p = 3, the excess rate is applicable;
SSH_p	Saved pth steam machine hours, when p = 1, the preferential rate is applicable; when p = 2, the basic preferential rate is applicable; when p = 3, the excess rate is applicable;
SEC_q	Saved qth electric power costs, when q = 1, the preferential rate is applicable; when q = 2, the basic preferential rate is applicable; when q = 3, the excess rate is applicable;
SGH_q	Saved qth generating machine hours, when q = 1, the preferential rate is applicable; when q = 2, the basic preferential rate is applicable; when q = 3, the excess rate is applicable;
β_{ij}	Saved activity driver demands of unit-level activity j (j ∈ U) for steam machine hours of one-unit product i;
γ_{ij}	Saved activity driver demand of unit-level activity j (j ∈ U) for generating machine hours of one-unit product i;
c_m	Unit cost of mth raw material;
a_{im}	Unit cost of mth raw material; mth raw material demand of one-unit product I;

Q_m	Available quantity of raw material Q.
d_j	Running activity cost per activity driver for activity j;
λ_{ij}	Activity driver demand of unit-level activity j ($j \in$ U) of one-unit product i;
LC_1	Total direct labor cost in LH_1 (see Figure 2);
LC_2	Total direct labor cost in LH_2 (see Figure 2);
TL	Total labor hours needed for the company;
LH_1	Upper limit of total direct labor hours of normal work (see Figure 2);
LH_2	Upper limit of total direct labor hours including overtime work (see Figure 2);
α_{ij}	The quantity of resource used by each batch-level activity j ($j \in$ B) for product i;
B_{ij}	The number of batches for batch-level activity j ($j \in$ B) used by product i;
μ_{ij}	The quantity of product i for each batch-level activity j ($j \in$ B);
T_j	The quantity available for the activity driver of batch-level activity j ($j \in$ B).
ρ_{ij}	Demand of activity driver needed by product-level activity j ($j \in$ P) for product i;
R_i	Production indicator of product i; If (Ri = 1), then product i will be produced. Otherwise, (Ri = 0);
V_i	Maximum demand for product i;
D_j	The quantity available for the activity driver of activity j ($j \in$ P).
MC_k	Total machine cost in MH_k (see Figure 3);
MH_k	Machine hours of kth level capacity (see Figure 3);
σ_k	SOS1 set of 0–1 variables (special order of the first kind), where one and only one variable must be nonzero; $\sigma_k = 1$ means that machine hour is expanded to MH_k.
δ_i	Machine hour demand for one unit of product i;
CE	Environmental management cost;
c_p	Processed gas p;
g_{ip}	The total quantity of gas p from product i.;
G	The quantity of carbon equivalent of various gases allowed to be emitted from the mill paper-making process.

Appendix A

Company A introduces ABC and TOC into the mathematical programming of the production process.

Table A1. The mathematical programming model with the example data of Table 1.

Z = [(income from main paper products) + (relevant byproducts) + (saved electric power cost) + (saved steam cost)] − [(unit-level activity cost: total material cost + total expense + total direct labor cost+ total machine cost) + (batch-level activity cost) + (product-level activity cost) + (total facility-level activity cost) + environmental management cost]

$$= (\sum_{i=1}^{n} p_i X_i + \sum_{i=1}^{n}\sum_{s=1}^{t} c_s b_{is} X_i + \sum_{q=0}^{n} SEC_q \varphi_q + \sum_{p=0}^{m} SSC_p \phi_p) - \left[\sum_{i=1}^{n}\sum_{m=1}^{s} c_m a_{im} X_i + \sum_{i=1}^{n}\sum_{j \in U} d_j \lambda_{ij} X_i + (LC_1\theta_1 + LC_2\theta_2) \right.$$

$$+ \sum_{i=1}^{n}\sum_{j \in B} d_j \mu_{ij} B_{ij} + \sum_{i=1}^{n}\sum_{j \in P} d_j \rho_{ij} R_i + \sum_{k=0}^{r} MC_k \sigma_k + CE \left] = (\sum_{i=1}^{3} p_i X_i + \sum_{i=1}^{3}\sum_{s=1}^{7} c_s b_{is} X_i + \sum_{p=0}^{2} SSC_p \phi_p + \sum_{q=0}^{2} SEC_q \varphi_q) - \left[\sum_{i=1}^{3}\sum_{m=1}^{5} c_m a_{im} X_i + \sum_{i=1}^{3}\sum_{j \in U} d_j \lambda_{ij} X_i + (LC_1\theta_1 + LC_2\theta_2) + \sum_{i=1}^{3}\sum_{j \in B} d_j \mu_{ij} B_{ij} + \sum_{i=1}^{3}\sum_{j \in P} d_j \rho_{ij} R_i + \sum_{k=0}^{2} MC_k \sigma_k + CE \right]$$

$=(320X_1 + 280X_2 + 250X_3) + \{[0.8^*(5X_1 + 6X_2 + 7X_3)] + [0.5^*(X_1 + 2X_2 + 3X_3)] + [0.035^*(10X_1 + 12X_2 + 15X_3)] + [0.02^*(10X_1 + 12X_2 + 15X_3)] + [0.2^*(23X_1 + 25X_2 + 30X_3)] + [0.05^*(23X_1 + 25X_2 + 30X_3)] + [0.7^*(23X_1 + 25X_2 + 30X_3)]\} + (92{,}000\phi_0 + 101{,}200\phi_1 + 108{,}100\phi_2) + (32{,}000\phi_0 + 35{,}200\phi_1 + 37{,}600\phi_2) - \{[(20^*3 + 15^*4 + 5^*2)X_1 + (20^*2 + 15^*3 + 5^*4)X_2 + (20^*1 + 15^*2 + 5^*7)X_3] + [(1^*10 + 2^*23)X_1 + (1^*12 + 2^*25)X_2 + (1^*15 + 2^*30)X_3] + (38{,}204\theta_1 + 62{,}082\theta_2) + [(6^*1)B_{11} + (6^*2)B_{21} + (6^*3)B_{31}] + [(14^*2)B_{12} + (14^*3)B_{22} + (14^*4)B_{32}] + (2^*3)B_{13} + (2^*4)B_{23} + (2^*5)B_{33}] + [(1^*2)B_{14} + (1^*2)B_{24} + (1^*3)B_{34}] + [(100^*15)R_1 + (100^*12)R_2 + (100^*18)R_3] + (85{,}600\sigma_0 + 98{,}975\sigma_1 + 115{,}025\sigma_2) + 20{,}000 = 160.9X_1 + 143.21X_2 + 126.425X_3 + 92{,}000\phi_0 + 101{,}200\phi_1 + 108{,}100\phi_2 + 32{,}000\phi_0 + 35{,}200\phi_1 + 37{,}600\phi_2 - 38{,}204\theta_1 - 62{,}082\theta_2 - 85{,}600\sigma_0 - 98{,}975\sigma_1 - 115{,}025\sigma_2 - 6B_{11} - 12B_{21} - 18B_{31} - 28B_{12} - 42B_{22} - 56B_{32} - 6B_{13} - 8B_{23} - 10B_{33} - 2B_{14} - 2B_{24} - 3B_{34} - 1500R_1 - 1200R_2 - 1800R_3 - 20{,}000$

Subject to:

Raw material and expense constraints:

$3X_1 + 2X_2 + X_3 \leq 11{,}310$

$4X_1 + 3X_2 + 2X_3 \leq 17{,}020$

$2X_1 + 4X_2 + 7X_3 \leq 24{,}750$

$10X_1 + 12X_2 + 15X_3 \leq 70{,}400$

$23X_1 + 25X_2 + 30X_3 \leq 147{,}960$

Batch-level treatment activity constraints:

$X_1 - 5B_{13} \leq 0$

$X_2 - 4B_{23} \leq 0$

$X_3 - 3B_{33} \leq 0$

$3B_{13} + 4B_{23} + 5B_{33} \leq 6100$

Stepwise facility-level machine hour constraints:

$6X_1 + 7X_2 + 8X_3 - 42{,}800\sigma_0 - 48{,}150\sigma_1 - 53{,}500\sigma_2 \leq 0$

$\sigma_0 + \sigma_1 + \sigma_2 = 1$

Direct labor constraints:

$4X_1 + 5X_2 + 6X_3 - 19{,}102\theta_1 - 28{,}653\theta_2 = 0$

$\theta_0 - \gamma_1 \leq 0$

$\theta_1 - \gamma_1 - \gamma_2 \leq 0$

$\theta_2 - \gamma_2 \leq 0$

$\theta_0 + \theta_1 + \theta_2 = 1$

$\gamma_1 + \gamma_2 = 1$

Batch-level cutting activity constraints:

$X_1 - 3B_{14} \leq 0$

$X_2 - 2B_{24} \leq 0$

$X_3 - B_{34} \leq 0$

$2B_{14} + 2B_{24} + 3B_{34} \leq 5400$

Product-level constraints:

$X_1 - 2000R_1 \leq 0$

$X_2 - 2500R_2 \leq 0$

$X_3 - 3000R_3 \leq 0$

$15R_1 + 12R_2 + 18R_3 \leq 50$

Saved electric power cost =

$92{,}000\phi_0 + 101{,}200\phi_1 + 108{,}100\phi_2$

Constraints:

$5X_1 + 6X_2 + 7X_3 - 36{,}800\phi_0 - 41{,}400\phi_1 - 46{,}000\phi_2 \leq 0$

$\phi_0 + \phi_1 + \phi_2 = 1$

Saved steam cost $= 32{,}000\varphi_0 + 35{,}200\varphi_1 + 37{,}600\varphi_2$

Constraints:

$X_1 + 2X_2 + 3X_3 - 12{,}800\varphi_0 - 14{,}400\varphi_1 - 16{,}000\varphi_2 \leq 0$

$\varphi_0 + \varphi_1 + \varphi_2 = 1$

Batch-level production activity constraints:

$X_1 - 3B_{11} \leq 0$

$X_2 - 2B_{21} \leq 0$

$X_3 - B_{31} \leq 0$

$B_{11} + 2B_{21} + 3B_{31} \leq 7750$

Batch-level preparation activity constraints:

$X_1 - 8B_{12} \leq 0$

$X_2 - 6B_{22} \leq 0$

$X_3 - 3B_{32} \leq 0$

$2B_{12} + 3B_{22} + 4B_{32} \leq 12{,}470$

Estimated data of different gases emitted from the mill paper-making

process: $\sum_{i=1}^{3}\sum_{p=1}^{8} c_p g_{ip} X_i \leq 80{,}000$

NO_x: $0.0006^*(23X_1 + 25X_2 + 30X_3)$

CO_2: $0.54^*(23X_1 + 25X_2 + 30X_3)$

SO_2: $0.0000024^*(23X_1 + 25X_2 + 30X_3)$

CO: $0.00012^*(23X_1 + 25X_2 + 30X_3)$

COD: $0.0009^*(23X_1 + 25X_2 + 30X_3)$

BOD: $0.00006^*(23X_1 + 25X_2 + 30X_3)$

SS: $0.000075^*(23X_1 + 25X_2 + 30X_3)$

AOX: $0.0000009^*(23X_1 + 25X_2 + 30X_3)$

In order to conform to the environmental protection policy, the carbon equivalent of the company should not exceed 80,000 units.

References

1. United Nations Framework on Climate Change (UNFCC). Adoption of the Paris Agreement. In the Report of the Conference of the Parties on its twenty-first session (held in Paris from 30 November to 13 December 2015), Addendum, Report No. FCCC/CP/2015/10/Add.1. United Nations, 29 January 2016. Available online: http://unfccc.int/resource/docs/2015/cop21/eng/10a01.pdf (accessed on 29 April 2020).
2. Tsai, W.-H.; Yang, C.-H.; Chang, J.-C.; Lee, H.-L. An Activity-Based Costing decision model for life cycle assessment in green building projects. *Eur. J. Oper. Res.* **2014**, *238*, 607–619. [CrossRef]
3. Georgiadis, P.; Vlachos, D. The effect of environmental parameters on product recovery. *Eur. J. Oper. Res.* **2004**, *157*, 449–464. [CrossRef]
4. Chen, W.Y.; Jim, C.Y. Resident valuation and expectation of the urban greening project in Zhuhai, China. *J. Environ. Plan. Manag.* **2011**, *54*, 851–869. [CrossRef]
5. Srivastava, S. Green Supply Chain Management: A State-of-The-Art Literature Review. *Int. J. Manag. Rev.* **2007**, *9*, 53–80. [CrossRef]
6. Liu, L.; Ma, X. Technology-based industrial environmental management: A case study of electroplating in Shenzhen, China. *J. Clean. Prod.* **2010**, *18*, 1731–1739. [CrossRef]
7. Christmann, P. Effects of "Best Practices" of Environmental Management on Cost Advantage: The Role of Complementary Assets. *Acad. Manag. J.* **2000**, *43*, 663–680.
8. Bansal, P.; Roth, K. Why Companies Go Green: A Model of Ecological Responsiveness. *Acad. Manag. J.* **2000**, *43*, 717–736.
9. Majumdar, S.K.; Marcus, A.A. Rules versus Discretion: The Productivity Consequences of Flexible Regulation. *Acad. Manag. J.* **2001**, *44*, 170–179.
10. Collado-Ruiz, D.; Ostad-Ahmad-Ghorabi, H. Comparing LCA results out of competing products: Developing reference ranges from a product family approach. *J. Clean. Prod.* **2010**, *18*, 355–364. [CrossRef]
11. Yu, S.; Zhang, R. Life Cycle Assessment in Different Product Design Stages—A Coffee Pot Case Study. *Appl. Mech. Mater.* **2010**, *34*, 988–994. [CrossRef]
12. Pati, R.K.; Vrat, P.; Kumar, P. Economic analysis of paper recycling vis-a-vis wood as raw material. *Int. J. Prod. Econ.* **2006**, *103*, 489–508. [CrossRef]
13. Berglund, C.; Söderholm, P. An econometric analysis of global waste paper recovery and utilization. *Environ. Resour. Econ.* **2003**, *26*, 429–456. [CrossRef]
14. Brandenburg, M.; Rebs, T. Sustainable supply chain management: A modeling perspective. *Ann. Oper. Res.* **2015**, *229*, 213–252. [CrossRef]
15. Ilgin, M.A.; Gupta, S.M. Environmentally conscious manufacturing and product recovery (ECMPRO): A review of the state of the art. *J. Environ. Manag.* **2010**, *91*, 563–591. [CrossRef] [PubMed]
16. Seuring, S. A review of modeling approaches for sustainable supply chain management. *Decis. Support Syst.* **2013**, *54*, 1513–1520. [CrossRef]
17. Chen, Y.-S. The Driver of Green Innovation and Green Image – Green Core Competence. *J. Bus. Ethics* **2008**, *81*, 531–543. [CrossRef]
18. Chen, Y.-S.; Lai, S.-B.; Wen, C.-T. The Influence of Green Innovation Performance on Corporate Advantage in Taiwan. *J. Bus. Ethics* **2006**, *67*, 331–339. [CrossRef]
19. Alsmadi, M.; Almani, A. Implementing an integrated ABC and TOC approach to enhance decision making in a Lean context. *Int. J. Qual. Reliab. Manag.* **2014**, *31*, 906–920. [CrossRef]
20. Hui, I.K.; Chan, A.H.S.; Pun, K.F. A study of the Environmental Management System implementation practices. *J. Clean. Prod.* **2001**, *9*, 269–276. [CrossRef]
21. Ambec, S.; Lanoie, P. Does it pay to be green? A systematic overview. *Acad. Manag. Perspect.* **2008**, *22*, 45–62.
22. Gavronski, I.; Klassen, R.D.; Vachon, S.; do Nascimento, L.F.M. A resource-based view of green supply management. *Transp. Res. Part E Logist. Transp. Rev.* **2011**, *47*, 872–885. [CrossRef]
23. Granek, F. Business value of toxics reduction and pollution prevention planning. *J. Clean. Prod.* **2011**, *19*, 559–560. [CrossRef]
24. Repoussis, P.P.; Paraskevopoulos, D.C.; Vazacopoulos, A.; Hupert, N. Optimizing emergency preparedness and resource utilization in mass-casualty incidents. *Eur. J. Oper. Res.* **2016**, *255*, 531–544. [CrossRef]
25. Russell, R.A.; Urban, T.L. Offsetting inventory replenishment cycles. *Eur. J. Oper. Res.* **2016**, *254*, 105–112. [CrossRef]

26. Lei, L.; Lee, K.; Dong, H. A heuristic for emergency operations scheduling with lead times and tardiness penalties. *Eur. J. Oper. Res.* **2016**, *250*, 726–736. [CrossRef]

27. Boonmee, A.; Sethanan, K. A GLNPSO for multi-level capacitated lot-sizing and scheduling problem in the poultry industry. *Eur. J. Oper. Res.* **2016**, *250*, 652–665. [CrossRef]

28. Esmaeilbeigi, R.; Charkhgard, P.; Charkhgard, H. Order acceptance and scheduling problems in two-machine flow shops: New mixed integer programming formulations. *Eur. J. Oper. Res.* **2016**, *251*, 419–431. [CrossRef]

29. Todosijević, R.; Benmansour, R.; Hanafi, S.; Mladenović, N.; Artiba, A. Nested general variable neighborhood search for the periodic maintenance problem. *Eur. J. Oper. Res.* **2016**, *252*, 385–396. [CrossRef]

30. Amorim, P.; Curcio, E.; Almada-Lobo, B.; Barbosa-Póvoa, A.P.; Grossmann, I.E. Supplier selection in the processed food industry under uncertainty. *Eur. J. Oper. Res.* **2016**, *252*, 801–814. [CrossRef]

31. Hästbacka, M.; Westerlund, J.; Westerlund, T. MISPT: A user friendly MILP mixed-time based production planning tool. In *Computer Aided Chemical Engineering*; Elsevier: Amsterdam, The Netherlands, 2007; Volume 24, pp. 637–642.

32. Björk, K.-M.; Carlsson, C. The effect of flexible lead times on a paper producer. *Int. J. Prod. Econ.* **2007**, *107*, 139–150. [CrossRef]

33. Korpunen, H.; Paltakari, J. Testing an activity-based costing model with a virtual paper mill. *Nord. Pulp Pap. Res. J.* **2013**, *28*, 146–155. [CrossRef]

34. Jiang, X.-Y.; Wu, H.-H. Optimization of setup frequency for TOC supply chain replenishment system with capacity constraints. *Neural Comput. Appl.* **2013**, *23*, 1831–1838. [CrossRef]

35. Andelkovic, P.; Andelkovic, A.; Dasic, P. The theory of constraints as a basis for production process improvement model. *АктуальніПроблеми Економіки* **2013**, *10*, 251–260.

36. Lockhart, J.; Taylor, A. Environmental considerations in product mix decisions using ABC and TOC. *Manag. Account. Q.* **2007**, *9*, 13.

37. Tsai, W.-H.; Chen, H.-C.; Leu, J.-D.; Chang, Y.-C.; Lin, T.W. A product-mix decision model using green manufacturing technologies under activity-based costing. *J. Clean. Prod.* **2013**, *57*, 178–187. [CrossRef]

38. Tsai, W.-H.; Chang, Y.-C.; Lin, S.-J.; Chen, H.-C.; Chu, P.-Y. A green approach to the weight reduction of aircraft cabins. *J. Air Transp. Manag.* **2014**, *40*, 65–77. [CrossRef]

39. Dessureault, S.; Benito, R. Data mining and activity based costing for equipment replacement decisions Part 1–establishing the information infrastructure. *Min. Technol.* **2012**, *121*, 73–82. [CrossRef]

40. Baptista, E.A.; Lucato, W.C.; Coppini, N.L.; Fortunato, F.A.D.S. Profit optimization in machining service providers using principles of the Theory of Constraints. *J. Braz. Soc. Mech. Sci. Eng.* **2013**, *35*, 347–355. [CrossRef]

41. Cannon, J.N.; Cannon, H.M.; Low, J.T. Modeling tactical product-mix decisions: A theory-of-constraints approach. *Simul. Gaming* **2013**, *44*, 624–644. [CrossRef]

42. Huang, S.-Y.; Chen, H.-J.; Chiu, A.-A.; Chen, C.-P. The application of the theory of constraints and activity-based costing to business excellence: The case of automotive electronics manufacture firms. *Total Qual. Manag. Bus. Excell.* **2014**, *25*, 532–545. [CrossRef]

43. Yang, H.-L.; Chen, Z.-J.; Ji, Y.-F.; Zhang, Y. Operation optimization of distribution center based on time-driven activity-based costing. *J. Dalian Marit. Univ.* **2012**, *4*, 26.

44. Tsai, W.-H. Green production planning and control for the textile industry by using mathematical programming and industry 4.0 techniques. *Energies* **2018**, *11*, 2072. [CrossRef]

45. Tsai, W.-H. Carbon Taxes and Carbon Right Costs Analysis for the Tire Industry. *Energies* **2018**, *11*, 2121. [CrossRef]

46. Tsai, W.-H.; Hung, S.-J. A fuzzy goal programming approach for green supply chain optimisation under activity-based costing and performance evaluation with a value-chain structure. *Int. J. Prod. Res.* **2009**, *47*, 4991–5017. [CrossRef]

47. Zeferino, J.A.; Antunes, A.P.; Cunha, M.C. Multi-objective model for regional wastewater systems planning. *Civ. Eng. Environ. Syst.* **2010**, *27*, 95–106. [CrossRef]

48. Arribas, C.A.; Blazquez, C.A.; Lamas, A. Urban solid waste collection system using mathematical modelling and tools of geographic information systems. *Waste Manag. Res.* **2010**, *28*, 355–363. [CrossRef]

49. Galante, G.; Aiello, G.; Enea, M.; Panascia, E. A multi-objective approach to solid waste management. *Waste Manag.* **2010**, *30*, 1720–1728. [CrossRef]

50. Tsai, W.-H. A Green Quality Management Decision Model with Carbon Tax and Capacity Expansion under Activity-Based Costing (ABC)—A Case Study in the Tire Manufacturing Industry. *Energies* **2018**, *11*, 1858. [CrossRef]
51. LINDO SYSTEMS INC. Archives—LINGO 15.0. Available online: https://www.lindo.com/index.php/2-uncategorised/160-archives-lingo-15-0 (accessed on 22 April 2020).
52. Kim, Y.-W.; Han, S.-H.; Yi, J.-S.; Chang, S. Supply chain cost model for prefabricated building material based on time-driven activity-based costing. *Can. J. Civ. Eng.* **2016**, *43*, 287–293. [CrossRef]
53. Mark, T.; Niraj, R.; Dawar, N. Uncovering customer profitability segments for business customers. *J. Bus. Bus. Mark.* **2012**, *19*, 1–32. [CrossRef]
54. Askarany, D.; Yazdifar, H. An investigation into the mixed reported adoption rates for ABC: Evidence from Australia, New Zealand and the UK. *Int. J. Prod. Econ.* **2012**, *135*, 430–439. [CrossRef]
55. Salim, I. Analyzing usage of activity based costing in commercial banking institutions of Tripoli Libya: A technological acceptance model approach. *J. Appl. Sci. Res.* **2012**, *8*, 414–419.
56. Dwyer, R.; Lamond, D.; Lee, K.H. Why and how to adopt green management into business organizations? *Manag. Decis.* **2009**, *47*, 1101–1121.
57. Liu, Z.L.; Anderson, T.D.; Cruz, J.M. Consumer environmental awareness and competition in two-stage supply chains. *Eur. J. Oper. Res.* **2012**, *218*, 602–613. [CrossRef]
58. Muriana, C. An EOQ model for perishable products with fixed shelf life under stochastic demand conditions. *Eur. J. Oper. Res.* **2016**, *255*, 388–396. [CrossRef]
59. Tanskanen, J.-H. Strategic planning of municipal solid waste management. *Resour. Conserv. Recycl.* **2000**, *30*, 111–133. [CrossRef]
60. Zhu, D.-S.; Lin, Y.-P.; Huang, S.-Y.; Lu, C.-T. The effect of competitive strategy, task uncertainty and organisation structure on the usefulness and performance of management accounting system (MAS). *Int. J. Bus. Perform. Manag.* **2009**, *11*, 336–363. [CrossRef]

Article

Research on the Impact of Various Emission Reduction Policies on China's Iron and Steel Industry Production and Economic Level under the Carbon Trading Mechanism

Ye Duan [1],*, Zenglin Han [2], Hailin Mu [3], Jun Yang [4] and Yonghua Li [1]

[1] College of Urban and Environmental, Liaoning Normal University, Dalian 116029, China; dydl@lnnu.edu.cn
[2] Center for Studies of Marine Economy and Sustainable Development, Liaoning Normal University, Dalian 116029, China; hzl@lnnu.edu.cn
[3] Key Laboratory of Ocean Energy Utilization and Energy Conservation of Ministry of Education, Dalian University of Technology, Dalian 116024, China; hailinmu@dlut.edu.cn
[4] Human Settlements Research Center, Liaoning Normal University, Dalian 116029, China; yangjun@lnnu.edu.cn
* Correspondence: dy200872083@mail.dlut.edu.cn

Received: 3 April 2019; Accepted: 23 April 2019; Published: 29 April 2019

Abstract: To study the emission reduction policies' impact on the production and economic level of the steel industry, this paper constructs a two-stage dynamic game model and analyzes various emission reduction policies' impact on the steel industry and enterprises. New results are observed in the study: (1) With the increasing emission reduction target (15%–30%) and carbon quota trading price (12.65–137.59 Yuan), social welfare and producer surplus show an increasing trend and emission macro losses show a decreasing trend. (2) Enterprises' reduction ranges in northwestern and southwestern regions are much higher than that of the other regions; the northeastern enterprise has the smallest reductions range. (3) When the market is balanced (0.8543–0.9320 billion tons), the steel output has decreased and the polarization in various regions has been alleviated to some extent. The model is the abstraction and assumption of reality, which makes the results have some deviations. However, these will provide references to formulate reasonable emissions reduction and production targets. In addition, the government needs to consider the whole and regional balance and carbon trading benchmark value when deciding the implementation of a single or mixed policy. Future research will be more closely linked to national policies and gradually extended to other high-energy industries.

Keywords: carbon trading mechanism; emission reduction policy; China's iron and steel industry; a two-stage dynamic game

---

1. Introduction

The iron and steel industry (hereafter referred to as the steel industry) is an important fundamental industry of China's economy. After the founding of New China, the development of steel industry has gone through three stages. The first stage was from 1949 to 1960. The steel industry achieved rapid development under the influence of a "steel-oriented" mindset and the large-scale steelmaking movement, and steel output rose sharply. In the second stage, from 1960 to 1980, the development of the steel industry during this period was in a sluggish state. Many factors had brought about a huge negative impact on the development of the steel industry, causing the steel industry to stagnate. In the third stage, after the 1980s, the steel industry developed rapidly. The rank of annual output jumped to the top in the world in 1996. In recent years, the annual steel output has ranked first in the world, achieving a leap in the development of the steel industry.

For a long time, the steel industry has provided important raw material guarantees for national construction, which has strongly supported the development of related industries, promoted the process of industrialization and modernization in China, and promoted the improvement of people's livelihood and social development.

Though the iron and steel industry has achieved remarkable achievements, it also faces many problems. The national steel industry equipment level is uneven, especially in terms of energy conservation and emission mitigation; a lot of debts of energy conservation and environmental protection investment were left over. Many enterprises have not achieved the comprehensive and stable discharge of pollutants; energy conservation and environmental protection facilities need to be further upgraded. Though the energy consumption and pollutant emissions of steel, per ton, have decreased in recent years, it cannot offset the increase in total energy consumption and pollutants caused by the increasing production. With the booming advancement of national industrialization and urbanization and the continuous upgrading of the consumption structure, the energy conservation and emission reduction work will become increasingly serious to the steel industry. At the same time, social development, ecological civilization construction, people's living needs, public opinion concerns, and more will impose new and more stringent requirements on energy conservation and emission reduction.

Obviously, for future development prospects, the steel industry still needs to make great efforts to adhere to green development, in addition to adhering to structural adjustment and innovation drive. It should aim to reduce energy consumption and pollutant emissions and to fully implement energy conservation and emission reduction. The government should keep on renovating, continuously optimizing the original fuel structure, vigorously developing the circular economy, actively researching and promoting green steel throughout the life cycle, and building a new pattern of steel manufacturing and social harmonious development.

At present, emission reduction policies such as the carbon trading mechanism or the carbon taxation mechanism have not been fully implemented. What are their impacts on the production level and economic profit of enterprises and the steel industry? Whether or not it is suitable for the actual situation of the steel industry is still unclear. Therefore, it is an urgent and difficult task for the steel industry to figure out the way to achieve green development in order to complete increasingly stringent emission reduction tasks.

In China, on 18th December, 2017, the National Development and Reform Commission announced the "National Carbon Emissions Trading Market Construction Plan (Power Generation Industry)," setting the threshold for China's power generation industry to be included in the carbon market: Annual emissions of 26,000 tons of carbon dioxide or above. It provides a reference for other industrial sectors, so this paper will study the carbon trading mechanism's impact on production level based on this program.

The remainder of this paper is organized as follows: In Section 2, a literature review focusing on the distribution mode and policy influence of carbon trading system is provided; in Section 3, a two-stage game theory modeling is constructed, and policy assumptions and data sources are provided; in Section 4, results and discussions are presented based on accounting data and statistical analysis; and in Section 5, conclusions and policy recommendations for China's iron and steel industry are drawn.

2. Literature Review of Distribution Mode and Policy Influence of Carbon Trading System

Dales [1] first proposed the concept of emissions trading, which aims to apply Coase's theorem of "efficiency in resource allocation through clear definition of rights" to water pollution control. In 2005, the European Union established the Emission Trading Scheme (ETS), covering more than 11,000 companies including manufacturing. The United States, the United Kingdom, and Australia have also established emission trading systems and exchanges. In China, although the "National Carbon Emissions Trading Market Construction Plan (Power Generation Industry)" was announced and the threshold for China's power generation industry to be included in the carbon market was set, the quota allocation method has not yet been announced. Moreover, the carbon market was only included in the power, cement, and electrolytic aluminum industries at the beginning, leaving the steel industry unincluded.

Scholars have studied the carbon trading mechanism from the perspectives of applicable industries, allocation, and influences. In the applicable industry research, most of the studies have been focused on the power industry where data and methods are relatively mature. Li and Colombier [2] analyzed the effects of different energy conservation and emission reduction policies on reducing carbon dioxide in the construction industry based on energy demand, population growth, and economic development trends. The results showed that the carbon emission trading mechanism has played a prominent role in improving the buildings' energy efficiency and energy-saving technologies' promotion. Anger [3], Chen and Tseng [4], and Alberola [5] discussed the application of carbon trading mechanisms in electricity, the manufacturing industry, and aviation. Considine and Larson [6] examined fuel switching in electricity production following the introduction of the European Union's Emissions Trading System (EU ETS) for greenhouse gas emissions. Shen et al. [7] used structural modeling methods to determine the factors affecting the carbon trading mechanisms' implementation in construction industry and explained the complex relationship between factors. In combination with the carbon footprint and the classic transportation model, Su [8] added carbon trading mechanism to the traditional three-level transportation network model and comprehensively analyzed the impact of carbon quotas and trading mechanisms on operating costs and transportation networks. Based on the Computable General Equilibrium (CGE) model, Zhang [9] studied the impact of China's establishment of carbon trading system on the construction industry.

Policy makers, economists, and researchers have different opinions on the issue of the allocation of quotas, due to their own situations. Ahman and Zetterberg [10] believed that, according to the different basic calculation data, the free distribution method can be divided into two distribution methods: Carbon emissions and production. However, the latter has higher data requirements than the former. Goeree et al. [11] believed that the free distribution method based on historical emissions as the basis for allocating quotas will lead to high carbon emission enterprises obtaining large quotas of carbon emission quotas due to historical data accounting, as well as raising quota prices in quota trading in the carbon market. Zetterberg et al. [12] suggested that the government should reward companies that have adopted emission reduction actions at the beginning of the carbon trading system, which will incentivize companies to obtain more quotas by increasing output. Grimm and Ilieva [13] believed that the allocation baseline for carbon emission rights will change over time and technology, so they cannot be allocated with reference to historical emissions data. For this reason, they proposed a scalable free allocation model. Wang and Wang [14] studied the carbon emission rights allocation in Beijing. They believed that distribution model combining free and auction is more suitable for China's national conditions. In the Emission Trading Scheme in Korea, Lee and Yu [15] observed the actual CO_2 emission levels in the first compliance year and estimated the maximum and minimum emission levels by conducting a sensitivity analysis. They also estimated the surplus or deficit of emission permits during the first phase by comparing the estimated emission levels and the permit supply. Finally, they explored the supply and future prospects of offset credits, as well as the allocated permits. Guo, Chen, and Long [16] studied the initial quota allocation of the consuming government (based on the perspective of government and family evolutionary game) and found that when the government's evolutionary stability strategy is "strict policy," the family's evolutionary stabilization strategy will be affected by carbon reduction costs and purchasing carbon emission rights costs. Zhang, Li and Jia [17] established a Computable General Equilibrium (CGE) model to analyze the impact of different ETS quota allocation schemes on the electricity industry and determined the best choice of a quota allocation scheme for the electricity industry in China. The research on China's carbon trading market may provide an important case for the global carbon trading market. In the European Union's Emissions Trading System (EU ETS), Duscha [18] analyzed the demand for certificates from a reserve of about 400 million allowances under different assumptions on production development as well as different design options for Phase IV. The analysis was built on freely available allocation data from Phase III along with projections of production trends from different time periods in the past.

The impact of implementing carbon emission trading policies is mainly reflected by the impact on macroeconomic factors such as government welfare, emission reduction effects, and prices. Chen and Wu [19] constructed a model for the free allocation and sale of carbon allowances. The results showed that China's implementation of carbon emission trading can reduce the economic losses caused by CO_2 emission reduction. Ellerman et al. [20] believed that developing countries can take full use of the new export chances brought about by carbon trading mechanisms to reap the benefits, while the countries with rich energy can have the largest revenues. Babiker M, Reilly J, and Viguier L [21] used CGE model to find that carbon emissions trading plays a negative role on social welfare. Wang, Chen, and Zou [22] found that carbon emission trading has a strong impact on prices. CO_2 emission reduction policies can stimulate energy efficiency in the energy sector to reduce carbon emissions effectively. However, China's GDP growth and labor efficiency will be negatively affected. Jamasb and Kohler [23] used learning curve theory to find that carbon emission trading could play a role in updating and promoting energy-saving and emission-reducing technologies; they also found that the continuous popularization of energy-saving and emission-reducing technologies would, in turn, reduce carbon emission reduction costs. Aviyonah and Uhlmann [24] believed that environmental externalities could be measured by carbon trading prices, and carbon trading mechanisms can generate profits for companies that reduce CO_2 emissions costs. Cong and Wei [25] studied the impact of power companies under different carbon trading allocation criteria and found that different allocations would also change carbon prices and electricity prices. Wu et al. [26] found that after the establishment of the carbon trading system, net exports of high-energy-consuming industries such as steel and electricity were greatly affected, and the power industry was even more affected. Li and Zhu [27] obtained the carbon emission trading's impact on high energy industries by constructing a partial equilibrium model of the three commodities in two countries and an emission reduction cost curve based on the micro level of technology. The results showed that in the carbon market with free quota allocation, carbon emission trading may cause distortion effects of non-backward capacity and backward production capacity. Wu, Fan, and Xia [28] used a multi-regional computable general equilibrium (CGE) model to analyze the economic impacts of ETS policy when combined with renewable energy sources (RES) policies in China. Dai, Xie, and Liu [29] used the CGE model to assess the carbon emissions trading's impact and renewable energy policies on economic contribution and to explore the influence of carbon trading volume, carbon price, and emission limits on the GDP in different sectors. Using the real trading data from the EU carbon market, Liu, Gao, and Guo [30] constructed a network model by integrating time windows with the network model; three types of network features were examined. The growth pattern of the carbon trading network was analyzed. Fernández et al. [31] analyzed the effectiveness of the carbon market as a basic tool in the reduction of emissions. The analysis also included other overlapping policies aimed at fighting climate change—the promotion of renewables, for example.

On the study of the comparison and selection of carbon tax and carbon trading, it could be found that domestic and foreign scholars have not reached a consensus. Aviyonah and Uhlmann [24] argued that the carbon tax mechanism is easier to implement than the carbon trading mechanism because carbon trading mechanisms face the challenge of setting emission reduction targets. Mann [32] contended that a carbon tax policy is more important because carbon tax has the advantages of a simple implementation, a favorable emission reduction path for enterprises, and a small space for local governments to implement local protectionism. Wang et al. [33] believed that in the short term, the carbon trading mechanism is more cost-effective because of the higher cost in CO_2 emission reduction technology.

For ease of reference, the references and summary information involved in this section are shown in Table 1.

Table 1. References and summary information about the carbon trading mechanism involved in this section.

Applicable industries		
Year	**Researcher**	**Industrial sector**
2009	Li and Colombier	construction
2010	Anger	electricity
2008	Chen and Tseng	manufacturing industry
2009	Alberola	aviation
2012	Considine and Larson	electricity
2016	Shen et al.	construction
2017	Su	transportation
2017	Zhang	construction
Allocation		
Year	**Researcher**	**Main ideas**
2007	Ahman and Zetterberg	Free distribution method can be divided into two distribution methods: Carbon emissions and production.
2010	Goeree et al.	Free distribution method based on historical emissions will lead to high carbon emission enterprises to obtain large quotas of carbon emission quotas and raise quota prices.
2012	Zetterberg et al.	The government should reward companies that have adopted emission reduction actions at the beginning of the carbon trading system.
2013	Grimm and Ilieva	Allocation baseline will change over time and technology, cannot be allocated with reference to historical emissions data.
2014	Wang	Distribution model combining free and auction is more suitable for China's national conditions.
2017	Lee and Yu	Permits were either in surplus or insufficient, depending on the sub-sector.
2018	Guo, Chen, and Long	When the government's evolutionary stability strategy is "strict policy," the family's evolutionary stabilization strategy will be affected by the cost of carbon reduction.
2018	Zhang, Li and Jia	Different quota allocation schemes have impacts on electricity price, and there are some spillover effects to other industries.
2018	Duscha	Amount of allowances foreseen for dynamic allocation is sufficient for Phase IV in the European Union's Emissions Trading System (EU ETS).
Impact of implementing carbon emission trading policies		
Year	**Researcher**	**Main effect**
1998	Chen and Wu	Reduce the economic losses caused by CO_2 emission reduction.
1998	Ellerman et al.	Gain new export opportunities to reap the benefits.
2004	Babiker M, Reilly J, and Viguier L	Negative impact on the social welfare.
2005	Wang, Chen, and Zou	A strong impact on production and prices. However, China's GDP growth and labor efficiency will be negatively affected.
2007	Jamasb and Kohler	Update and promote energy-saving and emission-reducing technologies.
2009	Aviyonah and Uhlmann	Generate profits for companies that reduce CO2 emissions costs.
2010	Cong and Wei	Change carbon prices and electricity prices.
2015	Wu, Fan, and Xia et al.	Net exports of high-energy-consuming industries were greatly affected.
2017	Li and Zhu	Cause distortion effects of non-backward capacity and backward production capacity.
2017	Wu, Fan and Xia	The combination of an ETS and a feed-in tariff (FIT) results in greater GDP cost and welfare loss in all Chinese regions
2018	Dai, Xie, and Liu et al.	Contributes to the achievement of emission reduction targets with less economic cost
2018	Liu, Gao and Guo	As the market grows, the geodesic distances become shorter and the clustering coefficients become larger.
2018	Fernández. et al.	EU-ETS is effective to reduce emissions, and each phase has a greater impact on the reduction.

Through the literature review of emission reduction policies, especially carbon trading mechanisms, we can find that the macroeconomic energy and economic model can be used to calculate carbon trading quotas, carbon prices, and other values, which can better analyze the impact on the economy, environment, and emission reduction. However, most of this research was conducted from a macro perspective, and there is less literature on using game theory to study the mutual decision-making and emission reduction mechanism between the industry and enterprises from both microscopic and macroscopic perspective. There is also relatively less research literature on the integration of policies such as mechanisms into the emission reduction model and the comparison of single and mixed emission reduction policies. With the continuous economy development and the constant attention of environmental concerns, emission mitigation pressure will increase continuously. An individual carbon emission reduction policy cannot meet the needs of national economic development and emission mitigation. Carbon capture and storage (CCS), carbon sinks, and other significant reductions methods are possible to be integrated into the overall mitigation system. There are few papers on relevant game theory, and, as such, more research needs to be introduced. In addition, China has not yet started the carbon tax mechanism, and the official has not formulated detailed implementation plans. Therefore, this paper uses the carbon trading mechanism as the background of emission reduction policies to conduct corresponding policy research and analysis.

Secondly, we find that researchers have usually focused on national level and entire sectors (industry, service industry, agriculture). Research on a provincial level or a single industry department are few. The iron and steel industry is a basic industry and one of the most CO_2-emitting heavy industrial sectors, which needs to do some further important emission reduction tasks research.

Thirdly, the model gives priority to the literature while giving less focus to empirical analysis in the previous research, and the parameters usually derive from the existing data of developed countries. The study of two typical enterprises is the most classic. In China, there are significant regional differences and economic development imbalances, and these areas' steel industry information and parameters need to be supplemented.

Above all, the reasonable CO_2 emission reduction policy will be the core issue in this paper. Therefore, this paper organizes the six main areas' steel industry's energy, environment, and economic data. Integrated with the government corresponding long-term plan, the carbon trading mechanism—including production subsidies, CCS, and CO_2 external losses—is included in this research; then, a two-stage dynamic game model is constructed. Using the background of different carbon emissions' decreased demand, we discuss game behavior in different regions under different reduction targets; and, finally, we investigate different emissions reduction scenarios' effect and economic impact.

3. Methods

3.1. Notations and Explanations

As shown in Figure 1, according to the regional division of China, China is divided into six regions: North China (i.e., ① Beijing, ② Tianjin, ③ Hebei, ④ Shanxi, and ⑤ Inner Mongolia), Northeast China (i.e., ⑥ Liaoning, ⑦ Jilin, and ⑧ Heilongjiang), East China (i.e., ⑨ Shanghai, ⑩ Jiangsu, ⑪ Zhejiang, ⑫ Anhui, ⑬ Fujian, ⑭ Jiangxi, ⑮ Shandong, and ㉜ Taiwan), South Central China (i.e., ⑯ Henan, ⑰ Hubei, ⑱ Hunan, ⑲ Guangdong, ⑳ Guangxi, ㉑ Hainan, ㉝ Hong Kong, and ㉞ Macau), Southwest China (i.e., ㉒ Chongqing, ㉓ Sichuan, ㉔ Guizhou, ㉕ Yunnan, and ㉛ Tibet), Northwest China (i.e., ㉖ Shaanxi, ㉗ Gansu, ㉘ Qinghai, ㉙ Ningxia, and ㉚ Xinjiang). For data reasons, Tibet, Hong Kong, Macau, and Taiwan are not included. In this paper, regions are presented by subscripts: 1 represents North China, 2 represents Northeast China, and so on.

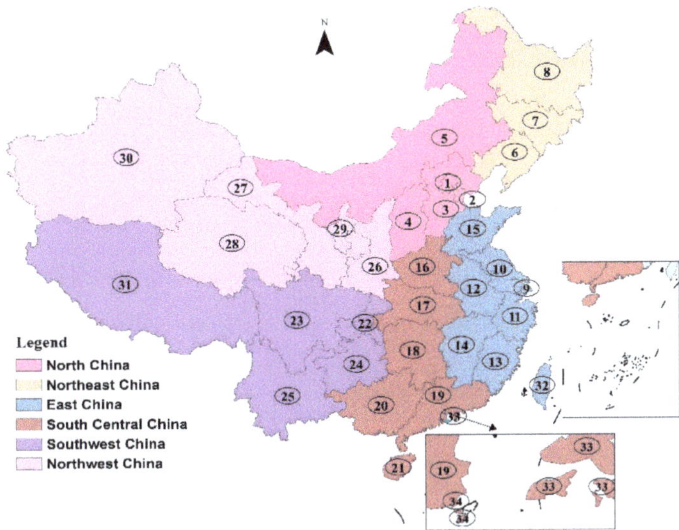

Figure 1. Regional division of China in this paper.

Consistent with the game sequence described in the literature of Duan et al. [34], the main research focus of this paper includes the government and the above six regions, and the regional steel industry data is regarded as a steel enterprise entity. The government emissions reduction policy is a double game problem. Firstly, there is a decomposition game between the government and regional enterprises: The government stipulates the emissions reduction target for a certain period, after which the regional enterprises should determine their respective reduction ranges according to their own cost curves. Secondly, there is a game of product output between the six regions, and the different targets of the regional enterprises will affect their respective output levels and market competitiveness. Therefore, the resultant game order is as follows: In the first stage, the government sets corresponding reduction targets and reduction policies (e.g., emissions allowances, product subsidies, CCS, independent or mixed); in the second stage, on the premise of guaranteeing profit maximization, the regional enterprises choose their respective reduction targets and output levels simultaneously. Based on this idea, this paper adopts the inverse method to solve the two-stage game problem. The notations used in this paper and their explanations are presented in Table 2. The main hypothesis of this paper is basically consistent with the hypothesis of Duan et al. [34].

Table 2. Notations and explanations used in this paper.

Notations	Explanations
Q	Steel production
P	The price of steel
α	The constant of the market inverse demand curve
β	The primary coefficient of the market inverse demand curve
q_i	Steel production of region i
$e_{2015,i}$	The region i CO_2 emission intensity of per ton steel in 2015
e_i	The region i CO_2 emission intensity of per ton steel at some stage
r_i	The decline range of CO_2 emission intensity of per ton steel in region i at some stage
R	The decline target of national CO_2 emission intensity of per ton steel at some stage
MAC	Marginal abatement cost curve in iron and steel industry
a_i	The quadratic coefficient of steel industry's MAC in region i
b_i	The primary coefficient of steel industry's MAC in region i

Table 2. *Cont.*

Notations	Explanations
C_i	The cost function of steel industry in region i
$C_{0,i}$	The production cost of steel industry in region i
c_i	The cost of base period emission reduction in region i
e_0	Carbon trading benchmarks
PP	Purchase price of carbon quota
SP	Selling price of carbon quota
CQ_i	Carbon quota of region i
W	Social welfare function
CS	Consumer surplus
PS	Producer surplus
$D(E)$	Total macro external environment loss of CO_2 emission
θ	The external loss parameter of CO_2
η	The production subsidies
m	The CO_2 emission reduced by CCS demonstration project
A	The primary coefficient of CCS demonstration project cost curve
B	The constant of CCS demonstration project cost curve
π_i	The profit function of steel industry in region i
E	The total CO_2 emissions in iron and steel industry
S	The total subsidy
M	The total cost of CCS demonstration project

3.2. Scenario Assumptions

For future development direction of steel industry and CO_2 emission reduction targets, the relevant development plans are mostly based on textual qualitative language, and the description of clear emission reduction policies is rather vague. Due to the uncertainty of the policy and the combing of the literature, this paper establishes a model of enterprise output selection under the carbon trading mechanism, studies the impact of future government (steel industry) emission reduction policies on the economic, environmental, and other factors of enterprises and the steel industry, and focuses on the trends of various economic and environmental indicators.

3.2.1. Case 1: Only Carbon Trading Scenarios

Recently, due to policy and technology constraints, the carbon trading mechanism is the only policy-based emission reduction measure to achieve emission reduction targets. The government (steel industry) sets CO_2 emission allowances for production enterprises, and enterprises buy or sell their quotas in the market. To prevent corporate speculation, this paper stipulates that the subsidy value is not greater than the enterprise purchase quota expenditure, and the time is set to 2020.

3.2.2. Case 2: Mixed Policy Scenario, Carbon Trading and Subsidy

When emission reduction targets continue to increase, steel companies that purchase carbon emission allowances will face increasing pressure to reduce emissions and purchase quotas, which will severely compress producers' profit margins. At this time, the rebate subsidy based on product output for enterprises purchasing carbon emission quota can improve their production enthusiasm and capacity. To prevent corporate speculation, this paper stipulates that the subsidy value is not greater than the enterprise purchase quota expenditure, and the time is set to 2025.

3.2.3. Case 3: Mixed Policy Scenario, Carbon Trading, Subsidy, and CCS

When the emission reduction targets gradually increase, the implementation of carbon trading and subsidies may not fully meet the CO_2 emission reduction target. CCS, which will play as a large-scale CO_2 emission reduction role, will be put into operation in the medium and long term. This research assumes that all subsidies are provided by the government (steel industry). To prevent corporate speculation, this paper stipulates that the subsidy value is not greater than the enterprise purchase quota expenditure, and the time is set to 2030. The government invests in the construction of one or two demonstration projects that can absorb millions of tons of CO_2.

3.3. The Two-Stage Dynamic Game Solving Process

Regional enterprises will face the given government emission mitigation targets and corresponding policies in the second stage, and select the decline in production output and emission intensity. The profit function in different situations is as follows:

$$
\begin{aligned}
\pi_{case1,i} &= P(Q)q_i - C_iq_i = (\alpha - \beta Q)q_i - q_iC_{0,i} - q_i\lambda\left(c_i + \int_0^{r_i} MAC_i(r)dr\right) + CT_i \\
&= (\alpha - \beta Q)q_i - q_iC_{0,case1,i} - q_i\lambda\left(c_i + \int_0^{r_i} MAC_i(r)dr\right) + CT_i
\end{aligned}
\tag{1}
$$

$$
\text{Among them, } CT_i = \begin{cases} SP \cdot [e_0 - e_{2015,i}(1 - r_i)] \cdot q_i, \text{ when } e_0 \geq e_{2015,i}(1 - r_i) \\ PP \cdot [e_0 - e_{2015,i}(1 - r_i)] \cdot q_i, \text{ when } e_0 < e_{2015,i}(1 - r_i) \end{cases}.
$$

$$
\begin{aligned}
\pi_{case2,i} &= P(Q)q_i - C_iq_i = (\alpha - \beta Q)q_i - q_iC_{0,i} - q_i\lambda\left(c_i + \int_0^{r_i} MAC_i(r)dr\right) + CT_i + \eta k_iq_i \\
&= (\alpha - \beta Q)q_i - q_iC_{0,case2,i} - q_i\lambda\left(c_i + \int_0^{r_i} MAC_i(r)dr\right) + CT_i + \eta k_iq_i
\end{aligned}
\tag{2}
$$

$$
\text{which, } CT_i = \begin{cases} SP \cdot [e_0 - e_{2015,i}(1 - r_i)] \cdot q_i, \text{when } e_0 \geq e_{2015,i}(1 - r_i) \\ PP \cdot [e_0 - e_{2015,i}(1 - r_i)] \cdot q_i, \text{when } e_0 < e_{2015,i}(1 - r_i) \end{cases}, k_i = \begin{cases} 0, \text{when } e_0 \geq e_{2015,i}(1 - r_i) \\ 1, \text{when } e_0 < e_{2015,i}(1 - r_i) \end{cases}.
$$

$$
\begin{aligned}
\pi_{case3,i} &= P(Q)q_i - C_iq_i = (\alpha - \beta Q)q_i - q_iC_{0,i} - q_i\lambda\left(c_i + \int_0^{r_i} MAC_i(r)dr\right) + CT_i + \eta k_iq_i \\
&= (\alpha - \beta Q)q_i - q_iC_{0,case3,i} - q_i\lambda\left(c_i + \int_0^{r_i} MAC_i(r)dr\right) + CT_i + \eta k_iq_i
\end{aligned}
\tag{3}
$$

$$
\text{which, } CT_i = \begin{cases} SP \cdot [e_0 - e_{2015,i}(1 - r_i)] \cdot q_i, \text{when } e_0 \geq e_{2015,i}(1 - r_i) \\ PP \cdot [e_0 - e_{2015,i}(1 - r_i)] \cdot q_i, \text{ when } e_0 < e_{2015,i}(1 - r_i) \end{cases}, k_i = \begin{cases} 0, \text{when } e_0 \geq e_{2015,i}(1 - r_i) \\ 1, \text{when } e_0 < e_{2015,i}(1 - r_i) \end{cases}.
$$

Each enterprise faces different profit functions because of different emission levels. In case 2 and case 3, the government subsidizes enterprises that purchase carbon quotas and does not subsidize enterprises which sell carbon quotas. When solving, it is necessary to comprehensively consider the relationship between the emission intensity and the reference value to determine whether the enterprise enjoys subsidies, and then determines the profit function of the enterprise.

Let $\dfrac{\partial \pi_i}{\partial q_i} = 0$ and $\dfrac{\partial \pi_i}{\partial r_i} = 0$, the corresponding reduction range of emission intensity r_i and output q_i of iron and steel enterprises in each region can be obtained.

In different cases, the social welfare function has been expanded, the specific form is as follows:

$$
\begin{aligned}
W_{Case1} &= CS + PS - D(E) = \int_0^Q P(q)dq - P(Q)Q + \sum_{i=1}^{6} \pi_{case1,i} - \theta E \\
&= \int_0^Q (\alpha - \beta q)dq - (\alpha - \beta \sum_{i=1}^{6} q_i) \sum_{i=1}^{6} q_i + \sum_{i=1}^{6} \pi_{case1,i} - \theta \sum_{i=1}^{6} e_{2015,i}(1 - r_i)q_i
\end{aligned}
\tag{4}
$$

$$W_{Case2} = CS + PS - S - D(E) = \int_0^Q P(q)dq - P(Q)Q + \sum_{i=1}^6 \pi_{case2,i} - \eta \sum_{i=1}^6 k_i q_i - \theta E$$
$$= \int_0^Q (\alpha - \beta q)dq - (\alpha - \beta \sum_{i=1}^6 q_i) \sum_{i=1}^6 q_i + \sum_{i=1}^6 \pi_{case2,i} - \eta \sum_{i=1}^6 k_i q_i - \theta \sum_{i=1}^6 e_{2015,i}(1 - r_i)q_i \tag{5}$$

$$W_{Case3} = CS + PS - S - D(E) - M = \int_0^Q P(q)dq - P(Q)Q + \sum_{i=1}^6 \pi_{case3,i} - \eta \sum_{i=1}^6 k_i q_i - \theta E - (Am + B)$$
$$= \int_0^Q (\alpha - \beta q)dq - (\alpha - \beta \sum_{i=1}^6 q_i) \sum_{i=1}^6 q_i + \sum_{i=1}^6 \pi_{case3,i} - \eta \sum_{i=1}^6 k_i q_i - \theta \sum_{i=1}^6 e_{2015,i}(1 - r_i)q_i - (Am + B) \tag{6}$$

In the first stage, 6 regions' enterprises obtain the corporate profit function by selecting the decline in their respective product output and emission intensity as a response to the government's corresponding emission reduction policy and emission reduction target R. The following relationship of government's decision can be obtained and expressed as:

Case 1

$$\max W$$
$$s.t. \begin{cases} \dfrac{\sum_{i=1}^6 e_{2015,i}(1 - r_i)q_i}{\sum_{i=1}^6 q_i} = e_{2010}(1 - R) \\ 0 < r_i < 1 \\ e_i > 0 \\ q_i > 0 \\ 0 \le SP \le PP \\ i = 1,2,3,4,5,6 \end{cases} \tag{7}$$

Case 2

$$\max W$$
$$s.t. \begin{cases} \dfrac{\sum_{i=1}^6 e_{2015,i}(1 - r_i)q_i}{\sum_{i=1}^6 q_i} = e_{2010}(1 - R) \\ 0 < r_i < 1 \\ e_i > 0 \\ q_i > 0 \\ 0 \le SP \le PP \\ \eta \ge 0 \\ 0 \le \dfrac{\eta \sum_{i=1}^6 k_i q_i}{-\sum_{i=1}^6 k_i \cdot PP \cdot [e_0 - e_{2015,i}(1 - r_i)]q_i} \le 1 \\ k = 0,1 \\ i = 1,2,3,4,5,6 \end{cases} \tag{8}$$

Case 3

$$\max W$$

$$s.t.\begin{cases} \dfrac{\sum\limits_{i=1}^{6} e_{2015,i}(1-r_i)q_i - m}{\sum\limits_{i=1}^{6} q_i} = e_{2010}(1-R) \\ 0 < r_i < 1 \\ e_i > 0 \\ q_i > 0 \\ 0 \leq SP \leq PP \\ \eta \geq 0 \\ 0 \leq \dfrac{\eta\sum\limits_{i=1}^{6} k_i q_i}{-\sum\limits_{i=1}^{6} k_i \cdot PP \cdot [e_0 - e_{2015,i}(1-r_i)]q_i} \leq 1 \\ k = 0,1 \\ 1 \times 10^6 \leq m \leq 2 \times 10^6 \\ i = 1,2,3,4,5,6 \end{cases} \tag{9}$$

The equilibrium of production and emissions mitigation can be described by SP, PP, and η. Then, we put the production and emission reduction results determined by SP, PP, and η into the first stage to get the expression formula of social economic welfare (W). In the game's first stage, the government maximizes social economic welfare determined by SP, PP, and η. Finally, we get W and other corresponding conclusions.

3.4. Data Sources

The statistics are from China Statistical Yearbook [35], China Industrial Statistical Yearbook [36], China Energy Statistical Yearbook [37], China Steel Yearbook [38], and the statistical yearbooks of various provinces. Relevant economic data are equivalent to comparable prices in 2010. The time span is from 2005 to 2016. In addition, CO_2 emissions in industrial production (IPPU CO_2), which also produces large amounts of CO_2, are included in this paper.

Because of the data availability, the steel industry's relevant energy consumption and economic data are derived from the ferrous metal smelting and calendaring processing industry in the China Statistical Yearbook. The CO_2 accounts of fossil energy consumption and IPPU refers to IPCC2006 [39] and Duan et al. [40].

4. Results and Discussions

4.1. The Parameter Fitting's Results

In this paper, referring to the research of Ma and Li [41], the market demand for steel industry is expressed by the apparent consumption, and the product price was obtained by dividing the industrial output value by the output. In the selection of the curve form, the inverse demand curve can be approximated as a straight line inclined to the lower right. The inverse demand curve fitting equation is as follows:

$$P = \alpha - \beta Q = 15769.56 - 1.13 \times 10^{-5}Q \tag{10}$$

In 2010, CO_2 emissions average level in steel industry was 3.1710 tons CO_2 per ton of steel (the same below, omitted). In the calculation below, the base period data of each region is the basic data for 2015 unless otherwise specified. According to the collection and calculation, the average level of CO_2 emission in 2015 was 2.8210. Correspondingly, the CO_2 emission levels of the six regions in 2015 were $e_1 = 2.3344$, $e_2 = 3.5698$, $e_3 = 2.9040$, $e_4 = 2.8779$, $e_5 = 3.2202$, and $e_6 = 4.5864$.

In this paper, China's steel industry CO_2 emissions from 2005 to 2016 were calculated by a non-parametric method, the CO_2 marginal abatement cost was estimated by shadow price, and then the marginal abatement cost curve (MACC) was obtained. The calculation method reference was made to the relevant literature of Duan et al. [42–44] (the data was updated to 2016 and the function form was slightly changed), and the quadratic form was selected as the regression equation of MACC. The relationship between the emission intensity reduction in each region and the CO_2 MACC is shown in Table 3. It should be pointed out that due to the estimation of shadow price, the error and uncertainty of the CO_2 marginal abatement cost curve are increased, which leads to a certain distance between the calculated abatement cost and the real abatement cost. This paper uses $\lambda\left(c_i + \int_0^{r_i} MAC_i(r)dr\right)$ to represent the actual emission reduction cost; λ is 0.5, in order to reduce the error.

Table 3. Some parameter values.

Notations	Unit		$i = 1$	$i = 2$	$i = 3$	$i = 4$	$i = 5$	$i = 6$
$e_{2015,i}$	t CO_2/t		2.3344	3.5698	2.9040	2.8779	3.2202	4.5864
a_i	-		11661	17208	16932	12952	6397.2	3485
b_i	-		−169.76	8876.7	−166.92	1483.6	502.52	421.13
c_i	Yuan		2168.2	3511.1	2165.4	3325.1	2368.7	3814.3
$C_{0,2015,i}$	Yuan	2015	2833.15	4898.47	3453.53	4153.15	3799.03	3832.38
		2020	2124.86	3918.77	2590.15	2491.89	3609.08	3640.76
		2025	1699.89	2743.14	2072.12	1868.92	3067.72	3094.64
		2030	1444.91	2194.51	1761.30	1588.58	2454.17	2475.71

Through the combing of relevant policy documents, the future economic development goals of China's iron and steel industry and target of energy conservation and CO_2 emission mitigation have been designated as mostly textual narratives, and there are basically no quantitative indicators of data. Accurately predicting future steel industry emission reduction targets, product yields, and development levels is impractical. Based on this, the focus of this paper is to explore the influence of implementation of different emission mitigation policies combination in China's steel industry's social welfare (W), consumer surplus (CS), producer surplus (PS), output, and total CO_2 emissions, as well as the enterprise's output, profits, CO_2 emissions, and other economic and environmental factors.

In the setting of the CO_2 emission intensity reduction target, this paper refers to target of reducing the comprehensive energy consumption of ton steel by 12 kgce in 2020 (Steel Industry Adjustment and Upgrade Plan (2016–2020)). The target of 2020 is about 85% of the energy consumption level in 2010. Therefore, this paper sets the CO_2 emission intensity in 2020 to be 15% lower than that in 2010. We will observe the reaction of output, profit, and emission intensity of iron and steel enterprises and industries when the emission reduction target under the condition of 15%–20%. We set the CO_2 emission intensity in 2025 to be 20% lower than that in 2010 and observed the reaction of output, profit, and emission intensity of iron and steel enterprises and industries when the emission reduction targe under the condition of 20%–25%. We set the CO_2 emission intensity in 2030 to be 25% lower than that in 2010 and observed the reaction of output, profit, and emission intensity of iron and steel enterprises and industries when the emission reduction target under the condition of 25%–30%.

Referring to Li's research [45], the average production cost of the base period can be obtained by calculating the business (regional) operating costs and steel production. This paper assumes that by 2020, the production costs in North China, East China, and South Central China will decrease significantly, and the production costs in Northeast China, Southwest China, and Northwest China will decrease less. Then, the gap will be gradually narrowed. By 2030, North China, East China, and South Central China will be basically in the same cost range, and the Northeast, Southwest, and Northwest regions will be located in another cost range. The external macro-environmental loss parameters refer to the study of Guenno and Tiezzi [46], $\theta = 14.55$ Yuan/ton CO_2. In 2030, it is assumed that the steel industry will plan to build 1–2 million tons of government-funded demonstration projects. The cost curve refers to [47–50], and the linear curve is $M = 204.96m + 11.53 \times 10^6$, $m \in \left[1 \times 10^6, 2 \times 10^6\right]$ ton.

The unit is *RMB*. Since the relationship between *PP* and *SP* is not easy to determine, let *PP* = *SP* in order to facilitate calculation; that is, when carbon trading is conducted between enterprises, the price used to purchase or sell unit quota is the same.

In the selection of the benchmark value of the carbon trading mechanism e_0, considering that China has just started to pilot the carbon trading mechanism, the steel industry emission benchmark value should not be set too high. After the system is mature, the benchmark value should be set strictly. Assume that the benchmark value for 2020 is 2.8210 tons of CO_2/per ton steel (the average CO_2 emission intensity in 2015). The benchmark value for 2025 is 2.3782 (25% lower than the national emission level in 2010). For 2030, it will be 2.2197 (30% lower than the national emissions level in 2010).

The specific data simulation parameter settings are shown in Table 3.

4.2. Empirical Analysis

4.2.1. Case 1, in 2020 Scenario

The carbon trading policy is the only emission mitigation policy. Detailed results are shown in Figures 2 and 3, Tables 4 and 5. When the emission reduction targets gradually increase from 15% to 20%, the transaction price of carbon emission allowance increases from 12.65 yuan/ton to 34.81 yuan/ton, which increases by 2.75 times. Total output, total social welfare (W), consumer surplus (CS), and producer surplus (PS) all show an upward trend, and the macro environmental damage caused by CO_2 emissions shows a decreasing trend. The total output maintains at around 854 million tons, which is a drop of nearly 25% compared to the 1.135 billion tons of steel production in 2016. This means that demands in the steel market declined significantly in 2020. As the emission reduction target gradually increases from 15% to 20%, total output, W, CS, PS, and emission loss increase by 0.05%, 0.21%, 0.10%, 0.34%, and −5.84%, respectively.

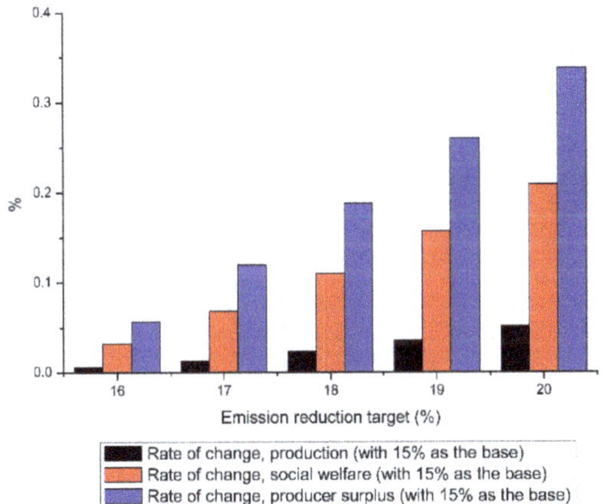

Figure 2. The rate of change of output, social welfare, and producer surplus under the carbon trading mechanism; take 15% as the baseline.

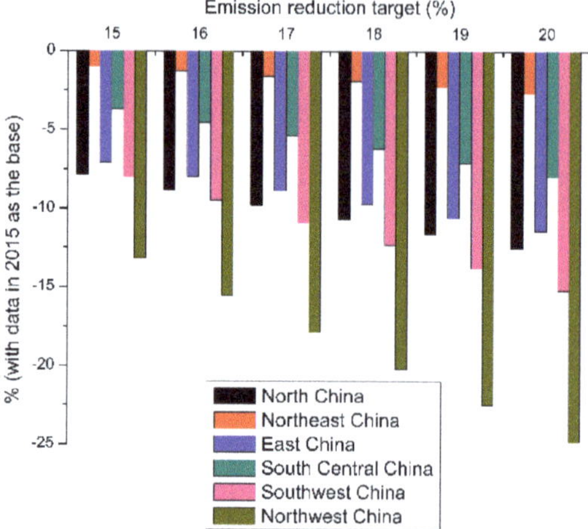

Figure 3. The change of regional emission intensity with the overall emission reduction target (15%–20%, carbon trade scenario only).

Table 4. The changes of the overall industry indicators under the carbon trade mechanism; the emission reduction target is 15%–20%.

Emission Reduction Target	15%	16%	17%	18%	19%	20%
Carbon quota transaction price (Yuan)	12.65	16.28	20.30	24.74	29.57	34.81
Production (100 million tons)	8.5427	8.5432	8.5439	8.5447	8.5457	8.5471
Rate of change, production (with 15% as the base)	-	0.01%	0.01%	0.02%	0.04%	0.05%
Rate of change, social welfare (with 15% as the base)	-	0.03%	0.07%	0.11%	0.16%	0.21%
Rate of change, producer surplus (with 15% as the base)	-	0.06%	0.12%	0.19%	0.26%	0.34%

Table 5. The changes of the various region enterprise indicators under the carbon trade mechanism; the emission reduction target is 15%–20%.

Emission Reduction Target		15%	16%	17%	18%	19%	20%
Production (100 million tons)	North China	2.5841	2.5854	2.5869	2.5884	2.5900	2.5916
	Northeast China	0.3872	0.3864	0.3856	0.3847	0.3837	0.3827
	East China	2.1673	2.1678	2.1684	2.1690	2.1696	2.1702
	South Central China	1.7405	1.7409	1.7413	1.7417	1.7422	1.7427
	Southwest China	1.1722	1.1723	1.1725	1.1726	1.1728	1.1730
	Northwest China	0.4915	0.4901	0.4886	0.4871	0.4856	0.4841
Rate of change, production (with 15% as the base)	North China	-	0.11%	0.23%	0.36%	0.50%	0.65%
	Northeast China	-	−0.42%	−0.91%	−1.47%	−2.11%	−2.82%
	East China	-	0.05%	0.10%	0.16%	0.22%	0.29%
	South Central China	-	0.05%	0.10%	0.15%	0.21%	0.27%
	Southwest China	-	0.02%	0.05%	0.09%	0.13%	0.18%
	Northwest China	-	−0.59%	−1.19%	−1.80%	−2.40%	−2.98%
The decline range of emission intensity (with data in 2015 as the base)	North China	−7.88%	−8.83%	−9.77%	−10.71%	−11.63%	−12.55%
	Northeast China	−1.00%	−1.28%	−1.58%	−1.92%	−2.28%	−2.66%
	East China	−7.10%	−7.98%	−8.85%	−9.72%	−10.58%	−11.43%
	South Central China	−3.71%	−4.53%	−5.36%	−6.22%	−7.09%	−7.97%
	Southwest China	−8.02%	−9.46%	−10.90%	−12.34%	−13.77%	−15.20%
	Northwest China	−13.18%	−15.52%	−17.85%	−20.18%	−22.50%	−24.82%

To achieve the 15%–20% emission reduction target, enterprises in each region choose the emission intensity and output. As the emission reduction target increases, the emission intensity of Southwest China and Northwest China decreases more. When the industry emission reduction target is 20%, the Northwest region's emission reduction rate reaches its maximum, which is 24.82%; the East China and North China range is moderate, with a range of 7%–13%; the emission reductions in the Northeast and South Central regions are small, and the Northeast is the smallest. For production, due to the different production costs and the impact of emission mitigation intensity, coupled with the implementation of the carbon trading policy, when the emission reduction targets grow, output of most regions has increased significantly, and only the steel production in Northeast China and Northwest China has declined. The output of the two regions has dropped significantly, which is approaching 3%.

Under the premise emission reduction target of 15%–20% and 2.8210 tons CO_2/ton steel as the benchmark value, the CO_2 emission intensity of North China, East China, and South Central China in the carbon trading market are lower than the benchmark value; they can choose to sell their carbon quota. The Northeast and Northwest regions have exceeded the benchmark value and need to purchase carbon quotas. For the Southwest region, when the emission reduction target is low (15%–18%), CO_2 emission intensity is more than the benchmark and the enterprise needs to be purchased for the quotas; when the emission reduction target is high (18%–20%) and CO_2 emission intensity is lower than the benchmark value, the enterprise can choose to sell its own carbon quotas.

4.2.2. Case 2, in 2025 Scenario

The mixed emission mitigation policy includes carbon trading and subsidy policies. Detailed results are shown in Figures 4 and 5, Tables 6 and 7. Due to the strict emission benchmark values, the CO_2 emission intensity of some regional enterprises cannot meet or fall below the benchmark value. The government needs to provide subsidies for enterprises with higher emission intensity. When the emission reduction targets gradually increase from 20% to 25%, the transaction price of carbon emission allowance increases from 47.21 yuan/ton to 84.43 yuan/ton, an increase of 1.79 times. The value is maintained at around 15.5 yuan to 17 yuan.

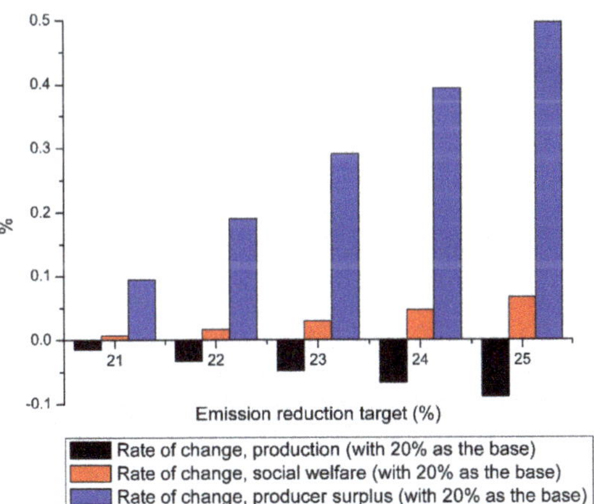

Figure 4. The rate of change of output, social welfare, and producer surplus under the carbon trading mechanism; take 20% as the baseline.

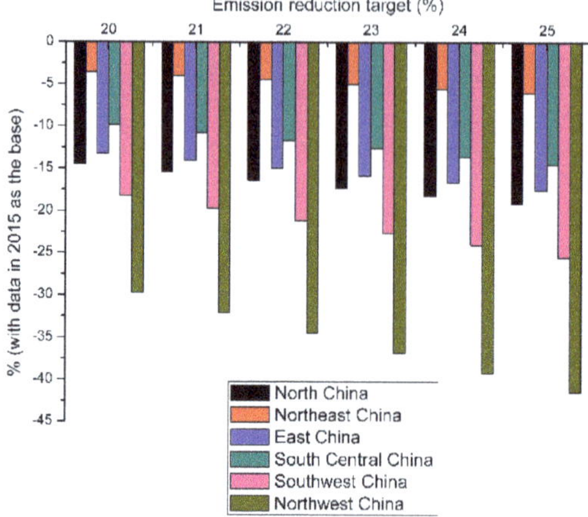

Figure 5. The change of regional emission intensity with the overall emission reduction target (20%–25%, carbon trade+ subsidy).

Table 6. The changes of the overall industry indicators under the carbon trade mechanism; the emission reduction target is 20%–25%.

Emission Reduction Target	20%	21%	22%	23%	24%	25%
Carbon quota transaction price (Yuan)	47.21	53.86	60.91	68.37	76.21	84.43
Carbon quota transaction price (Yuan)	9.0173	9.0159	9.0143	9.0129	9.0113	9.0093
Unit value of subsidy (Yuan)	17	17.5	17.5	17.5	17	15.5
Rate of change, production (with 20% as the base)	-	−0.02%	−0.03%	−0.05%	−0.07%	−0.09%
Rate of change, social welfare (with 20% as the base)	-	0.01%	0.02%	0.03%	0.05%	0.07%
Rate of change, producer surplus (with 20% as the base)	-	0.09%	0.19%	0.29%	0.39%	0.50%

Table 7. The changes of the various region enterprise indicators under the carbon trade mechanism; the emission reduction target is 20%–25%.

Emission Reduction Target		20%	21%	22%	23%	24%	25%
Production (100 million tons)	North China	2.4863	2.4883	2.4900	2.4921	2.4942	2.4963
	Northeast China	0.9253	0.9231	0.9209	0.9185	0.9160	0.9134
	East China	2.1506	2.1511	2.1517	2.1521	2.1526	2.1531
	South Central China	1.8143	1.8145	1.8149	1.8152	1.8154	1.8156
	Southwest China	1.1706	1.1707	1.1710	1.1712	1.1715	1.1717
	Northwest China	0.4702	0.4686	0.4673	0.4659	0.4646	0.4634
Rate of change, production (with 20% as the base)	North China	-	0.15%	0.32%	0.49%	0.68%	0.90%
	Northeast China	-	−0.48%	−1.01%	−1.58%	−2.21%	−2.91%
	East China	-	0.05%	0.09%	0.14%	0.17%	0.19%
	South Central China	-	0.03%	0.06%	0.09%	0.10%	0.10%
	Southwest China	-	0.04%	0.08%	0.12%	0.17%	0.20%
	Northwest China	-	−0.60%	−1.17%	−1.67%	−2.11%	−2.52%
The decline range of emission intensity (with data in 2015 as the base)	North China	−14.49%	−15.43%	−16.36%	−17.29%	−18.21%	−19.13%
	Northeast China	−3.55%	−4.02%	−4.51%	−5.01%	−5.54%	−6.08%
	East China	−13.23%	−14.09%	−14.96%	−15.81%	−16.67%	−17.52%
	South Central China	−9.85%	−10.77%	−11.69%	−12.62%	−13.55%	−14.47%
	Southwest China	−18.22%	−19.69%	−21.14%	−22.60%	−24.05%	−25.49%
	Northwest China	−29.72%	−32.09%	−34.45%	−36.81%	−39.15%	−41.48%

The total output, consumer surplus, and macro environmental damage caused by CO_2 emissions have decreased, and the overall social welfare and producer surplus have shown an upward trend. This shows that, although the emission reduction target has become increasingly severe, the behavior of returning subsidies to enterprises has increased the production enthusiasm of producers and helps to increase corporate profits. The total output has remained at around 900 million tons. Though the emission reduction target has increased, the total output has increased slightly compared to the 2020 emission reduction targets case. When emission reduction targets increase from 20% to 25%, total output, W, CS, PS, and emission loss decrease by 0.09%, −0.07%, 0.18%, −0.50%, and 6.33%, respectively.

For the sub-region, to achieve the emission mitigation target of 20%–25%, enterprises in each region choose the emission intensity and output. When the emission reduction targets gradually increased, the rate of decline in emission reduction intensity in various regions has gradually increased. In Southwest and Northwest China, the emission reduction rate has exceeded 20% in most cases. When the industry emission reduction target is 25%, the emission reduction rate in the northwest region has reaches the maximum, which is 41.48%, in East China and North China. When the emission range is in the middle, the decline range is about 13%–19%. The emission reduction range in the Northeast and South Central regions is small, and, in the Northeast, it is the smallest. Along with the effects of carbon trading policies and subsidies, the output changes show great regional differences. The output of North China, East China, South Central, and Southwest China present an upward trend; steel production in Northeast China and Northwest China show a downward trend, with a drop of 2.5%–3%.

Under the premise emission reduction target of 20%–25% and 2.3782 tons CO_2/ton steel as the carbon trading benchmark value, only the North China CO_2 emission intensity is lower than the benchmark value, and the enterprise can choose to sell its own carbon quotas; CO_2 emission intensity in other regions cannot be lower than the benchmark level, and carbon quotas need to be purchased. In addition, the government (steel industry) needs to return subsidies based on production to these regions.

4.2.3. Case 3, in 2030 Scenario

The mixed emission mitigation policies cover carbon trading policy, subsidies, and government-funded CCS demonstration projects. The results are shown in Figures 6 and 7, Tables 8 and 9. Due to the strict emission benchmark values, the CO_2 emission intensity of some regional enterprises cannot meet or fall below the benchmark value. The government needs to provide subsidies for enterprises with higher emission intensity. As the emission reduction target increases from 25% to 30%, the transaction price of carbon emission allowance increases from 91.70 yuan/ton to 137.59 yuan/ton—an increase of 1.79 times. The unit subsidy value is maintained at around 30–36 yuan; the CO_2 processed by the government-funded CCS demonstration project are always at the lowest value, $m = 1,000,000$ ton.

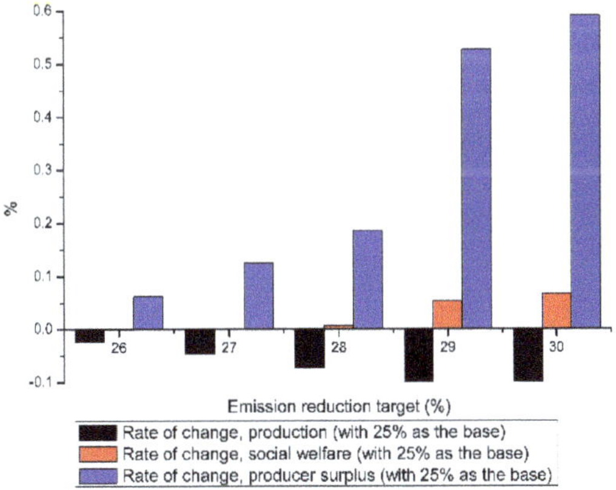

Figure 6. The rate of change of output, social welfare and producer surplus under the carbon trading mechanism; take 25% as the baseline.

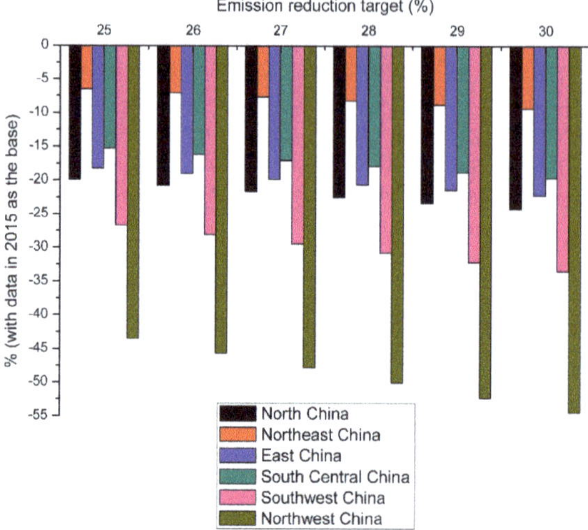

Figure 7. The change of regional emission intensity with the overall emission reduction target (25%–30%, carbon trade + subsidy + carbon capture and storage (CCS)).

Table 8. The changes of the overall industry indicators under the carbon trade mechanism; the emission reduction target is 25%–30%.

Emission Reduction Target	25%	26%	27%	28%	29%	30%
Carbon quota transaction price (Yuan)	91.70	100.24	108.95	117.98	127.83	137.59
Carbon quota transaction price (Yuan)	9.3391	9.3368	9.3348	9.3323	9.3283	9.3274
Unit value of subsidy (Yuan)	30.5	29.5	28.5	26.5	36.5	36
Rate of change, production (with 25% as the base)	-	−0.02%	−0.05%	−0.07%	−0.12%	−0.13%
Rate of change, social welfare (with 25% as the base)	-	0.00%	0.00%	0.01%	0.05%	0.07%
Rate of change, producer surplus (with 25% as the base)	-	0.06%	0.12%	0.19%	0.53%	0.59%

Table 9. The changes of the various region enterprise indicators under the carbon trade mechanism; the emission reduction target is 25%–30%.

Emission Reduction Target		25%	26%	27%	28%	29%	30%
Production (100 million tons)	North China	2.3949	2.3974	2.3999	2.4024	2.4049	2.4078
	Northeast China	1.0483	1.0449	1.0413	1.0377	1.0341	1.0302
	East China	2.1003	2.1005	2.1007	2.1008	2.1009	2.1009
	South Central China	1.7343	1.7342	1.7342	1.7341	1.7339	1.7337
	Southwest China	1.3864	1.3866	1.3869	1.3872	1.3875	1.3877
	Northwest China	0.6749	0.6743	0.6739	0.6736	0.6734	0.6732
Rate of change, production (with 25% as the base)	North China	-	0.21%	0.42%	0.66%	0.97%	1.17%
	Northeast China	-	−0.66%	−1.35%	−2.10%	−1.75%	−2.54%
	East China	-	0.02%	0.03%	0.03%	0.60%	0.61%
	South Central China	-	−0.01%	−0.02%	−0.05%	0.62%	0.60%
	Southwest China	-	0.04%	0.08%	0.11%	−1.28%	−1.18%
	Northwest China	-	−0.15%	−0.23%	−0.27%	−3.18%	−2.94%
The decline range of emission intensity (with data in 2015 as the base)	North China	−19.90%	−20.77%	−21.62%	−22.47%	−23.36%	−24.21%
	Northeast China	−6.55%	−7.09%	−7.63%	−8.19%	−8.79%	−9.37%
	East China	−18.24%	−19.04%	−19.83%	−20.62%	−21.44%	−22.22%
	South Central China	−15.26%	−16.14%	−17.01%	−17.88%	−18.79%	−19.66%
	Southwest China	−26.71%	−28.08%	−29.42%	−30.76%	−32.16%	−33.50%
	Northwest China	−43.45%	−45.68%	−47.85%	−50.01%	−52.27%	−54.44%

When the emission reduction targets gradually increase from 25% to 30%, the total output, consumer surplus, and macro-environmental losses caused by CO_2 emissions have decreased, and the overall social welfare and producer surplus have shown an upward trend. This suggests that, although the emission reduction target becomes increasingly severe, the behavior of returning subsidies to enterprises has increased the production enthusiasm of producers and helps to increase corporate profits. The total output has remained at around 933 million tons. Though the emission reduction target has increased, the total output has increased slightly compared to the 2025 emission reduction targets case. Total output, W, CS, PS, and emission loss decreased by 0.13%, −0.07%, 0.25%, −0.59%, and 6.81%, respectively.

For the sub-region, as the emission reduction target increases, and the rate of decline in emission reduction intensity in various regions has gradually increased. In most cases in Southwest China and Northwest China, the emission intensity decrease exceeded 30%. When the overall emission reduction target is 30%, the Northwest region's reduction rate reaches the maximum, which is 54.43%. In North China, East China, and South Central China, the emission reduction rate is in the middle; the rate in the Northeast is the smallest. Along with the effects of mixed carbon trading policies, the output changes show great regional differences. In North China, East China, and South Central China, it showed an upward trend; the output of steel in Northeast China, Southwest China, and Northwest China show a downward trend.

Under the premise emission reduction target of 25%–30% and 2.2197 tons CO_2/ton steel as the carbon trading benchmark value, only the North China CO_2 emission intensity is lower than the benchmark value, and the enterprise can choose to sell its own carbon quotas. CO_2 emission intensity in other regions cannot be lower than the benchmark value under most emission reduction targets; they need to purchase carbon quotas, and the government (steel industry) needs to return subsidies based on production. When the target is high (29%–30%), the CO_2 emission intensity in Southwest China and Northwest China is slightly lower than the benchmark value, so they can choose to sell their carbon quotas.

4.2.4. Comparison of Single Emission Reduction Policies and Mixed Emission Reduction Policies

Sections 4.2.2 and 4.2.3 discuss the impact of the implementation of the mixed emission reduction policy on the changes in the economic and environmental factors of steel companies and industries under the carbon trading mechanism. This section calculates the changes in the output and emission reduction intensity of enterprises and other economic factors in the implementation of the single policy under the same emission reduction targets, then compares and analyzes the single and mixed emission reduction policy.

(1) Comparison of only carbon trading policies and mixed policies in 2025.

After calculation and as shown in Table 10, Figures 8–10, the social welfare, total output, producer surplus, consumer surplus, carbon allowance transaction price, and the macro environmental damage caused by CO_2 emissions and the decline in CO_2 emission intensity under the mixed policy mechanism in various regions have increased more than that under the situation of the single policy. Under the mixed policy mechanism, the carbon price trading price and the regional emission intensity decrease than that in the single policy mechanism, but the gap is negligibly small.

Table 10. The indicators change under the carbon trade mechanism; the target in single policy is 20% to 25%.

Emission Reduction Target	20%	21%	22%	23%	24%	25%
Carbon quota transaction price (Yuan)	46.91	53.54	60.59	68.03	75.88	84.12
Production (100 million tons)	9.0066	9.0049	9.0033	9.0019	9.0006	8.9995
Rate of change, production (with 20% as the base)	-	−0.02%	−0.04%	−0.05%	−0.07%	−0.08%
Rate of change, social welfare (with 20% as the base)	-	0.01%	0.02%	0.03%	0.05%	0.07%
Rate of change, producer surplus (with 20% as the base)	-	0.09%	0.19%	0.29%	0.40%	0.50%

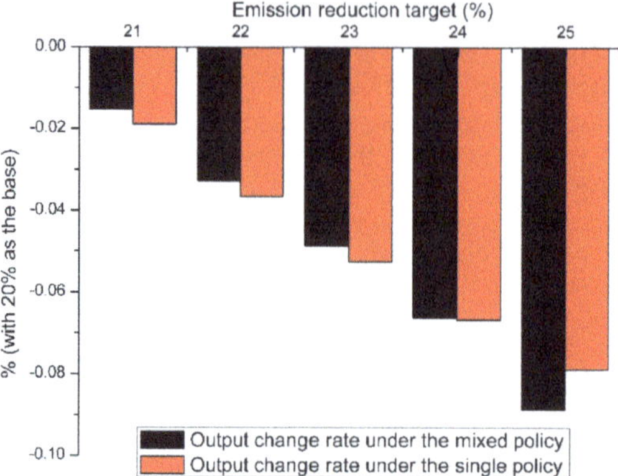

Figure 8. The output change rate compared under the single policy and the mixed policy in the carbon trade mechanism; take 20% as the baseline.

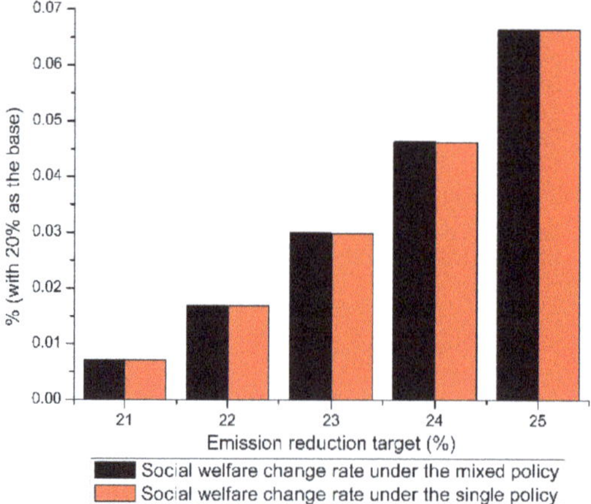

Figure 9. The social welfare change rate compared under the single policy and the mixed policy in the carbon trade mechanism; take 20% as the baseline.

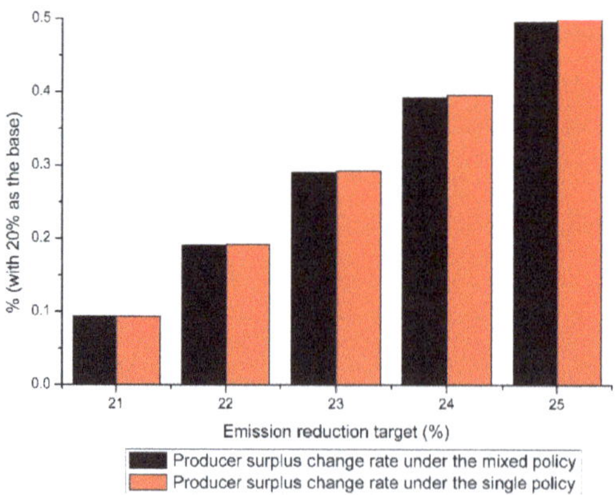

Figure 10. The producer surplus change rate compared under the single policy and the mixed policy in the carbon trade mechanism; take 20% as the baseline.

The rate of change reflects the intensity of changes in various indicators as the emission reduction target increases (compared to 20% reduction in 2010). From the figures, it can be found that in production change rates when emission reduction targets are low (20%–24%), steel production under a single policy changes significantly. When emission reduction targets are high (24%–25%), the change of steel production under the mixed policy is more obvious. In the change rate of producer surplus and social welfare, the two policy conditions are basically flat, and the single policy changes slightly larger.

(2) Comparison of only carbon trading policies and mixed policies in 2030.

Calculations have found that in the case of emission reduction targets of 25%–30%, the CO_2 emissions processed by CCS demonstration projects are at the lowest value; that is, under each emission reduction target, the CCS demonstration project is only at the lowest maintenance stage.

Similarly, after calculation and as shown in Table 11, Figures 11–13, the social welfare, total output, producer surplus, consumer surplus, carbon allowance transaction price, and the macro environmental damage caused by CO_2 emissions and the decline in CO_2 emission intensity under the mixed policy mechanism in various regions have increased more than that under the situation of the single policy. Under the mixed policy mechanism, the carbon price trading price and the regional emission intensity decrease more than that in the single policy mechanism, but the gap is so small that they can be ignored.

Table 11. The indicators change under the carbon trade mechanism; the target in single policy is 25% to 30%.

Emission Reduction Target	25%	26%	27%	28%	29%	30%
Carbon quota transaction price (Yuan)	91.17	99.72	108.45	117.51	126.89	136.59
Production (100 million tons)	9.3200	9.3183	9.3168	9.3156	9.3146	9.3138
Rate of change, production (with 25% as the base)	-	−0.02%	−0.03%	−0.05%	−0.06%	−0.07%
Rate of change, social welfare (with 25% as the base)	-	0.00%	0.00%	0.01%	0.02%	0.03%
Rate of change, producer surplus (with 25% as the base)	-	0.07%	0.13%	0.20%	0.27%	0.34%

The rate of change reflects the intensity of changes in various indicators as the emission reduction target increases (compared to 25% reduction in 2010). From the figures, it can be found that in production change rate, the effect of implementing a single policy on output change is significantly less than the impact of the implementation of the mixed policy. In the change rate of W and PS, when emission reduction targets are low (26%–28%), change rates are basically flat under the two policy conditions and slightly larger under a single policy. When emission reduction targets are higher (28%–30%), changes under mixed policies are more pronounced.

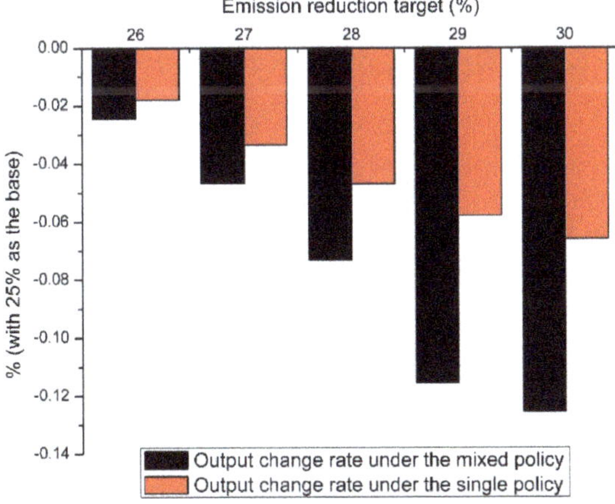

Figure 11. The output change rate compared under the single policy and the mixed policy in the carbon trade mechanism; take 25% as the baseline.

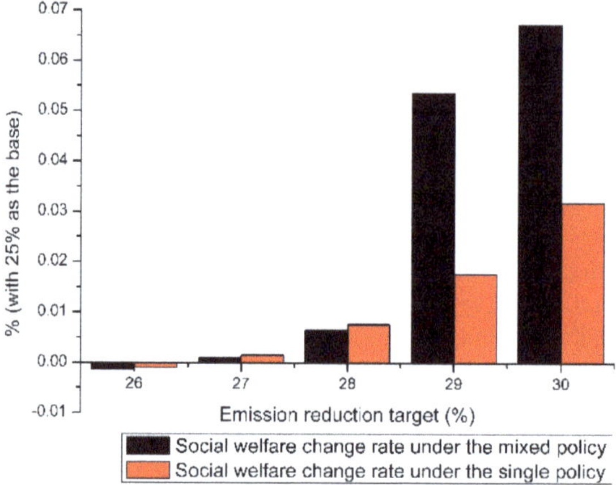

Figure 12. The social welfare change rate compared under the single policy and the mixed policy in the carbon trade mechanism; take 25% as the baseline.

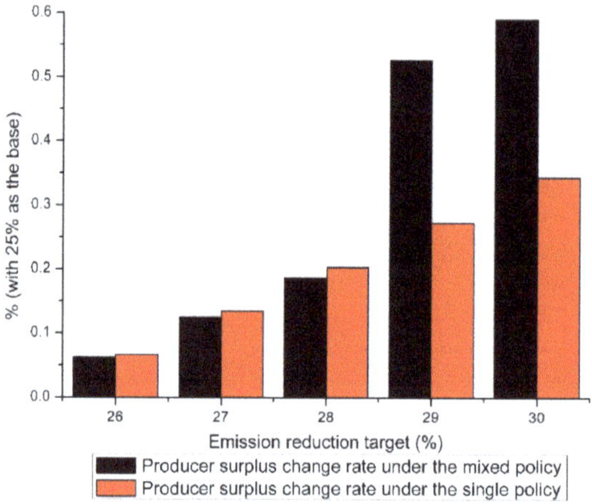

Figure 13. The producer surplus change rate compared under the single policy and the mixed policy in the carbon trade mechanism; take 25% as the baseline.

4.3. Discussions

Current research on the emissions mitigation targets is scarce. Therefore, this paper does not discuss optimal emissions intensity and methods. Only the above three scenarios (2020, 2025, 2030) have been analyzed. For China's industrial sector, the results of parameters and numerical calculations are few and lack certain references. Since the country has not announced the relevant details and trading methods of the carbon trading market, the carbon trading benchmark value is set based on the historical data. Moreover, due to these certain assumptions, there are some gaps between the calculations results and the actual situation, but some trends and rules can still be found and determined.

Since the carbon trading mechanism is based on the government's provision of a free carbon quota for each company, the benchmarks are set differently, and the results will vary. Under the conditions set in this paper, whether under the single carbon trading policy or the mixed policy, when the reduction target increases from 15% to 30%, the quota trading price steadily increases from 12.65 Yuan to about 137 Yuan (nearly 11 times), and the social welfare and producer surplus of the steel industry show an increasing trend—nearly 14% and 4%, respectively. While the macro-loss of CO_2 emissions shows a decreasing trend (decrease by more than 10%), other indicators increase or decrease. This shows that from the perspective of the implementation of the policy, the carbon trading policy and the mix with other emission reduction policies are conducive to the improvement of the economic level and emission reduction level of the steel industry.

For the sub-region, when the CO_2 emission reduction target gradually increases, the regional emission reduction pressure gradually increases as well. However, there are differences between regions. The decline range in emission reductions in Northwest China and Southwest China is much bigger than other four regions, and the range in the northeast region is the smallest.

Under the mixed policy mechanism, compared with the single carbon trading policy, the social welfare function, total output, producer surplus, consumer surplus, carbon quota trading price, macro-loss caused by CO_2 emissions, and emission reductions of the entire steel industry improve. Though under the same emission reduction target, the carbon allowance trading price in the single policy is less than that in the mixed policy, and the decline range in regional emissions reduction in the single policy is also lower than that under the mixed policies (though it is almost negligible—no more than 1 Yuan and 0.01 ton CO_2/ton). Therefore, under the carbon trading mechanism, it is necessary to comprehensively consider the balance between the whole industry and regions and to consider whether to use a single carbon trading mechanism or a mixed mechanism under the comprehensive evaluation.

Through numerical simulation, when the benchmark values in 2020, 2025, and 2030 are 2.8210 tons of CO_2/ton steel, 2.3782 tons of CO_2/ton steel, and 2.2197 tons of CO_2/ton steel, respectively, China's future steel output will be in the range of 854–934 million tons under the mixed policy mechanism. The carbon price trading price range is between 12.65–137.59 yuan. The subsidy unit value ranges from 15.5–36.5 yuan. The CCS demonstration projects are only in the minimum maintenance stage and have not played a significant role in social emission reduction. Under the single carbon trading policy, the future steel output will be in the range of 854–932 million tons, and the price of carbon trading quotas will be between 12.65 and 136.59 yuan.

Obviously, whether it is a single carbon trading policy or a mixed policy, when the market is balanced, the output is reduced by more than 200 million tons from the peak in 2016 (1.1346 billion tons). North China and East China have decreased a lot, while the northeast, southwest, and northwest regions have increased in production. This shows that the implementation of carbon trading policies and mixed policies have alleviated the polarization of production in various regions to some extent, which is conducive to optimizing the distribution of production between regions and enterprises so that enterprises can better achieve the best profit.

5. Conclusions

Through the introduction of carbon trading policy, subsidy policy, CCS, and other emission reduction policies, a two-stage multi-oligopoly enterprise production selection model including emission reduction policies is established. This paper makes a preliminary study on the changes of total production and social welfare indexes, which are under the restriction of the industry emission reduction target and emission reduction policy, and carries on the quantitative research with the operation data of iron and steel industry. The model is an abstraction and hypothesis of a practical problem, which causes the calculation results in this paper to potentially deviate a little from the actual situation. This paper emphasizes the change trends of corresponding indicators with the increase of governmental pressure on emissions reduction.

The main conclusions are as follows:

Under the carbon trading policy mechanism, the carbon quota trading price increases as the reduction targets increase, and the steel industry social welfare and producer surplus show an increasing trend. The CO_2 emission macro loss shows a decreasing trend, and the other indexes increase or decrease. When the emission reduction targets gradually increase, the pressure on emission reduction among regional oligarchic enterprises also gradually increases. The emission reduction rate of oligarchs in Northwest and Southwest China is much bigger than that in other regions. Oligarchy enterprises in

North China, East China and South Central China are second. Oligopolistic enterprises in Northeast China are the lowest. Since the benchmark values are set differently, the numerical results will be different, but the trends are basically similar.

Compared with the single carbon trading policy, the indicators under the mixed policy have increased. In the view of change rate in production, social welfare, and producer surplus, when under the lower emission reduction target, the change under a single policy is greater; when under the higher target, change caused by the mixed policy is more obvious.

Based on the above results and conclusions, the Chinese steel industry should focus on the following two aspects when formulating emission mitigation targets and auxiliary emission mitigation policies:

Firstly, the Chinese iron and steel industry at present is in the stage of excess production capacity, which means that the output capacity is far greater than the actual consumption level. Through the calculation of this paper, when the market reaches the equilibrium, the production and consumption are less than the present level, especially in 2016. As a result, for China's iron and steel industry (government) should alleviate the contradiction of excess production capacity and bring relief to the industry. It is immediate and necessary for the government to ban new production and push obsolete and backward-producing enterprises out of the market as soon as possible.

Secondly, at present, there is no clear-text regulation of the use conditions and industry scope of carbon trading policies in China. This paper uses the benchmark method to conduct a preliminary study on the carbon trading mechanism of the steel industry. In combination with the actual situation, we propose that in the use of single and mixed emission reduction policies under the carbon trading mechanism, the government should consider the whole industry sector, regional balance, and carbon trading benchmarks to determine the implementation of a single or mixed carbon trading policy. In the selection of the carbon trading benchmark value, it is necessary to carefully select according to the emission reduction target and the requirements for economic indicators.

This paper conducted a preliminary study on the carbon trade mechanism. With the gradual improvement of relevant national policies, laws, and regulations, this study will continue to explore the details of the carbon trading market of the steel industry and the game behavior between enterprises. Future research will be more closely linked to national policies and will be gradually extended to other high-energy industries such as cement industry and chemical industry.

Author Contributions: Both authors contributed equally to this work. In particular, H.M. and Y.D. had the original idea for the study, and both coauthors conceived of and designed the methodology. Y.D. drafted the manuscript, which was revised by Z.H., J.Y. and Y.L. All authors have read and approved the final manuscript.

Funding: This research was funded by the National Natural Science Foundation of China (41571122, 41771178).

Acknowledgments: The authors are grateful to the editor and anonymous referees who provided valuable comments and suggestions to significantly improve the quality of the paper.

Conflicts of Interest: The authors declare no conflict of interest.

References

1. Dales, J.H. *Pollution, Property & Prices: An Essay in Policy-Making and Economics*; Edward Elgar Publishing: Cheltenham, UK, 1968.
2. Li, J.; Colombier, M. Managing carbon emissions in China through building energy efficiency. *J. Environ. Manag.* **2009**, *90*, 2436–2447. [CrossRef] [PubMed]
3. Anger, A. Including aviation in the European emissions trading scheme: Impacts on the industry, CO emissions and macroeconomic activity in the EU. *J. Air Transp. Manag.* **2010**, *16*, 100–105. [CrossRef]
4. Chen, Y.; Tseng, C.L. Climate Policies and the Power Sector: Challenges and Issues. *J. Energy Eng.* **2008**, *134*, 31–32. [CrossRef]
5. Alberola, E.; Chevallier, J.; Chèze, B. Emissions Compliances and Carbon Prices under the EU ETS: A Country Specific Analysis of Industrial Sectors. *J. Policy Model.* **2009**, *31*, 446–462. [CrossRef]
6. Considine, T.; Larson, D.F. Short term electric production technology switching under carbon cap and trade. *Energies* **2012**, *5*, 4165–4185. [CrossRef]

7. Shen, L.; Song, X.; Wu, Y.; Liao, S.; Zhang, X. Interpretive Structural Modeling based factor analysis on the implementation of Emission Trading System in the Chinese building sector. *J. Clean. Prod.* **2016**, *127*, 214–227. [CrossRef]

8. Su, L. The Research on Transportaition Network Model Considering the Cap and Trade Policy. Master's Thesis, Beijing University of Chemical Technology, Beijing, China, 2017.

9. Zhang, Q. Impact Assessment of Carbon Trading System on the Construction Industry: Based on CGE Model. Master's Thesis, Chongqing University, Chongqing, China, 2017.

10. Ahman, M.; Zetterberg, L. Options for emission allowance allocation under the Eu Emissions Trading Directive. *Mitig. Adapt. Strateg. Glob. Chang.* **2007**, *12*, 1433. [CrossRef]

11. Goeree, J.K.; Palmer, K.; Holt, C.A.; Shobe, W.; Burtraw, D. An Experimental Study of Auctions Versus Grandfathering to Assign Pollution Permits. *J. Eur. Econ. Assoc.* **2010**, *8*, 514–525. [CrossRef]

12. Zetterberg, L.; Wråke, M.; Sterner, T.; Fischer, C.; Burtraw, D. Short-Run Allocation of Emissions Allowances and Long-Term Goals for Climate Policy. *Ambio* **2012**, *41*, 23–32. [CrossRef] [PubMed]

13. Grimm, V.; Ilieva, L. An experiment on emissions trading: The effect of different allocation mechanisms. *J. Regul. Econ.* **2013**, *44*, 308–338. [CrossRef]

14. Wang, J.; Wang, Y.Y. Study on the Initial Allocation Means of the Carbon Emission Rights in Beijing and International Experience Learning. *J. Beijing Univ. Technol.* **2014**, *14*, 19–26.

15. Lee, J.; Yu, J. Market Analysis during the First Year of Korea Emission Trading Scheme. *Energies* **2017**, *10*, 1974. [CrossRef]

16. Guo, D.Y.; Chen, H.; Long, R.Y. The allocation strategy of government for initial carbon allowance in downstream carbon trading market. *China Popul. Resour. Environ.* **2018**, *28*, 43–54.

17. Zhang, L.; Li, Y.; Jia, Z. Impact of carbon allowance allocation on power industry in China's carbon trading market: Computable general equilibrium based analysis. *Appl. Energy* **2018**, *229*, 814–827. [CrossRef]

18. Duscha, V. The EU ETS and Dynamic Allocation in Phase IV—An Ex-Ante Assessment. *Energies* **2018**, *11*, 409. [CrossRef]

19. Chen, W.Y.; Wu, Z.X. Carbon emission permit allocation and trading. *J. Tsinghua Univ.* **1998**, *38*, 15–18, 22.

20. Ellerman, A.D.; Jacoby, H.D.; Decaux, A. *The Effects on Developing Countries of the Kyoto Protocol and Carbon Dioxide Emissions Trading*; The World Bank Development Research Group: Washington, DC, USA, 1998.

21. Babiker, M.; Reilly, J.; Viguier, L. Is International Emissions Trading Always Beneficial? *Energy J.* **2004**, *25*, 33–56. [CrossRef]

22. Wang, C.; Chen, J.N.; Zou, J. Potential analysis of clean development mechanism put into effect in China. *China Environ. Sci.* **2005**, *25*, 310–314.

23. Jamaob, T.; Kohler, J. *Learning Curves for Energy Technology: A Critical Assessment*; Cambridge Univ., 2007; Available online: https://www.repository.cam.ac.uk/bitstream/handle/1810/194736/0752%26EPRG0723.pdf?sequence=1&isAllowed=y (accessed on 31 October 2007).

24. Aviyonah, R.S.; Uhlmann, D.M. Combating Global Climate Change: Why a Carbon Tax is a Better Response to Global Warming than Cap and Trade. *Stanf. Environ. Law J.* **2009**, *28*, 3–50.

25. Cong, R.G.; Wei, Y.M. Potential impact of (CET) carbon emissions trading on China's power sector: A perspective from different allowance allocation options. *Energy* **2010**, *35*, 3921–3931. [CrossRef]

26. Wu, J.; Fan, Y.; Xia, Y.; Liu, J.Y. Impacts of Initial Quota Allocation on Regional Macro-economy and Industry Competitiveness. *Manag. Rev.* **2015**, *27*, 18–26.

27. Li, Y.; Zhu, L. Impacts of ETS on Elimination of Backward Production Capacity in Energy-intensive Industries. *Green Pet. Petrochem.* **2017**, *2*, 6–15.

28. Wu, J.; Fan, Y.; Xia, Y. How can China achieve its nationally determined contribution targets combining emissions trading scheme and renewable energy policies? *Energies* **2017**, *10*, 1166. [CrossRef]

29. Dai, H.; Xie, Y.; Liu, J.; Masui, T. Aligning renewable energy targets with carbon emissions trading to achieve China's INDCs: A general equilibrium assessment. *Renew. Sustain. Energy Rev.* **2018**, *82*, 4121–4131. [CrossRef]

30. Liu, Y.P.; Gao, X.Y.; Guo, J.F. Network Features of the EU Carbon Trade System: An Evolutionary Perspective. *Energies* **2018**, *11*, 1501. [CrossRef]

31. Fernández Fernández, Y.; Fernández López, M.A.; González Hernández, D.; Olmedillas Blanco, B. Institutional Change and Environment: Lessons from the European Emission Trading System. *Energies* **2018**, *11*, 706. [CrossRef]

32. Mann, R.F. The Case for the Carbon Tax: How to Overcome Politics and Find Our Green Destiny. *Environ. Law Rep.* **2009**, *39*, 10118–10126.
33. Wang, W.J.; Xie, P.C.; Hu, J.L.; Wang, L.; Zhao, D.Q. Analysis of the Relative Mitigation Cost Advantages of Carbon Tax and ETS for the Cement Industry. *Clim. Chang. Res.* **2016**, *12*, 53–60.
34. Duan, Y.; Li, N.; Mu, H.L.; Gui, S.S. Research on CO_2 Emission Reduction Mechanism of China's Iron and Steel Industry under Various Emission Reduction Policies. *Energies* **2017**, *10*, 2026. [CrossRef]
35. National Bureau of Statistics of the People's Republic of China. *CSY, 2005–2017. China Statistical Yearbook*; National Bureau of Statistics of the People's Republic of China: Beijing, China, 2005–2017.
36. National Bureau of Statistics of the People's Republic of China. *CIESY, 2005–2017. China Industrial Statistical Yearbook*; National Bureau of Statistics of the People's Republic of China: Beijing, China, 2005–2017.
37. National Bureau of Statistics of the People's Republic of China. *CESY, 2005–2017. China Energy Statistical Yearbook*; National Bureau of Statistics of the People's Republic of China: Beijing, China, 2005–2017.
38. The Editorial Board of China Steel Yearbook. *CSY, 2005–2017. China Steel Yearbook*; The Editorial Board of China Steel Yearbook: Beijing, China, 2005–2017.
39. Intergovernmental Panel on Climate Change (IPCC). *IPCC Guidelines for National Greenhouse Gas Inventories*; United Kingdom Meteorological Office: Bracknell, UK, 2006.
40. Duan, Y.; Mu, H.L.; Li, N. Analysis of the Relationship between China's IPPU CO_2 Emissions and the Industrial Economic Growth. *Sustainability* **2016**, *8*, 426. [CrossRef]
41. Ma, W.J.; Li, M.G. The Systemic Calculation of Optimal Concentration Degree in China's Iron and Steel Industry:Empirical Study Based on Double Efficiency Goals Pursued by Firms and Industry and the Data of 2007. *J. Financ. Econ.* **2011**, *37*, 104–113.
42. Duan, Y.; Li, N.; Mu, H.; Li, L.X. Research on Provincial Shadow Price of Carbon Dioxide in China's Iron and Steel Industry. *Energy Procedia* **2017**, *142*, 2335–2340. [CrossRef]
43. Färe, R.; Grosskopf, S.; Pasurka Jr, C.A. Environmental production functions and environmental directional distance functions. *Energy* **2007**, *32*, 1055–1066. [CrossRef]
44. Lee, J.D.; Park, J.B.; Kim, T.Y. Estimation of the shadow prices of pollutants with production/environment inefficiency taken into account: A nonparametric directional distance function approach. *J. Environ. Manag.* **2002**, *64*, 365–375. [CrossRef]
45. Li, C.S. Mechanism Design of Emission Tax Based on Dynamic Game Theory. Ph.D. Thesis, University of Science and Technology of China, Hefei, China, 2012.
46. Guenno, G.; Tiezzi, S. *The Index of Sustainable Economics Welfare (ISEW) for Italy*; Fondazione Eni Enrico Mattei: Milano, Italy, 1998.
47. Dahowski, R.T.; Li, X.; Davidson, C.L.; Wei, N.; Dooley, J.J.; Gentile, R.H. A preliminary cost curve assessment of carbon dioxide capture and storage potential in China. *Energy Procedia* **2009**, *1*, 2849–2856. [CrossRef]
48. Liu, H.; Gallagher, K.S. Preparing to ramp up large-scale CCS demonstrations: An engineering-economic assessment of CO_2 pipeline transportation in China. *Int. J. Greenh. Gas Control* **2011**, *5*, 798–804. [CrossRef]
49. Zhu, L.; Fan, Y. Modeling the Investment of Coal-fired Power Plant Retrofit with CCS and Subsidy Policy Assessment. *China Popul. Resour. Environ.* **2014**, *24*, 99–105.
50. Wang, B.Q.; Li, H.Q.; Bao, W.J. A model of economy for overall process of CO_2 capture and saline storage. *CIESC J.* **2012**, *63*, 894–903.

Article

Flexible Options for Greenhouse Gas-Emitting Energy Producer

Andrey Krasovskii [1,*], Nikolay Khabarov [1], Ruben Lubowski [2] and Michael Obersteiner [1]

[1] Ecosystems Services and Management (ESM) Program, International Institute for Applied Systems Analysis (IIASA), 2351 Laxenburg, Austria; khabarov@iiasa.ac.at (N.K.); oberstei@iiasa.ac.at (M.O.)
[2] Environmental Defense Fund, Washington, DC 20009, USA; rlubowski@edf.org
* Correspondence: krasov@iiasa.ac.at

Received: 3 September 2019; Accepted: 29 September 2019; Published: 7 October 2019

Abstract: The reduction of emissions from deforestation and forest degradation (REDD) constitutes part of the international climate agreements and contributes to the Sustainable Development Goals. This research is motivated by the risks associated with the future CO_2 price uncertainty in the context of the offsetting of carbon emissions by regulated entities. The research asked whether it is possible to reduce these financial risks. In this study, we consider the bilateral interaction of a REDD supplier and a greenhouse gas (GHG)-emitting energy producer in an incomplete emission offsets market. Within this setting, we explore an innovative financial instrument—flobsion—a flexible option with benefit-sharing. For the quantitative assessment, we used a research method based on a two-stage stochastic technological portfolio optimization model established in earlier studies. First, we obtain an important result that the availability of REDD offsets does not increase the optimal emissions of the electricity producer under any future CO_2 price realization. Moreover, addressing concerns about a possible "crowding–out" effect of REDD-based offsets, we demonstrate that the emissions and offsetting cost will decrease and increase, respectively. Second, we demonstrate the flexibility of the proposed instrument by analyzing flobsion contracts with respect to the benefit-sharing ratio and strike price within the risk-adjusted supply and demand framework. Finally, we perform a sensitivity analysis with respect to CO_2 price distributions and the opportunity costs of the forest owner supplying REDD offsets. Our results show that flobsion's flexibility has advantages compared to a standard option, which can help GHG-emitting energy producers with managing their compliance risks, while at the same time facilitating the development of REDD programs. In this study we limited our analysis to the case of the same CO_2 price distributions foreseen by both parties; the flobsion pricing under asymmetric information could be considered in the future.

Keywords: optimal energy mix; CO_2 emissions; REDD offsets; risk-adjusted utility

---

1. Introduction

The 2015 Paris Climate Agreement encourages implementation and support of activities related to the reduction of emissions from deforestation and forest degradation (REDD+) [1], a climate change mitigation strategy based on the idea to reward countries for reducing their deforestation and forest degradation through financial benefits generated by carbon credits. However, the implementation of REDD+ is a complex international problem [2–4] despite it is considered as a relatively low-cost mitigation option [5,6], and its integration into the global mitigation strategy has the potential for larger emissions reductions to be made [7]. This integration can be done by linking REDD as an emission reduction credit program to major cap-and-trade programs [8]. In this context, credits that could be supplied by REDD projects are an attractive mitigation option; a range of literature is devoted to that topic, for example, References [9–11].

REDD principles, as part of the SDG 15 are contributing directly to the Sustainable Development Goals (SDG) [12]. However, there is an ongoing discussion related to uncertainties and risks in REDD implementation [13–15]. A substantial problem for a potential REDD investor is the missing legal background [16], which may mean that the future acceptance of emission credits generated by REDD projects (to be funded today) is not guaranteed. This uncertainty regarding acceptance and related conditions creates unacceptable risks for those potentially interested in funding/investing in REDD projects (e.g., energy companies in potential need for offsetting their emissions). To overcome this problem, establishment of the intermediaries such as the REDD Acceleration Fund [17] was recently suggested, along with approaches based on optionality in purchasing REDD-based offsets [10]. There are promising pioneering steps being made in California, heading towards a law on REDD acceptance for compliance purposes. The state of California has a placeholder in the suggested California Tropical Forest Standard allowing international credits of up to 2% of an entity's annual compliance obligation; however, it has not yet issued a detailed standard or introduced regulations to operationalize REDD application [18].

Accepting existing uncertainties, we explore bilateral interaction between a REDD supplier and a greenhouse gas (GHG)-emitting energy producer in the context of an incomplete REDD offsets market. We develop the FI-REDD model established in a series of publications [19,20]. FI-REDD is a two-stage stochastic technological portfolio optimization model describing an interaction between the REDD-offsets supplier, electricity producer and consumers. In this study, in order to contract REDD offsets in the model, we employ a novel financial instrument—a flexible option with a benefit-sharing mechanism called "flobsion" [21]. This instrument is different from the REDD offset contracts modeled in previous studies. In essence, a flobsion complements an option with a benefit-sharing mechanism. While the general idea of benefit-sharing is important within the REDD context [22], in our approach benefit-sharing stands for possible sharing of the profits stemming from flobsions. The FI-REDD model with exponential utility functions [23] allows the risk-averse behavior to be combined with benefit-sharing so that their impact on contracted amounts of flobsions can be analyzed.

Another modification consists of expanding the FI-REDD model by introducing the opportunity costs of the forest owner. Opportunity cost is the economic benefit forgone from the alternative land/forest use [24]. It sets a minimum amount to be paid to keep the land in forest. Thus, opportunity cost forms the basis for economic analyses of REDD [25]. In our model, the REDD supplier takes into account the opportunity cost curve, when making the decision about supplying REDD offsets.

The key driver of this research is the high uncertainty in future CO_2 prices and associated risks in the context of REDD offsetting. The research question is about the possible reduction of these risks using flobsion [21] and comparison of flobsion with the standard option in the REDD context. The elaboration of this instrument is an important step in the field of modeling financial instruments supporting REDD programs [6,10].

The structure of the paper is as follows. First, we analytically investigate the construction of flobsion in FI-REDD model. This allows the important result to be demonstrated, namely, that energy producers would not increase their emissions if they had acquired flobsions. Second, we present modeling results and sensitivity analysis with respect to CO_2 price distributions and opportunity costs. Finally, we discuss analytical and numerical results, as well as policy implications and possible future research directions.

2. Methodology

In this paper we further develop the FI-REDD model established in our previous work [19,20]. The model takes into account the potential market power of energy producers, which gives them flexibility in their decision-making under uncertain emission costs. The scheme of the model is shown in Figure 1. The model deals with optimization of the technological mix under market power, related to optimal scheduling of power systems [26] and market pricing in the power industry [27]. We propose an idea for setting a fair price of the REDD offsets, based on the indifference principle in two-stage problem setting. Utility-indifference pricing is a well-established approach to the valuation of derivative securities in incomplete markets [28]. The indifference price is defined as the price of the derivative toward which the investor is indifferent whether to use the derivative to maximize their expected utility or not to use it [29]. We use fair prices to evaluate REDD offsets [19] under future CO_2 price uncertainty. In the first stage (period), where details about the future REDD offsets market are uncertain, the parties (supplier and consumer of REDD offsets) assign their offset prices (buying and selling) in such a way that their profits or, generally, their utilities, stay the same in the second period (in which the REDD offsets price is revealed) no matter whether they have contracted REDD offsets in the first period or not.

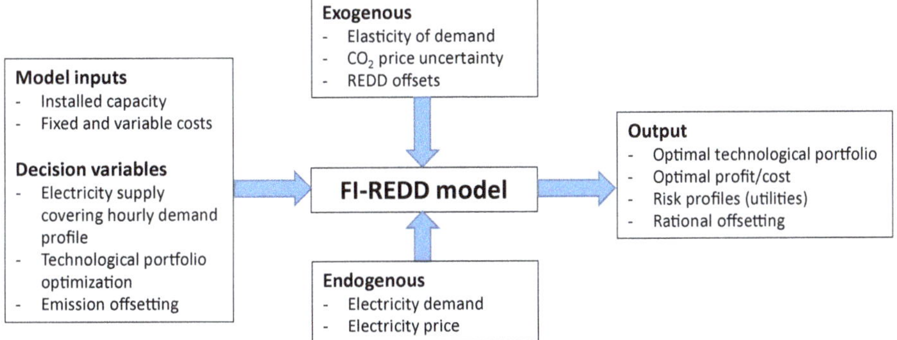

Figure 1. Scheme of the FI-REDD model.

We implement flobsion as the REDD offset in the model. Flobsion complements the standard option with a benefit-sharing mechanism. Methodologically, the idea is close to revenue-sharing contracts in supply chain coordination, under which a supplier receives a percentage of revenue generated by retailer [30]. Here we specifically consider a situation where benefits are shared between the REDD supplier and consumer. Therefore, we prefer the term "benefit-sharing" to distinguish our study from "revenue-sharing" in supply chain coordination, for example, Reference [31].

In this study we also advance the decision-making of the forest owner by implementing an opportunity costs curve in the model. Opportunity costs are foregone economic benefits from forest-uses, which are alternative to REDD. These include social-cultural costs and indirect costs [24]. In this study we use non-linearly increasing opportunity cost with respect to the amount of offsets supplied. We also perform sensitivity analysis with respect to the opportunity costs.

In summary, our methodology combines the following approaches:

1. Two-period technological portfolio optimization
2. Exponential risk-preferences
3. Utility-indifference pricing
4. Optionality of purchasing emission offsets
5. Benefit-sharing mechanism

In this section we provide the theoretical framework accompanied by some analytical results. In particular, we derive formulas for indifference prices, which form the basis of our analysis. These will allow us to construct the risk-adjusted fair prices of a seller and a buyer of REDD flobsions for all admissible benefit-sharing ratios and strike prices. The prices will, in turn, help in identifying the amounts and prices of contracted flobsions.

2.1. Decision-Making of the Electricity Producer

We consider two optimization problems of the electricity producer in the "second" period, which is when REDD offsets are traded on a market (as opposed to the "first" period when there is no such market). The first optimization problem is decision-making under the realized CO_2 price without REDD offsets, that is, flobsions in this study. The electricity producer can modify its technological mix to reduce emissions or raise the electricity price, thus making the consumer bear some of the costs [19]. The second problem is complemented by REDD-offsets flobsions. The electricity producer can use flobsions either to offset emissions or to sell on the market at the CO_2 price. In the latter case, the income from selling REDD offsets should be shared with the forest owner. The use of the flobsion also depends on the level of the strike price compared to the CO_2 price. Let us start with the first problem.

2.1.1. Optimization without Flobsions

For every CO_2 price realization, p_C, the electricity producer chooses technological mix, x, to maximize its profit Π_{EP}:

$$\underset{x(p_C)\in X}{\text{maximize}} \{\Pi_{EP}(x(p_C)) = \Pi_e(x(p_C)) - p_C E(x(p_C))\}, \tag{1}$$

where $x = x(p_C)$—technological mix, X—feasibility domain (all admissible technological portfolios), p_C—CO_2 price realization, Π_e—profit component without emission cost and E—emissions corresponding to technological mix x. In the problem formulation: profit is the objective function, technological mix is the control variable and emission prices are exogenous variables.

Let us denote optimal technological mix by $\hat{x}(p_C)$ and corresponding profit and emissions as

$$\hat{\Pi}_{EP}(p_C) = \Pi_{EP}(\hat{x}(p_C)), \tag{2}$$

$$\hat{E}(p_C) = E(\hat{x}(p_C)). \tag{3}$$

The solution of the problem (1) delivers an optimal response of the electricity producer in terms of profits and emissions to CO_2 prices.

2.1.2. Definition of Flobsion

A common "call" option for an asset (e.g., for an emission offset) or simply an option, implies that a buyer pays an amount p to the seller of the option for the future possibility of purchasing the asset at an agreed "strike" price p_{min}. The owner of the option decides in the future whether to make such a purchase or not, so that for them it is a possibility but not an obligation.

A flobsion is a generalized form of option. A buyer pays the amount p_δ to the seller of a flobsion for the future possibility of purchasing the asset at the agreed "strike" price p_{min} plus the discounted difference between the asset's market price p_C and p_{min}, if that difference is positive (or just p_{min} otherwise). If the flobsion holder decides to purchase an asset within the period of validity of a flobsion, they would pay the amount:

$$a_\delta = \begin{cases} p_{min} + (1-\delta) \times (p_C - p_{min}), & \text{if} \quad p_C > p_{min} \\ p_C, & \text{if} \quad p_C \le p_{min} \end{cases}, \tag{4}$$

meaning that an asset is being purchased at a market price p_C and the flobsion is not being executed if $p_C \le p_{min}$, where $0 \le \delta \le 1$ is a discount to the market price using p_{min} as a base. In the case of a flobsion, a future asset purchase is still optional for the buyer as in the case of a standard option. The price for the future purchase of an asset in the case of a flobsion, though not fixed, is still tied to the market price. For $\delta = 1$, meaning a 100% discount to the market price (retaining the strike price), a flobsion turns into an option.

2.1.3. Optimization with Flobsions

Let us consider the profit of the electricity producer $\Pi_{EP}^R(x)$ with flobsion at a CO_2 price realization in the second period:

$$\Pi_{EP}^R(x) = \begin{cases} \Pi_e(x) - p_C E(x) - p_E \mathcal{E}, & \text{if } p_C \leq p_{\min} \\ \Pi_e(x) - (p_{\min} + (1-\delta)(p_C - p_{\min}))\mathcal{E} + \\ + p_C[\mathcal{E} - E(x)]_+ - p_C[E(x) - \mathcal{E}]_+ - p_E \mathcal{E}, & \text{if } p_C > p_{\min} \end{cases} \tag{5}$$

where $\mathcal{E} \in [0, \mathcal{E}_{\max}]$ is the volume of offsets covered by flobsions and contracted in the first period, $\mathcal{E}_{\max}$—maximum amount of flobsions supplied by forest owner, $\delta \in [0,1]$—benefit-sharing ratio, $p_{\min}$—flobsion strike price, p_E—the price the electricity producer pays for flobsion in the first period, and

$$[Y]_+ = \begin{cases} Y, & \text{if } Y > 0 \\ 0, & \text{otherwise.} \end{cases} \tag{6}$$

Equation (5) can be interpreted as follows. When the CO_2 price realization is lower than the strike price, the electricity producer does not use flobsion. In the case where the price realization exceeds the strike price, the electricity producer pays the price $p_{\min}$ for flobsions to the forest owner and also shares their profit from the price difference (between the CO_2 price and the strike price) with the forest owner. The sharing is determined by ratio $\delta \in [0,1]$, such that the electricity producer gets a share of δ and share $(1-\delta)$ goes to the forest owner. Moreover, the electricity producer has two options: either to emit more CO_2 than the amount contracted through flobsions and pay the CO_2 price for the non-offset emissions or to emit less than the amount contracted through flobsions and sell the unused offsets on the market at respective CO_2 price. Additionally the price p_E is paid to the forest owner for flobsions in the first period, that is, that is the sunk cost.

Let us expand Equation (5) for the case $p_C > p_{\min}$ with respect to emissions in the second period.

Case 1. If $\mathcal{E} - E(x) > 0$, then

$$\begin{aligned} \Pi_{EP}^R(x) &= \Pi_e(x) - (p_{\min} + (1-\delta)(p_C - p_{\min}))\mathcal{E} + p_C(\mathcal{E} - E(x)) - p_E \mathcal{E} = \\ & \quad \Pi_e(x) - p_C E(x) + \delta(p_C - p_{\min})\mathcal{E} - p_E \mathcal{E}. \end{aligned} \tag{7}$$

Case 2. If $\mathcal{E} - E(x) \leq 0$, then

$$\begin{aligned} \Pi_{EP}^R(x) &= \Pi_e(x) - (p_{\min} + (1-\delta)(p_C - p_{\min}))\mathcal{E} - p_C(E(x) - \mathcal{E}) - p_e \mathcal{E} = \\ & \quad \Pi_e(x) - p_C E(x) + \delta(p_C - p_{\min})\mathcal{E} - p_E \mathcal{E}. \end{aligned} \tag{8}$$

Equations (7) and (8) are the same, showing the equivalence between offsetting emissions and selling offsets on the market for CO_2 price p_C. We can simplify Equation (5) as follows:

$$\Pi_{EP}^R(x) = \begin{cases} \Pi_e(x) - p_C E(x) - p_E \mathcal{E}, & \text{if } p_C \leq p_{\min} \\ \Pi_e(x) - p_C E(x) - p_E \mathcal{E} + \delta(p_C - p_{\min})\mathcal{E}, & \text{if } p_C > p_{\min} \end{cases} \tag{9}$$

Let us formulate the optimization problem with REDD flobsions in the second period.

Given the flobsion strike price $p_{\min}$, benefit-sharing ratio $\delta \in [0,1]$, amount of REDD offsets $\mathcal{E}$ contracted in the first period at price p_E, the electricity producer maximizes their profit at every CO_2 price realization p_C:

$$\underset{x(p_C) \in X}{\text{maximize}} \{\Pi_{EP}^R(x(p_C))\}, \tag{10}$$

where $\Pi_{EP}^R(x(p_C))$ is defined in Equation (9). Let us denote optimal technological mix by $\hat{x}^R(p_C)$ and corresponding profit by $\hat{\Pi}_{EP}^R(p_C) = \Pi_{EP}^R(\hat{x}^R(p_C))$.

Lemma 1. *For any amount $\mathcal{E}$, price p_E, benefit-sharing ratio δ and strike price p_{min}, technological mix $\hat{x}^R(p_C)$ solving the problem with flobsions (10) coincides with the optimal technological mix solving the problem without flobsions (1) at every CO_2 price realization, p_C.*

Proof. As terms $p_E\mathcal{E}, \delta(p_C - p_{min})\mathcal{E}$ in Equation (9) are independent of $x(p_C)$, then they are not part of the optimization problem with flobsions (Equation (5)) and are used only for calculating the resulting optimal profits. Therefore, the optimal mix $\hat{x}^R(p_C)$ solving problem (9) and (10) coincides with the mix $\hat{x}(p_C)$ solving problem (1). $\square$

Corollary 1. *Optimal profit with REDD flobsions, $\hat{\Pi}^R_{EP}(p_C)$, at every CO_2 price realization is calculated as follows:*

$$\hat{\Pi}^R_{EP}(p_C) = \begin{cases} \hat{\Pi}_{EP}(p_C) - p_E\mathcal{E}, & \text{if } p_C \leq p_{min} \\ \hat{\Pi}_{EP}(p_C) - p_E\mathcal{E} + \delta(p_C - p_{min})\mathcal{E}, & \text{if } p_C > p_{min} \end{cases} \tag{11}$$

Corollary 2. *Optimal emissions $\hat{E}(p_C)$ in the problem with REDD flobsions are the same as in the problem without REDD flobsions:*

$$\hat{E}(p_C) = E(\hat{x}^R(p_C)) = E(\hat{x}(p_C)).$$

Remark 1. *Corollary 2 shows that the optimal emissions of the electricity producer with REDD flobsion stay the same as in the case without REDD flobsion. This indicates that there is no risk that energy producers will change their production and emit more as compared to the case without offsets. This is explained by the fact that offsets can be sold at the CO_2 market price and that this opportunity is the highest profit, the energy producers can get from the offsets they possess.*

2.2. Decision-Making of the Forest Owner

We consider a forest owner, who decided to allocate part of their forest to REDD+ offsets and who assesses the value of the forest covering the offsets in amount $\mathcal{E} \in (0, \mathcal{E}_{max}]$, where $\mathcal{E}_{max}$ is the maximum available volume. There are two possibilities in the second period: the forest owner meets the CO_2 price either without participating in REDD or with an obligation corresponding to flobsions sold to the electricity producer.

Forest owner's profit without selling flobsions in the first period is calculated as follows:

$$\Pi_{FO} = \mathcal{E} \cdot \max\{p_{op}(\mathcal{E}), p_C\}, \tag{12}$$

where $p_{op} = p_{op}(\mathcal{E})$ is opportunity cost associated with forest values alternative to REDD+. If the forest owner did not engage in contracting flobsions in the first period, they still can sell the amount in the second period at the market CO_2 price or can take advantage of other opportunities (e.g., selling wood), whichever delivers a greater profit.

When forest owner engages in contracting flobsion in the first period, their profit at CO_2 price realization in the second period is calculated the following way:

$$\Pi^R_{FO}(p_C) = \begin{cases} \mathcal{E}p_F + \mathcal{E} \cdot \max\{p_{op}(\mathcal{E}), p_C\}, & \text{if } p_C \leq p_{min} \\ \mathcal{E}p_F + (p_{min} + (1-\delta)(p_C - p_{min}))\mathcal{E}, & \text{if } p_C > p_{min} \end{cases} \tag{13}$$

where p_F is the price paid to the forest owner for flobsions in the first period by the electricity producer. When CO_2 price realization is below strike price p_{min}, the forest owner has the income from selling flobsions in the first period, $\mathcal{E}p_F$. Moreover, as the electricity producer does not exercise flobsions in this case, the forest owner decides whether to sell that amount of flobsions on the market by comparing market price p_C with opportunity cost. When CO_2 price realization is higher than the strike price, the first term in Equation (13), the case $p_C > p_{min}$, stands for the income from selling the flobsion in the first period and the second term is the profit coming from the electricity producer (cf. Equation (9)) including the strike price p_{min} and shared benefits with parameter δ.

Let us note that we consider deterministic opportunity costs of the forest owner. This is based on the assumption that forest value is rather stable over time as compared to the offsets price.

2.3. Indifference Prices of the Forest Owner and Electricity Producer

To calculate indifference prices, we consider discrete distribution of CO_2 prices with probabilities w_i, $\sum_{i=1}^{N} w_i = 1$, corresponding to price realizations p_C^i, $i = 1, \ldots, N$, where N is the number of realizations and assume the growing sequence $p_C^i > p_C^{i-1}$. Let us consider expected utilities of the electricity producer $\mathbb{U}(\hat{\Pi}_{EP})$ and $\mathbb{U}(\hat{\Pi}_{EP}^R)$ without and with flobsion, respectively. Now we consider prices p_F and p_E as unknowns and find them from indifference equations. The fair price of the electricity producer is determined by the utility-indifference equation:

$$
\begin{aligned}
p_E &= p_E(\mathcal{E}, p_{min}, \delta, p_C^i, w_i) : \\
&\mathbb{U}(\hat{\Pi}_{EP}(p_C^i), w_i) = \mathbb{U}(\hat{\Pi}_{EP}^R(\mathcal{E}, p_E, p_{min}, \delta, p_C^i), w_i), \quad i = 1, \ldots, N,
\end{aligned}
$$

meaning that the expected utility stays the same, no matter if the electricity producer contracts flobsions in the first period or not. Where $\hat{\Pi}_{EP}(p_C^i)$ is the solution to problem without flobsions (1) and $\hat{\Pi}_{EP}^R(\mathcal{E}, p_E, p_{min}, \delta, p_C^i)$—with flobsions (10) at i-th CO_2 price realization.

Similarly, if we denote utilities of the forest owner by $\mathbb{U}(\hat{\Pi}_{FO})$ and $\mathbb{U}(\hat{\Pi}_{FO}^R)$, then their fair price is determined by equity:

$$
\begin{aligned}
p_F &= p_F(\mathcal{E}, p_{op}, p_{min}, \delta, p_C^i, w_i) : \\
&\mathbb{U}(\Pi_{FO}(\mathcal{E}, p_{op}, p_C^i), w_i) = \mathbb{U}(\Pi_{FO}^R(\mathcal{E}, p_F, p_{min}, \delta, p_C^i), w_i), \quad i = 1, \ldots, N,
\end{aligned}
$$

where $\Pi_{FO}(\mathcal{E}, p_{op}, p_C^i)$ and $\Pi_{FO}^R(\mathcal{E}, p_F, p_{min}, \delta, p_C^i)$ are profits (12) and (13), respectively, at i-th CO_2 price realization.

Indifference prices can be derived numerically for any distribution and utility. However, analytical derivation is not always possible [19]. Below we consider risk-neutral and exponential utilities, which allow for analytical derivation and modeling risk-preferences. When the indifference curves are constructed for a range of flobsions' amounts, we can check whether the amount can be contracted by comparing the prices of the electricity producer and forest owner. Namely, the amount $\mathcal{E}$ can be contracted if the buyer's price is not less than the seller's price: $p_E(\mathcal{E}) \geq p_F(\mathcal{E})$.

2.3.1. Risk-Neutral Utilities

In the case of risk-neutral (r.-n.) utilities the indifference prices are calculated according to equations (see Appendix A.1):

$$
p_E^{r.-n.} = p_E^{r.-n.}(\mathcal{E}, \delta, p_{min}, p_C^i, w_i) = \delta \cdot \sum_{i=i_*+1}^{N} (p_C^i - p_{min}) w_i, \tag{14}
$$

$$
\begin{aligned}
p_F^{r.-n.} &= p_F^{r.-n.}(\mathcal{E}, \delta, p_{min}, p_{op}, p_C^i, w_i) = \\
&\sum_{i=i_*+1}^{N} \left(\max\{p_{op}(\mathcal{E}), p_C^i\} - p_{min} - (1-\delta)(p_C^i - p_{min}) \right) w_i,
\end{aligned} \tag{15}
$$

where i_* is the largest number when $p_C^i \leq p_{min}$. In this case the price of the electricity producer does not depend on the flobsions' amount. The price of the forest owner depends on the quantity only via the opportunity cost $p_{op}(\mathcal{E})$.

Lemma 2. *In the risk-neutral case for $\delta > 0$, the volume of flobsions $\mathcal{E}$ can be contracted if the following inequality holds:*

$$\sum_{i=i_*+1}^{N} \max\{p_{op}(\mathcal{E}), p_C^i\}w_i \leq \sum_{i=i_*+1}^{N} p_C^i w_i. \tag{16}$$

Proof. Let us calculate the difference between p_E (14) and p_F (15), when $\delta > 0$:

$$p_F^{r.-n.}(\mathcal{E}) - p_E^{r.-n.} = \sum_{i=i_*+1}^{N} \left(\max\{p_{op}(\mathcal{E}), p_C^i\} - p_C^i \right)w_i.$$

Therefore, Equation (16) guarantees that seller's price exceeds buyer's price for amount $\mathcal{E}$, that is, $p_E^{r.-n.} \geq p_F^{r.-n.}(\mathcal{E})$. $\square$

Remark 2. *Lemma 2 shows that in the risk neutral case, whether the amount of flobsions is contracted or not, depends on the relationship between the opportunity costs of the forest owner and the CO_2 price distribution above the strike price $p_{\min}$ and it is independent of the benefit-sharing ratio δ.*

2.3.2. Exponential Utilities

In the case of exponential utilities (risk preferences, r.-p.) the indifference prices are calculated according to equations (see Appendix A.2):

$$
\begin{aligned}
p_E^{r.-p.} &= p_E^{r.-p.}(\mathcal{E}, \alpha, \delta, p_{\min}, p_C^i, w_i) = \frac{1}{\alpha\mathcal{E}} \cdot \left(\ln\left(\sum_{i=1}^{N} e^{-\alpha\hat{\Pi}_{EP}(p_C^i)}w_i \right) - \right.\\
&\qquad \left. \ln\left(\sum_{i=1}^{i_*} e^{-\alpha\hat{\Pi}_{EP}(p_C^i)}w_i + \sum_{i=i_*+1}^{N} e^{-\alpha(\hat{\Pi}_{EP}(p_C^i)+\delta(p_C^i-p_{\min})\mathcal{E})}w_i \right) \right),
\end{aligned}
\tag{17}
$$

$$
\begin{aligned}
p_F^{r.-p.} &= p_F^{r.-p.}(\mathcal{E}, \alpha, \delta, p_{op}, p_{\min}, p_C^i, w_i) = \\
&\frac{1}{\alpha\mathcal{E}} \cdot \left(\ln\left(\sum_{i=1}^{i_*} e^{-\alpha\mathcal{E}\cdot\max\{p_{op}(\mathcal{E}),p_C^i\}}w_i + \sum_{i=i_*+1}^{N} e^{-\alpha\mathcal{E}(p_{\min}+(1-\delta)(p_C^i-p_{\min}))}w_i \right) - \right.\\
&\left. \ln\left(\sum_{i=1}^{N} e^{-\alpha\cdot\mathcal{E}\cdot\max\{p_{op}(\mathcal{E}),p_C^i\}}w_i \right) \right),
\end{aligned}
\tag{18}
$$

where $\alpha \neq 0$ is the parameter of risk preferences [23]. When $\alpha \to 0$ the risk-adjusted prices converge to the risk-neutral ones.

Remark 3. *When $\delta = 0$ the price $p_E^{r.-p.}$ (Equation (17)) equals to zero—the same as in the risk neutral case (Equation (14)). This means that when there is no benefit-sharing (i.e., no discount to a market price), the electricity producer is indifferent to contracting the offsets at zero price.*

3. Modeling Results

In this section we present modeling results making use of the methodology described above. We employ the FI-REDD model calibrated in previous studies. Basically, to calculate the indifference prices, we need information only about optimal profits of the electricity producer (see Corollary 1). In the example we take the following distribution:

$$p_C^i = 10 \cdot (i-1), \quad w_i = 1/9, \quad i = 1,\ldots,9, \tag{19}$$

where CO_2 price varies from 0 US\$/tCO$_2$ to 80 US\$/tCO$_2$ with the step 10 US\$/tCO$_2$. Here we consider the uniform distribution, that is, each price realization has the same probability equal to 1/9. Profits of the electricity producer at each price realization are shown in Figure 2.

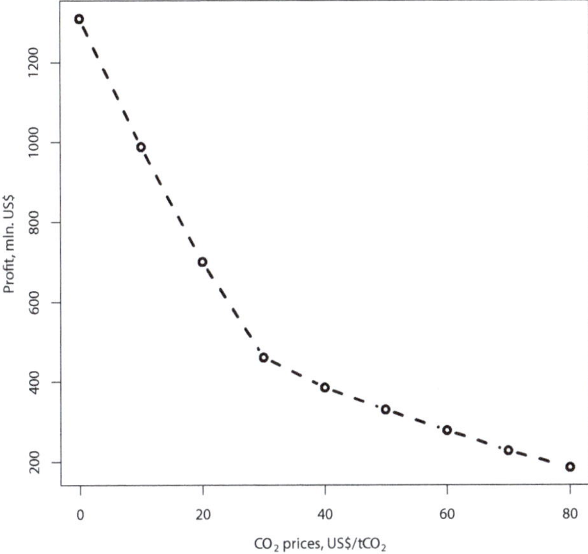

Figure 2. Annual profit of the electricity producer in the FI-REDD model with respect to growing CO_2 price, adapted from Reference [19].

In our experiments we vary offsets' amount in the range from 1 to 100 $MtCO_2$. For the forest owner we assume the following opportunity cost function:

$$p_{op}(\mathcal{E}) = K \cdot (0.4 \cdot \mathcal{E} + 0.005 \cdot \mathcal{E}^2), \tag{20}$$

where $\mathcal{E}$ is measured in tons of CO_2 and K is a scaling coefficient. Opportunity costs increase with the amount of flobsions, equivalent to the forest area allocated for offsets. The more forest is allocated to flobsion, the higher is the opportunity cost. We choose the price range consistent with the CO_2 price distribution. Below we consider the case, $K = 0.9$, when opportunity cost varies between 0 US\$/tCO_2 (for zero offsets) and 81 US\$/tCO_2 (for 100 $MtCO_2$). This range is consistent with some empirical studies (e.g., Reference [32]).

3.1. Impacts of Risk-Aversion on Contracted Amounts and Equilibrium Prices

In our experiments we compare risk-neutral case with risk-averse cases. In Figure 3 we show risk-neutral indifference curves (dashed) lines based on Equations (14) and (15) and risk-averse (solid curves) based on Equations (17) and (18), corresponding to coefficient $\alpha = 0.001$ for the strike price $p_{\min} = 20$ US\$/tCO_2 and benefit-sharing ratio $\delta = 0.5$. We also show curves for two intermediate values of the risk-preference parameter $\alpha = 0.0001$ and $\alpha = 0.0005$.

In the study we use the following notations for the cases with considered risk-preferences:

- N—risk-neutral ($\alpha = 0$);
- A1—risk-aversion parameter $\alpha = 0.0001$;
- A2—risk-aversion parameter $\alpha = 0.0005$;
- A3—risk-aversion parameter $\alpha = 0.001$.

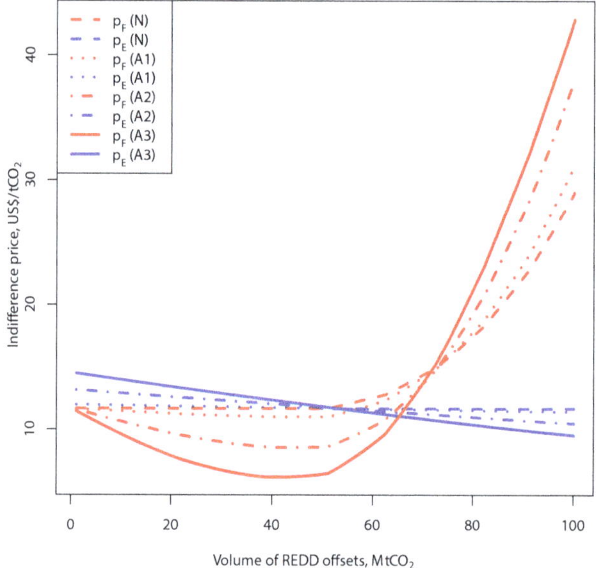

Figure 3. Indifference prices in the FI-REDD model with flobsions. Comparison of risk-neutral case (N) with risk-averse cases: A1, A2 and A3, for benefit-sharing parameter $\delta = 0.5$, strike price $p_{min} = 20$ US\$/MtCO$_2$, CO$_2$ distribution (19) and opportunity cost (20) with $K = 0.9$. Blue curves correspond to indifference prices of the electricity producer and red curves of the forest owner.

Blue curves correspond to indifference prices of the electricity producer and red curves—of the forest owner. In the risk neutral case the price of the electricity producer is constant. The price of the forest owner (15) stays the same until the amount of offsets 50.9 MtCO$_2$ and increases afterwards. Therefore, for the fixed parameter $\delta = 0.5$ and strike price $p_{min} = 20$ US\$/tCO$_2$, the contracted amount is 50.9 MtCO$_2$ at the price 11.67 US\$/tCO$_2$ contracted via flobsion (and assuming an additional future payment). For small values of parameter α (case A1) the indifference curves are close to the risk-neutral lines. The figure shows how the risk-aversion (cases A1, A2 and A3) transforms the indifference curves of the parties. The price of the electricity producer is a monotonically declining with respect to amount of flobsions, while the price of the forest owner becomes rather U-shaped as demonstrated in the figure. This shape can be explained by relatively low opportunity costs for the smaller amounts of flobsions and, therefore, the risk-averse forest owner prefers to sell those via flobsion to have a "guaranteed" higher income. However, when opportunity costs are high, a rational forest owner would avoid entering into a REDD-offsetting contract. In the particular case indicated in Figure 3, risk-aversion increases the contracted amount of flobsions, that is, the maximum amount 65.3 MtCO$_2$ can be contracted at the intersection of solid lines (A3) at the equilibrium price 11.09 US\$/tCO$_2$.

3.2. Contracted Amounts and Equilibrium Prices with Respect to Benefit-Sharing Ratio

Let us fix the strike price as in Figure 3, $p_{min} = 20$ US\$/tCO$_2$ and calculate the contracted amounts for all possible benefit-sharing ratios. According to Remark 3, we consider the case of zero discount (and hence zero-purchase price) as degraded and, therefore, stick to the range of $\delta \in [0.05, 1]$. Contracted amounts are shown in Figure 4a. In the risk-neutral case (dashed blue line) the contracted amount is constant and equals 50.95 MtCO$_2$. As the plots show risk-aversion increases the contracted amounts for this strike price. An interesting feature is that there is a nonlinear dependence of the contracted amount with respect to benefit-sharing ratio δ in the risk-averse case. Red curves are concave with respect to benefit-sharing ratio, meaning that there is a maximum amount of contracted flobsions for every risk-aversion parameter. Here for A1 the maximum contracted amount is 57.1 MtCO$_2$, that is reached at the benefit-sharing ratio, $\delta = 0.5$. For A2 the maximum amount is 64.4 MtCO$_2$ at $\delta = 0.4$ and for A3—65.65 MtCO$_2$ at $\delta = 0.4$. Note, that here we considered discrete values of δ with the step 0.05.

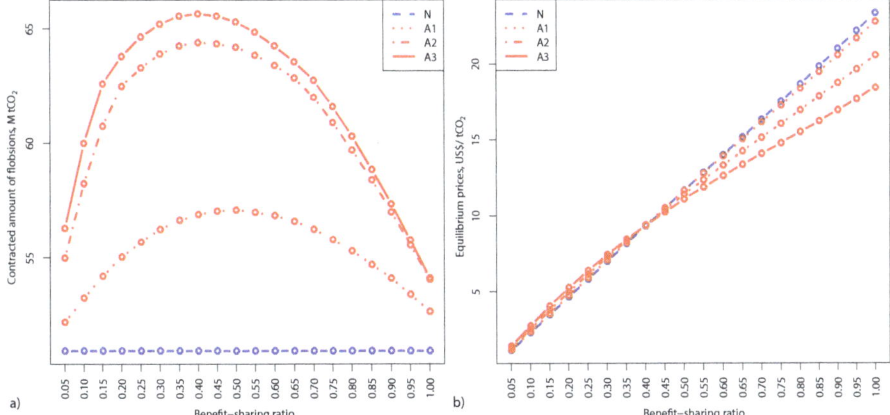

Figure 4. Contracted amounts (**a**) and equilibrium prices (**b**) of flobsion with respect to benefit-sharing parameter $\delta \in [0.05, 1]$ for the fixed strike price $p_{\min} = 20$ US\$/tCO$_2$, CO$_2$ distribution (19) and opportunity cost (20) with $K = 0.9$. Comparison of risk-neutral case N (blue line) with risk-averse cases A1, A2 and A3 (red curves).

In Figure 4b we show the equilibrium prices corresponding to contracted amounts in Figure 4a. The prices are increasing with respect to growing benefit-sharing ratio. Note, that the maximum price is achieved when $\delta = 1$. This case corresponds to standard option. Thus, flobsion decreases the equilibrium price compared to option due to additional flexibility in choosing the benefit-sharing ratio. We also find out that for lower benefit-sharing ratios the equilibrium prices in the risk-averse case (red curves) are higher as compared to the prices in the risk-neutral case (blue line) but it is opposite for higher benefit-sharing ratios. This can be explained by the risk-aversion of the electricity producers, who are more comfortable with higher ratios. However, together with the price increasing at higher ratios, the amounts of contracts decline as shown in Figure 4a.

3.3. Impacts of Strike Prices on Contracted Amounts and Equilibrium Prices

To be consistent with CO$_2$ price distribution (19), we consider 8 strike prices varying in the range from 0 to 70 US\$/tCO$_2$ with the step 10 US\$/tCO$_2$. In Figure 5a we show how contracted amounts of flobsions change with respect to $p_{\min}$ for the fixed benefit-sharing ratio $\delta = 0.7$. One can see that contracted volumes increase with the growing strike price in all cases. The smallest amounts in all cases can be contracted at zero strike price: 21.85 MtCO$_2$ (N), 33.75 MtCO$_2$ (A1), 46.45 MtCO$_2$ (A2), 50.05 MtCO$_2$ (A3).

Moreover, risk aversion increases the contracted amounts for small values of the strike price. This can be explained by the relatively high opportunity costs of the forest owner compared to the low strike price and at the same time by higher CO$_2$ prices expected by the electricity producer. This situation changes when the strike price is relatively high; for larger values of the strike price the risk-averse amounts start to converge to risk-neutral case. At final rather extreme strike price, $p_{\min} = 70$ US\$/tCO$_2$, which is close to maximum CO$_2$ price, the amount contracted in risk-neutral case exceeds the amount in risk-averse cases. This can be explained by the high value of the benifit-sharing ratio, $\delta = 0.7$ considered in this case.

Figure 5a shows that maximum contracted amounts for $\delta = 0.7$ take place at maximum strike price $p_{\min} = 70$ US\$/tCO$_2$: 99.25 MtCO$_2$ (N), 96.85 MtCO$_2$ (A1), 96.65 MtCO$_2$ (A2), 96.45 MtCO$_2$ (A3).

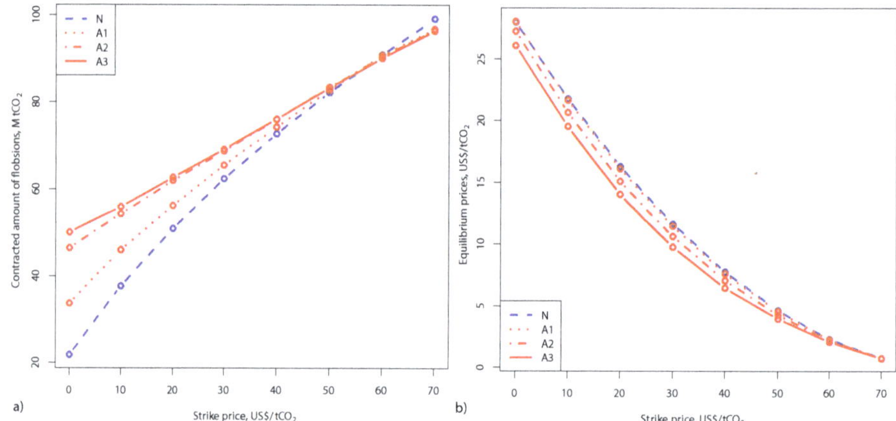

Figure 5. Contracted amounts (**a**) and equilibrium prices (**b**) of flobsions with respect to strike price $p_{\min}$ for benefit-sharing ratio $\delta = 0.7$, CO_2 distribution (19) and opportunity cost (20) with $K = 0.9$. Comparison of risk-neutral case N with risk-averse cases A1, A2 and A3.

The equilibrium prices with respect to $p_{\min}$ are shown in Figure 5b for $\delta = 0.7$. The case A1 is very close to N in terms of prices. All equilibrium prices decrease with respect to growing $p_{\min}$, as a higher strike price implies less possibilities for the energy producer to "earn" on a discount as compared to the market price. In the risk-neutral case the price varies from 0.78 to 28 US\$/tCO$_2$, while in the risk-averse case A3—from 0.79 to 26.12 US\$/tCO$_2$.

3.4. Full Flexibility Of Flobsion

Let us now consider all combinations of benefit-sharing ratios and strike prices and their impacts on contracted amounts of flobsions and their prices. Figure 6a shows the contracted amount with respect to $\delta \in [0.05, 1]$ and $p_{\min} \in [0, 70]$ in the risk-neutral case N. We see that the surface has a concave shape, increasing with respect to growing $p_{\min}$. We also note, that the surface levels are constant with respect to δ. However, looking at the corresponding equilibrium prices in Figure 7a we observe an evident decrease of the prices with respect to declining δ. Thus, even in the risk-neutral case benefit-sharing has a positive effect of decreasing the equilibrium price. Figure 7a shows that highest prices at every $p_{\min}$ correspond to largest $\delta = 1$, that is, the case of standard option.

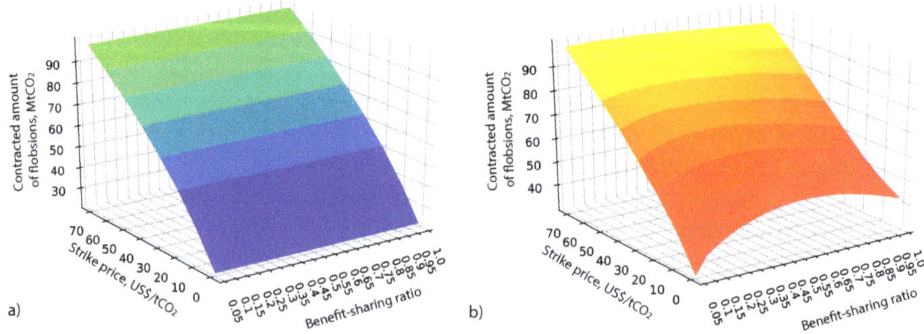

Figure 6. Contracted amounts of flobsions with respect to benefit-sharing ratio, $\delta \in [0.05, 1]$ and strike price $p_{\min} \in [0, 70]$: (**a**) risk-neutral case N, (**b**) risk-averse case A3. CO_2 distribution (19) and opportunity cost (20) with $K = 0.9$.

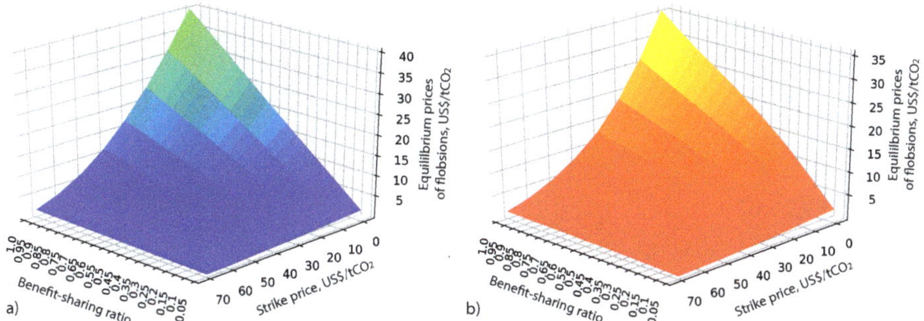

Figure 7. Equilibrium prices of flobsions with respect to benefit-sharing ratio, $\delta \in [0.05, 1]$ and strike price, $p_{\min} \in [0, 70]$: (**a**) risk-neutral case N, (**b**) risk-averse case A3. CO_2 distribution in Equation (19) and opportunity cost in Equation (20), where $K = 0.9$.

Figure 6b shows how the risk-aversion case A3 transforms the risk-neutral surface (cf. Figure 6a). The minimum value in case A3, 30.45 $MtCO_2$, corresponds to $\delta = 0.05$ and $p_{\min} = 0$, while the maximum is 99.45 $MtCO_2$ corresponds to $\delta = 0.15$ and $p_{\min} = 70$. Both minimum and maximum values in case A3 are larger than in case N, where the values are 21.85 $MtCO_2$ and 99.25 $MtCO_2$, respectively.

Equilibrium prices in case A3 are depicted in Figure 7b. Although the shape of the surface is similar to case N (cf. Figure 7a), we observe lower prices in case A3, particularly, for lower strike prices and higher benefit-sharing ratios.

3.5. Sensitivity Analysis

In this section we perform sensitivity analysis of modeling results with respect to opportunity cost of the forest owner and CO_2 price distribution envisioned by both decision-makers. For illustration we take the risk-averse case A3 and compare the contracted amount of flobsions to those depicted in Figure 6b.

3.5.1. Non-Uniform CO_2 Price Distributions

The analysis above was performed for the uniform distribution. However, analytical formulas are valid for arbitrary distribution. In order to check how modeling results change with respect to distributions, we analyze two non-uniform cases. In the first case, CO_2 prices are the same as in (19) but the probabilities are shifted to lower price realizations, meaning that decision-makers envision smaller CO_2 prices more likely to happen. We consider the following distribution:

$$
\begin{aligned}
p_C^i &= 10 \cdot (i-1), \quad i = 1, \dots, 9, \\
w_1 &= 0.11, \ w_2 = 0.22, \ w_3 = 0.33, \ w_4 = 0.22, \ w_5 = 0.11, \ w_j = 0, \quad j = 6, \dots, 9.
\end{aligned}
\tag{21}
$$

The mean CO_2 price in this case is 20 US\$/tCO_2, which is half less than 40 US\$/tCO_2 in (19). The contracted amounts are shown in Figure 8a. We see that they are becoming zero after the strike price reaches the highest CO_2 price with positive weight $p_C^5 = 40$ US\$/tCO_2, as the electricity producer would not buy any offsets. For $p_{\min} \geq 40$ the surface stays at zero level for all benefit-sharing ratios. Before this threshold, determined by distribution (21) the surface is qualitatively similar to the one with uniform distribution (cf. Figure 6b) but the values are lower due to lower expected values in the right side of the distribution. The value at $\delta = 0.05$ and $p_{\min} = 0$ is 23.5 $MtCO_2$, which is 30.45 $MtCO_2$ in Figure 6b. The largest contracted amount, 62.95 $MtCO_2$, is achieved when $\delta = 0.25$ and $p_{\min} = 30$, that is the maximum feasible strike price in this case. Qualitatively the outcomes stay similar to the case with uniform distribution.

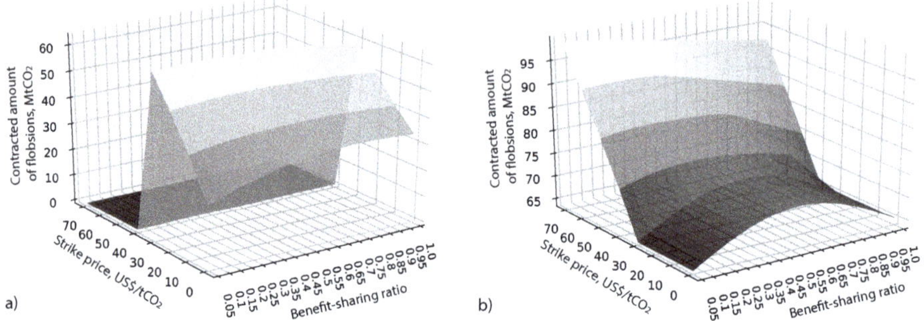

Figure 8. Contracted amounts of flobsions with respect to benefit-sharing ratio, $\delta \in [0.05, 1]$ and strike price, $p_{\min} \in [0, 70]$. We consider risk-averse case A3 with opportunity cost in Equation (20), where $K = 0.9$ and CO_2 distributions in: (**a**) Equation (21), (**b**) Equation (22).

The second alternative distribution is opposite to the previous one as it assumes more weight put on higher CO_2 price realizations. We consider the following distribution:

$$
\begin{aligned}
p_C^i &= 10 \cdot (i-1), \quad i = 1, \ldots, 9, \\
w_j &= 0, \, w_5 = 0.11, \, w_6 = 0.22, \, w_7 = 0.33, \, w_8 = 0.22, \, w_9 = 0.11, \quad j = 1, \ldots, 4.
\end{aligned}
\tag{22}
$$

The mean CO_2 price in this case is 60 US\$/tCO_2. The results are depicted in Figure 8b. For $p_{\min} \leq 30$, which correspond to the range of CO_2 prices with zero probability in the distribution, the surface has steady shape. For this range the contracted amounts does not change with respect to strike price but still have concave shape with respect to δ. Note that the contracted amounts are larger compared to the case with uniform distribution (cf. Figure 6b). This is explained by the fact that parties put more value on flobsions in the situation of the foreseen higher CO_2 prices. The minimum value in Figure 8b is 63.95 MtCO_2. For strike prices higher than $p_{\min} = 40$, we see the increase in contracted amounts. Comparing the figure to Figure 6b, we observe a sharper incline towards smaller benefit-sharing ratios. This can be interpreted as the risk-averse electricity producer is ready to pay higher prices in the face of high CO_2 price realizations and forest owner is fine with providing a larger discount. The maximum 99.25 MtCO_2 is achieved at $p_{\min} = 70$ and $\delta = 0.05$. Let us note that this maximum does not coincide with the one in the case of uniform distribution 99.45 MtCO_2 although the expected prices $p_C^9 = 80$ and probability $w_9 = 1/9$ coincide. This is due to the fact that a risk-averse decision maker takes into account the entire CO_2 price distribution while calculating indifference prices as stated in Equations (17) and (18).

3.5.2. Sensitivity to Opportunity Costs

For illustration, we are not changing the shape of the opportunity cost curve but vary the scaling parameter K in Equation (20) that was set to $K = 0.9$ above. We consider lower opportunity cost curve by setting $K = 0.5$, meaning that the maximum cost is $p_{op}(100) = 45$ US\$/tCO_2. The case of lower opportunity cost curve is shown in Figure 9a for the uniform distribution. In this case flobsions are very attractive to forest owner and they are willing to contract larger amounts compared to the case of higher opportunity costs (cf. Figure 6b). When the strike price is high, the maximum amount of flobsions 100 MtCO_2 is contracted. This happens when $p_{\min} = 30$ and $p_{\min} = 40$ for a set of benefit-sharing ratios and when $p_{\min} \geq 50$ for all δ. For $p_{\min} \leq 30$ surface conserves the concavity properties with respect to $p_{\min}$ and δ. The lower opportunity costs increase the minimum contracted amount, which constitutes 47.3 MtCO_2 in this case.

In Figure 9b we show the case of higher opportunity cost curve with parameter $K = 1.3$ in Equation (16), meaning that the highest cost is $p_{op}(100) = 117$ US\$/tCO_2. In this case the shape of the surface is similar to the one in Figure 6b but quantitatively the contracted amounts decrease. The minimum contracted amount is 22.55 MtCO$_2$, while the maximum—is 78.15 MtCO$_2$. This is due to the fact that for each flobsions' amount the opportunity cost curve becomes higher, while CO$_2$ price distribution stays the same. Here the maximum value is achieved at the highest strike price for the range of benefit-sharing ratios $\delta \in [0.15, 0.30]$, indicating that at that highest strike price, the contracted amounts start to reach some saturation level similar to Figure 9a.

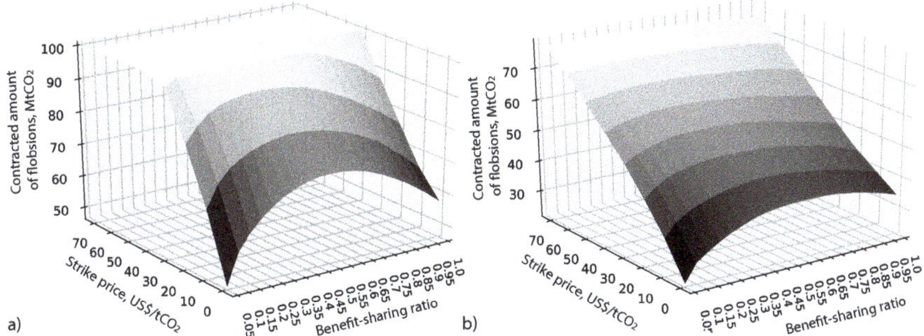

Figure 9. Contracted amounts of flobsions with respect to benefit-sharing ratio, $\delta \in [0.05, 1]$ and strike price, $p_{min} \in [0, 70]$. We consider risk-averse case A3 with CO$_2$ distribution (19) and opportunity cost in Equation (20), where: (**a**) $K = 0.5$, (**b**) $K = 1.3$.

4. Discussion

In this paper the FI-REDD model was elaborated by implementing a flobsion and an opportunity cost of the forest owner (REDD supplier). The results showed (Lemma 1) that REDD offsets provided to the energy producer would not change the optimal emissions compared to the case of no offsets. This delivers an important signal to policy makers. Even if the offsets are provided to the electricity producer at no cost, it is not rational to change the technological portfolio from the optimal one (determined solely by the carbon price) and emit more. When the market for offsets is established, it is more profitable to sell the offsets on the market. Moreover, in the situation of higher emission costs, the optimal emissions will decrease compared to the situation with zero emissions costs as shown in Reference [19].

The impact of risk-preferences on the volume of contracted flobsions was analyzed. In the risk-neutral case, the contracted amounts at every strike price are the same for both the flobsion and the standard option. However, the price in the case of the standard option coincides with flobsion's price only at the largest benefit-sharing ratio, which is equal to one. For lower benefit-sharing ratios, the equilibrium prices of the flobsion are lower compared to the standard option. They decrease together with decreasing benefit-sharing ratio. This price decrease is more vivid for lower strike prices.

In the risk-averse case, the situation is similar in terms of prices and,moreover, the benefit-sharing ratio allows the maximum contracted amount of flobsions to be found for every strike price. In particular, for relatively small strike prices compared to the maximum range of CO$_2$ prices, the choice of benefit-sharing ratio allows for a considerable increase in the contracted amounts. That fact is of a potential interest in the REDD context, larger contracted volumes mean more forest being allocated for the generation of carbon offset sand that it is hence protected under the REDD umbrella. The contracted amounts also increase when there is stronger risk-aversion.

The findings show that contracted amounts increase as the strike price grows. This is quite a natural result, as the higher strike prices are advantageous for both parties—forest owner and electricity producer. Thereafter, the equilibrium prices also decrease with growing strike prices. The full set of benefit-sharing ratios and strike prices and how they impact the contracted amount and the equilibrium prices is illustrated in Figures 6 and 7. Interestingly, the risk-averse case shows that at every strike price, there are values of the benefit-sharing ratio that deliver the maximum of contracted amounts.

As the research was based on the FI-REDD model, a similar range of offset amounts and CO_2 prices was used. The opportunity cost curve was chosen to be consistent with these data for illustrative purposes. Nevertheless, to check how the results react to a different setup a sensitivity analysis was performed. The analysis shows that qualitative features remain valid. The contracted amounts are higher when higher CO_2 prices are more likely and lower when lower CO_2 prices are more likely, in both cases compared to uniform distribution. We also show that lower opportunity costs facilitate the contracting of more offsets, while higher ones decrease the contracted amounts. The results of the sensitivity analysis did not provide any irregularities, as the situation of symmetric information of a buyer and a seller with respect to future CO_2 price distribution was considered.

5. Conclusions

While the flobsion construct apparently has more universal applications, we see how good its fit is in the REDD-offsetting context. This is because it supports the provision of up-front financing for the development of offset-generating projects, while at the same time providing enough flexibility for a balance of interest to be found between the offsets buyer and seller in the face of uncertainties associated with future terms of offsetting. The problem of the acceptance and fungibility of REDD-based offsets with emission allowances is still open at both national and international level [33,34]. This situation leads to necessity to alleviate corresponding risks, and flobsion might be helpful in this regard.

Despite technical details of potential policies remaining uncertain, there is progress towards inclusion of REDD-based offsetting for compliance at a legal level (e.g., California). Although policy changes create opportunities, at the same time they can create risks for certain market players. The flobsion's properties prove it to be a potential candidate to accommodate and alleviate those risks, and, therefore, accelerate the implementation of new policies.

The numerical results on the flobsion's properties presented in this paper cover a rich set of cases with respect to the varied parameters—the strike price and the benefit-sharing ratio (discount)— and therefore make this work complete for practical applications. Such instruments could be considered in terms of providing flexibility in the future through the benefit-sharing mechanism, thereby reducing the initial investment. However, it is important for decision-makers to clearly formulate their policies and in particular to formalize the legal aspects of the acceptance of the flobsion (issued today) in the future, as well as the sharing agreements. Although it is beyond the scope of this paper, we would like to draw attention to the legal aspects of community-based human rights [35] and consideration of the needs of indigenous peoples [36], in the implementation of REDD.

Further research could be directed towards exploration of flobsion applications beyond REDD. The asymmetric information of a buyer and seller in terms of future uncertainties would also be an interesting analysis.

Author Contributions: Conceptualization, A.K., N.K., R.L. and M.O.; methodology, A.K. and N.K.; software, formal analysis, writing—original draft preparation, A.K.; writing—review and editing, N.K., R.L. and M.O.

Funding: This work was supported by the project "Delivering Incentives to End Deforestation: Global Ambition, Private/ Public Finance and Zero-Deforestation Supply Chains" funded by the Norwegian Agency for Development Cooperation under agreement number QZA-0464, QZA-16/0218.

Conflicts of Interest: The authors declare no conflict of interest. The funders had no role in the design of the study; in the collection, analyses, or interpretation of data; in the writing of the manuscript, or in the decision to publish the results.

Abbreviations

The following abbreviations are used in this manuscript:

GHG   Greenhouse Gas
REDD   Reducing emissions from deforestation and forest degradation
SDGs   Sustainable Development Goals
FI-REDD   Financial instruments for REDD (model)

Appendix A. Derivation of Indifference Prices for Flobsions

Appendix A.1. Indifference Prices for Risk-Neutral Utilities

For the utility of the electricity producer without REDD we have:

$$\mathbb{U}(\hat{\Pi}_{EP}) = \sum_{i=1}^{N} \hat{\Pi}_{EP}(p_C^i)w_i = \sum_{i=1}^{i_*} \hat{\Pi}_{EP}(p_C^i)w_i + \sum_{i=i_*+1}^{N} \hat{\Pi}_{EP}(p_C^i)w_i,$$

where i_* is the largest index when $p_C^i \leq p_{\min}$.

Using Equation (11), we get the expected mean utility with REDD flobsions:

$$\begin{aligned} \mathbb{U}(\hat{\Pi}_{EP}^R) &= \sum_{i=1}^{i_*}(\hat{\Pi}_{EP}(p_C^i) - p_E\mathcal{E})w_i + \\ &\quad \sum_{i=i_*+1}^{N} (\hat{\Pi}_{EP}(p_C^i) + \delta(p_C^i - p_{\min})\mathcal{E} - p_E\mathcal{E})w_i = \\ &\quad \sum_{i=1}^{i_*}\hat{\Pi}_{EP}(p_C^i)w_i + \sum_{i=i_*+1}^{N}\hat{\Pi}_{EP}(p_C^i)w_i + \mathcal{E}\cdot\sum_{i=i_*+1}^{N}\delta(p_C^i - p_{\min})w_i - p_E\mathcal{E}. \end{aligned}$$

By applying indifference principle $\mathbb{U}(\hat{\Pi}_{EP}) = \mathbb{U}(\hat{\Pi}_{EP}^R)$ we get the price:

$$p_E^{r,-n.} = p_E^{r,-n.}(\mathcal{E}, \delta, p_{\min}, p_C^i, w_i) = \delta\cdot\sum_{i=i_*+1}^{N}(p_C^i - p_{\min})w_i.$$

Utility of the forest owner without flobsion (12):

$$\mathbb{U}(\Pi_{FO}) = \mathcal{E}\cdot\sum_{i=1}^{N}\max\{p_{op}(\mathcal{E}), p_C^i\}w_i.$$

Using Equation (13), we can calculate utility of the forest owner with flobsion:

$$\begin{aligned} \mathbb{U}(\Pi_{FO}^R) &= \sum_{i=1}^{i_*}(\mathcal{E}p_F + \mathcal{E}\cdot\max\{p_{op}(\mathcal{E}), p_C^i\})w_i + \\ &\quad \sum_{i=i_*+1}^{N} ((p_{\min} + (1-\delta)(p_C^i - p_{\min}))\mathcal{E} + \mathcal{E}p_F + p_{op}(\mathcal{E})\cdot\mathcal{E})w_i = \\ &\quad \mathcal{E}\cdot\left(p_F + \sum_{i=1}^{i_*}\max\{p_{op}(\mathcal{E}), p_C^i\}w_i + \sum_{i=i_*+1}^{N}(p_{\min} + (1-\delta)(p_C^i - p_{\min}))w_i\right). \end{aligned}$$

By applying the indifference condition $\mathbb{U}(\hat{\Pi}_{FO}) = \mathbb{U}(\hat{\Pi}_{FO}^{R})$, we derive the following indifference price of the forest owner:

$$
p_F^{r.-n.} = p_F^{r.-n.}(\mathcal{E}, \delta, p_{\min}, p_{op}, p_C^i, w_i) = \\
\sum_{i=i_*+1}^{N} \left(\max\{p_{op}(\mathcal{E}), p_C^i\} - p_{\min} - (1-\delta)(p_C^i - p_{\min}) \right) w_i.
$$

Appendix A.2. Indifference Prices for Exponential Utilities

For the electricity producer without REDD flobsions we get the following utility:

$$
\mathbb{U}(\hat{\Pi}_{EP}) = \frac{1}{\alpha} \cdot \left(1 - \sum_{i=1}^{N} e^{-\alpha \hat{\Pi}_{EP}(p_C^i)} w_i \right), \tag{A1}
$$

where α is the parameter of risk preferences. Expected utility of the electricity producer with REDD flobsions is calculated as follows:

$$
\mathbb{U}(\hat{\Pi}_{EP}^{R}) = \frac{1}{\alpha} - \frac{1}{\alpha} \cdot \sum_{i=1}^{i_*} e^{-\alpha(\hat{\Pi}_{EP}(p_C^i) - p_E \mathcal{E}))} w_i - \tag{A2}
$$

$$
\frac{1}{\alpha} \cdot \sum_{i=i_*+1}^{N} e^{-\alpha(\hat{\Pi}_{EP}(p_C^i) + \delta(p_C^i - p_{\min})\mathcal{E} - p_E \mathcal{E}))} w_i =
$$

$$
\frac{1}{\alpha} - \frac{1}{\alpha} \cdot e^{\alpha p_E \mathcal{E}} \cdot \left(\sum_{i=1}^{i_*} e^{-\alpha \hat{\Pi}_{EP}(p_C^i)} w_i + \sum_{i=i_*+1}^{N} e^{-\alpha(\hat{\Pi}_{EP}(p_C^i) + \delta(p_C^i - p_{\min})\mathcal{E})} w_i \right).
$$

Using Equations (A1) and (A2) we derive the indifference price:

$$
p_E^{r.-p.} = \frac{1}{\alpha \mathcal{E}} \cdot \left(\ln \left(\sum_{i=1}^{N} e^{-\alpha \hat{\Pi}_{EP}(p_C^i)} w_i \right) - \right.
$$

$$
\left. \ln \left(\sum_{i=1}^{i_*} e^{-\alpha \hat{\Pi}_{EP}(p_C^i)} w_i + \sum_{i=i_*+1}^{N} e^{-\alpha(\hat{\Pi}_{EP}(p_C^i) + \delta(p_C^i - p_{\min})\mathcal{E})} w_i \right) \right).
$$

For the forest owner without flobsion we have:

$$
\mathbb{U}(\hat{\Pi}_{FO}) = \frac{1}{\alpha} \cdot \left(1 - \sum_{i=1}^{N} e^{-\alpha \cdot \mathcal{E} \cdot \max\{p_{op}(\mathcal{E}), p_C^i\}} w_i \right),
$$

and with flobsion:

$$
\mathbb{U}(\hat{\Pi}_{FO}^{R}) = \frac{1}{\alpha} - \frac{1}{\alpha} \cdot \sum_{i=1}^{i_*} e^{-\alpha(\mathcal{E} p_F + \mathcal{E} \cdot \max\{p_{op}(\mathcal{E}), p_C^i\})} w_i -
$$

$$
\frac{1}{\alpha} \cdot \sum_{i=i_*+1}^{N} e^{-\alpha((p_{\min} + (1-\delta)(p_C^i - p_{\min}))\mathcal{E} + \mathcal{E} p_F)} w_i =
$$

$$
\frac{1}{\alpha} - \frac{1}{\alpha} \cdot e^{-\alpha \cdot \mathcal{E} p_F} \cdot \left(\sum_{i=1}^{i_*} e^{-\alpha \mathcal{E} \cdot \max\{p_{op}(\mathcal{E}), p_C\}} w_i + \sum_{i=i_*+1}^{N} e^{-\alpha \mathcal{E}(p_{\min} + (1-\delta)(p_C^i - p_{\min}))} w_i \right).
$$

We get the following indifference price:

$$
p_F^{r.-p.} = \frac{1}{\alpha \mathcal{E}} \cdot \ln \left(\frac{\sum_{i=1}^{i_*} e^{-\alpha \mathcal{E} \cdot \max\{p_{op}(\mathcal{E}), p_C^i\}} w_i + \sum_{i=i_*+1}^{N} e^{-\alpha \mathcal{E}(p_{\min} + (1-\delta)(p_C^i - p_{\min}))} w_i}{\sum_{i=1}^{N} e^{-\alpha \cdot \mathcal{E} \cdot \max\{p_{op}(\mathcal{E}), p_C^i\}} w_i} \right).
$$

References

1. United Nations Framework Convention on Climate Change. *Adoption of the Paris Agreement*; United Nations Office at Geneva: Geneva, Switzerland, 2015.
2. Corbera, E.; Schroeder, H. Governing and implementing REDD+. *Environ. Sci. Policy* **2011**, *14*, 89–99. [CrossRef]
3. Law, E.A.; Thomas, S.; Meijaard, E.; Dargusch, P.J.; Wilson, K.A. A modular framework for management of complexity in international forest-carbon policy. *Nat. Clim. Chang.* **2012**, *2*, 155. [CrossRef]
4. McDermott, C.L. REDDuced: From sustainability to legality to units of carbon—The search for common interests in international forest governance. *Environ. Sci. Policy* **2014**, *35*, 12–19. [CrossRef]
5. Busch, J.; Strassburg, B.; Cattaneo, A.; Lubowski, R.; Bruner, A.; Rice, R.; Creed, A.; Ashton, R.; Boltz, F. Comparing climate and cost impacts of reference levels for reducing emissions from deforestation. *Environ. Res. Lett.* **2009**, *4*, 044006. [CrossRef]
6. Lubowski, R.N.; Rose, S.K. The potential for REDD+: Key economic modeling insights and issues. *Rev. Environ. Econ. Policy* **2013**, *7*, 67–90. [CrossRef]
7. Koch, N.; Reuter, W.H.; Fuss, S.; Grosjean, G. Permits vs. offsets under investment uncertainty. *Resour. Energy Econ.* **2017**, *49*, 33–47. [CrossRef]
8. Angelsen, A.; Rudel, T.K. Designing and implementing effective REDD+ policies: A forest transition approach. *Rev. Environ. Econ. Policy* **2013**, *7*, 91–113. [CrossRef]
9. Bosetti, V.; Lubowski, R.; Golub, A.; Markandya, A. Linking reduced deforestation and a global carbon market: Implications for clean energy technology and policy flexibility. *Environ. Dev. Econ.* **2011**, *16*, 479–505. [CrossRef]
10. Golub, A.A.; Fuss, S.; Lubowski, R.; Hiller, J.; Khabarov, N.; Koch, N.; Krasovskii, A.; Kraxner, F.; Laing, T.; Obersteiner, M.; et al. Escaping the climate policy uncertainty trap: Options contracts for REDD+. *Clim. Policy* **2018**, *18*, 1227–1234. [CrossRef]
11. Houghton, R.A.; Duffy, P.B.; Nassikas, A. Forests: The Bridge to a Fossil-Free Future. 2016. Available online: http://whrc.org/wp-content/uploads/2018/01/PB_Forests.pdf (accessed on 1 October 2019).
12. United Nations Development Programme. Sustainable Development Goals—Goal 15 Life on Land. 2016. Available online: http://www.undp.org/content/undp/en/home/sustainable-development-goals/goal-15-life-on-land.html (accessed on 1 October 2019).
13. Plugge, D.; Baldauf, T.; Köhl, M. The global climate change mitigation strategy REDD: Monitoring costs and uncertainties jeopardize economic benefits. *Clim. Chang.* **2013**, *119*, 247–259. [CrossRef]
14. Fletcher, R.; Dressler, W.; Büscher, B.; Anderson, Z.R. Questioning REDD+ and the future of market-based conservation. *Conserv. Biol.* **2016**, *30*, 673–675. [CrossRef] [PubMed]
15. Boer, H.J. The role of government in operationalising markets for REDD+ in Indonesia. *For. Policy Econ.* **2018**, *86*, 4–12. [CrossRef]
16. Laing, T.; Taschini, L.; Palmer, C. Understanding the demand for REDD+ credits. *Environ. Conserv.* **2016**, *43*, 389–396. [CrossRef]
17. The Rockefeller Foundation. REDD+ Acceleration Fund. 2017. Available online: https://assets.rockefellerfoundation.org/app/uploads/20170706180732/Redd-Acceleration-Fund.pdf (accessed on 1 October 2019).
18. Keohane, N. California Sets Stage for Global Forest Protection through State's Cap-and-Trade Program. 2018. Available online: https://www.edf.org/media/california-sets-stage-global-forest-protection-through-states-cap-and-trade-program (accessed on 1 October 2019).
19. Krasovskii, A.; Khabarov, N.; Obersteiner, M. Fair pricing of REDD-based emission offsets under risk preferences and benefit-sharing. *Energy Policy* **2016**, *96*, 193–205. [CrossRef]
20. Krasovskii, A.; Khabarov, N.; Obersteiner, M. CO_2-intensive power generation and REDD-based emission offsets with a benefit-sharing mechanism. *Energy Syst.* **2017**, *8*, 857–883. [CrossRef]
21. Khabarov, N.; Lubowski, R.; Krasovskii, A.; Obersteiner, M. Flobsion—Flexible Option with Benefit Sharing. *Int. J. Financ. Stud.* **2019**, *7*, 22. [CrossRef]
22. Dunlop, T.; Corbera, E. Incentivizing REDD+: How developing countries are laying the groundwork for benefit-sharing. *Environ. Sci. Policy* **2016**, *63*, 44–54. [CrossRef]

23. Raiffa, H. *Decision Analysis: Introductory Lectures on Choices und Uncertainty*; Addison-Wesley: Boston, MA, USA, 1968.
24. White, D.; Minang, D. *Estimating the Opportunity Costs of REDD+: A Training Manual*; World Bank Institute: Washington, DC, USA, 2010; 200p.
25. Boucher, D.H. *Out of the Woods: A Realistic Role for Tropical Forests in Curbing Global Warming*; Union of Concerned Scientists: Cambridge, MA, USA, 2008.
26. Che, Q.; Lou, S.; Wu, Y.; Zhang, X.; Wang, X. Optimal Scheduling of a Multi-Energy Power System with Multiple Flexible Resources and Large-Scale Wind Power. *Energies* **2019**, *12*, 3566. [CrossRef]
27. Lisin, E.; Kurdiukova, G.; Okley, P.; Chernova, V. Efficient Methods of Market Pricing in Power Industry within the Context of System Integration of Renewable Energy Sources. *Energies* **2019**, *12*, 3250. [CrossRef]
28. Staum, J. Incomplete markets. *Handb. Oper. Res. Manag. Sci.* **2007**, *15*, 511–563.
29. Musiela, M.; Zariphopoulou, T. An example of indifference prices under exponential preferences. *Financ. Stoch.* **2004**, *8*, 229–239. [CrossRef]
30. Cachon, G.P.; Lariviere, M.A. Supply chain coordination with revenue-sharing contracts: Strengths and limitations. *Manag. Sci.* **2005**, *51*, 30–44. [CrossRef]
31. Li, S.; Zhu, Z.; Huang, L. Supply chain coordination and decision making under consignment contract with revenue sharing. *Int. J. Prod. Econ.* **2009**, *120*, 88–99. [CrossRef]
32. Graham, V.; Laurance, S.G.; Grech, A.; McGregor, A.; Venter, O. A comparative assessment of the financial costs and carbon benefits of REDD+ strategies in Southeast Asia. *Environ. Res. Lett.* **2016**, *11*, 114022. [CrossRef]
33. Dooley, K.; Gupta, A. Governing by expertise: The contested politics of (accounting for) land-based mitigation in a new climate agreement. *Int. Environ. Agreem. Polit. Law Econ.* **2017**, *17*, 483–500. [CrossRef]
34. Beltran, A.M.; Angelsen, A.; den Elzen, M.; Gierløff, C.W.; Böttcher, H. *Analysing the Options and Impacts of Including REDD Credits in Carbon Markets*; Technical Report; PBL Working Paper 7; PBL: The Hague, The Netherlands, 2013.
35. Raftopoulos, M. REDD+ and human rights: Addressing the urgent need for a full community-based human rights impact assessment. *Int. J. Hum. Rights* **2016**, *20*, 509–530. [CrossRef]
36. Aguilar-Støen, M. Better Safe than Sorry? Indigenous Peoples, Carbon Cowboys and the Governance of REDD in the Amazon. *Forum Dev. Stud.* **2017**, *44*, 91–108. [CrossRef]

Review

A Review of CO_2 Storage in View of Safety and Cost-Effectiveness

Cheng Cao [1,2], Hejuan Liu [3,*], Zhengmeng Hou [1,2], Faisal Mehmood [2,4], Jianxing Liao [1,2,*] and Wentao Feng [1,2]

1   Research Center of Energy Storage Technologies, Clausthal University of Technology, 38640 Goslar, Germany; caochengcn@163.com (C.C.); hou@tu-clausthal.de (Z.H.); wentao.feng@tu-clausthal.de (W.F.)
2   Institute of Petroleum Engineering, Clausthal University of Technology, 38678 Clausthal-Zellerfeld, Germany; faisal.mehmood@tu-clausthal.de
3   State Key Laboratory of Geomechanics and Geotechnical Engineering, Institute of Rock and Soil Mechanics, Chinese Academy of Sciences, Wuhan 430071, China
4   Department of Petroleum & Gas Engineering, University of Engineering & Technology Lahore, Lahore 54890, Pakistan
*   Correspondence: hjliu@whrsm.ac.cn (H.L.); jianxing.liao@tu-clausthal.de (J.L.)

Received: 21 November 2019; Accepted: 19 January 2020; Published: 29 January 2020

Abstract: The emissions of greenhouse gases, especially CO_2, have been identified as the main contributor for global warming and climate change. Carbon capture and storage (CCS) is considered to be the most promising strategy to mitigate the anthropogenic CO_2 emissions. This review aims to provide the latest developments of CO_2 storage from the perspective of improving safety and economics. The mechanisms and strategies of CO_2 storage, focusing on their characteristics and current status, are discussed firstly. In the second section, the strategies for assessing and ensuring the security of CO_2 storage operations, including the risks assessment approach and monitoring technology associated with CO_2 storage, are outlined. In addition, the engineering methods to accelerate CO_2 dissolution and mineral carbonation for fixing the mobile CO_2 are also compared within the second section. The third part focuses on the strategies for improving economics of CO_2 storage operations, namely enhanced industrial production with CO_2 storage to generate additional profit, and co-injection of CO_2 with impurities to reduce the cost. Moreover, the role of multiple CCS technologies and their distribution on the mitigation of CO_2 emissions in the future are summarized. This review demonstrates that CO_2 storage in depleted oil and gas reservoirs could play an important role in reducing CO_2 emission in the near future and CO_2 storage in saline aquifers may make the biggest contribution due to its huge storage capacity. Comparing the various available strategies, CO_2-enhanced oil recovery (CO_2-EOR) operations are supposed to play the most important role for CO_2 mitigation in the next few years, followed by CO_2-enhanced gas recovery (CO_2-EGR). The direct mineralization of flue gas by coal fly ash and the pH swing mineralization would be the most promising technology for the mineral sequestration of CO_2. Furthermore, by accelerating the deployment of CCS projects on large scale, the government can also play its role in reducing the CO_2 emissions.

Keywords: CO_2 storage; CO_2 utilization; security assessment; cost-effectiveness; CO_2 storage projects

---

1. Introduction

In the last few centuries, the CO_2 concentration in the atmosphere has already risen above 410 ppm from a level of below 300 ppm in pre-industrial times [1,2]. As shown in Figure 1, the continuous rise in the Earth's surface temperature appears to be strongly linked with atmospheric concentration of

CO_2, which suggests that CO_2 may be the main contributor to global warming and climate change. In addition, it makes up an estimated 77% of greenhouse gases [3,4].

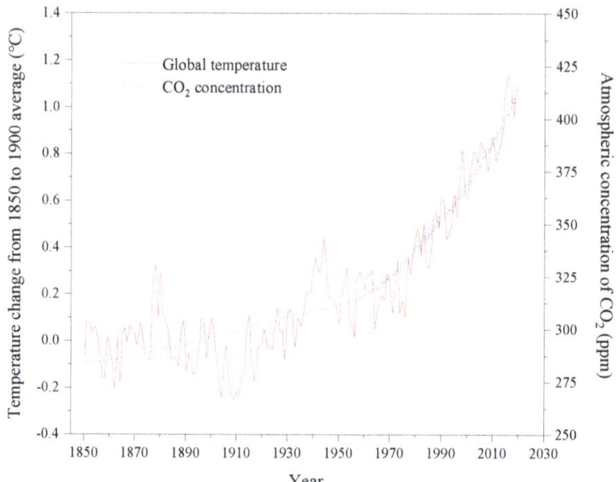

Figure 1. Correlation between atmospheric concentration of CO_2 and the global temperature since 1850s (Data from [1,2]).

In addition, the CO_2 emission may increase the frequency of extreme extratropical cyclones. Unless the greenhouse gas emissions are efficiently mitigated, the extratropical cyclones are projected to more than triple in number by the end of this century in both Europe and North America [5]. In response to such intense global climate change, the Intergovernmental Panel on Climate Change's (IPCC) assessment suggested that the average global warming should be limited less than 2 °C within this century on the basis of estimated results from integrated assessment models (IAMs) [6].

To achieve this goal, carbon capture and storage (CCS) needs to be promoted, as it is currently the most effective technology for slowing down atmospheric CO_2 enrichment and extenuating associated potential climate problems [7]. According to the estimation by the IEA [8], CCS alone could undertake almost a 19% reduction in global CO_2 emissions by 2050 (Figure 2). Furthermore, the overall cost of achieving the same emission reduction targets will increase by 70% without CCS [9], which highlights the importance of CCS on the mitigation of CO_2 emission from the economic point of view as well.

There are 51 CCS engineering projects across the world, mainly scheduled in North America, Australia, China, and Western Europe, although only 19 CCS projects are currently in operation (Figure 3) [10]. Although CCS has been proven to be technically feasible, the risks and economics associated with CCS challenges its large-scale application. Consequently, the contribution of CCS is still very limited in attenuating climate change because of the high cost and safety risk associated with CO_2 leakage [11,12]. More efforts on improving the safety and economics are required to develop the CCS technology, gaining support from government, and improving public acceptance to accelerate the application of large-scale CCS.

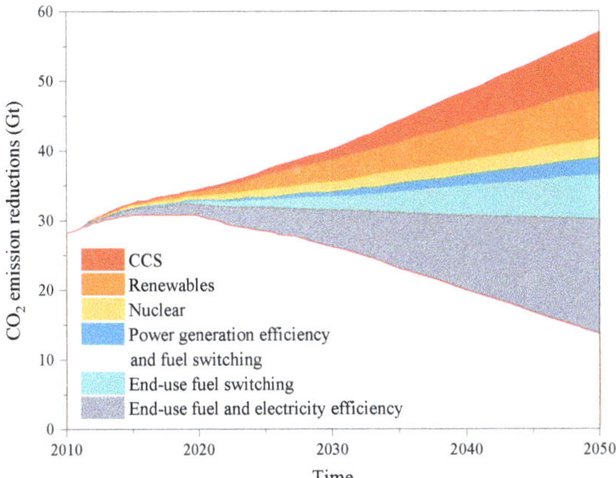

Figure 2. International Energy Agency (IEA) forecasts of key technologies for CO_2 emission reductions (modified from [8]).

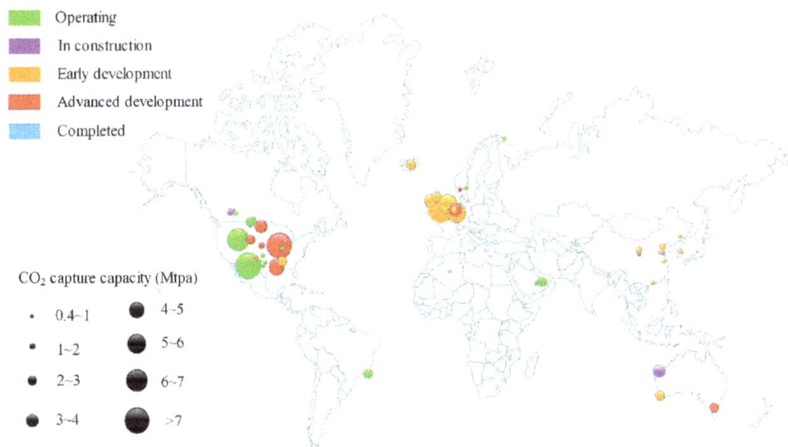

Figure 3. Commercial-scale integrated carbon capture and storage (CCS) projects around the world. Circle size is proportional to the CO_2 capture capacity, and the color indicates different stages of the lifecycle of the project (data from [10]).

In the last decade, nearly every aspect of CCS has been discussed extensively [13–44], see Table 1. However, the strategies for improving the safety and economics have not been reviewed in detail. In addition, the CCS technology is developing rapidly, and the recent progress needs to be reviewed and analyzed.

This review attempts to discuss the most recent developments on addressing challenges associated with assessing and decreasing the risks of leakage, cutting the cost of CO_2 storage and promoting the developments of commercial-scale CCS projects. Firstly, different mechanisms and strategies of CO_2 storage are summarized and discussed. Then, the risk assessment of CO_2 storage and strategies for accelerating CO_2 dissolution and mineral carbonation are reviewed. Finally, the strategies for improving cost-effectiveness, including enhanced industrial production with CO_2 storage and co-injection of CO_2 with impurities, are examined.

Table 1. Summary of review literature on CCS technology.

Research Fields	Ref.	Review Scope
CO$_2$ capture and utilization	[13]	Review of the application of CO$_2$ for enhanced oil and gas recovery
	[14]	Review of CO$_2$ capture and reuse technologies, highlighting the strategies of CO$_2$ capture in variety of scenarios, and the state of the art for CO$_2$ utilization
	[15]	Review of CO$_2$ capture, utilization, and storage (CCUS) in Chinese Academy of Sciences, highlighting the strategies for CCUS in China
	[16]	Review of the property impacts of CCS, highlighting the effect of uncertainties in thermal–physical properties on the design of components and processes in CCS
	[17]	Review of CCS highlighting the CO$_2$ capture technologies, the pilot plants, and the economic and legal aspects of CCS
	[18]	Review of CO$_2$ enhanced coal bed methane recovery, highlighting the CO$_2$ storage trials in the San Juan Basin in USA, and the estimation of CO$_2$ storage capacity in coal seams
	[19]	Review of CCUS technologies highlighting the engineering projects and their developments in China
	[20]	Review of CCS highlighting the findings obtained in CCS operational projects including the technologies of CO$_2$ capture, separation, transport, and storage
Options for CO$_2$ storage and CCS projects	[21]	Review of CCS highlighting the options for CO$_2$ storage, the evaluation criteria for CO$_2$ storage sites, and the major CO$_2$ storage projects
	[22]	Review of biomass with CCS (Bio-CCS), highlighting the economics and global status of Bio-CCS, and the role of Bio-CCS in the food–water–energy–climate nexus
	[23]	Review of CO$_2$ storage in saline aquifers, highlighting the geological and operation parameters, and the monitoring technologies for existing saline aquifers storage operations
	[24]	Review of the CCS in a coal-fired plant in Malaysia, highlighting the choices of coal plants and the capture technologies
	[25]	Review of CO$_2$ storage in saline formations, highlighting the modeling of solubility trapping
	[26]	Review of mineral carbonation (MC) technologies for CO$_2$ sequestration, highlighting the mechanisms of MC technologies and their contribution in decreasing the cost of CCS
	[27]	Review of CCS projects and future opportunities, highlighting the technical details and business plan for CCS projects
	[28]	Review of CO$_2$ storage projects in China, highlighting the CO$_2$ source, and CO$_2$ storage strategies in China
	[29]	Review of CO$_2$ mineralization product forms, highlighting the mineralization process for CO$_2$ storage
	[30]	Review of CCS by using coal fly ash, highlighting the feasibility and prospects of CCS using coal fly ash
CO$_2$-brine-rock systems	[31]	Review of the relative permeability and residual trapping in CO$_2$ storage systems, highlighting the estimating and measuring methods
	[32]	Review of the geochemical aspects of CO$_2$ storage in saline aquifers, highlighting the advantages of CO$_2$ storage in saline aquifers, and the CO$_2$–brine–rock interactions in the aquifers
	[33]	Review of geomechanical modeling of CO$_2$ storage, highlighting the numerical methods and their application in the modeling of ground deformation, faults, and fracture propagation
	[34]	Review of CO$_2$ sequestration highlighting the trapping mechanisms and the flow of CO$_2$ brine in porous media system
Well integrity and risk assessment	[35]	Review of the cement degradation in CO$_2$-rich conditions of CCS projects, highlighting the degradation of Portland cement
	[36]	Review of the risk assessment of CO$_2$ storage, highlighting the regulations and strategies of risk assessment for CO$_2$ storage
	[37]	Review of the isotopic composition of CO$_2$ for leakage monitoring in CCS project, highlighting the stable isotopes as a tracer for injected CO$_2$
	[38]	Review of the integrity of existing wells for CCS, highlighting the mechanical well failure and chemical issue due to cement carbonation
	[39]	Review of well integrity of CCS, highlighting the corrosion of metallic and cement, and the remedial measures
	[40]	Review of caprock sealing mechanisms for CO$_2$ storage, highlighting the problems associated with CO$_2$ leakage, the leakage paths, and the factors that affect leakage
	[41]	Review of CO$_2$ storage highlighting the capacity estimation of storage sites, the monitoring technologies, and the simulation tools for CCS
	[42]	Review of CO$_2$ storage and caprock integrity, highlighting the major CCS project in operation and CO$_2$ migration in the reservoirs
Storage efficiency and environmental considerations	[43]	Review of CO$_2$ storage efficiency in saline aquifers, highlighting the factors that affect CO$_2$ plume migration and the methods to estimate the storage capacity
	[44]	Review of environmental considerations for CO$_2$ storage in a sub-seabed, highlighting the potential ecological impacts

2. Mechanisms and Strategies of CO_2 Storage

2.1. Mechanisms of CO_2 Storage

As shown in Figure 4a, there are four main trapping mechanisms of CO_2 storage involving (1) structural and stratigraphic, (2) residual, (3) solubility, and (4) mineral trapping [42]. With structural and stratigraphic trapping, which is the most dominant trapping mechanism, once CO_2 is injected subsurface, it will rise to the top of geological structures due to the buoyancy effect but stay below the impermeable caprock. In residual trapping, the injected CO_2 displaces formation fluid when it flows through the formation rock. The displaced fluid disconnects and traps the remaining CO_2 within the pores of rocks due to the capillary force [45]. In the residual trapping mechanism, the saturation of trapped CO_2 is at least 10% and can reach more than 30% of the pore volume in some reservoir rocks [46,47]. In solubility trapping, CO_2 dissolves in formation fluids and becomes immobile, thereby decreasing the volume of free CO_2 [48]. This dissolved CO_2 will slightly increase the density of formation fluid by around 1%. It is sufficient to promote the convection flow with such a small density difference [49], which is also in favor of the trapping of CO_2. The solubility of CO_2 in groundwater ranges from 2% to 6%, and it decreases with the rising temperature and salinity [47]. For the mineral trapping mechanism, CO_2 is trapped by geochemical reactions in reservoir, usually precipitating as carbonate, which can trap the CO_2 in immobile secondary phases effectively [50].

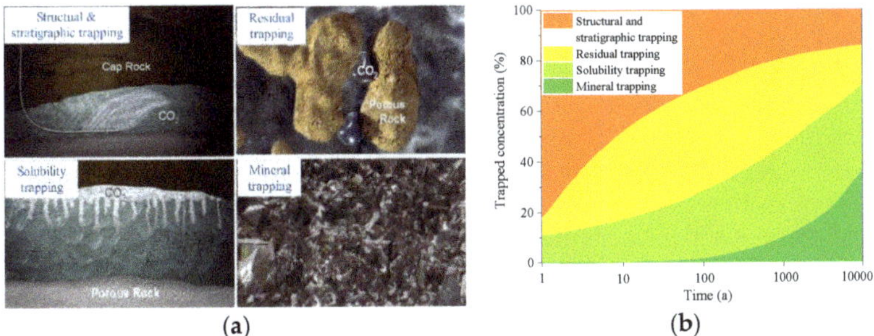

Figure 4. (**a**) The four main CO_2 trapping mechanisms [51]; (**b**) the contribution of four CO_2 trapping mechanisms with time (modified from [52]).

As shown in Figure 4b, these trapping mechanisms play different roles on CO_2 storage in the time scale from 1 to 10,000 years. Clearly, structural trapping plays a vital role in the initial stage of CO_2 storage, and its effect weakens gradually. The residual trapping and solubility trapping show a significant impact in tens of years and lock up a certain amount of CO_2 for thousands of years. With respect to mineral trapping, it begins to show a significant effect almost around a hundred years and plays a key role at a geological timescale.

2.2. Strategies of CO_2 Storage

2.2.1. CO_2 Storage in Saline Aquifers

CO_2 storage in saline aquifers is one of the most important options due to the huge amount of storage capacity, which is estimated to be sufficient for the sequestration of 10,000 Gt of CO_2, namely the emissions from large stationary sources for more than 100 years [32,53,54]. Compared with the other storage sites, the saline aquifers usually possess more wide distribution and greater regional coverage. Therefore, it has a better chance to be located near the CO_2 emission sources, which could reduce the cost of CO_2 transportation [47,55]. The most crucial issue brought by the sequestration of CO_2 in saline aquifers is pressure build up and CO_2 plume migration in formation, which has the potential to lead to the fracturing of formation and reactivation of faults and leakage of CO_2, which should receive more attention [56]. Birkholzer et al. [57] conducted a numerical simulation to determine the influence of large-scale CO_2 storage with an injection rate of 1.52 million tons per year (Mtpa) in an open saline aquifer. Their results indicated that there is significant pressure build up in the formation more than 100 km away from the injection zone, but the CO_2 plume migration is rather small—that is, around 2 km—and is concentrated on the top of saline aquifer due to the buoyancy effect. They also showed that the pressure perturbation may reach shallow groundwater formation when there is a caprock with relatively high permeability (higher than 10^{-18} m^2) between the saline aquifer and the shallow layers. However, the migration of reservoir fluids into groundwater formation is extremely unlikely. This demonstrates the safety of large-scale CO_2 storage in saline aquifers.

There are mainly five commercial-scale CCS projects in saline aquifers, including the Sleipner project [58–60], the Snøhvit project [61], the In Salah project [62,63], the Gorgon project [64], and the Quest project [65]. Detailed geological and engineering data are shown in Table 2.

Table 2. Large-scale CCS project in saline aquifers.

Num.	Project	Injection Rate (t/d)	Permeability (mD)	Depth (m)	Thickness of Reservoir (m)	Thickness of Caprock (m)	Reservoir Temperature (°C)	Reservoir Pressure (MPa)	Ref.
1	Snøhvit	2000	450	2550	60	30	95	28.5	[23,61]
2	Sleipner	2700	3000	1000	250	75	37	10.3	[23,58,66]
3	In Salah	3500	13	1800	20	900	90	17.9	[23,63]
4	Gorgon	10,410	25	2300	280	250	100	22	[23,64]
5	Quest	2960	100	2000	40	70	55	18.9	[65–68]

Regarding the Sleipner project, CO_2 was separated from the methane produced at the Sleipner field in the North Sea. Then, the CO_2 was injected into a regional saline aquifer within the Utsira Sand formation, and a total of 18 million tons were injected by 2018 since its initiation in 1996 [60]. Based on the engineering experiences from the Sleipner project, the CO_2 separated from the liquefied natural gas (LNG) project was injected to the deeper Tubåen Formation at a rate of 2000 tons per day in the Snøhvit CCS project, which is located in the Barents Sea. The project was launched in 2008 with a total amount of 1600 ktons of CO_2 injected until August 2012, and it is expected that about 23 million tons of CO_2 will be stored there based on the projected lifetime of the Snøhvit LNG project [61,69].

The CCS project at In Salah, Algeria, is one of the world's pioneering CCS projects. More than 3.8 million tons of CO_2 have been injected to the Carboniferous sandstone at the Krechba field since 2004 [62]. This CCS project is unique due to the diversity of monitoring methods, including satellite monitoring and 4D seismic data, which have been used to monitor the response of formation to CO_2 injection. Meanwhile, the accessibility of these monitoring data to the public is very high [62,63,70–77], so it could be a commendable case to study the CCS in saline aquifers.

The Gorgon CCS project located in the northwest of Australia, and it owns a Jurassic saline reservoir in the Dupuy Formation. In the lifetime of the Gorgon project, more than 120 million tons of CO_2 is planned to be injected at a rate of approximately 3.8 Mtpa [64].

The Quest CCS project began in 2015 and is designed to store the CO_2 from an existing facility for upgrading heavy oil in Scotford of Alberta, Canada. It is expected that approximately 27 million tons of CO_2 can be injected to the Basal Cambrian Sands formation at an injection rate of 1.08 Mtpa through three to eight vertical wells [65].

Aside from the previously mentioned large-scale CCS projects, there are some small-scale projects as well. These include the Ketzin pilot site [78,79], the Illinois Basin-Decatur Project [80], and the Shenhua CCS demonstration project [81]. These projects have been conducted with detailed modeling and monitoring during operation, which demonstrates the safety of this technology and helps increase the public acceptance.

Although the storage capacity of saline aquifers is huge, the overall progress of CO_2 storage in such aquifers throughout the world is still slow due to the lack of financial incentives. Therefore, some policies related to the taxes on carbon emission with a higher price might need to be formulated, which highlights the important role of the government on the application of CCS at a large scale.

2.2.2. CO_2 Storage in Depleted Oil and Gas Reservoirs

There are many advantages of CO_2 storage in depleted oil and gas reservoirs. Firstly, oil and gas reservoirs have a large amount of existing equipment installed on the surface and underground, which could be reused for CO_2 storage with only minor modification. Secondly, the seal quality and integrity of the caprock are guaranteed and have been comprehensively characterized during the exploration and production process [56]. Thirdly, the extent of pressure perturbations and induced stress changes is much smaller in depleted oil and gas reservoirs compared with aquifers because of the long-term extraction of oil and gas [56]. Compared with depleted oil reservoirs, the depleted gas reservoirs are more favorable for CCS due to higher ultimate recovery and compressibility of gas, a larger storage capacity per pore volume is available [82–84]. Comparing the types of reservoirs used in this form of storage, condensate gas reservoirs are more advantageous over wet and dry gas reservoirs resulting from the little remaining gas, the phase behavior of the mixture of condensate gas and CO_2, as well as the good injectivity of it [85]. Furthermore, the sequestrated CO_2 per pore volume in depleted condensate reservoirs is very high: approximately 13 times higher than that of the equivalent aquifer [82]. However, attention should be paid as the phase change may occur in depleted condensate reservoirs, while not in dry and wet gas reservoirs.

There are some traits associated with the long-term trapping mechanisms of CO_2 in natural gas fields. The noble gas and carbon isotope traces results show that solubility trapping in formation water is dominant, while mineral trapping is limited in the natural gas reservoirs with siliciclastic or carbonate lithologies [86]. It should be mentioned that the residual gas in the depleted reservoirs has an effect on the CO_2 sequestration. Generally, capillary trapping capacity exhibits a positive relation with the remaining gas, while structural trapping capacity, dissolution trapping capacity, and the total storage capacity are inversely related with it [87].

The most important issue related to CO_2 storage in depleted gas reservoirs is the low reservoir pressure, it's sometimes below 20 bar at the initial stage of injection, which may lead to a strong Joule–Thomson cooling effect, probably reducing the reservoir temperature, further forming hydrate, freezing the residual water, and even compromising the well injectivity, especially when cold CO_2 is injected [88–90]. When the temperature of the reservoir is over 40 °C, the Joule–Thomson cooling effect is not noticeable in permeable reservoirs, even though the initial formation pressure is as low as 2 MPa. However, the Joule–Thomson cooling may lead to the formation of hydrate, as the initial formation pressure is 6 MPa and the reservoir temperature is less than 20 °C [88]. In order to avoid the Joule–Thomson cooling and ensure a relatively high temperature at a low reservoir pressure, a high temperature of injected CO_2 or a high mass flow rate should be applied. It should be noted that the high mass flow rate may lead to other problems at the beginning and shut-in of the injection. The system in the ROAD project connects a CO_2 capture system at the Masvlakte Power Plant with an offshore depleted gas field that has a depletion pressure below 2 MPa [91]. It is a single source and sink system that allows pressure and temperature control at the shoreline inlet of the offshore pipeline by adjusting the level of after cooling at the compressor. It is used to ensure a high downhole temperature and ease the Joule–Thomson cooling effect. That is, a high temperature of injected CO_2 is applied in the reservoir with low pressure, whereas a low temperature of injected fluids can be acceptable in the reservoir with higher pressure, to keep the injection pressure requirement at a low level [90]. Furthermore, the co-injection of SO_2 and CO_2 is an alternative method to suppress Joule–Thomson cooling and shows a beneficial thermal consequence [92]. In addition, the presence of methane can potentially reduce the Joule–Thomson cooling effect [93].

A few projects dedicated to CO_2 storage have been implemented in depleted gas reservoirs. The first demonstration project in Australia named the CO2CRC Otway Project is well-known [94], in which the CO_2 was injected into the Waarre C Formation at a depth of about 2050 m. This project commenced in March 2008 and ended on August 2009, with a total storage capacity of 65,445 tons [95]. It is worth mentioning that a community led "stakeholder reference group" has been set up in this project to communicate with the public and help increase their acceptance about CCS technology, which could be a demonstration for other CCS projects. Overall, the CO2CRC Otway Project demonstrates that the sequestration of CO_2 in depleted gas fields can be achieved safely [95], and it provides a basis for the large-scale CO_2 sequestration in depleted oil and gas fields. According to the experience gained in this project, the suitability and storage capacity of similar depleted gas reservoirs has been evaluated. For example, the depleted P18-4 gas field on the offshore of Netherlands [96] and the DF-1 South China Sea Gas field [97], owning a potential capacity of 1 Gt and 8 Mt respectively, are identified as suitable sites for the sequestration of CO_2.

Generally, due to the advantages of low risk and cost-effectiveness, CCS in depleted reservoirs can play an important role in the mitigation of global warming, before the wide application of large-scale CCS in saline aquifers [98].

2.2.3. CO_2 Storage in Coal Beds

CO_2 injection into coal beds is another attractive strategy for CO_2 storage. Most of the suitable coal beds for CO_2 storage are located at a depth ranging from 300 to 900 m [99]. The sequestration of CO_2 in coal beds possesses the major advantage that the potential coal beds are usually located nearby the existing or planned coal-fired power plants. Therefore, the transportation cost could be reduced significantly. However, CO_2 storage in coal beds is still an immature technology, and only some pilot studies have been conducted on its suitability and storage capacity. The evaluated effective storage capacity of Cretaceous–Tertiary coal beds in Alberta, Canada is 6.4 Gt [99], and the potential storage capacity for the coal beds in China is about 142.67 Gt [100], which signifies the potential contribution of coal beds on the mitigation of CO_2 emissions.

2.2.4. CO_2 Storage in Deep Ocean

The CO_2 can also be directly injected into the deep ocean at water depth of more than 2700 m [101,102], where the liquid CO_2 can sink downward to the seafloor, because the CO_2 is denser than seawater under high pressure and low temperature [102,103]. The storage capacity is extremely large due to the enormous volume of the ocean. However, this CCS technology cannot be applied widely because it may affect the marine environment.

2.2.5. CO_2 Storage in Deep-Sea Sediments

The option of CO_2 storage in deep-sea sediments not only combines the merits of geologic storage and ocean storage, but it also avoids many shortcomings [104–107]. For example, it is free from the potential harm to the ocean ecosystems as the CO_2 is injected into the sediment deep beneath the ocean rather than directly into ocean. The storage mechanisms in terrestrial sequestration such as dissolution trapping, residual trapping, and mineral trapping still play a positive role. In addition, new storage mechanisms, including gravitational trapping and hydrate trapping, also work in the sequestration. The gravitational trapping comes from the fact that the higher density drives the CO_2 into the deep sea [103] to the so-called negative buoyancy zone (NBZ). The depth at which the density of CO_2 is identical to the salinity and temperature-dependent density of seawater is approximately 2700 m [52]. The hydrate trapping works because of the formation of CO_2 hydrate under the condition of high pressure and low temperature [104]. Figure 5 shows the long-term evolution of injected CO_2 in the deep-sea sediments. At the initial stage of injection, a little of hydrates form at the bottom of hydrate formation zone (HFZ), which is beneficial to reduce the permeability of the caprock. The area of hydrate caprock expands along with more CO_2, reaching the bottom of the HFZ and limiting the CO_2 below it. Meanwhile, the aqueous saturated with CO_2 will sink downward because of the buoyancy-driven advection. Finally, the hydrate CO_2 and liquid CO_2 will dissolve in seawater and change into CO_2 aqueous solution through diffusion, and permanent storage occurs.

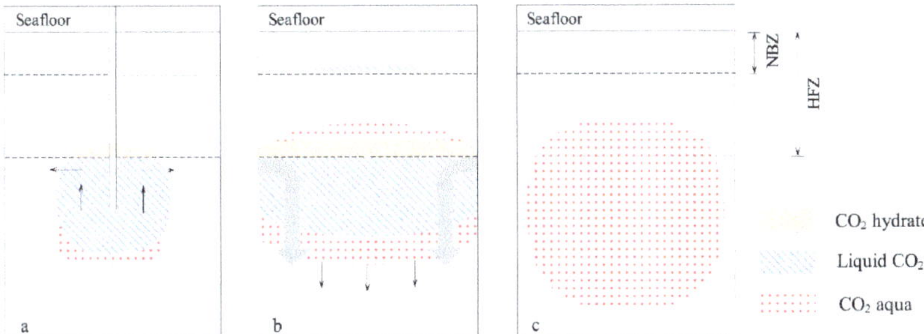

Figure 5. The long-term evolution of the injected CO_2 in deep-sea sediments (modified from [104]).

Despite the enormous capacity and feasibility this technology shares, it is still in the formulation technology readiness level. In addition, CO_2 storage in deep-sea sediments is far more expensive than onshore methods. In addition, it may take a long time to increase the public acceptance of this method [106,108].

In a summary of this part, there are several options for the underground CO_2 storage, including the saline aquifers, depleted oil and gas reservoirs, coal beds, deep ocean, and the deep-sea sediments. The pros and cons of these storage strategies are summarized in Table 3.

Table 3. The pros and cons of a variety of storage strategies.

Option	Pros	Cons
Saline aquifers	Huge amount of storage capacity, wide distribution, commercial technology readiness level	No economic benefit
Depleted oil and gas reservoirs	Existing installed equipment, guaranteed caprock integrity, characterized geological conditions, small pressure perturbations and induced stress changes, additional oil and gas recovery	Demonstration technology readiness level
Coal beds	Low transportation cost due to its potential location near the coal-fired power plants, additional coal bed methane recovery	Pilot plant technology readiness level
Deep ocean	Large storage capacity	Formulation technology readiness level, no economic benefit, may affect the marine environment
Deep-sea sediments	Enormous storage capacity, free from the potential harm to the ocean ecosystems	Formulation technology readiness level, no economic benefit, far more expensive than onshore methods

3. Security Assessment of Underground CO_2 Storage

3.1. Numerical Methods for the Security Assessment of CO_2 Storage

The brine migration caused by CO_2 injection may affect the ground water resources [57]. In addition, the chemical reaction between CO_2 with the cement and well string [79,109,110], and the formation response such as the reactivation of faults and the shear failure of caprock, may lead to the failure of well integrity and caprock integrity, resulting in the leakage of CO_2 [11,42]. Therefore, before the implementation of a CCS project, it is very important to assess risks through predicting the CO_2 injection-induced formation responses, including the formation pressure change, formation deformation, and migration of CO_2 plume, etc. Generally, such temporal and spatial responses of formation can be predicted both analytically and numerically. However, due to the physical and geological complexities in CCS projects, only a few semi-analytical models have been developed to estimate risks related with the migration of CO_2 plume and leakage along abandoned wells [111–115]. For example, Nordbotten et al. [112] derived the solution for CO_2 plume evolution during injection in dimensionless form as follows:

$$\left(Q_{\text{well}} \left(p(r_{\text{well}}, t) - p_{\text{init}} \right) + \Delta E_{\text{p}} \right) \left(\frac{2\pi \lambda_{\text{w}} kB}{Q_{\text{well}}^2} \right) \\ = -\int_0^1 \frac{(\lambda-1)\ln r'(b')}{((\lambda-1)b'+1)^2} db' + \Gamma \int_0^1 b'[r'(b')]^2 db' + \frac{1}{2}\frac{\lambda-1}{\lambda} \ln\left(\frac{\pi B \varphi}{V(t)} \right) \tag{1}$$

where Q_{well} presents the volumetric injection rate; $p(r_{\text{well}}, t)$ denotes the fluid pressure at the injection well; p_{init} is the initial pressure; ΔE_{p} presents the potential energy required to submerge the CO_2 into the denser water; λ_{w} presents the mobility of water; k is the permeability; B is the reservoir thickness; λ denotes the total mobility; φ is the porosity; $V(t)$ denotes the total injected volume; and b presents the thickness in the CO_2 plume profile, which is a function of radial distance from the injection well (r) and time (t).

The numerical simulation methods are more popular in assessment of the risks associated with CO_2 storage. Many thermal–hydraulic (TH) coupled simulators have been developed for multi-component and multi-phase flow in CO_2 storage, such as TOUGH [57,116,117], ECLIPSE [118,119], CMG-GEM [120–123], STOMP [124,125], MRST [126], COMET3 [127], IPARS [128], MUFTE-UG [129], FLUENT [130], and Tempest [131,132]. The hydro–thermal–mechanical (THM) coupled simulators are mostly constructed based on the coupling framework of fluid flow simulator (TH) and mechanical simulator (M). Specifically, the THM simulators used in CCS mainly include TOUGH-FLAC3D [63,133,134], TOUGH2-RDCA [135], TOUGH-RBSN [136], Sierra Arpeggio (Aria-Adagio) [73], OpenGeoSys-ECLIPSE [137], ABAQUS-ECLIPSE [138], and ECLIPSE-VISAGE [76]. Among them, the TOUGH-FLAC has been well tested and applied in many simulations of CCS [63,139], and it was developed by the Lawrence Berkeley National Laboratory [140,141].

In recent years, some software has been developed for the risk assessment in CCS, such as the National Risk Assessment Partnership (NRAP) Toolset and Leakage Assessment and Cost Estimation (PyLACE). The NRAP is designed to evaluate the environmental risks associated with CCS operation by the U.S. Department of Energy's National Energy Technology Laboratory [142]. In NARP, the geological system of CCS is divided into several subsystems firstly, and each subsystem is characterized by a reduced-order model [143]. Then, the reduced-order models are linked by an integrated assessment model based on the system modeling approach [142]. Finally, the whole system model can be used to evaluate the risk performance. The framework of the NARP is shown in Figure 6.

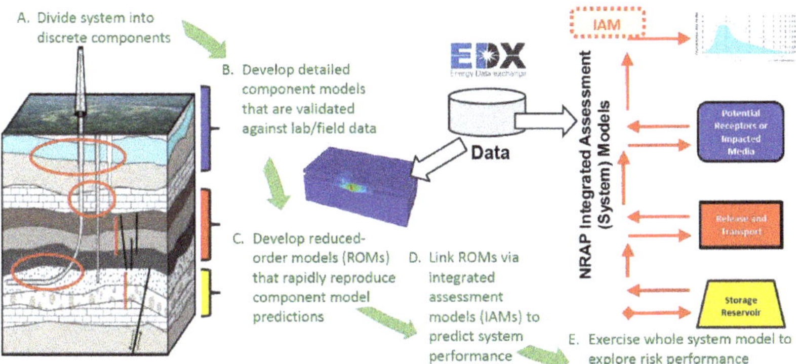

Figure 6. The framework of the National Risk Assessment Partnership (NARP) [144].

By using NARP, two major types of risk, including CO_2 leakage and induced seismicity, can be simulated. Additionally, the behavior of several important components in the CCS systems, including reservoirs, seals, wells, and ground water aquifers, could be modeled using corresponding tools. For example, the Wellbore Leakage Analysis Tool (WLAT) could be used for the assessment of the leakage potential of existing wells [145], and the Design for Risk Evaluation and Monitoring (DREAM) is developed for the evaluation and optimization of monitoring designs for long-term CO_2 storage operation.

The Python-based web application PyLACE is designed to quantify the financial risks associated with potential CO_2 leakage in a CCS system [146]. There are two major functional blocks in PyLACE: one of them is metamodel development, and the other one is metamodel-based decision support. It can convert the process-level risk assessment models into high-fidelity metamodels for the purpose of online assessment by using the high-performance computing and cloud computing infrastructures.

Recently, a deep neural network inversion was applied on 4D seismic data for estimating saturation and pressure [147], proving the availability of deep neutral network on the application of data-based inversion. To make the assessment of CCS more efficient and effective, the machine learning technology is encouraged to be used. It can be forecasted that the evaluation of CCS will be more and more intelligent with the development of machine learning technology.

3.2. Monitoring Technologies in the Assessment of the CCS Risks

The injected CO_2 will be retained in underground for a long period, and the CO_2 plume may affect the surrounding environment and the groundwater in particular [148]. Since it is difficult to predict with reasonable accuracy the key issues or risks in CCS by utilizing only simulation tools [149], the monitoring and history matching is very important in CCS assessment. The monitoring is used in most of the field development plans and routine field operations [62]. The most common monitoring technology used in CCS includes 3D seismic, 4D seismic, microseismic, vertical seismic profiling, gravimetry, cross-hole electromagnetic, pressure and temperature monitoring, geochemical sampling, soil and gas sampling analysis, tracers, atmospheric monitoring, microbiology, core analysis, satellite monitoring, and distributed temperature sensing technology.

The 3D seismic can provide a tri-dimensional image of the formation structures and the CO_2 plume. The quality of the 3D seismic is affected by the medium. In off-shore monitoring, the 3D seismic monitoring data with high quality could be obtained, and the CO_2 bodies above 106 kg at the depth of 1–2 km could identified due to the enhanced penetration of seismic waves in water [148].

The 4D seismic involves repeating 3D seismic in time-lapse mode to image the CO_2 plume in the reservoir over time, which is beneficial for the monitoring of the migration process of CO_2. A major challenge for the 4D signal to reflect the field data with high accuracy is the non-repeatable noise level in the data. This is on account of that the seismic imaging experiment is difficult to be repeated from one survey to the next due to the variations in the sources' receiver positioning and geometry, the soil moisture content, and the formation water properties [150].

The microseismic activity levels show a correlation with the CO_2 injection periods. Thus, it is beneficial for the understanding of the subsurface CO_2 injection and migration process. In this procedure, the one-dimensional array that consists of several three-component downhole geophones will be deployed in the vertical well. Afterwards, the waveform data will be detected by the geophones and then transferred to the digitizers for recording [151].

Apart from conventional surface seismic acquisition, the hydrophone arrays, buried geophone, and the fiber optic cables are permanently installed in the vertical seismic profiling systems to achieve long-term monitoring and obtain the geological structure details [152]. This monitoring technology has been successfully used in the Ketzin pilot site and the MRCSP project.

Gravimetry testing can detect the variations of fluid density due to CO_2 injection and thus provide the information of the location of CO_2. The limit of this method is that the testing result is affected by the shape of the CO_2 plume [148,153].

Cross-hole electromagnetic is a non-invasive method on the determination of the subsurface physical and chemical properties. It can also provide the information for the detection and monitoring of the location of CO_2. In this domain, the electrical conductivity before and after the CO_2 injection is obtained firstly. Then, it can be converted to the CO_2 saturation with the help of inversion algorithms and appropriate rock-physics models [154,155]. It should be mentioned that this monitoring technology is only suitable for small-scale areas such as the area between wells [154].

Monitoring the formation fluid pressure and the variation of temperature can help in evaluating the risks associated with the failure of the caprock integrity and identifying the flow path of the injected CO_2, respectively. It is important to point out that the wellhead pressure and temperature cannot provide enough information for the CO_2 injection process, which has been certified at the Ketzin pilot site [156]. Therefore, the downhole pressure and temperature monitoring are recommended in the CO_2 storage operation.

The geochemical sampling analysis could reveal the chemical variations, such as the drop of pH and natural variations in water chemistry, which is crucial for establishing a useful baseline for groundwater hydrology [157]. In addition, it can provide the information of the variation of the concentration of minerals that may be induced by the dissolution of carbonates and precipitation of anhydrite. It should be mentioned that the chemical reaction is a slow process in the sandstone formation [158], which is difficult to be detected in a short-term CO_2 storage process.

Soil and gas monitoring can provide the data of CO_2 concentration, which is beneficial for the definition of the baseline before the injection [148]. In addition, it can provide more data on natural CO_2 variations in different environments and associated seasonal fluctuations.

Tracers monitoring is a cost-effective method for monitoring the origin of CO_2 observations at wells and in the storage complex. The mechanism involves the co-injection of some specific compounds that can be detected in a very small concentration such as SF_6, SF_5CF_3, and the isotope ^{14}C together with CO_2 [95,159]. Thus, the trail of the injected CO_2 could be reflected by the traces.

Atmospheric monitoring means detecting the atmospheric concentration of CO_2 that may be changed by the leakage of CO_2 from the underground, which is beneficial for the identification of the anomalies above the natural base line [160]. The reliability of this technology may be affected by the significant natural variation of atmospheric concentration of CO_2 induced by the organic matter decomposition and the soil respiration [148].

Microbiology monitoring can be conducted on the samples of reservoir fluids and minerals before and after the CO_2 injection, which could be defined as a baseline and the modification caused by the presence of CO_2, respectively [161]. Specifically, the biocenosis such as the sulfate-reducing bacteria (SRB) in the rock substrate and fluid samples could be analyzed by the molecular biological method, polymerase chain reaction–single strand conformation polymorphism (PCR-SSCP) method, and fluorescent in situ hybridization method [162]. This information related to the microbiology is valuable for the identification of biogeochemical process that affect the diffusion of CO_2 in the reservoirs.

Core analysis is essential for the acquisition of the petrophysical and rock mechanical properties. Some measuring methods including the SEM imaging, XRD, and X-ray elemental analysis are usually included to obtain the micro morphological and mineralogical properties of the core [62].

The satellite-borne synthetic aperture radar (SAR) monitoring can provide the information related to the change of the ground surface caused by CO_2 injection operation, which can be used to modify the model of CO_2 distribution in the underground. By using SAR, the amplitude and the phase can be obtained firstly. Then, the phase difference between two observations can be converted into the displacement of the surface through the platform altitude and look angle [163]. Comparing the geophysical surveys methods, the satellite-borne SAR monitoring is considered as a cost-effective monitoring tool, and it has been successfully used in the In Salah project [163].

The distributed temperature sensing (DTS) technology is developed to overcome the limitations of conventional temperature monitoring that cannot provide the high vertical spatial resolution and real-time data. The fiber optic cable is used as a distributed sensor in DTS, which offers the possibility to measure the temperature from the surface to the bottomhole along the extension of the fiber [164,165].

The advantages and applications of the monitoring technologies in the CCS demonstration projects are summarized in Tables 4 and 5, respectively. It shows that the geochemical sampling analyses, 3D seismic, microseismic, and pressure and temperature logs have gained the most popularity among the monitoring technologies due to their high performance in acquiring the characteristics of geological structures and formation fluids.

Table 4. Main monitoring technologies in CCS.

Monitoring Technology	Advantages	Ref.
3D seismic	Provides a tridimensional image of geological structures and the plume migration of CO_2.	[62]
4D seismic	Significant benefits for overburden imaging and time-lapse responses with improved acquisition plan.	[62]
Microseismic	It is very useful for monitoring geomechanical response to injection.	[151]
Vertical seismic profiling	Valuable information on the geological structure details.	[152]
Gravimetry	Beneficial for the evaluation of formation fluids density and CO_2 plume.	[153]
Cross-hole electromagnetic	Advantageous for the detection and monitoring of the location of CO_2.	[155]
Pressure and temperature monitoring	Direct information for the evaluation of the stability of the reservoir.	[156]
Geochemical sampling	Natural variations in water chemistry are crucial for establishing a useful baseline for groundwater hydrology.	[157]
Soil and gas sampling	More data on natural CO_2 variations in different environments and associated seasonal fluctuations is needed.	[62]
Tracers	Valuable and cost-effective method for monitoring the origin of CO_2 observations at wells and in the storage complex.	[62]
Atmospheric monitoring	Useful data to identity the anomalies above the natural baseline.	[160]
Microbiology	Valuable data to identify biogeochemical process that affect the diffusion of CO_2 in the reservoirs.	[161]
Core analysis	Good petrophysical data and rock mechanical properties are essential.	[62]
Satellite monitoring	Valuable and cost-effective monitoring data for onshore CO_2 injection operation.	[163]
Distributed temperature sensing technology	It can provide high-resolution information on the migration of CO_2 in the reservoir.	[164]

Table 5. Application of the main monitoring technologies in some CCS demonstration projects (modified from [148]).

Monitoring Technology	Sleipner	Frio	Nagaoka	Ketzin	In-Salah	Otway	Weyburn	MRCSP
3D seismic	×		×	×	×	×		
4D seismic				×	×			
Microseismic	×		×		×		×	×
Vertical seismic profiling		×						×
Gravimetry	×				×		×	×
Cross-hole electromagnetic		×		×	×			
Pressure and temperature logs		×	×	×	×			×
Geochemical sampling		×	×		×	×	×	×
Soil and gas sampling		×			×		×	
Tracers		×			×	×		
Atmospheric monitoring						×		
Microbiology				×				
Core analysis					×		×	
Satellite monitoring					×			×
Distributed temperature sensing technology								×

3.3. Generating CO_2-in-Water Foams

The injected CO_2 exhibits much lower viscosity and density compared with oil and brine in the reservoir conditions and leads to a poor displacement efficiency. Generating high viscosity CO_2-in-water foams with large gas volume fraction would be an effective strategy to address this issue [166], which has some advantages for both oil reservoirs and saline aquifers. For oil reservoirs, CO_2-in-water foams can improve oil recovery and the economics of CCUS in petroleum systems [167]. Guo and Aryana [168] used a glass microfluidic device to investigate the flow behavior of foam in oil-saturated heterogeneous porous medium, and it resulted that the foam injection can improve the oil recovery through the improvement of the sweep efficiency. A field test of CO_2 foam flooding was conducted in the North Ward-Estes field in Texas, demonstrating that the foam can notably improve the sweep efficiency and be economically successful [169]. For saline aquifers, CO_2-in-water foams can also improve the sweep efficiency, storage capacity, and economics. Guo et al. [170] used a glass-fabricated microfluidic device to investigate the effect of various factors on the CO_2 storage capacity in aquifers. Their results showed that the CO_2 storage capacity would be increased by over 30% with the foam injection compared with CO_2 injection cases, demonstrating the superiority of CO_2 foam on the improvement of CO_2 storage capacity. It should be mentioned that the foam can also reduce the risk of leakage in the underground CO_2 storage unit result from the reduction of fluid mobility. For instance, due to the significant shear rates differences between the flowing in the reservoir rock and leakage pathway, the foam may become gel inside the leak and reduce the leakage through the shear-induced gelation, i.e., the particles break the interaction barrier of forming clusters in high shear stress condition [171].

To achieve stable CO_2 foams, the surfactants are used traditionally. The hydrophilic/CO_2-philic balance (HCB) is considered as the principle for the designing of surfactants for CO_2 foams, which can characterize the balance of surfactant and solvent interactions [172]. In recent years, the nanoparticles have been used combining with the surfactant solutions to improve the stability of CO_2 foams, and Worthen et al. developed the concept of HCB to be nanoparticle HCB [173]. Worthen et al. [174] generated viscous and stable CO_2-in-water foams with the mixture of nanoparticles (bare colloidal silica) and surfactant (caprylamidopropyl betaine). They suggested that the formation of foams was caused by the reduction of interfacial tension through the surfactant, while the stability of foams may be improved by the adsorption of nanoparticles at the CO_2–water interface. The behavior of silica nanoparticles on the reduction of carbon footprint was also demonstrated by Rognmo et al. [167]. In addition, some other nanoparticles such as nano lauramidopropyl betaine with alpha-olefin sulfonate also have been demonstrated to perform well in maintaining a high foam quality [170], showing the superiority of nanoparticles on the stabilizing of CO_2 foams. Overall, the generating of CO_2-in-water foams especially combining with the nanoparticles are supposed to be a candidate in the designing of CO_2 storage, while the overall cost should be considered.

3.4. Accelerating CO_2 Dissolution Process

As mentioned in above sections, the free CO_2 can remain more than 1000 years underneath the caprock, which increases the uncertainties in the long-term fate of injected CO_2 and increases the cost of long-term monitoring operation. To address this issue, accelerating the dissolution process of CO_2 and minimizing the free CO_2 in the underground is an effective strategy [175]. Cameron and Durlofsky [176] used the Hooke–Jeeves direct search algorithm to optimize the locations of CO_2 injection wells and the injection rate to minimize the mobile CO_2 in the CCS system within 1000 years, and the results showed that the fraction of mobile CO_2 would decrease from 0.220 to 0.072 in the optimal case, highlighting the importance of well location optimization. Anchliya et al. [175] proposed an engineered injection method to promote the dissolution and trapping of CO_2. Figure 7 shows the schematic of the engineered injection process. Compared with conventional CO_2 injection scenarios, there is another brine injection well located exactly over the horizontal CO_2 injection well and near the top of reservoir. In the engineered injection system, another two brine production wells are placed at either side of the CO_2 injection well. The brine injection well is used to limit the upward movement and impel the horizontal flows of CO_2 under a lateral pressure gradient provided by brine injection, which increases the sweep efficiency and enhances the CO_2 dissolution and trapping [175]. Numerical simulation studies show that about 90% of the injected CO_2 could be immobilized within 20 years after CO_2 injection ceases due to the fast dissolution trapping and residual trapping. It should be pointed out that the potential risk of pressurization due to CO_2 injection could also be addressed by the engineered injection method through controlling the brine injection and production rates. In addition, it can enhance the storage capacity of CO_2 in a bounded aquifer formation. Additional drilled wells are needed for the engineered injection system, which increases the cost of this CCS operation. Consequently, the adaptability of the engineered injection method should be quantified and site specific.

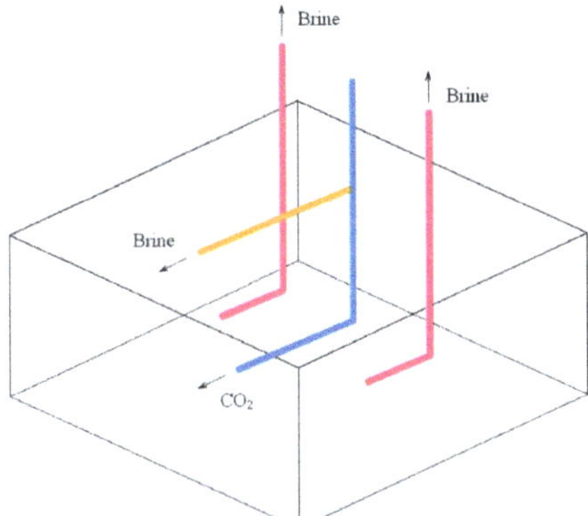

Figure 7. Engineered injection method to accelerate CO_2 dissolution and trapping (adapted from [175]).

Another injection scheme called water-alternating-gas (WAG) injection was firstly proposed in the petroleum industry in the late 1950s to improve the sweep efficiency of reservoirs. The different types of WAG injection are illustrated in Figure 8. Zhang and Agarwal investigated the potential of the WAG injection scheme with the goal of improving the efficiency of CO_2 storage [177,178]. The results showed that the optimized WAG scheme could accelerate CO_2 dissolution and decrease the impact zone up to 14% comparing with that of the constant gas injection scheme. However, the WAG injection scheme may decrease the total storage capacity of the reservoirs due to the large amount of injected water. In addition, it may increase the cost in the injection process, thus it has not been applied widely in the CCS operation.

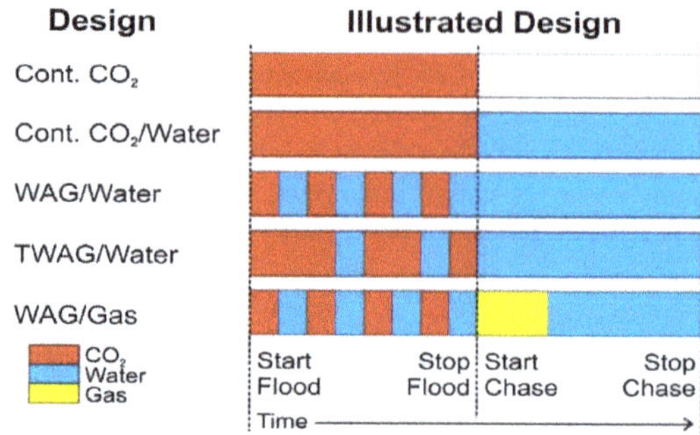

Figure 8. Schematic of various schemes of water-alternating-gas (WAG) injection [179].

To eliminate the adverse effects of WAG injection schemes on storage capacity, an injection scheme combining the intermittent injection method and brine production was proposed by Tanaka et al. [180]. The schematic diagram of intermittent injection is shown in Figure 9, which suggests that a diagonal pair of wells were used for CO_2 injection alternately. In this injection scheme, a diagonal pair of wells (Well 1 and Well 3) are used for CO_2 injection, while another pair of wells (Well 2 and Well 4) are used for producing brine, which will be re-injected to the reservoir through Well 1 and Well 3. Numerical simulation results show that both the dissolved and residual CO_2 are increased. Specifically, the ratio of the trapped CO_2 increases by 20% compared with the base case, which has only one well with continuous injection. Another advantage of this injection scheme is that it could mitigate the pressure buildup through intermittent injection and water production, which highlights the importance of the CO_2 injection design and brine management.

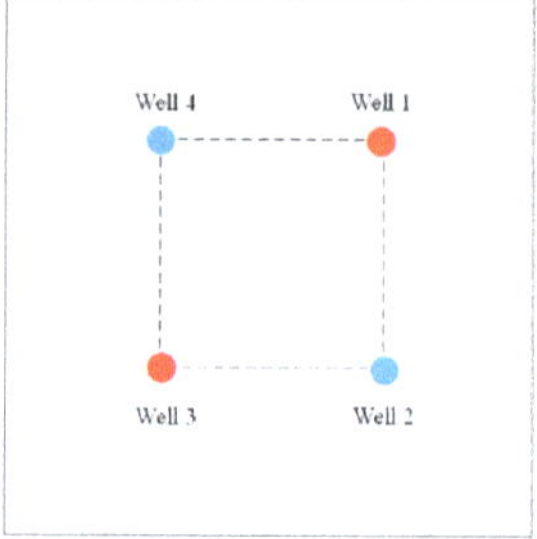

Figure 9. Schematic of the intermittent injection method (modified from [180]).

In terms of the above methods for accelerating the dissolution of CO_2, they may increase the cost in the injection process due to water injection. In addition, additional wells are needed for the engineered injection and intermittent injection with four wells, which results high costs on the operation. However, the cost on the monitoring would be reduced due to the relative low leakage risk caused by the rapid dissolution of CO_2. The overall economic costs of the CCS unit and the storage capacity are supposed to be taken into account for optimization when considering the utilization of these technologies.

3.5. Accelerating Mineral Carbonation Process

The mineralization of CO_2 is an effective way to fix the injected CO_2 and guarantee its security permanently. However, for conventional CCS in saline aquifers, it takes tens of thousands of years for the mineral trapping because of the low reactivity of silicate minerals in sedimentary rocks with CO_2 [52]. To accelerate the mineral trapping of CO_2, some novel methods have been developed, such as CO_2 storage in basalt rock formation [12], CO_2 storage in peridotite formation [181], direct mineralization of flue gas by coal fly ash [182], direct aqueous mineral carbonation [29], and pH swing mineralization [183].

3.5.1. CO_2 Storage in Basalt Rock Formation

Applying the sequestration of CO_2 in basalt formations is proposed by Gislason and Oelkers [12]. Basalt contains approximately 25% of magnesium, calcium, and iron oxides, and it is far more reactive with carbonic water compared with sedimentary silicate rock [184]. Another merit of this method is that abounding basaltic rocks exist on the Earth's surface [185], which offers the possibility of large application of CCS in basalt formation. The field test of CCS in basalt was conducted in the CarbFix pilot project in Iceland in 2012 [159]. In this project, 175 tons of pure CO_2 were injected with water firstly. Then, 73 tons of CO_2–H_2S mixture (55 tons CO_2) were fully dissolved in water and injected. The measured dissolved inorganic carbon of ^{14}C and the monitoring well shows that more than 95% of injected CO_2 was mineralized to carbonate minerals within two years, demonstrating the efficiency of mineral trapping of CO_2 in basalt. However, large-scale application of this technology requires substantial quantities of water during the CO_2 injection process. In addition, the cost of storing and transporting CO_2 for the CarbFix project is about twice that of storage in typical sedimentary basins. Therefore, the application prospect of this technology is not very promising in the near future.

3.5.2. CO_2 Storage in Peridotite Formation

The mantle peridotite is mainly composed of olivine and pyroxene, which can react with CO_2 and H_2O to form hydrous silicate, Fe-oxides, and carbonates. Isotope analysis and reconnaissance mapping indicate that about 10^4 to 10^5 tons of CO_2 per year are trapped to solid minerals via peridotite weathering effect in Oman [181]. The main reaction can be formulated as:

$$2Mg_2SiO_4 + Mg_2Si_2O_6 + 4H_2O = 2Mg_3Si_2O_5(OH)_4 \tag{2}$$

$$Mg_2SiO_4 + 2CO_2 = 2MgCO_3 + SiO_2 \tag{3}$$

$$Mg_2SiO_4 + CaMgSi_2O_6 + 2CO_2 + 2H_2O = Mg_3Si_2O_5(OH)_4 + CaCO_3 + MgCO_3. \tag{4}$$

As shown in Figure 10, the carbonation rate could be enhanced more than 1 million times compared with the natural rate with the three-step operation process, beginning with drilling and fracturing, followed by injection of hot CO_2 (approximately 185 °C) at a rapid rate to heat the fractured peridotite, and then injecting CO_2 with normal temperature. In this case, the system maintains a temperature of 185 °C and high carbonation rate due to the exothermic carbonation of peridotites. It is estimated that the peridotite in Oman alone could trap more than 1 billion tons of CO_2 per year into carbonate minerals, showing a huge amount of capacity.

However, except for the peridotite exposing through large thrust faults, the mantle peridotite is generally more than 6 km below the seafloor and 40 km below the land surface [181], making it difficult for the sequestration of CO_2 in these formations. Thus, the peridotite in shallow areas could be used effectively for the sequestration of CO_2.

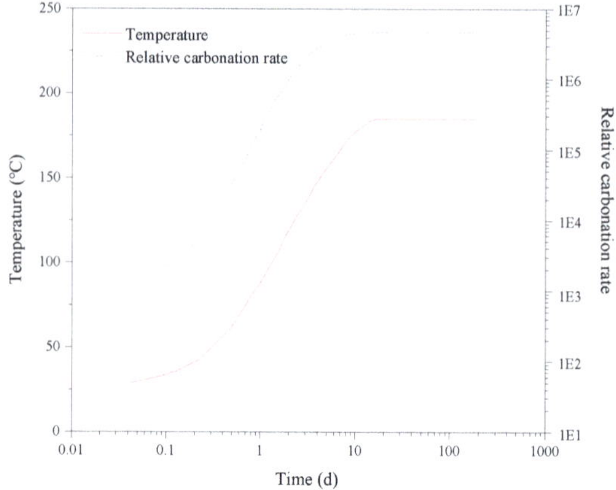

Figure 10. Calculated temperature and the carbonation rate relative to the rate for CO_2 in surface water at 25 °C and 0.1 MPa in the three-step injection operation (adapted from [181]).

3.5.3. Direct Mineralization of Flue Gas by Coal Fly Ash

Reddy et al. [182] conducted a preliminary experiment to study the reaction between flue gas and coal fly ash in a fluidized bed reactor. This experimental setup is shown in Figure 11. The results show that the concentrations of CO_2 and SO_2 in flue gas dropped from 13.0% to 9.6%, from 107.8 to 15.1 ppmv, respectively in 2 min. In addition, the Hg in flue gas was also mineralized by fly ash particles. Reddy et al. [182] conducted a pilot-scale study with a fly ash content of 100–300 kg to investigate the feasibility of this technology. In the pilot studies, the fly ash particles were fluidized by the flowing of flue gas in a fluidized bed reactor to ensure sufficient mixing and contact between them, and the reaction occurred under a fixed pressure of 115.1 kPa. Experimental results show that the content of $CaCO_3$ produced by the reaction of flue gas and fly ash ranged from 2.5% to 4% in 10 min. Meanwhile, the contents of S and Hg in the fly ash increased from non-detectable to 0.45 and 0.5 mg/kg, respectively. This confirmed that the flue gas components can be captured without separation and mineralized by fly ash particles using an accelerated mineral carbonation process. However, the treatment of the carbonated fly ash produced by using this method is still an important problem that needs be addressed [30].

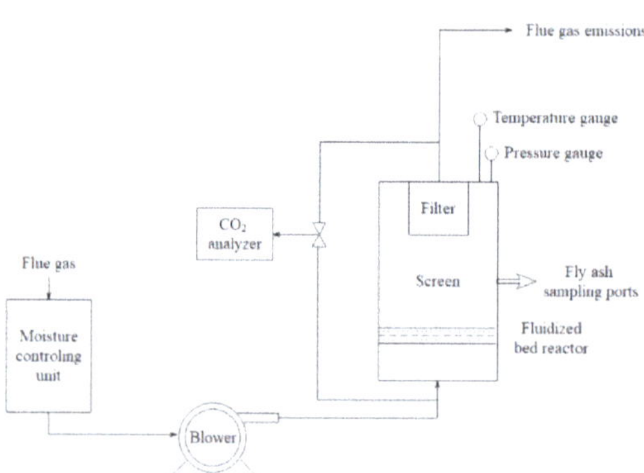

Figure 11. Preliminary experimental setup for CO_2 capture and mineralization (modified from [182]).

3.5.4. Direct Aqueous Mineral Carbonation

Direct aqueous mineral carbonation is a process that uses a bicarbonate-bearing solution mixed with reactant minerals such as magnesium and calcium silicate rocks to convert gaseous CO_2 into solid form [186]. The magnesium and calcium silicate rocks are distributed all over the world. For instance, the magnesium silicate rock in Eastern Finland has the ability to store 10 million tons of CO_2 per year for 200 to 300 years [187].

The overall chemical reaction of the carbonation of serpentine can be described as:

$$Mg_3Si_2O_5(OH)_4 + 3CO_2 = 3MgCO_3 + 2SiO_2 + 2H_2O. \tag{5}$$

In the process, the serpentine was treated at 630 °C to attain an active mineral. Then, the stoichiometric conversion of 78% of the silicate to carbonate was observed by using the bicarbonate-bearing solution at the conditions of 150 °C and 18.5 MPa, with 15% solids. This occurred within 30 min, showing an extremely high reaction rate for the carbon mineralization. To reduce the energy consumption on the heat reactivation and increase the effective reaction area, mechanical activation was proposed by adding an attrition grinding step. In this case, the reaction condition was optimized to 25 °C and 1 MPa, which still led to up to 65% carbonation within 1 h [186]. For instance, the combination of grinding and heat activation was used to attain a better carbonation mineralization performance, as shown in Figure 12 [29]. Based on this concept, the feasibility of this technology in the mineralization of flue gas was investigated by Verduyn et al. [29]. As shown in Figure 12, the relative high pH due to the lower solubility of flue gas makes it difficult for the leaching of cations. To eliminate this, the contact of the mineral and flue gas is done in a slurry mill and a leaching basin.

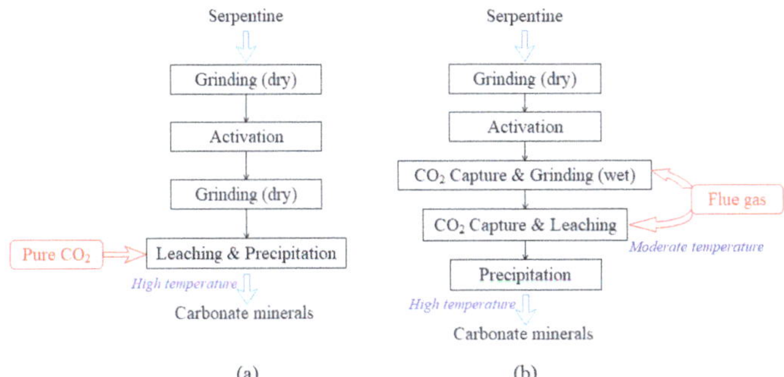

Figure 12. Mineralization process concept for (**a**). pure CO_2, (**b**). flue gas (modified from [29]).

3.5.5. pH Swing Mineralization

To increase the conversion efficiency in the CO_2 mineral trapping process, the pH swing approach was proposed by Park and Fan [188]. They combined the internal grinding in acidic solvent for a rapid dissolution of a serpentine sample. Three solid products including SiO_2-rich solids, iron oxide, and magnesium carbonate were produced by controlling the pH. Teir et al. [187] used HCl and HNO_3 to dissolute serpentinite and then transformed the serpentinite to hydromagnesite with the help of CO_2. In their results, the pure hydromagnesite, which is thermally stable at 300 °C, was produced. However, the additional amount of chemicals used in this work increase the costs and make it infeasible for CCS operations. To reduce the cost, a pH swing CO_2 mineralization method with a recyclable reaction solution was proposed by Kodama et al. [189]. They selectively extracted the alkaline-earth metal from steelmaking slag in an ammonium chloride solution. The reacted solution was used for CO_2 absorbent and produced ammonium carbonate. Then, the calcium carbonate was precipitated in another reactor with the recovery of ammonium chloride. The results showed that the selectivity of the calcium extraction reaction reached 60%, and pure $CaCO_3$ was produced with an energy consumption of 300 kWh/t-CO$_2$. Wang and Maroto-Valer [183] proposed a modified carbon mineralization process as shown in Figure 13. Through experimental studies, they concluded that (NH$_4$)$_2$CO$_3$ is more beneficial for increasing the efficiency of carbon fixation compared with NH$_4$HCO$_3$, and the optimal efficiency of CO$_2$ mineralization reaches 46.6%. Furthermore, the pH swing mineralization process was optimized by Sanna et al. [190] under different temperature conditions. The results showed that the total CO$_2$ trapping efficiency was 62.6% at the condition of 80 °C, with the molar ratio of 1:4:3 for Mg:NH$_4$ salts:NH$_3$. However, the energy consumption and overall economic cost need to be lowered before any large-scale application.

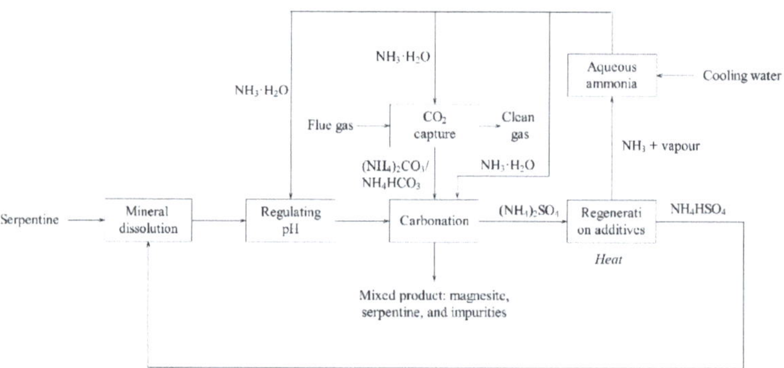

Figure 13. The schematic of carbon mineralization process using recyclable ammonium salts (modified from [183]).

To summarize this section, CO$_2$ storage in basalt rock formation and peridotite formation is limited by the distribution of the particular rock. In addition, the substantial quantities of water and energy consumption for heating increase the cost a lot. The direct mineralization of flue gas by coal fly ash would be a promising technology result since the gas could be mineralized without separation, which decreases the overall cost a lot. The pH swing mineralization may be another promising technology for the sequestration of CO$_2$ due to the sustainability. The recyclable and cheap chemical reagents ought to be introduced into the mineralization process to make this technology more cost-effective.

4. Strategies for Improving the Cost-Effectiveness of CO$_2$ Storage

4.1. Enhanced Industrial Production with CO$_2$ Storage

Resources production during CO$_2$ storage is an effective method to partly cover the cost of CCS and obtain additional economic benefits, which is called CO$_2$ capture, utilization, and storage (CCUS) [19]. In the process of CCUS, CO$_2$ usually works as a working fluid to enhance the recovery of underground resources through displacement, dissolution, thermal conductivity, and reactive transport. The potential geological formations for CCUS include oil reservoirs, gas reservoirs, saline aquifers, shale formation, un-mineable coal seams, hot dry rock, uranium deposit formation, and natural gas hydrate reservoirs [191]. The corresponding CCUS technologies are CO$_2$-enhanced oil recovery (CO$_2$-EOR), CO$_2$-enhanced gas recovery (CO$_2$-EGR), CO$_2$-enhanced water recovery (CO$_2$-EWR), CO$_2$-enhanced shale gas recovery (CO$_2$-ESGR), CO$_2$-enhanced coal bed methane recovery (CO$_2$-ECBM), CO$_2$-enhanced geothermal systems (CO$_2$-EGS), CO$_2$-enhanced in situ uranium leaching (CO$_2$-IUL), and CH$_4$–CO$_2$ replacement from natural gas hydrates, respectively [191,192].

4.1.1. CO$_2$-EOR

In CO$_2$-EOR technology, the CO$_2$ is injected into oil reservoirs to enhance the recovery of crude oil, which is the most successful and promising technology combining the utilization and sequestration of CO_2 [27,193,194]. The displacement of oil by CO_2 can be classified as multicontact miscible and immiscible processes, depending on the properties of CO_2 and the reservoir fluids at the condition of reservoir pressure and temperature [195,196]. In the multicontact miscible displacement procedure, the minimum miscibility pressure (MMP) is required for multicontact miscible displacement. The immiscible displacement occurs when the pressure is lower than the MMP, with less components exchange between CO_2 and oil in the reservoir [197].

There are three CO_2 injection schemes for the operation of CO_2-EOR, including continuous injection, water-alternating-gas (WAG) injection, and cyclic injection. In the continuous injection scheme, CO_2 injection and oil production are running continuously, which has been applied in the North Cross Devonian Unit for enhanced oil recovery [198]. The multicontact process can be achieved through vaporizing and condensing [199]. However, this injection scheme has not gained much popularity in the field application compared with the WAG injection and cyclic injection. WAG injection is the widely used form, because it can decrease the mobility ratio between the injection fluids with oil and lead to late gas breakthrough and high oil recovery. In the design of WAG injection, the optimization algorithm such as the Lagrangian and stochastic simplex approximate gradient algorithm could be used to obtain the maximum net present value [200]. Although the WAG injection is beneficial for improving the oil recovery, it may cause the gas to flow upward, while the water and oil flow downward due to the large density differences, resulting in early gas breakthrough, especially in the reservoirs with highly permeable channels and large vertical heterogeneity [199]. To address this issue, the cyclic injection process, i.e., gas huff-n-puff process was proposed, which is composed of three stages, including the gas injection stage, well shutting state, and the oil production stage.

For conventional oil reservoirs, CO_2 flow dominated throng the rock matrix. The mechanism of CO_2-EOR is due to the solubility of CO_2 in oil under the supercritical phase condition, which can decrease the oil density and viscosity, leading to enhancing the oil recovery [199]. For the unconventional tight oil reservoirs such as shale oil reservoirs, fracturing is an essential technic for the exploitation. In this scenario, CO_2 flowing is dominated by fracture flow instead of rock matrix flow. The process of the CO_2-EOR in fractured oil reservoirs can be divided into four steps, as shown in Figure 14. In the initial stage (Step 1), the injected CO_2 flows rapidly through the fracture. Then, the CO_2 starts to permeate into the rock matrix under the displacement effect (Step 2). During this stage, the permeating CO_2 may carry oil into the rock and decrease the oil production. Simultaneously, the permeated CO_2 would lead to the swelling of oil and then mitigating out of the matrix, which is beneficial for the oil production. The oil continues to swell and lower viscosity by the permeated CO_2, and moves to the fracture in the follow stage (Step 3), which corresponds to the well shutting stage in the huff-n-puff process. Finally, the pressure equilibrium inside of the matrix is approached, thus the migrating of the miscible or immiscible oil from the matrix to the fracture is dominated by diffusion effects. The oil in the bulk CO_2 is swept through the fractures to the production well by the production pressure [201]. The cyclic injection scheme can also promote the propagating of reservoir pressure due to CO_2 injection near the injection well, especially for the reservoir with ultralow permeability. Specifically, the CO_2 huff-n-puff performs better on the oil recovery when the reservoir permeability is lower than 0.03 mD [202]. Characterization of the flow behavior of CO_2 and oil in the low permeability formation with complex natural and hydraulically created fractures under the in situ conditions are supposed to be emphasized, which is beneficial for improving the efficiency of CO_2-EOR.

Apart from increasing oil recovery, CO_2-EOR provides an additional advantage of CO_2 sequestration, which could be an important economic incentive for early CO_2 storage projects [203]. Typically, 3 tons of CO_2 injection can produce approximately 1 bbl of incremental oil. It is shown that about a 5% to 15% enhancement in oil production can be achieved by using CO_2-EOR [204]. In the largest discovered fields over the world, it is estimated that approximately 470 billion barrels of incremental oil could be produced simultaneously with 140 billion metric tons of stored CO_2 by using CO_2-EOR [205].

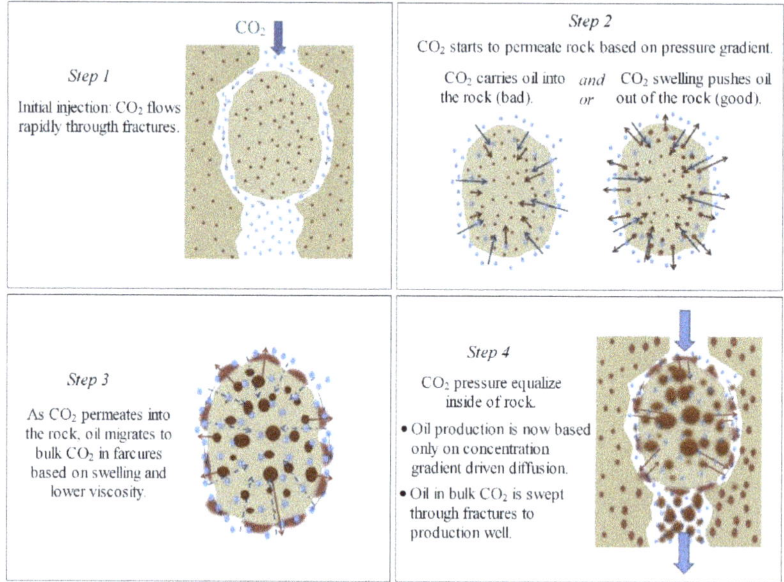

Figure 14. Conceptual steps of CO_2-enhanced oil recovery (CO_2-EOR) in fractured tight oil reservoirs ([201]).

The first CO_2-EOR pilot project was implemented at the SACROC oil field in 1972 [206], in which CO_2 foam was implemented to alter the mobility and improve the sweep efficiency [207,208]. At present, CO_2-EOR is a relatively mature technology that has been widely used in the petroleum industry to enhance oil recovery for tens of years, with the capacity of more than 1000 million tons of CO_2 stored subsurface [209,210]. This technology has gained great success in North America. In the USA, more than 260,000 bb/d are produced due to the application of CO_2-EOR technology [211]. In the Weyburn oilfield in Canada, the CO_2-EOR project was conducted to extend the life of an oilfield. About 20 million tons of CO_2 is planned to be sequestrated in the oil reservoir [212,213]. In recent years, the feasibility of CO_2-EOR in China has been massively studied. The first CO_2-EOR project in China, i.e., the Jilin Oilfield, injected nearly 217,000 tons of CO_2 with a storage efficiency of 96% by April 2013 [214], and the CO_2 capacity is about 600,000 tons [10]. The technology of CO_2-EOR has application prospects in the Shengli Oilfield and Bohai Bay Basin, with 6.7% incremental oil recovery and 683 million tons of incremental oil production, respectively [215,216]. CO_2-EOR also attracted much attention in Europe. In Poland, the potential utilization of anthropogenic CO_2 for CO_2-EOR was studied in the B8 oilfields in Baltic Sea and Brage on the Norwegian Continental Shelf [217], which is a part of the ongoing PRO_CCS project funded by Norway Grants. The simulation results showed that the total storage capacity of Brage and B8 oilfields are 33 and 4.8 million tons in 17 years of injection, with an expected incremental oil production of 98 and 14.6 million bbls, respectively.

To co-optimize the CO_2 storage and enhanced oil recovery, Ampomah et al. [218] proposed an objective function (Equation (6)) considering both CO_2 storage and oil production, which can be optimized by the neural network and genetic algorithm. In the optimal case of the Farnsworth field unit, more than 94% of CO_2 could be sequestrated with approximately 80% of the oil produced, which provides a guideline for the co-optimization of CO_2 storage and EOR.

$$f = w_1 \times FOPT + w_2 \times FGIT \tag{6}$$

where w is weight assigned to vector, *FOPT* is the cumulative produced gas, and *FGIT* is the cumulative injected gas.

Similarly, a framework to co-optimize the oil production and CO_2 storage was developed by Jahangiri and Zhang [219]. In the framework, the net present value (NPV) is treated as the optimization objection function, which was solved by the ensemble-based optimization algorithm.

$$NPV = \sum_{t=1}^{T} \frac{C}{(1+r)^t} - C_0 \tag{7}$$

where t is the time step, T is the operation period, r is the periodic discount rate, C is the cash flow in the time step that is determined by the price, injection, and production volume of CO_2 and oil, and C_0 is initial investment.

By using this method, the well injection patterns and injection rates for the maximum NPV can be determined. In addition, the discrete time optimization model could be used for maximizing the total profit in CO_2-EOR operations, with consideration of both enhanced oil recovery and geological CO_2 sequestration [220].

Artificial neural network models can be used to predict and optimize the performances of CO_2-EOR, as shown in Figure 15 during the multi-cycled water-alternating-gas process [221]. There are four neurons in the input layer corresponding to water-to-gas injection time ratios (WAG), temperature, permeability ratio, and initial water saturation, respectively. Subsequently, the oil recovery, oil production rate, gas and oil ratio (GOR), and net CO_2 storage amounts are set as the targets, which correspond with four neurons in the output layer and with 10 neurons in the hidden layer. The oil recovery and net CO_2 storage can be accurately predicted in this framework. For instance, the optimal injection scheme could be obtained for a maximum economic profit in various reservoir conditions by using this method.

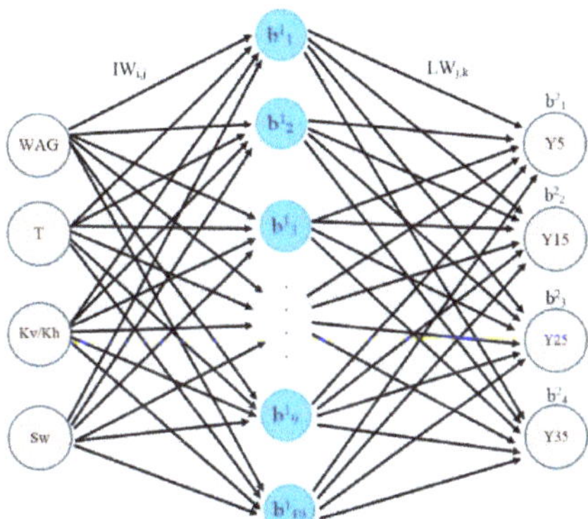

Figure 15. Artificial neural network structure of the models, Y represents oil recovery, oil production rate, gas and oil ratio (GOR), and net CO_2 storage amount [221].

The machine learning approach can also be applied to optimize oil recovery as well as CO_2 storage [222]. As shown in Figure 16, a history matching model was developed based on production history data in the CO_2-EOR process under this optimization framework. Then, the hybridized multi-layer and radial basis function neural network method were utilized to train a proxy model, which is beneficial for increasing computational effectiveness in the optimization process. After a proxy model with reliable accuracy was realized, the machine learning optimization algorithm was used to obtain the optimal solution of the objective function that incorporates the role of parameters such as oil recovery and CO_2 storage. This work highlights the adaptability of a robust machine learning approach for optimizing the CO_2-EOR process. Considering the mature technology and huge market demand, it can be concluded that CO_2-EOR may play a more important role on the mitigation of CO_2 than other strategies of CO_2 utilization in the next few years.

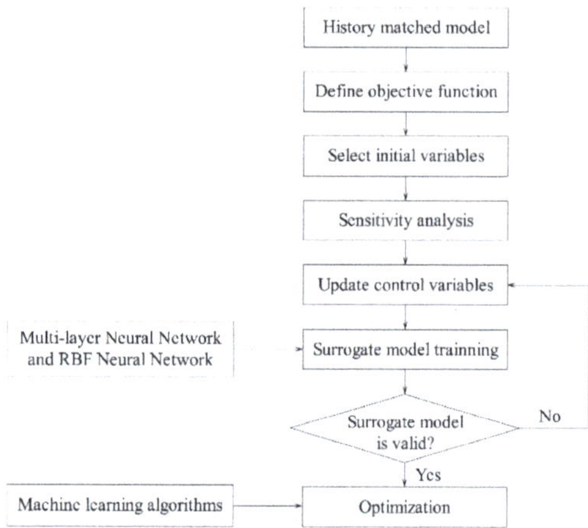

Figure 16. Flowchart of the optimization framework (modified from [222]).

4.1.2. CO_2-EGR

The technology of CO_2-EGR means that it enhances the gas recovery by CO_2 injection. Enhanced gas recovery is realized by both the displacement and re-pressurization of the remaining gas in a depleting or depleted reservoir [223], especially for sour gas reservoirs in which CO_2 is produced mixed with the natural gas. The separated CO_2 from the produced gas could be injected back into the reservoir to enhance gas recovery. Additionally, CO_2 has the potential to reduce the dew point pressure of reservoir fluids in wet gas reservoirs, which is favorable for eliminating condensate blockage and improving CH_4 production [224,225]. It is estimated that up to 11% incremental gas recovery can be achieved by CO_2 displacement [223].

The feasibility of CO_2-EGR has been investigated by many experimental and numerical simulation studies [83,131,132,211,224,226–230]. Some typical displacement experiments in a variety of temperature and pressure conditions, revealing the mechanism of CO_2-EGR and providing a guide line for the application of this technology, are summarized in Table 6.

Table 6. Typical displacement experiments on the CO_2-enhanced gas recovery (CO_2-EGR) process.

Rock Type	Saturated Fluids	T (°C)	P (MPa)	Key Observations	Ref.
Carbonate core	CH_4	20–60	3.55–20.79	Whether CO_2 is in the gas, liquid, or supercritical phase, it could enhance the recovery of CH_4.	[83]
Carbonate core	Saturated with methane with or without water	20–80	3.55–20.79	The coefficient of CO_2 increases with temperature and decreases with pressure.	[231]
Berea sandstone core	Dry core, initial saturation of 10% water, and initial saturation of 10% brine (20 wt %), respectively	40	8.96	The salinity of connate water will decrease the dispersion of CO_2 in CH_4.	[232]
Sandstone and carbonate core	CH_4	60–80	10–12	The residual water narrows the pore and consequently increases the dispersion of supercritical CO_2 and CH_4.	[233]
Sandstone core	CH_4 and simulate natural gas (90% CH_4 + 10% CO_2) respectively	40–55	10–14	The dispersion coefficient of CO_2 in the simulate natural gas is larger than that of CH_4.	[234]
Sandstone core	Formation water and N_2	50	21	The gravity segregation effect is notable in the porous and permeable core, while the heterogeneity effect becomes dominant in the low permeability of the core.	[235]
Bandera sandstone core	CH_4	50	8.96	The gravity has significant effects on the flow behavior of SCO_2 at lower flow rates.	[236]

The most critical hurdle in CO_2-EGR is the breakthrough of CO_2 in the reservoir producing CO_2-contaminated gas [131,237]. Actually, the preferential pathway has a significant impact on CO_2 breakthrough and ultimate CH_4 recovery [238], so the geological formations for CO_2-EGR, especially the microstructures, are supposed to be characterized in detail. Irreducible water in reservoirs also has an impact on the mixing of CO_2 and CH_4 [239]. The dispersion increases with irreducible water, because the pores occupied by irreducible water lead to much narrower pores and more tortuous flow paths.

In addition to the above-mentioned geological parameters, engineering parameters also have a significant influence on gas mixing and CO_2-EGR performance [131,240,241]. CO_2 injection with a horizontal well in the lower parts and CH_4 production in the upper parts of reservoirs could mitigate the breakthrough of CO_2 at the production well [131,242]. CO_2 injection during the early decline phase of natural gas production is beneficial for ensuring the displacement in supercritical phase and achieving a high CH_4 recovery [131]. On the contrary, it may cause the trapping of CH_4 in unswept areas under high pressure. CO_2 injection in the late phase could avert this shortcoming and improve the performance of CCS, which is more attractive when the CO_2 sequestration is considered [131,243]. On the whole, it can be concluded that the time of CO_2 injection to obtain a maximum incremental recovery is highly affected by the allowable produced CO_2 concentration at the production well, which is determined by the economics of CCS projects [244].

Whether the CO_2 is injected in the early or later stage, it is recommended to inject CO_2 at a relatively high pressure to ensure the supercritical phase in the displacement process. In this case, the distribution of CO_2 is dominated by gravity forces [245]. As the CO_2 is much more dense than CH_4, CO_2 will occupy the smaller space and spread at a slower rate, which could mitigate CO_2 breakthrough. However, if the injected CO_2 is in gas phase in the reservoir, the CO_2 will occupy a large volume and mix with the CH_4 more easily, which can lead to an early CO_2 breakthrough [245].

Regarding the injection rate, of course, a high injection rate can increase the gas recovery [241]. However, a high injection rate also brings excessive gas mixing, which is harmful for methane production. It is suggested that the injection rate should be lower than the production rate to avoid an early breakthrough of CO_2 [237]. In Al-Hasami's study [223], 9% incremental methane recovery can be achieved when the CO_2 injection rate is only 13% of the production rate. Instead of the constant injection rate, a constant pressure injection scheme was proposed to avoid potential risks related to high pressure [246]. The optimal injection strategy could be achieved by an optimization code based on genetic algorithm and multi-phase simulator TOUGH2 (GA-TOUGH2).

Geological parameters also greatly affect the performance of CO_2-EGR. The viscous and gravity force affecting parameters e.g., permeability, formation dip, and thickness, play a vital role in the stability of displacement. The fluid properties such as the diffusion coefficient and water salinity take second place on affecting the CO_2 breakthrough [244]. Specifically, the connate water in reservoirs has a positive impact on CO_2-EGR performance. As a result, the dissolution of CO_2 in reservoir fluids is favorable for enhancing the storage capacity and mitigating the CO_2 breakthrough in the production well [223,237,247].

There are several CO_2-EGR projects around the world, including the Alberta gas field project, the K12-B field project, and the CLEAN project. The Alberta gas field project is located in Canada. In this project, impure CO_2 with less than 2% of H_2S has been injected into the depleted Long Coulee Glauconite F gas Pool in southeastern Alberta since 2002, but the operation was terminated in 2005 because of the breakthrough of acid gas [245]. The K12-B gas field is located in the Dutch continental shelf in the North Sea, with a reservoir depth of about 3800 m below the sea level. The reservoir pressure has dropped from 40 MPa to 4 MPa with a production of 90% of the initial gas in place. The initial reservoir temperature is 128 °C [248]. Over 0.1 million tons of CO_2, which is separated from the produced gas directly at the offshore platform, has been injected over a period of 13 years since 2004. Monitoring data shows that the well integrity has remained stable [249]. Furthermore, no major complications occurred in the lifetime of this project, which proves that the safety of CO_2-EGR can be ensured [250]. The CLEAN project was conducted between 2008 and 2011 to inject CO_2 into the Altmark natural gas field in Germany. The risk assessment of this project has been conducted based on digital databases. The results showed that the safety and efficiency of EGR technology based on CO_2 injection could be achieved. Meanwhile, the borehole integrity could be achieved without any intervention, providing a guideline on CO_2-EGR [251].

Generally, the technology of CO_2-EGR is still immature and needs more efforts to address the problems, such as mitigating the CO_2 breakthrough and achieving a favorable performance in both enhancing CH_4 recovery and CO_2 sequestration. The economic success is largely dependent on the political developments in the next years and decades [251].

4.1.3. CO_2-EWR

Similar to CO_2-EOR and CO_2-EGR, CO_2-EWR is a methodology combining CO_2 sequestration and saline water production [252–254], which is developed from the technology of CCS in saline aquifers. Figure 17 shows a diagram of CO_2-EWR technology. The operation of CO_2-EWR could decrease formation pressure and avoid potential leakage through extracting formation water, thus it could further improve the storage efficiency and achieve higher security and stability of large-scale geological CO_2 sequestration [255]. Besides, the produced saline water could be used for drinking, industrial, and agricultural utilization after desalination treatment such as using a high-efficiency reverse osmosis system [252]. Meanwhile, the deep brine resources obtained through the cascade extraction may create economic profit and fill the cap of cost in the operation of CCS technology [253]. Kobos et al. [252] proposed a numerical simulation model to investigate the feasibility of CO_2-EWR based on a hypothetical case study from a representative power plant and saline formation in the southwestern part of the United States. In their work, the extracted saline water was treated with a high-efficiency reverse osmosis system and then used as power plant cooling water. The results showed that the coupled technology of CO_2 storage and saline water extraction and treatment is feasible for tens to hundreds of years.

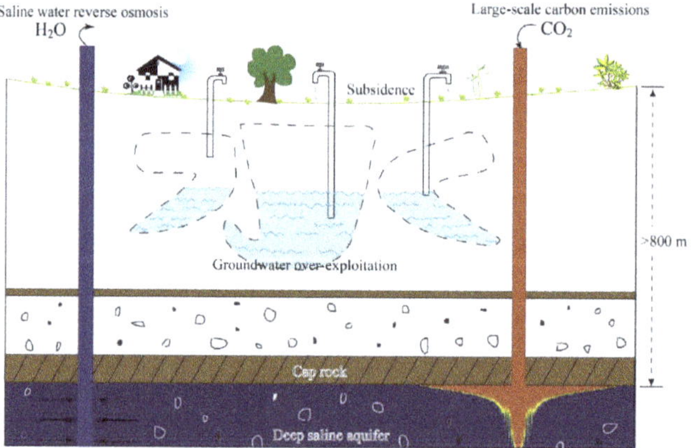

Figure 17. Depiction of the CO_2-enhanced water recovery (CO_2-EWR) technology [253].

Unfortunately, the added cost of extraction wells is considered a shortcoming of CO_2-EWR [252]. Furthermore, the production of brine must be ceased once the breakthrough of CO_2 occurs [256]. In general, under effective engineering design, the CO_2-EWR technology has application prospects.

4.1.4. CO_2-ESGR

In regard to CO_2-ESGR technology, the CO_2 is injected into shale gas reservoirs to replace and displace shale gas, for the ultimate goal of enhancing the shale gas recovery, with a side benefit of

CO_2 sequestration synchronously [257]. The dominate mechanism of CO_2-ESGR is the competitive absorption of CO_2 by shale matrix [199], such as with a CO_2 sorption capacity up to 1 mmol per gram for the Muderong Shale [258]. In addition, the pressure gradient displacement plays an important role [259]. Liu et al. [259] conducted a numerical simulation to evaluate the feasibility of CO_2-ESGR, and results showed that over 95% of the injected CO_2 was instantaneously adsorbed and sequestrated in the reservoirs. However, only limited ESGR performance was detected due to the limited communication between the wells in this study. The feasibility of CO_2-ESGR on the Devonian Gas Shale Play of eastern Kentucky was investigated by Schepers et al. [127]. They found that the huff-n-puff scenario was not suitable, while the full-field continuous CO_2 injection was a good option. About 300 tons of CO_2 were injected within one and half months. A significantly increased recovery was attained, and approximately half of the injected CO_2 was sequestrated. Therefore, it can be concluded that there remains a long way before the application of CO_2-ESGR, and the contribution to the mitigation of CO_2 emission is still limited.

4.1.5. CO_2-ECBM

In CO_2-ECBM, the CO_2 is injected into un-mineable coal seams to displace and replace coal bed methane, simultaneously achieving CO_2 sequestration in the coal seams. Similar to the CO_2-ESGR, CO_2 works as displacing fluid and is competitive in the process of CO_2-ECBM [260,261]. The potential ECBM recovery in China is estimated to be over 3.751 Tm^3 [100], highlighting the superiority of this technology. However, the injected CO_2 in CO_2-ECBM projects is usually less than 1 million tons per year, and many coal seams usually with low permeability such as those in Western Europe are not suitable for the application of this technology [148]. Herein, the role of CO_2-ECBM on mitigating the emission of CO_2 is limited.

4.1.6. CO_2-EGS

Geothermal energy is regarded as a clean, renewable, and reliable energy for its advantages of sustainability and environment-friendly characteristics [262]. It is extracted through water traditionally. Brown [263] firstly proposed the concept of using supercritical CO_2 instead of water as the heat exchange fluid in an EGS. It has been proven that the heat extraction efficiency of CO_2-based systems is superior to water-based systems. If this concept is popularized, more regions worldwide with relatively low temperature can be used for electricity production in an economically beneficial manner [254,262,264]. In addition, the mobility of CO_2 is better than that of water, which is beneficial for the production of fluids and the extraction of geothermal energy.

In recent years, the technology combining geothermal extraction and CO_2 storage has gained more attention [264,265], which can achieve an efficient geothermal energy extraction as well as mitigating CO_2 emission. For example, the regional energy deficit could decrease by 22.1% and the CO_2 emissions could decrease by 31.3% in the Latium Region in Central Italy if the CO_2-EGS was applied [266]. However, the technology of the CO_2 plume geothermal system is still at the conceptual stage and pre-feasibility studies phase, and many efforts need be devoted toward its study before its application.

4.1.7. CO_2-IUL

CO_2-IUL is a technology that injects CO_2 and leaches uranium ore out of geological formation through reaction with ore and minerals in the ore deposits [191]. CO_2-IUL could increase the recovery of uranium and simultaneously be favorable for CO_2 storage, especially for sandstone-type uranium mining [191]. However, the global annual demand of natural uranium is only around 0.1 million tons [267], thus it may be difficult for CO_2-IUL to reduce CO_2 emission significantly due to its limits and demands.

4.1.8. CH$_4$-CO$_2$ Replacement from Natural Gas Hydrates

The technology of CH$_4$-CO$_2$ replacement from natural gas hydrates (NGH) is regarded as a win–win method for exploring NGH and simultaneously storing CO$_2$ in the form of CO$_2$ hydrates formation [268–271]. As shown in Figure 18, the mechanisms of this technology can be divided into four steps. Firstly, the CO$_2$ molecule diffuses into the surface of CH$_4$ hydrate and decreases the stability of CH$_4$ hydrate structure (Figure 18a). Secondly, due to CH$_4$ hydrate dissociation, the CH$_4$ molecule escapes from the hydrate cage (Figure 18b). In the next stage, the hydrate is re-formed. As shown in Figure 18c, the CO$_2$ molecules mainly occupy the large cage, while CH$_4$ molecules occupy the small cage. Finally, the CH$_4$ molecules diffuse from the surface of hydrate and change into gas, while the CO$_2$ molecule diffuse into deeper hydrate layer to continue replacing the CH$_4$ in hydrate (Figure 18d) [270]. To improve the performance of CH$_4$–CO$_2$ replacement from natural gas hydrates, a thermal stimulation approach was proposed [271]. By using this approach, the CH$_4$ replacement exhibits an upper limit of 64.63%, and the maximum CO$_2$ storage efficiency can reach up to 78.40%–96.73% [271].

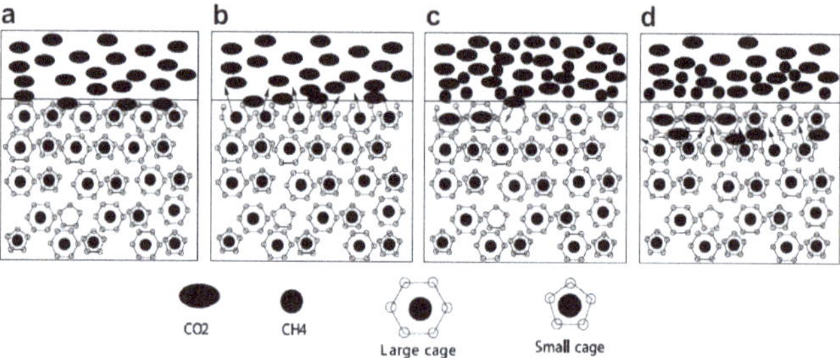

Figure 18. Schematic diagram for CH$_4$–CO$_2$ replacement in hydrates [270].

Based on the concept of thermal stimulation to CH$_4$–CO$_2$ replacement, a geothermal-assisted CO$_2$ replacement method (GACR) was proposed by Liu et al. [272]. As described in Figure 19, the CO$_2$ with ambient temperature was injected into geothermal reservoir for heating, then the heated CO$_2$ flows upward into the hydrate-bearing layer (HBL) to enhance the NGH dissociation. Numerical simulation results showed that the GACR method can significantly accelerate NGH dissociation and increase CH$_4$ recovery. However, the application of this method is limited by a strict precondition that a thermal reservoir exists below the methane hydrate reservoir.

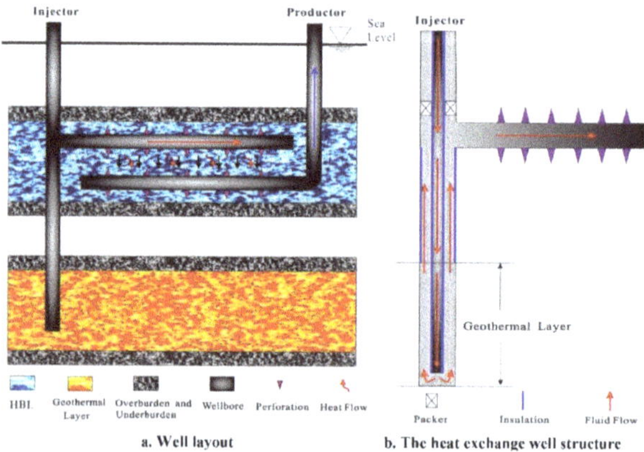

Figure 19. Schematic well group configuration diagram of the geothermal-assisted CO$_2$ replacement method (GACR) [272].

The research on CH_4–CO_2 replacement from natural gas hydrates is still in the preliminary experimental and numerical study phase [271,273,274]. However, it is expected that great progress will be made in the near future under the stimulation of methane hydrate production and CO_2 sequestration.

In a short summary of this part, the utilization of CO_2 for resources production and CO_2 storage are encouraged to be designed for the whole process of engineering operation to optimize its performance. To achieve this goal, artificial intelligence may play an important role. Although CO_2-EOR has been used commercially, the other technologies are in the pilot plant phase in terms of technology readiness level [204]. CO_2-EOR would be the most promising technology combining the utilization and sequestration of CO_2 in the next few years. CO_2-EGR is another promising technology, whose application is limited by the mixing of CO_2 and CH_4 in the reservoir. Considering that the mixing behavior is affected by the geological parameters, i.e., porosity, permeability, residual water saturation, and engineering parameters, i.e., injection rate, injection pressure, production rate, a site selection system for CO_2-EGR project is encouraged to be developed.

4.2. Co-Injection of CO_2 with Impurities

The biggest obstacle for large-scale CO_2 storage is the lack of financial incentives [12]. Most strategies of CO_2 storage could not generate profit, so the measures to decrease the cost of CO_2 storage is very important and beneficial for the large-scale application of this technology. It is estimated that the cost of carbon capture and storage is dominated by the process of capture and gas separation, which costs \$55 to \$112 per ton of CO_2 [12]. Therefore, co-injection of CO_2 with impurities can be a cost-effective option for CO_2 sequestration [275].

The impurities may be co-injected with CO_2 including CH_4, H_2S, SO_2, N_2, and O_2. Among them, H_2S and CH_4 are usually mixed with CO_2 in produced acid gas, which can be used in CO_2 storage. Other gases are the main components of flue gas, which is captured from the major CO_2 emission sources such as power plants [276].

The impure CO_2 can also be used for CCUS especially for CO_2-EOR with multicontact miscible CO_2 flooding [195]. In the multicontact miscible displacement procedure, the MMP is a key control variable due to it having a notable impact on the design and development of assets and being closely related to the economically feasibility, which is affected by the impurities in CO_2 a lot. In general, the presence of H_2S and SO_2 in CO_2 can reduce the MMP [195,277], while the presence of CH_4, N_2, and O_2 can increase the MMP of CO_2 [278,279], which is disadvantageous for the CO_2-EOR operation and may increase the risks for the fracturing of the formation due to the higher injection pressure required.

The presence of impurities may change the other CO_2 thermophysical properties and phase behavior [280], and affect the performance of CCS. N_2 would lead to a delay for CO_2 breakthrough when it is co-injected, because the solubility of CO_2 in irreducible water is much higher than that of N_2 [281]. However, the N_2 would decrease the density of the dissolved phase and increase the risk in the long term [282]. Generally, the storage capacity of reservoirs decreases proportionally to the concentration and the compressibility factor of impurities when N_2 is co-injected with CO_2 [82]. For instance, the reduced storage capacity may be even higher than the volume fraction of impurities when O_2 is included. However, the negative impact of impurities on capacity could be alleviated by storing the impure CO_2 in a reservoir with high temperature [276].

There is no significant effect of H_2S, with a fraction of less than 30%, as impurity on the dissolution of CO_2 [275]. However, when it was co-injected with CO_2 under the condition of 20 MPa and 45 °C, the H_2S with a concentration over 20% has a potential to decrease the interfacial tension and increase the contact angle, leading to a low capillary force [283]. This means that H_2S may increase the risks of gas leakage, which should receive attention. The impure CO_2 with H_2S can be trapped by hematite even in a dry system driven by the reduction of ferric iron in hematite by sulfide species, verifying the feasibility of co-injection of CO_2 with H_2S [284].

CH_4 produced from an acid gas reservoir may also serve as impure gas and be used in CCS projects. There is no significant negative influence of CH_4 on the interfacial tension and wettability even with the concentration of up to 20% [283]. The storage capacity of reservoirs also decreases proportionally to the concentration and compressibility factor of CH_4 [82]. As the concentration of CH_4 in injected CO_2 is very low, its effect on the CCS is minor and can be neglected.

The presence of SO_2 usually controls the acid-induced reactions with calcium-rich minerals when it is co-injected into reservoirs as an impurity, but the quantitative effect is very minor and could be negligible based on the German Bunter Sandstone [285]. Generally, the porosity in sandstone increases under the impact of SO_2, while the porosity of the intermediate shale layer decreases because of the conversion of dominant calcite to anhydrite [286]. For instance, the conversion of Ca^{2+} bearing carbonate to anhydrite is observed when the SO_2 was co-injected with CO_2 into the German Bunter Sandstone [287]. A field study was conducted to investigate the geochemical impacts of SO_2 and O_2 as impurities on the reactions of minerals and fluids in a siliciclastic reservoir [288]. The CO_2-saturated water with impurities was injected into reservoirs and allowed to interact with minerals for three weeks. The results showed that the pyrite dissolved due to the O_2 acting as an oxidizing agent. However, the concentrations of SO_2 and O_2 are 67 ppm and 6150 ppm respectively, which is too low to lead to a significant effect on fluid–rock interaction. It could be inferred that the impact of impurities on the interaction with formation rock is highly dependent on the composition of minerals, which should be analyzed site specifically. It should be mentioned that the co-injection of SO_2 and CO_2 could suppress Joule–Thomson cooling, which is a beneficial thermal consequence for CCS [92].

In short, co-injection of CO_2 with impurities is an effective strategy to reduce the cost of CO_2 storage. However, the interaction of the impurities with formation rock, and the effect of impurities on thermophysical properties of reservoir fluids need to be further studied to reduce the uncertainties in CO_2 storage process.

4.3. Prospects of CCS/CCUS Technologies

The economic factor for the CCS projects is believed to be one of the most important incentives for the industry. The price for CO_2 emissions at the first major carbon market and also the biggest one, i.e., the European Union Emission Trading System, is approximately $7 per ton of CO_2, which is much lower than the cost of CCS [12]. Therefore, there are no financial incentives of CCS for industries unless a higher price of carbon emission is set, demonstrating the important role that should be played by the government in the mitigation of CO_2 emission.

Table 7 shows the large-scale CCS projects (more than 0.4 Mtpa) throughout the world from 1972 until the end of the 2020s. It can be seen that the CCUS for EOR and CCS in saline formations have made major contributions toward CO_2 storage, which is in accordance with the prediction model formulated by Mac Dowell et al. [203]. Nearly half of 51 large-scale CCS projects scheduled by the end of the 2020s are designed for EOR, which shows its economic viability in EOR operations. In addition, 21 projects of CCS in saline formations are also planned, since its CO_2 storage capacity may up to 4 Mtpa. Moreover, the average CO_2 capture capacity of CCS in depleted gas fields is much greater than that of EOR and may reach to 2.8 Mtpa, showing the potential of the mitigating of CO_2 emissions. However, on account of the great extent of the mixing between CO_2 and CH_4, which is an obstacle for enhancing additional recovery of CH_4, there are only three large-scale CCS projects in depleted gas fields in the near future. With the development of technology to address this issue, CCS in depleted gas fields can play a more important role in CO_2 storage.

Table 7. Large-scale CCS projects (more than 0.4 Mtpa) from 1972 to the end of the 2020s (data from [10]).

Strategy	Under Evaluation	EOR	CCS		Total
			Saline Formation	Depleted Gas Fields	
Quality of project	24	3	21	3	51
Capture capacity (Mtpa)	42.11–43.41	8.1–8.6	40.35–85.1	7.5–8.5	98.06–145.61
Average capture capacity (Mtpa)	1.75–1.81	2.7–2.87	1.92–4.05	2.5–2.83	1.92–2.86

5. Conclusions

The status of the strategies for CO_2 storage has been discussed in view of assessing the security as well as improving the cost-effectiveness. In addition, the role of CCS technologies and their potential contribution on the mitigation of CO_2 emissions in future are summarized. Based on the studies carried out in this review, the following conclusions have been obtained.

Firstly, the sequestration of CO_2 in depleted oil and gas reservoirs could play an important role in reducing CO_2 emissions in the near future, because the existing installed equipment and comprehensively characterized reservoir integrity will significantly reduce the cost of CCS. The leakage of CO_2 through abandoned wells is an obstacle for the application of this technology. To address this issue, the long-term experiments and molecular dynamic simulations are needed to figure out the kinetics between CO_2 with the well string, cement, as well as formation minerals under the relevant conditions. Secondly, if implemented on a large scale, CO_2 storage in saline aquifers may make the biggest contribution in reducing CO_2 emissions due to its huge storage capacity. Moreover, the scientifically proven technologies such as CO_2 storage in coal beds, deep ocean, and deep-sea sediments are still immature technologies and do not appear to be capable of making a great contribution to the mitigation of CO_2 emissions in the foreseeable future.

Another point is the need to investigate accurate risk assessment associated with CO_2 storage and provide a guideline for the design of CCS projects. Attempting to make the CCS assessment more intelligential, the machine learning technology ought to be used.

It has also been demonstrated that the direct mineralization of flue gas by coal fly ash would be a promising technology result since the gas could be mineralized without separation. In addition, the pH swing mineralization may be another promising technology for the sequestration of CO_2 due to the sustainability. The recyclable and cheap chemical reagents ought to be introduced into the mineralization process to make this technology more cost-effective.

Among the variety of CCUS strategies, CO_2-EOR followed by CO_2-EGR is supposed to play the most important role in the mitigation of CO_2 in the next few years. The utilization of other strategies seems to be negligible in the near future. The co-injection of impurities with CO_2 is an effective methodology to decrease the overall cost of CO_2 storage. The physical and chemical effects of the impurities on reservoir fluids and formation rock should be studied site specific, to reduce the uncertainties in CO_2 storage.

The government is supposed to play a major role in mitigating CO_2 emission, a higher tax on CO_2 emissions and financial subsidy on CO_2 storage is encouraged to accelerate the deployment of CCS projects at a large-scale.

Author Contributions: Conceptualization, C.C., H.L. and J.L.; methodology, Z.H. and F.M.; data, C.C. and W.F.; writing—original draft preparation, C.C.; writing—review and editing, C.C., H.L., J.L. and F.M. All authors have read and agreed to the published version of the manuscript.

Funding: This research was funded by the National Natural Science Foundation of China (NSFC), grant number 51809259; the CAS Pioneer Hundred Talents Program in China; the China Scholarship Council, grant number 201808080067, 201708080145.

Conflicts of Interest: The authors declare no conflict of interest.

References

1. Met Office. Our Changing World—Global Indicators. Available online: https://www.metoffice.gov.uk/climate-guide (accessed on 9 June 2019).
2. Scripps CO_2 Program. CO_2 Concentration at Mauna Loa Observatory, Hawaii. Available online: http://scrippsco2.ucsd.edu (accessed on 9 June 2019).
3. MacDowell, N.; Florin, N.; Buchard, A.; Hallett, J.; Galindo, A.; Jackson, G.; Adjiman, C.S.; Williams, C.K.; Shah, N.; Fennell, P. An overview of CO_2 capture technologies. *Energy Environ. Sci.* **2010**, *3*, 1645–1669. [CrossRef]
4. Rahman, F.A.; Aziz, M.M.A.; Saidur, R.; Bakar, W.A.W.A.; Hainin, M.R.; Putrajaya, R.; Hassan, N.A. Pollution to solution: Capture and sequestration of carbon dioxide (CO_2) and its utilization as a renewable energy source for a sustainable future. *Renew. Sustain. Energy Rev.* **2017**, *71*, 112–126. [CrossRef]
5. Hawcroft, M.; Walsh, E.; Hodges, K.; Zappa, G. Significantly increased extreme precipitation expected in Europe and North America from extratropical cyclones. *Environ. Res. Lett.* **2018**, *13*, 124006. [CrossRef]
6. Edenhofer, O.; Pichs-Madruga, R.; Sokona, Y.; Kadner, S.; Minx, J.; Brunner, S. *Change 2014: Mitigation of Climate Change*; Contribution of Working Group III to the Fifth Assessment Report of the Intergovernmental Panel on Climate Change; Cambridge University Press: Cambridge, UK and New York, USA, 2014.
7. Brinckerhoff, P. *Accelerating the Uptake of CCS: Industrial Use of Captured Carbon Dioxide*; Global CCS Institute, 2011. Available online: https://www.globalccsinstitute.com/resources/publications-reports-research/?search=Accelerating+the+Uptake+of+CCS (accessed on 11 June 2019).
8. IEA. Energy Technology Perspectives, Scenarios and Strategies to 2050. Available online: https://www.iea.org/publications/freepublications/publication/etp2010.pdf (accessed on 9 June 2019).
9. IEA. Technology Roadmap Carbon Capture and Storage. Available online: https://insideclimatenews.org/sites/default/files/IEA-CCS%20Roadmap.pdf (accessed on 13 June 2019).
10. GCCSI. Status CCS Project Database. Available online: https://co2re.co/FacilityData (accessed on 15 November 2019).
11. Pawar, R.J.; Bromhal, G.S.; Carey, J.W.; Foxall, W.; Korre, A.; Ringrose, P.S.; Tucker, O.; Watson, M.N.; White, J.A. Recent advances in risk assessment and risk management of geologic CO_2 storage. *Int. J. Greenh. Gas Control* **2015**, *40*, 292–311. [CrossRef]
12. Gislason, S.R.; Oelkers, E.H. Carbon storage in basalt. *Science* **2014**, *344*, 373–374. [CrossRef]
13. Atia, A.; Mohammedi, K. A review on the application of enhanced oil/gas recovery through CO_2 sequestration. In *Carbon Dioxide Chemistry, Capture and Oil Recovery*, 1st ed.; Karamé, I., Shaya, J., Srour, H., Eds.; IntechOpen: London, UK, 2018; pp. 241–253.
14. Koytsoumpa, E.I.; Bergins, C.; Kakaras, E. The CO_2 economy: Review of CO_2 capture and reuse technologies. *J. Supercrit. Fluids* **2018**, *132*, 3–16. [CrossRef]
15. Li, L.; Zhao, N.; Wei, W.; Sun, Y. A review of research progress on CO_2 capture, storage, and utilization in Chinese Academy of Sciences. *Fuel* **2013**, *108*, 112–130. [CrossRef]
16. Tan, Y.; Nookuea, W.; Li, H.; Thorin, E.; Yan, J. Property impacts on Carbon Capture and Storage (CCS) processes: A review. *Energy Convers. Manag.* **2016**, *118*, 204–222. [CrossRef]
17. Boot-Handford, M.E.; Abanades, J.C.; Anthony, E.J.; Blunt, M.J.; Brandani, S.; Mac Dowell, N.; Fernández, J.R.; Ferrari, M.-C.; Gross, R.; Hallett, J.P.; et al. Carbon capture and storage update. *Energy Environ. Sci.* **2014**, *7*, 130–189. [CrossRef]
18. Godec, M.; Koperna, G.; Gale, J. CO_2-ECBM: A review of its status and global potential. *Energy Procedia* **2014**, *63*, 5858–5869. [CrossRef]
19. Liu, H.J.; Were, P.; Li, Q.; Gou, Y.; Hou, Z. Worldwide status of CCUS technologies and their development and challenges in China. *Geofluids* **2017**, 1–25. [CrossRef]
20. Pires, J.C.M.; Martins, F.G.; Alvim-Ferraz, M.C.M.; Simões, M. Recent developments on carbon capture and storage: An overview. *Chem. Eng. Res. Des.* **2011**, *89*, 1446–1460. [CrossRef]
21. Aminu, M.D.; Nabavi, S.A.; Rochelle, C.A.; Manovic, V. A review of developments in carbon dioxide storage. *Appl. Energy* **2017**, *208*, 1389–1419. [CrossRef]
22. Kemper, J. Biomass and carbon dioxide capture and storage: A review. *Int. J. Greenh. Gas Control* **2015**, *40*, 401–430. [CrossRef]

23. Michael, K.; Golab, A.; Shulakova, V.; Ennis-King, J.; Allinson, G.; Sharma, S.; Aiken, T. Geological storage of CO_2 in saline aquifers—A review of the experience from existing storage operations. *Int. J. Greenh. Gas Control* **2010**, *4*, 659–667. [CrossRef]

24. Oh, T.H. Carbon capture and storage potential in coal-fired plant in Malaysia—A review. *Renew. Sustain. Energy Rev.* **2010**, *14*, 2697–2709. [CrossRef]

25. Riaz, A.; Cinar, Y. Carbon dioxide sequestration in saline formations: Part I—Review of the modeling of solubility trapping. *J. Petrol. Sci. Eng.* **2014**, *124*, 367–380. [CrossRef]

26. Sanna, A.; Uibu, M.; Caramanna, G.; Kuusik, R.; Maroto-Valer, M.M. A review of mineral carbonation technologies to sequester CO_2. *Chem. Soc. Rev.* **2014**, *43*, 8049–8080. [CrossRef]

27. Singh, P.; Haines, M. A review of existing carbon capture and storage cluster projects and future opportunities. *Energy Procedia* **2014**, *63*, 7247–7260. [CrossRef]

28. Tang, Y.; Yang, R.; Bian, X. A review of CO_2 sequestration projects and application in China. *Sci. World J.* **2014**, 381854. [CrossRef]

29. Verduyn, M.; Geerlings, H.; Mossel, G.V.; Vijayakumari, S. Review of the various CO_2 mineralization product forms. *Energy Procedia* **2011**, *4*, 2885–2892. [CrossRef]

30. Wee, J.-H. A review on carbon dioxide capture and storage technology using coal fly ash. *Appl. Energy* **2013**, *106*, 143–151. [CrossRef]

31. Burnside, N.M.; Naylor, M. Review and implications of relative permeability of CO_2/brine systems and residual trapping of CO_2. *Int. J. Greenh. Gas Control* **2014**, *23*, 1–11. [CrossRef]

32. De Silva, G.P.D.; Ranjith, P.G.; Perera, M.S.A. Geochemical aspects of CO_2 sequestration in deep saline aquifers: A review. *Fuel* **2015**, *155*, 128–143. [CrossRef]

33. Pan, P.; Wu, Z.; Feng, X.; Yan, F. Geomechanical modeling of CO_2 geological storage: A review. *J. Rock Mech. Geotech. Eng.* **2016**, *8*, 936–947. [CrossRef]

34. Abidoye, L.K.; Khudaida, K.J.; Das, D.B. Geological carbon sequestration in the context of two-phase flow in porous media: A review. *Crit. Rev. Environ. Sci. Technol.* **2015**, *45*, 1105–1147. [CrossRef]

35. Abid, K.; Gholami, R.; Choate, P.; Nagaratnam, B.H. A review on cement degradation under CO_2-rich environment of sequestration projects. *J. Nat. Gas Sci. Eng.* **2015**, *27*, 1149–1157. [CrossRef]

36. Li, Q.; Liu, G. Risk assessment of the geological storage of CO_2: A review. In *Geologic Carbon Sequestration*; Vishal, V., Singh, T.N., Eds.; Springer: Basel, Switzerland, 2016; pp. 249–284. [CrossRef]

37. Mayer, B.; Humez, P.; Becker, V.; Dalkhaa, C.; Rock, L.; Myrttinen, A.; Barth, J.A.C. Assessing the usefulness of the isotopic composition of CO_2 for leakage monitoring at CO_2 storage sites: A review. *Int. J. Greenh. Gas Control* **2015**, *37*, 46–60. [CrossRef]

38. Zhang, M.; Bachu, S. Review of integrity of existing wells in relation to CO_2 geological storage: What do we know? *Int. J. Greenh. Gas Control* **2011**, *5*, 826–840. [CrossRef]

39. Bai, M.; Zhang, Z.; Fu, X. A review on well integrity issues for CO_2 geological storage and enhanced gas recovery. *Renew. Sustain. Energy Rev.* **2016**, *59*, 920–926. [CrossRef]

40. Song, J.; Zhang, D. Comprehensive review of caprock-sealing mechanisms for geologic carbon sequestration. *Environ. Sci. Technol.* **2013**, *47*, 9–22. [CrossRef]

41. Zahid, U.; Lim, Y.; Jung, J.; Han, C. CO_2 geological storage: A review on present and future prospects. *Korean J. Chem. Eng.* **2011**, *28*, 674–685. [CrossRef]

42. Shukla, R.; Ranjith, P.; Haque, A.; Choi, X. A review of studies on CO_2 sequestration and caprock integrity. *Fuel* **2010**, *89*, 2651–2664. [CrossRef]

43. Bachu, S. Review of CO_2 storage efficiency in deep saline aquifers. *Int. J. Greenh. Gas Control* **2015**, *40*, 188–202. [CrossRef]

44. Carroll, A.G.; Przeslawski, R.; Radke, L.C.; Black, J.R.; Picard, K.; Moreau, J.W.; Haese, R.R.; Nichol, S. Environmental considerations for subseabed geological storage of CO_2: A review. *Cont. Shelf Res.* **2014**, *83*, 116–128. [CrossRef]

45. Bradshaw, J.; Bachu, S.; Bonijoly, D.; Burruss, R.; Holloway, S.; Christensen, N.P.; Mathiassen, O.M. CO_2 storage capacity estimation: Issues and development of standards. *Int. J. Greenh. Gas Control* **2007**, *1*, 62–68. [CrossRef]

46. Krevor, S.; Blunt, M.J.; Benson, S.M.; Pentland, C.H.; Reynolds, C.; Al-Menhali, A.; Niu, B. Capillary trapping for geologic carbon dioxide storage—From pore scale physics to field scale implications. *Int. J. Greenh. Gas Control* **2015**, *40*, 221–237. [CrossRef]

47. Zhang, Z.; Huisingh, D. Carbon dioxide storage schemes: Technology, assessment and deployment. *J. Clean. Prod.* **2017**, *142*, 1055–1064. [CrossRef]

48. Bian, X.Q.; Xiong, W.; Kasthuriarachchi, D.T.K.; Liu, Y.B. Phase equilibrium modeling for carbon dioxide solubility in aqueous sodium chloride solutions using an association equation of state. *Ind. Eng. Chem. Res.* **2019**, *58*, 10570–10578. [CrossRef]

49. Zhang, K.; Wu, Y.S.; Pruess, K. *User's Guide for TOUGH2-MP-a Massively Parallel Version of the TOUGH2 Code (No. LBNL-315E)*; Lawrence Berkeley National Laboratory: Berkeley, CA, USA, 2008.

50. Sundal, A.; Hellevang, H.; Miri, R.; Dypvik, H.; Nystuen, J.P.; Aagaard, P. Variations in mineralization potential for CO_2 related to sedimentary facies and burial depth—A comparative study from the North Sea. *Energy Procedia* **2014**, *63*, 5063–5070. [CrossRef]

51. Zhao, X.; Liao, X.; Wang, W.; Chen, C.; Rui, Z.; Wang, H. The CO_2 storage capacity evaluation: Methodology and determination of key factors. *J. Energy Inst.* **2014**, *87*, 297–305. [CrossRef]

52. Metz, B.; Davidson, O.; De Coninck, H.; Loos, M.; Meyer, L. *Carbon Dioxide Capture and Storage*; IPCC Special Report. New York, NY, USA, 2005. Available online: https://www.researchgate.net/publication/239877190_IPCC_Special_Report_on_Carbon_dioxide_Capture_and_Storage (accessed on 15 June 2019).

53. Celia, M.A.; Bachu, S.; Nordbotten, J.M.; Bandilla, K.W. Status of CO_2 storage in deep saline aquifers with emphasis on modeling approaches and practical simulations. *Water Resour. Res.* **2015**, *51*, 6846–6892. [CrossRef]

54. Davison, J.; Freund, P.; Smith, A. *Putting Carbon Back in the Ground*; IEA Greenhouse Gas R & D Programme. Available online: https://www.osti.gov/etdeweb/biblio/20204888 (accessed on 22 June 2019).

55. Cooper, C. A technical basis for carbon dioxide storage. *Energy Procedia* **2009**, *1*, 1727–1733. [CrossRef]

56. Orlic, B. Geomechanical effects of CO_2 storage in depleted gas reservoirs in the Netherlands: Inferences from feasibility studies and comparison with aquifer storage. *J. Rock Mech. Geotech. Eng.* **2016**, *8*, 846–859. [CrossRef]

57. Birkholzer, J.T.; Zhou, Q.; Tsang, C.F. Large-scale impact of CO_2 storage in deep saline aquifers: A sensitivity study on pressure response in stratified systems. *Int. J. Greenh. Gas Control* **2009**, *3*, 181–194. [CrossRef]

58. Audigane, P.; Gaus, I.; Pruess, K.; Xu, T. A long term 2D vertical modelling study of CO_2 storage at Sleipner (North Sea) using TOUGHREACT. In Proceedings of theTOUGH Symposium, Berkeley, CA, USA, 15–17 May 2006.

59. Audigane, P.; Gaus, I.; Czernichowski-Lauriol, I.; Pruess, K.; Xu, T. Two-dimensional reactive transport modeling of CO_2 injection in a saline aquifer at the Sleipner site, North Sea. *Am. J. Sci.* **2007**, *307*, 974–1008. [CrossRef]

60. Williams, G.; Chadwick, A. Chimneys and channels: History matching the growing CO_2 plume at the Sleipner storage site. In Proceedings of the Fifth CO_2 Geological Storage Workshop, Utrecht, The Netherlands, 21–23 November 2018.

61. Hansen, O.; Gilding, D.; Nazarian, B.; Osdal, B.; Ringrose, P.; Kristoffersen, J.B.; Eiken, O.; Hansen, H. Snøhvit: The history of injecting and storing 1 Mt CO_2 in the fluvial Tubåen Fm. *Energy Procedia* **2013**, *37*, 3565–3573. [CrossRef]

62. Ringrose, P.S.; Mathieson, A.S.; Wright, I.W.; Selama, F.; Hansen, O.; Bissell, R.; Saoula, N.; Midgley, J. The In Salah CO_2 storage project: Lessons learned and knowledge transfer. *Energy Procedia* **2013**, *37*, 6226–6236. [CrossRef]

63. Rutqvist, J.; Vasco, D.W.; Myer, L. Coupled reservoir-geomechanical analysis of CO_2 injection and ground deformations at In Salah, Algeria. *Int. J. Greenh. Gas Control* **2010**, *4*, 225–230. [CrossRef]

64. Flett, M.A.; Beacher, G.J.; Brantjes, J.; Burt, A.J.; Dauth, C.; Koelmeyer, F.M.; Lawrence, R.; Leigh, S.; McKenna, J.; Gurton, R.; et al. Gorgon project: Subsurface evaluation of carbon dioxide disposal under Barrow Island. In Proceedings of the SPE Asia Pacific Oil and Gas Conference and Exhibition, Perth, Australia, 20–22 October 2008.

65. Bourne, S.; Crouch, S.; Smith, M. A risk-based framework for measurement, monitoring and verification of the Quest CCS Project, Alberta, Canada. *Int. J. Greenh. Gas Control* **2014**, *26*, 109–126. [CrossRef]

66. Arts, R.J.; Chadwick, A.; Eiken, O.; Thibeau, S.; Nooner, S. Ten years' experience of monitoring CO_2 injection in the Utsira Sand at Sleipner, offshore Norway. *First Break* **2008**, *26*, 65–72.

67. Huang, X.; Bandilla, K.W.; Celia, M.A.; Bachu, S. Basin-scale modeling of CO_2 storage using models of varying complexity. *Int. J. Greenh. Gas Control* **2014**, *20*, 73–86. [CrossRef]

68. Bachu, S. Drainage and imbibition CO_2/brine relative permeability curves at in situ conditions for sandstone formations in western Canada. *Energy Procedia* **2013**, *37*, 4428–4436. [CrossRef]

69. Hermanrud, C.; Eiken, O.; Hansen, O.R.; Nordgaard Bolaas, H.M.; Simmenes, T.; Teige, G.M.G.; Hansen, H.; Johansen, S. Importance of pressure management in CO_2 storage. In Proceedings of the Offshore Technology Conference, Houston, TX, USA, 6–9 May 2013.

70. Bjørnarå, T.I.; Bohloli, B.; Park, J. Field-data analysis and hydromechanical modeling of CO_2 storage at In Salah, Algeria. *Int. J. Greenh. Gas Control* **2018**, *79*, 61–72. [CrossRef]

71. Eiken, O.; Ringrose, P.; Hermanrud, C.; Nazarian, B.; Torp, T.A.; Høier, L. Lessons learned from 14 years of CCS operations: Sleipner, In Salah and Snøhvit. *Energy Procedia* **2011**, *4*, 5541–5548. [CrossRef]

72. Gemmer, L.; Hansen, O.; Iding, M.; Leary, S.; Ringrose, P. Geomechanical response to CO_2 injection at Krechba, In salah, Algeria. *First Break* **2012**, *30*, 79–84.

73. Newell, P.; Yoon, H.; Martinez, M.J.; Bishop, J.E.; Bryant, S.L. Investigation of the influence of geomechanical and hydrogeological properties on surface uplift at In Salah. *J. Petrol. Sci. Eng.* **2017**, *155*, 34–45. [CrossRef]

74. Rinaldi, A.P.; Rutqvist, J. Modeling of deep fracture zone opening and transient ground surface uplift at KB-502 CO_2 injection well, In Salah, Algeria. *Int. J. Greenh. Gas Control* **2013**, *12*, 155–167. [CrossRef]

75. Rinaldi, A.P.; Rutqvist, J.; Finsterle, S.; Liu, H.H. Inverse modeling of ground surface uplift and pressure with iTOUGH-PEST and TOUGH-FLAC: The case of CO_2 injection at In Salah, Algeria. *Comput. Geosci.* **2017**, *108*, 98–109. [CrossRef]

76. Shi, J.Q.; Smith, J.; Durucan, S.; Korre, A. A coupled reservoir simulation-geomechanical modelling study of the CO_2 injection-induced ground surface uplift observed at Krechba, in Salah. *Energy Procedia* **2013**, *37*, 3719–3726. [CrossRef]

77. Stork, A.L.; Verdon, J.P.; Kendall, J.M. The microseismic response at the In Salah Carbon Capture and Storage (CCS) site. *Int. J. Greenh. Gas Control* **2015**, *32*, 159–171. [CrossRef]

78. Martens, S.; Kempka, T.; Liebscher, A.; Lüth, S.; Möller, F.; Myrttinen, A.; Norden, B.; Schmidt-Hattenberger, C.; Zimmer, M.; Kühn, M. Europe's longest-operating on-shore CO_2 storage site at Ketzin, Germany: A progress report after three years of injection. *Environ. Earth Sci.* **2012**, *67*, 323–334. [CrossRef]

79. Opedal, N. Ensuring integrity of CO_2 storage: An overview of ongoing experimental activity. In Proceedings of the Fifth CO_2 Geological Storage Workshop, Utrecht, The Netherlands, 21–23 November 2018.

80. Finley, R.J.; Frailey, S.M.; Leetaru, H.E.; Senel, O.; Couëslan, M.L.; Scott, M. Early operational experience at a one-million tonne CCS demonstration project, Decatur, Illinois, USA. *Energy Procedia* **2013**, *37*, 6149–6155. [CrossRef]

81. Yang, G.; Li, Y.; Atrens, A.; Liu, D.; Wang, Y.; Jia, L.; Lu, Y. Reactive transport modeling of long-term CO_2 sequestration mechanisms at the Shenhua CCS demonstration project, China. *J. Earth Sci.* **2017**, *28*, 457–472. [CrossRef]

82. Barrufet, M.A.; Bacquet, A.; Falcone, G. Analysis of the storage capacity for CO_2 sequestration of a depleted gas condensate reservoir and a saline aquifer. *J. Can. Pet. Technol.* **2010**, *49*, 23–31. [CrossRef]

83. Mamora, D.D.; Seo, J.G. Enhanced gas recovery by carbon dioxide sequestration in depleted gas reservoirs. In Proceedings of the SPE Annual Technical Conference and Exhibition, San Antonio, TX, USA, 29 September–2 October 2002.

84. Stein, M.H.; Ghotekar, A.L.; Avasthi, S.M. CO_2 sequestration in a depleted gas field: A material balance study. In Proceedings of the SPE EUROPEC/EAGE Annual Conference and Exhibition, Barcelona, Spain, 14–17 June 2010.

85. Raza, A.; Gholami, R.; Rezaee, R.; Rasouli, V.; Bhatti, A.A.; Bing, C.H. Suitability of depleted gas reservoirs for geological CO_2 storage: A simulation study. *Greenh. Gases Sci. Technol.* **2018**, *8*, 876–897. [CrossRef]

86. Gilfillan, S.M.; Lollar, B.S.; Holland, G.; Blagburn, D.; Stevens, S.; Schoell, M.; Cassidy, M.; Ding, Z.; Zhou, Z.; Lacrampe-Couloume, G.; et al. Solubility trapping in formation water as dominant CO_2 sink in natural gas fields. *Nature* **2009**, *458*, 614–618. [CrossRef]

87. Raza, A.; Gholami, R.; Rezaee, R.; Bing, C.H.; Nagarajan, R.; Hamid, M.A. CO_2 storage in depleted gas reservoirs: A study on the effect of residual gas saturation. *Petroleum* **2018**, *4*, 95–107. [CrossRef]

88. Mathias, S.A.; Gluyas, J.G.; Oldenburg, C.M.; Tsang, C.F. Analytical solution for Joule–Thomson cooling during CO_2 geo-sequestration in depleted oil and gas reservoirs. *Int. J. Greenh. Gas Control* **2010**, *4*, 806–810. [CrossRef]

89. Oldenburg, C.M. Joule-Thomson cooling due to CO_2 injection into natural gas reservoirs. *Energy Convers. Manag.* **2007**, *48*, 1808–1815. [CrossRef]

90. Twerda, A.; Belfroid, S.; Neele, F. CO_2 injection in low pressure depleted reservoirs. In Proceedings of the Fifth CO_2 Geological Storage Workshop, Utrecht, The Netherlands, 21–23 November 2018.

91. Böser, W.; Belfroid, S. Flow assurance study. *Energy Procedia* **2013**, *37*, 3018–3030. [CrossRef]

92. Ziabakhsh-Ganji, Z.; Kooi, H. Sensitivity of Joule-Thomson cooling to impure CO_2 injection in depleted gas reservoirs. *Appl. Energy* **2014**, *113*, 434–451. [CrossRef]

93. Loeve, D.; Hofstee, C.; Maas, J.G. Thermal effects in a depleted gas field by cold CO_2 injection in the presence of methane. *Energy Procedia* **2014**, *63*, 5378–5393. [CrossRef]

94. Sharma, S.; Cook, P.; Berly, T.; Lees, M. The CO2CRC Otway Project: Overcoming challenges from planning to execution of Australia's first CCS project. *Energy Procedia* **2009**, *1*, 1965–1972. [CrossRef]

95. Jenkins, C.R.; Cook, P.J.; Ennis-King, J.; Undershultz, J.; Boreham, C.; Dance, T.; de Caritat, P.; Etheridge, D.M.; Freifeld, B.M.; Hortle, A.; et al. Safe storage and effective monitoring of CO_2 in depleted gas fields. *Proc. Natl. Acad. Sci. USA* **2012**, *109*, E35–41. [CrossRef]

96. Arts, R.J.; Vandeweijer, V.P.; Hofstee, C.; Pluymaekers, M.P.D.; Loeve, D.; Kopp, A.; Plug, W.J. The feasibility of CO_2 storage in the depleted P18-4 gas field offshore the Netherlands (the ROAD project). *Int. J. Greenh. Gas Control* **2012**, *11*, S10–S20. [CrossRef]

97. Zhang, L.; Niu, B.; Ren, S.; Zhang, Y.; Yi, P.; Mi, H.; Ma, Y. Assessment of CO_2 storage in DF1-1 gas field South China Sea for a CCS demonstration. *J. Can. Pet. Technol.* **2010**, *49*, 9–14. [CrossRef]

98. Hannis, S.; Lu, J.; Chadwick, A.; Hovorka, S.; Kirk, K.; Romanak, K.; Pearce, J. CO_2 storage in depleted or depleting oil and gas fields: What can we learn from existing projects? *Energy Procedia* **2017**, *114*, 5680–5690. [CrossRef]

99. Bachu, S. Carbon dioxide storage capacity in uneconomic coal beds in Alberta, Canada: Methodology, potential and site identification. *Int. J. Greenh. Gas Control* **2007**, *1*, 374–385. [CrossRef]

100. Yu, H.; Zhou, G.; Fan, W.; Ye, J. Predicted CO_2 enhanced coalbed methane recovery and CO_2 sequestration in China. *Int. J. Coal Geol.* **2007**, *71*, 345–357. [CrossRef]

101. Brewer, P.G.; Friederich, G.; Peltzer, E.T.; Orr, F.M. Direct experiments on the ocean disposal of fossil fuel CO_2. *Science* **1999**, *284*, 943–945. [CrossRef] [PubMed]

102. Fer, I.; Haugan, P.M. Dissolution from a liquid CO_2 lake disposed in the deep ocean. *Limnol. Oceanogr.* **2003**, *48*, 872–883. [CrossRef]

103. Levine, J.S.; Matter, J.M.; Goldberg, D.; Cook, A.; Lackner, K.S. Gravitational trapping of carbon dioxide in deep sea sediments: Permeability, buoyancy, and geomechanical analysis. *Geophy. Res. Lett.* **2007**, *34*. [CrossRef]

104. House, K.Z.; Schrag, D.P.; Harvey, C.F.; Lackner, K.S. Permanent carbon dioxide storage in deep-sea sediments. *Proc. Natl. Acad. Sci. USA* **2006**, *103*, 12291–12295. [CrossRef]

105. Koide, H.; Shindo, Y.; Tazaki, Y.; Iijima, M.; Ito, K.; Kimura, N.; Omata, K. Deep sub-seabed disposal of CO_2—The most protective storage. *Energy Convers. Manag.* **1997**, *38*, S253–S258. [CrossRef]

106. Schrag, D.P. Storage of carbon dioxide in offshore sediments. *Science* **2009**, *325*, 1658–1659. [CrossRef]

107. Teng, Y.; Zhang, D. Long-term viability of carbon sequestration in deep-sea sediments. *Sci. Adv.* **2018**, *4*, eaao6588. [CrossRef]

108. Adams, E.E.; Caldeira, K. Ocean Storage of CO_2. *Elements* **2008**, *4*, 319–324. [CrossRef]

109. Gaus, I.; Audigane, P.; André, L.; Lions, J.; Jacquemet, N.; Durst, P.; Czernichowski-Lauriol, I.; Azaroual, M. Geochemical and solute transport modelling for CO_2 storage, what to expect from it? *Int. J. Greenh. Gas Control* **2008**, *2*, 605–625. [CrossRef]

110. Gawel, K.; Todorovic, J.; Liebscher, A.; Wiese, B.; Opedal, N. Study of materials retrieved from a Ketzin CO_2 monitoring well. *Energy Procedia* **2017**, *114*, 5799–5815. [CrossRef]

111. Celia, M.A.; Nordbotten, J.M.; Court, B.; Dobossy, M.; Bachu, S. Field-scale application of a semi-analytical model for estimation of CO_2 and brine leakage along old wells. *Int. J. Greenh. Gas Control* **2011**, *5*, 257–269. [CrossRef]

112. Nordbotten, J.M.; Celia, M.A.; Bachu, S. Injection and storage of CO_2 in deep saline aquifers: Analytical solution for CO_2 plume evolution during injection. *Transp. Porous Media* **2005**, *58*, 339–360. [CrossRef]

113. Xu, Z.; Fang, Y.; Scheibe, T.D.; Bonneville, A. A fluid pressure and deformation analysis for geological sequestration of carbon dioxide. *Comput. Geosci.* **2012**, *46*, 31–37. [CrossRef]

114. Wang Huimin, W.J.G. A fully coupled mathematical model with geochemical reaction for caprock sealing efficiency in CO_2 geosequestration. In Proceedings of the 15th China Rock Mechanics and Engineering Academic Annual Meeting, Beijing, China, 19–22 September 2018.

115. Bao, J.; Chu, Y.; Xu, Z.; Tartakovsky, A.M.; Fang, Y. Uncertainty quantification for the impact of injection rate fluctuation on the geomechanical response of geological carbon sequestration. *Int. J. Greenh. Gas Control* **2014**, *20*, 160–167. [CrossRef]

116. Lengler, U.; De Lucia, M.; Kühn, M. The impact of heterogeneity on the distribution of CO_2: Numerical simulation of CO_2 storage at Ketzin. *Int. J. Greenh. Gas Control* **2010**, *4*, 1016–1025. [CrossRef]

117. Wasch, L.J.; Wollenweber, J.; Tambach, T.J. Intentional salt clogging: A novel concept for long-term CO_2 sealing. Greenh. *Gases Sci. Technol.* **2013**, *3*, 491–502. [CrossRef]

118. Iogna, A.; Guillet-Lhermite, J.; Wood, C.; Deflandre, J.P. CO_2 storage and enhanced gas recovery: Using extended black oil modelling to simulate CO_2 injection on a North Sea depleted gas field. In Proceedings of the SPE Europec featured at 79th EAGE Conference and Exhibition, Paris, France, 12–15 June 2017.

119. Rafiee, M.M.; Ramazanian, M. Simulation study of enhanced gas recovery process using a compositional and a black oil simulator. In Proceedings of the SPE Enhanced Oil Recovery Conference, Kuala Lumpur, Malaysia, 19–21 July 2011.

120. Jia, W.; McPherson, B.J.; Pan, F.; Xiao, T.; Bromhal, G. Probabilistic analysis of CO_2 storage mechanisms in a CO_2-EOR field using polynomial chaos expansion. *Int. J. Greenh. Gas Control* **2016**, *51*, 218–229. [CrossRef]

121. Mishra, S.; Ganesh, P.R.; Schuetter, J. Developing and validating simplified predictive models for CO_2 geologic sequestration. *Energy Procedia* **2017**, *114*, 3456–3464. [CrossRef]

122. Ren, B. Local capillary trapping in carbon sequestration: Parametric study and implications for leakage assessment. *Int. J. Greenh. Gas Control* **2018**, *78*, 135–147. [CrossRef]

123. Wriedt, J.; Deo, M.; Han, W.S.; Lepinski, J. A methodology for quantifying risk and likelihood of failure for carbon dioxide injection into deep saline reservoirs. *Int. J. Greenh. Gas Control* **2014**, *20*, 196–211. [CrossRef]

124. Bao, J.; Hou, Z.; Fang, Y.; Ren, H.; Lin, G. Uncertainty quantification for evaluating impacts of caprock and reservoir properties on pressure buildup and ground surface displacement during geological $CO2$ sequestration. Greenh. *Gases Sci. Technol.* **2013**, *3*, 338–358. [CrossRef]

125. Hou, Z.; Bacon, D.H.; Engel, D.W.; Lin, G.; Fang, Y.; Ren, H.; Fang, Z. Uncertainty analyses of CO_2 plume expansion subsequent to wellbore CO_2 leakage into aquifers. *Int. J. Greenh. Gas Control* **2014**, *27*, 69–80. [CrossRef]

126. Allen, R.; Nilsen, H.M.; Lie, K.A.; Møyner, O.; Andersen, O. Using simplified methods to explore the impact of parameter uncertainty on CO_2 storage estimates with application to the Norwegian Continental Shelf. *Int. J. Greenh. Gas Control* **2018**, *75*, 198–213. [CrossRef]

127. Schepers, K.C.; Nuttall, B.C.; Oudinot, A.Y.; Gonzalez, R.J. Reservoir modeling and simulation of the Devonian gas shale of eastern Kentucky for enhanced gas recovery and CO_2 storage. In Proceedings of the SPE International Conference on CO_2 Capture, Storage, and Utilization, San Diego, CA, USA, 10–11 November 2009.

128. Jung, H.; Singh, G.; Espinoza, D.N.; Wheeler, M.F. Quantification of a maximum injection volume of CO_2 to avert geomechanical perturbations using a compositional fluid flow reservoir simulator. *Adv. Water Resour.* **2018**, *112*, 160–169. [CrossRef]

129. Ebigbo, A.; Class, H.; Helmig, R. CO_2 leakage through an abandoned well: Problem-oriented benchmarks. *Comput. Geosci.* **2006**, *11*, 103–115. [CrossRef]

130. Luo, F.; Xu, R.N.; Jiang, P.X. Numerical investigation of the influence of vertical permeability heterogeneity in stratified formation and of injection/production well perforation placement on CO_2 geological storage with enhanced CH_4 recovery. *Appl. Energy* **2013**, *102*, 1314–1323. [CrossRef]

131. Khan, C.; Amin, R.; Madden, G. Economic modelling of CO_2 injection for enhanced gas recovery and storage: A reservoir simulation study of operational parameters. *Energy Environ. Res.* **2012**, *2*. [CrossRef]

132. Khan, C.; Amin, R.; Madden, G. Carbon dioxide injection for enhanced gas recovery and storage (reservoir simulation). *Egypt. J. Pet.* **2013**, *22*, 225–240. [CrossRef]

133. Rinaldi, A.P.; Rutqvist, J. Modeling ground surface uplift during CO_2 sequestration: The case of in Salah, Algeria. *Energy Procedia* **2017**, *114*, 3247–3256. [CrossRef]

134. Rutqvist, J. Status of the TOUGH-FLAC simulator and recent applications related to coupled fluid flow and crustal deformations. *Comput. Geosci.* **2011**, *37*, 739–750. [CrossRef]

135. Pan, P.Z.; Rutqvist, J.; Feng, X.T.; Yan, F. An approach for modeling rock discontinuous mechanical behavior under multiphase fluid flow conditions. *Rock Mech. Rock Eng.* **2014**, *47*, 589–603. [CrossRef]

136. Liu, H.H.; Houseworth, J.; Rutqvist, J.; Zheng, L.; Asahina, D.; Li, L.; Vilarrasa, V.; Chen, F.; Nakagawa, S.; Finsterle, S.; et al. *Report on THMC Modeling of the Near Field Evolution of a Generic Clay Repository: Model Validation and Demonstration*; Lawrence Berkeley National Laboratory: Berkeley, CA, USA, 2013.

137. Benisch, K.; Graupner, B.; Bauer, S. The coupled OpenGeoSys-eclipse simulator for simulation of CO_2 storage—Code comparison for fluid flow and geomechanical processes. *Energy Procedia* **2013**, *37*, 3663–3671. [CrossRef]

138. Fei, W.B.; Li, Q.; Liu, X.H.; Wei, X.C.; Jing, M.; Song, R.R.; Li, X.C.; Wang, Y.S. Coupled analysis for interaction of coal mining and CO_2 geological storage in Ordos Basin, China. In Proceedings of the 8th Asian Rock Mechanics Symposium, Sapporo, Japan, 14–16 October 2014.

139. Gou, Y.; Hou, Z.; Liu, H.; Zhou, L.; Were, P. Numerical simulation of carbon dioxide injection for enhanced gas recovery (CO_2-EGR) in Altmark natural gas field. *Acta Geotechnica* **2014**, *9*, 49–58. [CrossRef]

140. Rutqvist, J.; Tsang, C.F. TOUGH-FLAC: A numerical simulator for analysis of coupled thermal-hydrologic-mechanical processes in fractured and porous geological media under multi-phase flow conditions. In Proceedings of the TOUGH Symposium 2003, Berkeley, CA, USA, 12–14 May 2003.

141. Rutqvist, J.; Tsang, C.F. A study of caprock hydromechanical changes associated with CO_2-injection into a brine formation. *Environ. Geol.* **2002**, *42*, 296–305. [CrossRef]

142. Pawar, R.J.; Bromhal, G.S.; Chu, S.; Dilmore, R.M.; Oldenburg, C.M.; Stauffer, P.H.; Zhang, Y.; Guthrie, G.D. The National Risk Assessment Partnership's integrated assessment model for carbon storage: A tool to support decision making amidst uncertainty. *Int. J. Greenh. Gas Control* **2016**, *52*, 175–189. [CrossRef]

143. Namhata, A.; Zhang, L.; Dilmore, R.M.; Oladyshkin, S.; Nakles, D.V. Modeling changes in pressure due to migration of fluids into the Above Zone Monitoring Interval of a geologic carbon storage site. *Int. J. Greenh. Gas Control* **2017**, *56*, 30–42. [CrossRef]

144. NETL. U.S. DOE's National Risk Assessment Partnership: Assessing Carbon Storage Risk Performance to Support Decision Making Amidst Uncertainty. Available online: https://www.cslforum.org/cslf/sites/default/files/documents/AbuDhabi2017/Bromhal-NRAP-TG-AbuDhabi0517.pdf (accessed on 12 May 2019).

145. Doherty, B.; Vasylkivska, V.; Huerta, N.J.; Dilmore, R. Estimating the leakage along wells during geologic CO_2 Storage: Application of the Well Leakage Assessment Tool to a hypothetical storage scenario in Natrona County, Wyoming. *Energy Procedia* **2017**, *114*, 5151–5172. [CrossRef]

146. Sun, A.Y.; Jeong, H.; González-Nicolás, A.; Templeton, T.C. Metamodeling-based approach for risk assessment and cost estimation: Application to geological carbon sequestration planning. *Comput. Geosci.* **2018**, *113*, 70–80. [CrossRef]

147. Dramsch, J.S.; Corte, G.; Amini, H.; Lüthje, M.; MacBeth, C. Deep learning application for 4D pressure saturation inversion compared to Bayesian inversion on North Sea data. In Proceedings of the Second EAGE Workshop Practical Reservoir Monitoring, Amsterdam, The Netherlands, 1–4 April 2019.

148. Leung, D.Y.C.; Caramanna, G.; Maroto-Valer, M.M. An overview of current status of carbon dioxide capture and storage technologies. *Renew. Sustain. Energy Rev.* **2014**, *39*, 426–443. [CrossRef]

149. Nordbotten, J.M.; Flemisch, B.; Gasda, S.E.; Nilsen, H.M.; Fan, Y.; Pickup, G.E.; Wiese, B.; Celia, M.A.; Dahle, H.K.; Eigestad, G.T.; et al. Uncertainties in practical simulation of CO_2 storage. *Int. J. Greenh. Gas Control* **2012**, *9*, 234–242. [CrossRef]

150. Lumley, D. 4D seismic monitoring of CO_2 sequestration. *Lead. Edge* **2010**, *29*, 150–155. [CrossRef]

151. Oye, V.; Aker, E.; Daley, T.M.; Kühn, D.; Bohloli, B.; Korneev, V. Microseismic monitoring and interpretation of injection data from the in Salah CO_2 storage site (Krechba), Algeria. *Energy Procedia* **2013**, *37*, 4191–4198. [CrossRef]

152. Götz, J.; Lüth, S.; Henninges, J.; Reinsch, T. Vertical seismic profiling using a daisy-chained deployment of fibre-optic cables in four wells simultaneously—Case study at the Ketzin carbon dioxide storage site. *Geophys. Prospect.* **2018**, *66*, 1201–1214. [CrossRef]

153. Kabirzadeh, H.; Sideris, M.G.; Shin, Y.J.; Kim, J.W. Gravimetric Monitoring of confined and unconfined geological CO_2 reservoirs. *Energy Procedia* **2017**, *114*, 3961–3968. [CrossRef]

154. Böhm, G.; Carcione, J.M.; Gei, D.; Picotti, S.; Michelini, A. Cross-well seismic and electromagnetic tomography for CO_2 detection and monitoring in a saline aquifer. *J. Petrol. Sci. Eng.* **2015**, *133*, 245–257. [CrossRef]

155. Carcione, J.M.; Gei, D.; Picotti, S.; Michelini, A. Cross-hole electromagnetic and seismic modeling for CO_2 detection and monitoring in a saline aquifer. *J. Petrol. Sci. Eng.* **2012**, *100*, 162–172. [CrossRef]
156. Liebscher, A.; Möller, F.; Bannach, A.; Köhler, S.; Wiebach, J.; Schmidt-Hattenberger, C.; Weiner, M.; Pretschner, C.; Ebert, K.; Zemke, J. Injection operation and operational pressure–temperature monitoring at the CO_2 storage pilot site Ketzin, Germany—Design, results, recommendations. *Int. J. Greenh. Gas Control* **2013**, *15*, 163–173. [CrossRef]
157. Boreham, C.; Underschultz, J.; Stalker, L.; Kirste, D.; Freifeld, B.; Jenkins, C.; Ennis-King, J. Monitoring of CO_2 storage in a depleted natural gas reservoir: Gas geochemistry from the CO2CRC Otway Project, Australia. *Int. J. Greenh. Gas Control* **2011**, *5*, 1039–1054. [CrossRef]
158. Gaus, I. Role and impact of CO_2–rock interactions during CO_2 storage in sedimentary rocks. *Int. J. Greenh. Gas Control* **2010**, *4*, 73–89. [CrossRef]
159. Matter, J.M.; Stute, M.; Snæbjörnsdottir, S.Ó.; Oelkers, E.H.; Gislason, S.R.; Aradottir, E.S.; Sigfusson, B.; Gunnarsson, I.; Sigurdardottir, H.; Gunnlaugsson, E.; et al. Rapid carbon mineralization for permanent disposal of anthropogenic carbon dioxide emissions. *Science* **2016**, *352*, 1312–1314. [CrossRef]
160. Etheridge, D.; Luhar, A.; Loh, Z.; Leuning, R.; Spencer, D.; Steele, P.; Zegelin, S.; Allison, C.; Krummel, P.; Leist, M.; et al. Atmospheric monitoring of the CO2CRC Otway Project and lessons for large scale CO_2 storage projects. *Energy Procedia* **2011**, *4*, 3666–3675. [CrossRef]
161. Morozova, D.; Zettlitzer, M.; Let, D.; Würdemann, H. Monitoring of the microbial community composition in deep subsurface saline aquifers during CO_2 storage in Ketzin, Germany. *Energy Procedia* **2011**, *4*, 4362–4370. [CrossRef]
162. Schilling, F.; Borm, G.; Würdemann, H.; Möller, F.; Kühn, M. Status report on the first European on-shore CO_2 storage site at Ketzin (Germany). *Energy Procedia* **2009**, *1*, 2029–2035. [CrossRef]
163. Onuma, T.; Okada, K.; Otsubo, A. Time series analysis of surface deformation related with CO_2 injection by satellite-borne SAR interferometry at In Salah, Algeria. *Energy Procedia* **2011**, *4*, 3428–3434. [CrossRef]
164. Mawalkar, S.; Brock, D.; Burchwell, A.; Kelley, M.; Mishra, S.; Gupta, N.; Pardini, R.; Shroyer, B. Where is that CO_2 flowing? Using Distributed Temperature Sensing (DTS) technology for monitoring injection of CO_2 into a depleted oil reservoir. *Int. J. Greenh. Gas Control* **2019**, *85*, 132–142. [CrossRef]
165. Shatarah, I.S.; lbrycht, R. Distributed temperature sensing in optical fibers based on Raman scattering: Theory and applications. *Meas. Autom. Monit.* **2017**, *63*, 41–44.
166. Worthen, A.J.; Parikh, P.S.; Chen, Y.; Bryant, S.L.; Huh, C.; Johnston, K.P. Carbon dioxide-in-water foams stabilized with a mixture of nanoparticles and surfactant for CO_2 storage and utilization applications. *Energy Procedia* **2014**, *63*, 7929–7938. [CrossRef]
167. Rognmo, A.U.; Heldal, S.; Fernø, M.A. Silica nanoparticles to stabilize CO_2-foam for improved CO_2 utilization: Enhanced CO_2 storage and oil recovery from mature oil reservoirs. *Fuel* **2018**, *216*, 621–626. [CrossRef]
168. Guo, F.; Aryana, S.A. Improved sweep efficiency due to foam flooding in a heterogeneous microfluidic device. *J. Petrol. Sci. Eng.* **2018**, *164*, 155–163. [CrossRef]
169. Chou, S.I.; Vasicek, S.L.; Pisio, D.L.; Jasek, D.E.; Goodgame, J.A. CO_2 foam field trial at north Ward-Estes. In Proceedings of the SPE Annual Technical Conference and Exhibition, Washington, DC, USA, 4–7 October 1992.
170. Guo, F.; Aryana, S.A.; Wang, Y.; McLaughlin, J.F.; Coddington, K. Enhancement of storage capacity of CO_2 in megaporous saline aquifers using nanoparticle-stabilized CO_2 foam. *Int. J. Greenh. Gas Control* **2019**, *87*, 134–141. [CrossRef]
171. Pizzocolo, F.; Peters, E.; Loeve, D.; Hewson, C.W.; Wasch, L.; Brunner, L.J. Feasibility of novel techniques to mitigate or remedy CO_2 leakage. In Proceedings of the SPE Europec featured at 79th EAGE Conference and Exhibition, Paris, France, 12–15 June 2017.
172. Johnston, K.P.; Rocha, S.R.P.d. Colloids in supercritical fluids over the last 20 years and future directions. *J. Supercrit. Fluids* **2009**, *47*, 523–530. [CrossRef]
173. Worthen, A.J.; Bagaria, H.G.; Chen, Y.; Bryant, S.L.; Huh, C.; Johnston, K.P. Nanoparticle-stabilized carbon dioxide-in-water foams with fine texture. *J. Colloid Interface Sci.* **2013**, *391*, 142–151. [CrossRef]
174. Worthen, A.J.; Bryant, S.L.; Huh, C.; Johnston, K.P. Carbon dioxide-in-water foams stabilized with nanoparticles and surfactant acting in synergy. *AIChE J.* **2013**, *59*, 3490–3501. [CrossRef]
175. Anchliya, A.; Ehlig-Economides, C.A.; Jafarpour, B. Aquifer management to accelerate CO_2 dissolution and trapping. *SPE J.* **2012**, *17*, 805–816. [CrossRef]

176. Cameron, D.A.; Durlofsky, L.J. Optimization of well placement, CO_2 injection rates, and brine cycling for geological carbon sequestration. *Int. J. Greenh. Gas Control* **2012**, *10*, 100–112. [CrossRef]

177. Zhang, Z.; Agarwal, R.K. Numerical simulation and optimization of CO_2 sequestration in saline aquifers for vertical and horizontal well injection. *Comput. Geosci.* **2012**, *16*, 891–899. [CrossRef]

178. Zhang, Z.; Agarwal, R. Numerical simulation and optimization of CO_2 sequestration in saline aquifers. *Comput. Fluids* **2013**, *80*, 79–87. [CrossRef]

179. Harris, J.; Kovscek, A.R.; Orr, F.M.; Zoback, M.D. *Geologic Storage of CO_2 in Coal Beds*; Global Climate and Energy Project (GCEP) Technical Report; Stanford University: Stanford, CA, USA, 2009.

180. Tanaka, K.; Vilcáez, J.; Sato, K. Improvement of CO_2 geological storage efficiency by injection and production well design. *Energy Procedia* **2013**, *37*, 4591–4597. [CrossRef]

181. Kelemen, P.B.; Matter, J. In situ carbonation of peridotite for CO_2 storage. *Proc. Natl. Acad. Sci. USA* **2008**, *105*, 17295–17300. [CrossRef]

182. Reddy, K.J.; John, S.; Weber, H.; Argyle, M.D.; Bhattacharyya, P.; Taylor, D.T.; Christensen, M.; Foulke, T.; Fahlsing, P. Simultaneous capture and mineralization of coal combustion flue gas carbon dioxide (CO_2). *Energy Procedia* **2011**, *4*, 1574–1583. [CrossRef]

183. Wang, X.; Maroto-Valer, M.M. Optimization of carbon dioxide capture and storage with mineralisation using recyclable ammonium salts. *Energy* **2013**, *51*, 431–438. [CrossRef]

184. Schaef, H.T.; McGrail, B.P.; Owen, A.T. Carbonate mineralization of volcanic province basalts. *Int. J. Greenh. Gas Control* **2010**, *4*, 249–261. [CrossRef]

185. Goldberg, D.S.; Takahashi, T.; Slagle, A.L. Carbon dioxide sequestration in deep-sea basalt. *Proc. Natl. Acad. Sci. USA* **2008**, *105*, 9920–9925. [CrossRef] [PubMed]

186. O'Connor, W.K.; Dahlin, D.C.; Nilsen, D.N.; Gerdemann, S.J.; Rush, G.E.; Walters, R.P.; Turner, P.C. Research status on the sequestration of carbon dioxide by direct aqueous mineral carbonation. In Proceedings of the 18th Annual International Pittsburgh Coal Conference, Newcastle, Australia, 3–7 December 2001.

187. Teir, S.; Eloneva, S.; Fogelholm, C.J.; Zevenhoven, R. Fixation of carbon dioxide by producing hydromagnesite from serpentinite. *Appl. Energy* **2009**, *86*, 214–218. [CrossRef]

188. Park, A.H.A.; Fan, L.S. CO_2 mineral sequestration: Physically activated dissolution of serpentine and pH swing process. *Chem. Eng. Sci.* **2004**, *59*, 5241–5247. [CrossRef]

189. Kodama, S.; Nishimoto, T.; Yamamoto, N.; Yogo, K.; Yamada, K. Development of a new pH-swing CO_2 mineralization process with a recyclable reaction solution. *Energy* **2008**, *33*, 776–784. [CrossRef]

190. Sanna, A.; Dri, M.; Maroto-Valer, M. Carbon dioxide capture and storage by pH swing aqueous mineralisation using a mixture of ammonium salts and antigorite source. *Fuel* **2013**, *114*, 153–161. [CrossRef]

191. Wei, N.; Li, X.; Fang, Z.; Bai, B.; Li, Q.; Liu, S.; Jia, Y. Regional resource distribution of onshore carbon geological utilization in China. *J. CO_2 Util.* **2015**, *11*, 20–30. [CrossRef]

192. Burton, E.; Beyer, J.; Bourcier, W.; Mateer, N.; Reed, J. Carbon utilization to meet California's climate change goals. *Energy Procedia* **2013**, *37*, 6979–6986. [CrossRef]

193. Azzolina, N.A.; Peck, W.D.; Hamling, J.A.; Gorecki, C.D.; Ayash, S.C.; Doll, T.E.; Nakles, D.V.; Melzer, L.S. How green is my oil? A detailed look at greenhouse gas accounting for CO_2—nhanced oil recovery (CO_2-EOR) sites. *Int. J. Greenh. Gas Control* **2016**, *51*, 369–379. [CrossRef]

194. Bian, X.Q.; Han, B.; Du, Z.M.; Jaubert, J.N.; Li, M.J. Integrating support vector regression with genetic algorithm for CO_2-oil minimum miscibility pressure (MMP) in pure and impure CO_2 streams. *Fuel* **2016**, *182*, 550–557. [CrossRef]

195. Metcalfe, R.S. Effects of impurities on minimum miscibility pressures and minimum enrichment levels for CO_2 and rich-gas displacements. *SPE J.* **1982**, *22*, 219–225. [CrossRef]

196. Brush, R.M.; Davitt, H.J.; Aimar, O.B.; Arguello, J.; Whiteside, J.M. Immiscible CO_2 flooding for increased oil recovery and reduced emissions. In Proceedings of the SPE/DOE Improved Oil Recovery Symposium, Tulsa, OK, USA, 3–5 April 2000.

197. Sahin, S.; Kalfa, U.; Celebioglu, D. Bati Raman field immiscible CO_2 application-status quo and future plans. In Proceedings of the Latin American and Caribbean Petroleum Engineering Conference, Buenos Aires, Argentina, 15–18 April 2007.

198. Aryana, S.A.; Barclay, C.; Liu, S. North cross devonian unit - a mature continuous CO_2 flood beyond 200% HCPV injection. In Proceedings of the SPE Annual Technical Conference and Exhibition, Amsterdam, The Netherlands, 27–29 October 2014.

199. Jia, B.; Tsau, J.S.; Barati, R. A review of the current progress of CO_2 injection EOR and carbon storage in shale oil reservoirs. *Fuel* **2019**, *236*, 404–427. [CrossRef]

200. Chen, B.; Reynolds, A.C. Optimal control of ICV's and well operating conditions for the water-alternating-gas injection process. *J. Petrol. Sci. Eng.* **2017**, *149*, 623–640. [CrossRef]

201. Hawthorne, S.B.; Gorecki, C.D.; Sorensen, J.A.; Steadman, E.N.; Harju, J.A.; Melzer, S. Hydrocarbon mobilization mechanisms from upper, middle, and lower Bakken reservoir rocks exposed to CO_2. In Proceedings of the SPE Unconventional Resources Conference Canada, Calgary, AB, Canada, 5–7 November 2013.

202. Zuloaga, P.; Yu, W.; Miao, J.; Sepehrnoori, K. Performance evaluation of CO_2 huff-n-puff and continuous CO_2 injection in tight oil reservoirs. *Energy* **2017**, *134*, 181–192. [CrossRef]

203. Mac Dowell, N.; Fennell, P.S.; Shah, N.; Maitland, G.C. The role of CO_2 capture and utilization in mitigating climate change. *Nat. Clim. Chang.* **2017**, *7*, 243–249. [CrossRef]

204. Bui, M.; Adjiman, C.S.; Bardow, A.; Anthony, E.J.; Boston, A.; Brown, S.; Fennell, P.S.; Fuss, S.; Galindo, A.; Hackett, L.A.; et al. Carbon capture and storage (CCS): The way forward. *Energy Environ. Sci.* **2018**, *11*, 1062–1176. [CrossRef]

205. Carpenter, S.M.; Koperna, G. Development of the first internationally accepted standard for geologic storage of carbon dioxide utilizing Enhanced Oil Recovery (EOR) under the International Standards Organization (ISO) Technical Committee TC-265. *Energy Procedia* **2014**, *63*, 6717–6729. [CrossRef]

206. Kane, A.V. Performance Review of a large-scale CO_2-WAG enhanced recovery project, SACROC Unit Kelly-Snyder field. *J. Petrol. Technol.* **1979**, *31*, 217–231. [CrossRef]

207. Sanders, A.W.; Jones, R.M.; Linroth, M.A.; Nguyen, Q.P. Implementation of a CO_2 foam pilot study in the SACROC field: Performance evaluation. In Proceedings of the SPE Annual Technical Conference and Exhibition, San Antonio, TX, USA, 8–10 October 2012.

208. Langston, M.V.; Hoadley, S.F.; Young, D.N. Definitive CO_2 flooding response in the SACROC unit. In Proceedings of the SPE/DOE Enhanced Oil Recovery Symposium, Tulsa, OK, USA, 16–21 April 1988.

209. Gozalpour, F.; Ren, S.R.; Tohidi, B. CO_2 EOR and storage in oil reservoir. *Oil Gas Sci. Technol.* **2006**, *60*, 537–546. [CrossRef]

210. Marston, P.M. Incidentally speaking: A systematic assessment and comparison of incidental storage of CO_2 during EOR with other Near-term storage options. *Energy Procedia* **2017**, *114*, 7422–7430. [CrossRef]

211. Clemens, T.; Secklehner, S.; Mantatzis, K.; Jacobs, B. Enhanced gas recovery-challenges shown at the example of three gas fields. In Proceedings of the SPE EUROPEC/EAGE Annual Conference and Exhibition, Barcelona, Spain, 14–17 June 2010.

212. Hattenbach, R.P.; Wilson, M.; Brown, K.R. Capture of carbon dioxide from coal combustion and its utilization for enhanced oil recovery. In Proceedings of the GHGT-4 Conference, Interlaken, Switzerland, 30 August–2 September 1998.

213. Preston, C.; Monea, M.; Jazrawi, W.; Brown, K.; Whittaker, S.; White, D.; Law, D.; Chalaturnyk, R.; Rostron, B. IEA GHG Weyburn CO_2 monitoring and storage project. *Fuel Process. Technol.* **2005**, *86*, 1547–1568. [CrossRef]

214. Lui, L.C.; Leamon, G. Developments towards environmental regulation of CCUS projects in China. *Energy Procedia* **2014**, *63*, 6903–6911. [CrossRef]

215. Lv, G.; Li, Q.; Wang, S.; Li, X. Key techniques of reservoir engineering and injection–production process for CO_2 flooding in China's SINOPEC Shengli Oilfield. *J. CO_2 Util.* **2015**, *11*, 31–40. [CrossRef]

216. Yang, W.; Peng, B.; Liu, Q.; Wang, S.; Dong, Y.; Lai, Y. Evaluation of CO_2 enhanced oil recovery and CO_2 storage potential in oil reservoirs of Bohai Bay Basin, China. *Int. J. Greenh. Gas Control* **2017**, *65*, 86–98. [CrossRef]

217. Mathisen, A.; Skagestad, R. Utilization of CO_2 from Emitters in Poland for CO_2 -EOR. *Energy Procedia* **2017**, *114*, 6721–6729. [CrossRef]

218. Ampomah, W.; Balch, R.S.; Cathar, M.; Will, R.; Lee, S.Y.; Dai, Z. Performance of CO_2-EOR and storage processes under uncertainty. In Proceedings of the SPE Europec featured at 78th EAGE Conference and Exhibition, Vienna, Austria, 30 May–2 June 2016.

219. Jahangiri, H.R.; Zhang, D. Ensemble based co-optimization of carbon dioxide sequestration and enhanced oil recovery. *Int. J. Greenh. Gas Control* **2012**, *8*, 22–33. [CrossRef]

220. Tapia, J.F.D.; Lee, J.Y.; Ooi, R.E.H.; Foo, D.C.Y.; Tan, R.R. CO_2 allocation for scheduling enhanced oil recovery (EOR) operations with geological sequestration using discrete-time optimization. *Energy Procedia* **2014**, *61*, 595–598. [CrossRef]

221. Le Van, S.; Chon, B.H. Evaluating the critical performances of a CO_2–Enhanced oil recovery process using artificial neural network models. *J. Petrol. Sci. Eng.* **2017**, *157*, 207–222. [CrossRef]

222. You, J.; Ampomah, W.; Kutsienyo, E.J.; Sun, Q.; Balch, R.S.; Aggrey, W.N.; Cather, M. Assessment of enhanced oil recovery and CO_2 storage capacity using machine learning and optimization framework. In Proceedings of the SPE Europec featured at 81st EAGE Conference and Exhibition, London, UK, 3–6 June 2019.

223. Al-Hasami, A.; Ren, S.; Tohidi, B. CO_2 injection for enhanced gas recovery and geo-storage: Reservoir simulation and economics. In Proceedings of the SPE Europec/EAGE Annual Conference, Madrid, Spain, 13–16 June 2005.

224. Odi, U. Analysis and potential of CO_2 huff-n-puff for near wellbore condensate removal and enhanced gas recovery. In Proceedings of the SPE Annual Technical Conference and Exhibition, San Antonio, TX, USA, 8–10 October 2012.

225. Odi, U. Optimal Process Design for Coupled CO_2 Sequestration and Enhanced Gas Recovery in Carbonate Reservoirs. Ph.D. Thesis, Texas A&M University, College Station, TX, USA, 2013.

226. Eliebid, M.; Mahmoud, M.; Shawabkeh, R.; Elkatatny, S.; Hussein, I.A. Effect of CO_2 adsorption on enhanced natural gas recovery and sequestration in carbonate reservoirs. *J. Nat. Gas Sci. Eng.* **2018**, *55*, 575–584. [CrossRef]

227. Klimkowski, Ł.; Nagy, S.; Papiernik, B.; Orlic, B.; Kempka, T. Numerical simulations of enhanced gas recovery at the Załęcze gas field in Poland confirm high CO_2 storage capacity and mechanical integrity. *Oil Gas Sci. Technol.* **2015**, *70*, 655–680. [CrossRef]

228. Narinesingh, J.; Alexander, D. CO_2 enhanced gas recovery and geologic sequestration in condensate reservoir: A simulation study of the effects of injection pressure on condensate recovery from reservoir and CO_2 storage efficiency. *Energy Procedia* **2014**, *63*, 3107–3115. [CrossRef]

229. Seo, J.G.; Mamora, D.D. Experimental and simulation studies of sequestration of supercritical carbon dioxide in depleted gas reservoirs. *J. Energy Resour. Technol.* **2005**, *127*, 1–6. [CrossRef]

230. Zangeneh, H.; Safarzadeh, M.A. Enhanced gas recovery with carbon dioxide sequestration in a water-drive gas condensate reservoir: A case study in a real gas Field. *J. Petrol. Sci. Eng.* **2017**, *7*, 3–11.

231. Seo, J.G. Experimental and Simulation Studies of Sequestration of Supercritical Carbon Dioxide in Depleted Gas Reservoirs. Ph.D. Thesis, Texas A&M University, College Station, TX, USA, 2004.

232. Abba, M.K.; Abbas, A.J.; Nasr, G.G. Enhanced gas recovery by CO_2 injection and sequestration: Effect of connate water salinity on displacement efficiency. In Proceedings of the SPE Abu Dhabi International Petroleum Exhibition & Conference, Abu Dhabi, UAE, 13–16 November 2017.

233. Abdoulghafour, H.; Gouze, P.; Luquot, L.; Leprovost, R. Characterization and modeling of the alteration of fractured class-G Portland cement during flow of CO_2 -rich brine. *Int. J. Greenh. Gas Control* **2016**, *48*, 155–170. [CrossRef]

234. Liu, S.; Song, Y.; Zhao, C.; Zhang, Y.; Lv, P.; Jiang, L.; Liu, Y.; Zhao, Y. The horizontal dispersion properties of CO_2-CH_4 in sand packs with CO_2 displacing the simulated natural gas. *J. Nat. Gas Sci. Eng.* **2018**, *50*, 293–300. [CrossRef]

235. Liu, K.; Yu, Z.; Saeedi, A.; Esteban, L. Effects of permeability, heterogeneity and gravity on supercritical CO_2 displacing gas under reservoir conditions. In Proceedings of the SPE Enhanced Oil Recovery Conference, Kuala Lumpur, Malaysia, 11–13 August 2015.

236. Abba, M.; Abbas, A.; Saidu, B.; Nasr, G.; Al-Otaibi, A. Effects of gravity on flow behaviour of supercritical CO_2 during enhanced gas recovery (EGR) by CO_2 injection and sequestration. In Proceedings of the Fifth CO_2 Geological Storage Workshop, Utrecht, The Netherlands, 21–23 November 2018.

237. Zangeneh, H.; Jamshidi, S.; Soltanieh, M. Coupled optimization of enhanced gas recovery and carbon dioxide sequestration in natural gas reservoirs: Case study in a real gas field in the south of Iran. *Int. J. Greenh. Gas Control* **2013**, *17*, 515–522. [CrossRef]

238. Wang, J.G.; Liu, J.; Liu, J.; Chen, Z. Impact of rock microstructures on the supercritical CO_2 enhanced gas recovery. In Proceedings of the CPS/SPE International Oil and Gas Conference and Exhibition in China, Beijing, China, 8–10 June 2013.

239. Honari, A.; Zecca, M.; Vogt, S.J.; Iglauer, S.; Bijeljic, B.; Johns, M.L.; May, E.F. The impact of residual water on CH_4-CO_2 dispersion in consolidated rock cores. *Int. J. Greenh. Gas Control* **2016**, *50*, 100–111. [CrossRef]

240. Dou, X.; Liao, X.; Wang, H.; Zhao, T.; Cheng, Z.; Ren, W.; Zhang, R. The study of CO_2 flooding and sequestration in tight gas reservoir: Different completion measures. In Proceedings of the SPE Nigeria Annual International Conference and Exhibition, Lagos, Nigeria, 4–6 August 2015.

241. Khan, C.M.H. Techno-Economic Reservoir Simulation Model for CO_2 Sequestration Evaluation. Ph.D. Thesis, Curtin University, Perth, Australia, 2013.

242. Narinesingh, J.; Alexander, D. Injection well placement analysis for optimizing CO_2 enhanced gas recovery coupled with sequestration in condensate reservoirs. In Proceedings of the SPE Trinidad and Tobago Section Energy Resources Conference, Port of Spain, Trinidad and Tobago, 13–15 June 2016.

243. Feather, B.; Archer, R. Enhanced natural gas recovery by carbon dioxide injection for storage purposes. In Proceedings of the 17th Australasian Fluid Mechanics Conference, Auckland, New Zealand, 5–9 December 2010.

244. Regan, M.L.M. A Numerical Investigation into the Potential to Enhance natural Gas Recovery in Water-Drive Gas Reservoirs Through the Injection of CO_2. Ph.D. Thesis, The University of Adelaide, Adelaide, Australia, 2010.

245. Pooladi-Darvish, M.; Hong, H.; Theys, S.O.P.; Stocker, R.; Bachu, S.; Dashtgard, S. CO_2 injection for enhanced gas recovery and geological storage of CO_2 in the long Coulee Glauconite, F. pool, Alberta. In Proceedings of the SPE Annual Technical Conference and Exhibition, Denver, CO, USA, 21–24 September 2008.

246. Biagi, J.; Agarwal, R.; Zhang, Z. Simulation and optimization of enhanced gas recovery utilizing CO_2. *Energy* **2016**, *94*, 78–86. [CrossRef]

247. Patel, M.J.; May, E.F.; Johns, M.L. Inclusion of connate water in enhanced gas recovery reservoir simulations. *Energy* **2017**, *141*, 757–769. [CrossRef]

248. Geel, C.R.; Arts, R.J.; Van Eijs, R.M.H.E.; Kreft, E.; Hartman, J.; D'Hoore, D. Geological site characterization of the nearly depleted K12-B gas field, offshore the Netherlands. In Proceedings of the International Symposium on Site Characterization for CO_2 Geological Storage, Berkeley, CA, USA, 20–22 March 2006.

249. Vandeweijer, V.; van der Meer, B.; Hofstee, C.; Mulders, F.; D'Hoore, D.; Graven, H. Monitoring the CO_2 injection site: K12-B. *Energy Procedia* **2011**, *4*, 5471–5478. [CrossRef]

250. Vandeweijer, V.; Hofstee, C.; Graven, H. 13 Years of safe CO_2 injection at K12-B. In Proceedings of the Fifth CO_2 Geological Storage Workshop, Utrecht, The Netherlands, 21–23 November 2018.

251. Kühn, M.; Tesmer, M.; Pilz, P.; Meyer, R.; Reinicke, K.; Förster, A.; Kolditz, O.; Schäfer, D. CLEAN: Project overview on CO_2 large-scale enhanced gas recovery in the Altmark natural gas field (Germany). *Environ. Earth Sci.* **2012**, *67*, 311–321. [CrossRef]

252. Kobos, P.H.; Cappelle, M.A.; Krumhansl, J.L.; Dewers, T.A.; McNemar, A.; Borns, D.J. Combining power plant water needs and carbon dioxide storage using saline formations: Implications for carbon dioxide and water management policies. *Int. J. Greenh. Gas Control* **2011**, *5*, 899–910. [CrossRef]

253. Li, Q.; Wei, Y.N.; Liu, G.; Shi, H. CO_2-EWR: A cleaner solution for coal chemical industry in China. *J. Clean. Prod.* **2015**, *103*, 330–337. [CrossRef]

254. Liu, H.; Hou, Z.; Were, P.; Sun, X.; Gou, Y. Numerical studies on CO_2 injection–brine extraction process in a low-medium temperature reservoir system. *Environ. Earth Sci.* **2015**, *73*, 6839–6854. [CrossRef]

255. Liu, G.; Gorecki, C.D.; Saini, D.; Bremer, J.M.; Klapperich, R.J.; Braunberger, J.R. Four-site case study of water extraction from CO_2 storage reservoirs. *Energy Procedia* **2013**, *37*, 4518–4525. [CrossRef]

256. Dewers, T.; Eichhubl, P.; Ganis, B.; Gomez, S.; Heath, J.; Jammoul, M.; Kobos, P.; Liu, R.; Major, J.; Matteo, E.; et al. Heterogeneity, pore pressure, and injectate chemistry: Control measures for geologic carbon storage. *Int. J. Greenh. Gas Control* **2018**, *68*, 203–215. [CrossRef]

257. Dahaghi, A.K. Numerical simulation and modeling of enhanced gas recovery and CO_2 sequestration in shale gas reservoirs: A feasibility study. In Proceedings of the SPE International Conference on CO_2 Capture, Storage, and Utilization, New Orleans, LA, USA, 10–12 November 2010.

258. Busch, A.; Alles, S.; Gensterblum, Y.; Prinz, D.; Dewhurst, D.; Raven, M.; Stanjek, H.; Krooss, B. Carbon dioxide storage potential of shales. *Int. J. Greenh. Gas Control* **2008**, *2*, 297–308. [CrossRef]

259. Liu, F.; Ellett, K.; Xiao, Y.; Rupp, J.A. Assessing the feasibility of CO_2 storage in the New Albany Shale (Devonian–Mississippian) with potential enhanced gas recovery using reservoir simulation. *Int. J. Greenh. Gas Control* **2013**, *17*, 111–126. [CrossRef]

260. Baran, P.; Zarębska, K.; Krzystolik, P.; Hadro, J.; Nunn, A. CO_2-ECBM and CO_2 sequestration in Polish Coal Seam—Experimental study. *J. Sustain. Mining* **2014**, *13*, 22–29. [CrossRef]

261. Fang, Z.; Li, X.; Hu, H. Gas mixture enhance coalbed methane recovery technology: Pilot tests. *Energy Procedia* **2011**, *4*, 2144–2149. [CrossRef]

262. Randolph, J.B.; Saar, M.O. Impact of reservoir permeability on the choice of subsurface geothermal heat exchange fluid: CO_2 versus water and native brine. In Proceedings of the Annual Meeting of the Geothermal Council (Geothermal 2011), San Diego, CA, USA, 23–26 October 2011.

263. Brown, D.W. A hot dry rock geothermal energy concept utilizing supercritical CO_2 instead of water. In Proceedings of the twenty-fifth workshop on geothermal reservoir engineering, Stanford, CA, USA, 24–26 January 2000.

264. Randolph, J.B.; Saar, M.O. Combining geothermal energy capture with geologic carbon dioxide sequestration. *Geophy. Res. Lett.* **2011**, *38*, L10401. [CrossRef]

265. Liu, H.; Hou, Z.; Li, X.; Wei, N.; Tan, X.; Were, P. A preliminary site selection system for a CO_2-AGES project and its application in China. *Environ. Earth Sci.* **2015**, *73*, 6855–6870. [CrossRef]

266. Procesi, M.; Cantucci, B.; Buttinelli, M.; Armezzani, G.; Quattrocchi, F.; Boschi, E. Strategic use of the underground in an energy mix plan: Synergies among CO_2, CH_4 geological storage and geothermal energy. Latium Region case study (Central Italy). *Appl. Energy* **2013**, *110*, 104–131. [CrossRef]

267. Underhill, D.H. *Analysis of Uranium Supply to 2050*; No. IAEA-SM-362; 2000. Available online: https://inis.iaea.org/search/search.aspx?orig_q=RN:31054412 (accessed on 16 October 2019).

268. Chong, Z.R.; Yang, S.H.B.; Babu, P.; Linga, P.; Li, X.S. Review of natural gas hydrates as an energy resource: Prospects and challenges. *Appl. Energy* **2016**, *162*, 1633–1652. [CrossRef]

269. Ota, M.; Abe, Y.; Watanabe, M.; Smith, R.L.; Inomata, H. Methane recovery from methane hydrate using pressurized CO_2. *Fluid Phase Equilib.* **2005**, *228–229*, 553–559. [CrossRef]

270. Yuan, Q.; Sun, C.Y.; Yang, X.; Ma, P.C.; Ma, Z.W.; Liu, B.; Ma, Q.L.; Yang, L.Y.; Chen, G.J. Recovery of methane from hydrate reservoir with gaseous carbon dioxide using a three-dimensional middle-size reactor. *Energy* **2012**, *40*, 47–58. [CrossRef]

271. Zhang, L.; Yang, L.; Wang, J.; Zhao, J.; Dong, H.; Yang, M.; Liu, Y.; Song, Y. Enhanced CH_4 recovery and CO_2 storage via thermal stimulation in the CH_4/CO_2 replacement of methane hydrate. *Chem. Eng. J.* **2017**, *308*, 40–49. [CrossRef]

272. Liu, Y.; Hou, J.; Zhao, H.; Liu, X.; Xia, Z. A method to recover natural gas hydrates with geothermal energy conveyed by CO_2. *Energy* **2018**, *144*, 265–278. [CrossRef]

273. Lee, Y.; Choi, W.; Shin, K.; Seo, Y. CH_4-CO_2 replacement occurring in sII natural gas hydrates for CH_4 recovery and CO_2 sequestration. *Energy Convers. Manag.* **2017**, *150*, 356–364. [CrossRef]

274. Xu, C.G.; Cai, J.; Yu, Y.S.; Chen, Z.Y.; Li, X.S. Research on micro-mechanism and efficiency of CH_4 exploitation via CH_4-CO_2 replacement from natural gas hydrates. *Fuel* **2018**, *216*, 255–265. [CrossRef]

275. Jafari Raad, S.M.; Hassanzadeh, H. Prospect for storage of impure carbon dioxide streams in deep saline aquifers—A convective dissolution perspective. *Int. J. Greenh. Gas Control* **2017**, *63*, 350–355. [CrossRef]

276. Wang, J.; Ryan, D.; Anthony, E.J.; Wigston, A.; Basava-Reddi, L.; Wildgust, N. The effect of impurities in oxyfuel flue gas on CO_2 storage capacity. *Int. J. Greenh. Gas Control* **2012**, *11*, 158–162. [CrossRef]

277. Sayegh, S.G.; Krause, F.F.; Fosti, J.E. Miscible displacement of crude oil by CO_2/SO_2 mixtures. *SPE Reserv. Eng.* **1987**, *2*, 199–208. [CrossRef]

278. Zhang, P.Y.; Huang, S.; Sayegh, S.; Zhou, X.L. Effect of CO_2 impurities on gas-injection EOR processes. Proceedings of SPE/DOE Fourteenth Symposium on Improved Oil Recovery, Tulsa, OK, USA, 17–21 April 2004.

279. Yang, F.; Zhao, G.B.; Adidharma, H.; Towler, B.; Radosz, M. Effect of oxygen on minimum miscibility pressure in carbon dioxide flooding. *Ind. Eng. Chem. Res.* **2007**, *46*, 1396–1401. [CrossRef]

280. Hajiw, M.; Corvisier, J.; El Ahmar, E.; Coquelet, C. Impact of impurities on CO_2 storage in saline aquifers: Modelling of gases solubility in water. *Int. J. Greenh. Gas Control* **2018**, *68*, 247–255. [CrossRef]

281. Turta, A.T.; Sim, S.S.; Singhal, A.K.; Hawkins, B.F. Basic investigations on enhanced gas recovery by gas-gas displacement. In Proceedings of the Petroleum Society's 8th Canadian International Petroleum Conference, Calgary, AB, Canada, 12–14 June 2007.

282. Li, D.; Jiang, X.; Meng, Q.; Xie, Q. Numerical analyses of the effects of nitrogen on the dissolution trapping mechanism of carbon dioxide geological storage. *Comput. Fluids* **2015**, *114*, 1–11. [CrossRef]

283. Chen, C.; Chai, Z.; Shen, W.; Li, W. Effects of impurities on CO_2 sequestration in saline aquifers: Perspective of interfacial tension and wettability. *Ind. Eng. Chem. Res.* **2017**, *57*, 371–379. [CrossRef]

284. Alpermann, T.; Dietrich, M.; Ostertag-Henning, C. Mineral trapping of a CO_2/H_2S mixture by hematite under initially dry hydrothermal conditions. *Int. J. Greenh. Gas Control* **2016**, *51*, 346–356. [CrossRef]

285. Fischer, S.; Wolf, L.; Fuhrmann, L.; Gahre, H.; Rütters, H. Simulated fluid-rock interactions during storage of temporally varying impure CO_2 streams. In Proceedings of the Fifth CO_2 Geological Storage Workshop, Utrecht, The Netherlands, 21–23 November 2018.

286. Wolf, J.L.; Niemi, A.; Bensabat, J.; Rebscher, D. Benefits and restrictions of 2D reactive transport simulations of CO_2 and SO_2 co-injection into a saline aquifer using TOUGHREACT V3.0-OMP. *Int. J. Greenh. Gas Control* **2016**, *54*, 610–626. [CrossRef]

287. Wolf, J.L.; Fischer, S.; Rütters, H.; Rebscher, D. Reactive transport simulations of impure CO_2 injection into saline aquifers using different modelling approaches provided by TOUGHREACT V3.0-OMP. *Proc. Earth Planet. Sci.* **2017**, *17*, 480–483. [CrossRef]

288. Vu, H.P.; Black, J.R.; Haese, R.R. The geochemical effects of O_2 and SO_2 as CO_2 impurities on fluid-rock reactions in a CO_2 storage reservoir. *Int. J. Greenh. Gas Control* **2018**, *68*, 86–98. [CrossRef]

Article

A Cradle-to-Grave Multi-Pronged Methodology to Obtain the Carbon Footprint of Electro-Intensive Power Electronic Products

Giovanni Andrés Quintana-Pedraza [1], Sara Cristina Vieira-Agudelo [1,*] and Nicolás Muñoz-Galeano [2]

[1] Grupo de Investigación en Ingeniería y Gestión Ambiental (GIGA), Departamento de Ingeniería Civil, Universidad de Antioquia (UdeA), Calle 70 No. 52-21, Medellín 050010, Colombia
[2] Grupo en Manejo Eficiente de la Energía (GIMEL), Departamento de Ingeniería Eléctrica, Universidad de Antioquia (UdeA), Calle 70 No. 52-21, Medellín 050010, Colombia
* Correspondence: sara.vieira@udea.edu.co; Tel.: +57-4-2195511

Received: 31 July 2019; Accepted: 26 August 2019; Published: 30 August 2019

Abstract: This paper proposes the application of a cradle-to-grave multi-pronged methodology to obtain a more realistic carbon footprint (CF) estimation of electro-intensive power electronic (EIPE) products. The literature review shows that methodologies for establishing CF have limitations in calculation or are not applied from the conception (cradle) to death (grave) of the product; therefore, this paper provides an extended methodology to overcome some limitations that can be applied in each stage during the life cycle assessment (LCA). The proposed methodology is applied in a cradle-to-grave scenario, being composed of two approaches of LCA: (1) an integrated hybrid approach based on an economic balance and (2) a standard approach based on ISO 14067 and PAS 2050 standards. The methodology is based on a multi-pronged assessment to combine conventional with hybrid techniques. The methodology was applied to a D-STATCOM prototype which contributes to the improvement of the efficiency. Results show that D-STATCOM considerably decreases CF and saves emissions taken place during the usage stage. A comparison was made between Sweden and China to establish the environmental impact of D-STATCOM in electrical networks, showing that saved emissions in the life cycle of D-STATCOM were 5.88 and 391.04 ton CO_2eq in Sweden and China, respectively.

Keywords: carbon footprint (CF); life cycle assessment (LCA); hybrid life cycle assessment; electro-intensive power electronic (EIPE) products; energy efficiency

---

1. Introduction

In recent decades, the electronic industry has grown as well as its greenhouse gas (GHG) emissions, generating emissions comparable with the airline industry. On the other hand, governments and organizations have created standards such as the Kyoto protocol [1] and Cop 21 [2], in which countries that are involved should follow the guidelines to reduce GHG emissions in the short, medium, and long terms. However, a concrete method to estimate carbon footprint (CF) in the electronic industry that is comparable with other CF industry estimations has not yet been developed [3].

At the present time, there is no standardized legislation to control GHG emissions of the electronic industry. However, governments and organizations around the world have created policies to regulate GHG emissions of industries. In the United States, the environmental protection agency (EPA) promulgated the "Mandatory reporting of greenhouse gases" (74 FR 56260) in 2009, requiring facilities that emit more than 25,000 metric tons of greenhouse gases per year to report annually on their emissions [4]. EPA mandatory reporting of GHG data collected by the Congressional Budget

Office in 2010 helped to develop a study titled "Effects of the carbon tax on the economy and the environment" [5–7]. South Korea approved a law to establish a commercial system of GHG emissions in 2015 and other laws to obtain 11% of the electricity from renewable sources in order to reduce its CF. China also approved laws to use renewable energy to supply electricity and to install electro-intensive power electronic (EIPE) products to improve efficiency [8,9].

One of the COP 21 objectives is to shift the energy sector onto a low-carbon path that supports economic growth and energy access, making the energy sector more resilient to climate change [10]. EIPE products are devices that can contribute to the improvement of energy efficiency and consequently to CF reduction. Therefore, researchers around the world have focused their efforts on the measurement and reduction of CF when EIPE products are used: Mostert et al. [11] compared several electrical energy storage technologies regarding their material and CF when renewable energy technologies are used. Due to renewable energy sources being intermittent, batteries were sized in order to obtain the minimum CF and, at the same time, to guarantee the electricity service. Siraganyan et al. [12] designed a tool to evaluate the environmental feasibility of solar photovoltaics, solar thermal panels, and energy storage technologies using CO_2eq estimation. Xie et al. [13] performed a planning exercise of an electric power system management with carbon emission constraints, concluding that the use of wind power and hydropower would be the best choices in terms of CF reduction.

In this paragraph are the techniques or methodologies that are conventionally used for CF estimation. There are three scopes that define the quality of CF estimation: scope 1 considers direct emissions from operations controlled by manufacturers, scope 2 considers indirect emissions associated with energy consumption, and scope 3 considers other indirect emissions detailed in downstream transportation and distribution of the product. After calculating emissions associated with scope 1, 2, and 3, the sum of these is divided by the total production to get the emission factor (*EF*) [14]. Life cycle assessment (LCA) is the guide to estimate CF of any product due to LCA-defined stages that are consecutive and interlinked. The LCA stages are raw material extraction, manufacturing, distribution, usage, and final disposal. There are different LCA techniques to estimate CF [15]. Conventional LCA approaches are process-based (PA) LCA and input–output-based (IOA) LCA, which allow for estimating the environmental impact. PA LCA uses a bottom-up approach to address CF during the entire life cycle of a process or a product; nevertheless it defines boundary selection per activity of a stage that can be ambiguous or can generate a truncation error, which can be as high as 50% [16]. IOA LCA considers an economic assessment to estimate CF using high-level aggregations per economic sector. Moreover, it underestimates the emissions of the involved sectors in the supply chain [17]. The PA and IOA approaches only cover scopes 1 and 2, generating a CF estimation that only includes 25% of the emissions. Standards used to estimate CF are based on these approaches [18]. For example, PAS 2050 [19] specifies the requirements for LCA of GHG emissions for goods and services, covering all activities from the acquisition of raw materials until its management as waste. ISO 14067 [20] developed an analysis for CF, providing requirements and guidance to quantify and report an inventory of GHG emissions associated with a specific product.

There are hybrid LCA techniques that combine PA and IOA approaches to generate a more accurate CF estimation because they take into account all the scopes. These type of approaches search for reduction in ambiguities in the estimations, integrating the largest amount of information from the supply chain. There are three types of hybrid analysis: (1) tiered hybrid analysis, (2) input–output hybrid analysis, and (3) integrated hybrid analysis [3]. In the following paragraph, the first two will be described, while the third one is part of our approach and will be described in the methodology section.

Tiered hybrid analysis is a detailed process-based analysis, which is carried out when environmental impact data is available for processes associated with the development of a product. If part of the data related to the development processes is unavailable, it can be covered by other hybrid approaches. Krishnan et al. [21] estimate the environmental impacts of the high-purity specialty chemicals and materials used in semiconductor manufacturing through this hybrid approach. LCA data on such specialty is private because chemical producers are reserved with the processes and materials used in the development of their product. In this case, processes that are not available are studied using an input–output hybrid analysis. Input–output hybrid analysis depends on the information available by each sector. In this analysis, it is advisable to have detailed input–output data of the processes and to discompose the information by sectors to increase the precision of the estimation. Nakamura et al. [22] applied an input–output hybrid approach to study environmental impacts from scraps and byproducts during the production processes of metallic elements for electronic products such as fuel cells, LEDs, and solar cells. The production of metals such as gold, silver, bismuth, and indium generates metal subproducts such as copper, lead, and zinc.

In this study, the application of an extended methodology is proposed, which is based on existing techniques used to estimate CF. This methodology includes raw material extraction, manufacturing, distribution, usage, and final disposal stages. For the raw material extraction and manufacturing stages, the proposed methodology implements an integrated hybrid analysis for the cradle-to-gate scenario. For posterior stages, to complete the cradle-to-grave scenario, the application of ISO 14067 and PAS 2050 standards is proposed. Due to EIPE products being used to improve the efficiency of the system, a usage stage is proposed that involves the environmental implications of the incorporation of an EIPE device in an electrical grid. Last, in the final disposal stage, three end-of-life methods are presented that depend on the level of the technology and the capacity of the solid waste management center. The contribution of this study is to propose a detailed and clear methodology to estimate CF, to propose a methodology that can be applied to the life cycle of the product under the cradle to grave scenario, and to include the usage stage of the product in the estimation of CF. This paper is organized as follows: Section 2 includes the details of the proposed methodology, Section 3 presents the results applied to a D-STATCOM of 30 kvar, and Section 4 highlights the main aspects and conclusions of the paper.

2. Proposed Methodology: Cradle-to-Grave Multi-Pronged Methodology

CF estimation of an EIPE product requires a comprehensive multi-pronged approach that includes the use of product group-oriented standards, hybrid LCA techniques, and integration of CF into the supply chain considering GHG emissions from cradle-to-grave. Commonly, methodologies to estimate CF are based on PA or IOA LCA. However, these LCA approaches only cover scopes 1 and 2. Hybrid LCA considers the three scopes and uses high-level aggregation of the product or the data that is available. In this paper, an integrated hybrid analysis is used to estimate CF in the cradle-to-gate scenario because there is data available from the producers. The later stages are analyzed based on PAS 2050 [19] and ISO 14067 [20] standards because there is no control over these stages and they are subject to assumptions depending on where the device is installed. Figure 1 shows a scheme of the methodology applied to perform the CF estimation.

Figure 1 shows the entire methodology to estimate CF using an LCA approach per stage. Estimation in (a) is based on the hybrid LCA method from cradle to gate. Estimations in (b), (c), and (d) are based on standards ISO 14067 [20] and PAS 2050 [19], allowing the estimation of CF in the distribution, usage, and end-of-life stages. Finally, in the final step (e), the CF estimations from previous stages are added up to generate an approximate estimate of CF in the life cycle of an EIPE product.

Stage (a) takes into account processes related to raw extraction material and to manufacturing of an LCA. In this stage, an economic balance model is used to estimate CF similar to the estimation of CF for electronic products proposed by References [3,23]. In stage (a), it is required to divide the analysis into two sections. The first section has inputs as the bill of components, the emission factor related with extraction per material used in the product, and the emission factor related with the assembly of components. The next step considers the disassembly of the EIPE product which comprises material content data and information about the processes involved. Then, with the weight of the materials, CO_2eq emissions are estimated using the corresponding *EF*.

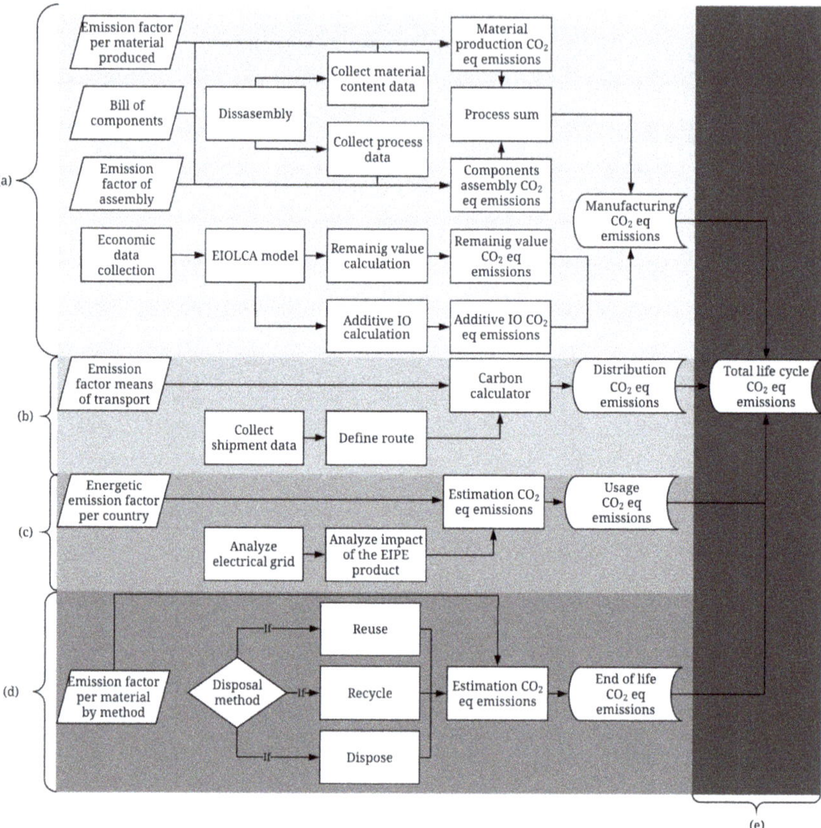

Figure 1. Methodology diagram for electro-intensive power electronic (EIPE) carbon footprint (CF) estimation: The stages considered for CF estimation are (**a**) manufacturing, (**b**) distribution, (**c**) usage, (**d**) end of life, and (**e**) total cycle life emissions. Processes considered for each stage are also shown.

In stage (a), the second section takes into account CO_2eq emissions that are not considered in the bill of components using an economic balance. This method uses economic data which is processed in the Economic Input-Output Life Cycle Assessment (EIOLCA) model [24]. This model estimates two economic correction factors: remaining value and additive input–output (IO). After, these correction values are extrapolated to CO_2eq emissions. Finally, results from the economic data and the material data are added up in order to get manufacturing CO_2eq emissions [3,23].

Stage (b) estimates CO_2eq emissions related to the distribution of a product. It starts collecting shipment data related to the product as weight, volume, and packaging. In the next step, a route should be defined from the factory to the customer, specifying the type of transport. All this information feeds a carbon calculator [25], which gives the emissions associated with the distribution stage.

Stage (c) estimates the CO_2eq emissions related to usage of the product. EIPE products have a direct impact on the electrical grid. Thus, the analysis requires definition of the electrical grid where the device will be used. After this definition, the impact of the product on the electrical grid is analyzed in terms of the energy consumed during its useful life. It is important to highlight that a calculation of saved energy (SE) must be included for a more realistic CF estimation. Many EIPE products are installed in power networks to improve processes or to solve some limitations. In general terms, they are directly or indirectly used to increase the electric power system efficiency so that EIPE products facilitate the reduction of the energy consumption; thus CF emissions are significantly reduced in most cases.

Stage (d) estimates the CO_2eq emissions related to end-of-life of the product. This stage is different from the previous ones because it takes into account three methods of disposal: reuse, recycle, and disposal. Each one has its own implications with a direct relationship with the location of the product [26]. Estimation of CO_2eq emissions uses an *EF* factor per material per method of disposal in a specific place.

Finally, the results of manufacturing, distribution, usage, and end-of-life stages are added up to obtain total life cycle CO_2eq emissions.

2.1. Hybrid LCA to Estimate CF in Cradle-to-Gate Scenario

The integrated hybrid analysis, described in stage (a) of Figure 1, is the most appropriated method to estimate CF of an EIPE product. This method uses the computational mathematical structure of LCA, along with input–output analysis and an economic balance. It collects data on materials and components directly from the dismantling of the product rather than relying on general databases or published material data. In addition, it collects data on energy consumption at the process level through industrial reports and literature review for stages of extraction and production of bulk materials, production and assembly of the components of the product, and its final manufacture. To assure the minimum cut-off error, this approach adds an Economic Input–Output (EIO) correction to check inputs in the supply chain that are not involved in the process-sum inventory. The EIO correction is composed of two factors: additive and remaining value (RV) [27]. The additive factor accounts for product components with specific economic data on requirements per product, omitting involved processes. The RV estimates the contribution from all the remaining, unaccounted sectors based on the available economic value of the product [23]. Equation (1) depicts a simplification of the model:

$$E = E^{Process} + E^{Additive} + E^{RV} \tag{1}$$

where $E^{Process}$ represents energy used in processes considered in the analysis (for example, semiconductor fabrication and board circuits assembly). $E^{Additive}$ represents the involved portion of other industries that are not directly considered in the $E^{Process}$ because the process data are restricted (for example, using the price of the chemicals used in the assembly due to the lack of knowledge of the process of its manufacture). Also, economic input–output models are formulated in terms of the producer or purchase prices. The production cost is the price of the product when it leaves the factory, while the customer price considers the cost of manufacturing, transportation, distribution, and sales margins [28].

Finally, E^{RV} estimates the contributions of the processes using an EIO model to generate the economic contribution of the sectors involved. This approach considers the RV in the manufacturing process that is not covered in the additive or process analysis [29].

2.2. Distribution Stage

In the shipment of merchandise, CF refers to direct consumption of energy and/or fuel. Companies move their merchandises by air, land, and sea transport. Because of this diversity, a simplified equation based on ISO 14067 [20], PAS 2050 [19], and the GHG Protocol [30] is implemented. Stage (b) of Figure 1 describes in detail how to estimate the CF related to distribution. This estimation considers measurements of different types of transport to ship the merchandise until its final destination.

$$TE = \sum M \times D \times EF \tag{2}$$

where *TE* is the total GHG emissions in kg CO_2eq related to the shipment and *M* refers to the mass (kg) of the merchandise when it is transported by air or land. When considering sea transportation, *M* refers to the occupation volume (m^3). *D* is the distance traveled during transport, and *EF* is a specific emission factor (g CO_2eq/km) that considers relevant load factors to allocate CO_2eq emissions.

2.3. Usage Stage

The estimation of CF in the usage stage of an EIPE device starts as an estimate for an electronic device based on PAS 2050 [19] and ISO 14067 [20]. The equation proposed in these standards estimates the annual average of energy emission factors for specific countries, considering the average energy consumption of the device in the country. Equation (3) shows the expression for the estimation:

$$TE = T \times C \times EF \tag{3}$$

where TE is the total GHG emissions during their useful life (g CO_2eq); T is the average time the device is working (h); C is the average electric power consumption (kWh), which depends strictly on the designer; and EF is the specific emission factor that depends on how electricity is generated in the country where the device is located (g CO_2eq/kWh).

However, to determine the impact of using the EIPE device on the electric grid, it is necessary to account for generated emissions with and without the device in the electric network. This estimation is performed by means of Equation (3) as described in stage (c) of Figure 1. Thus, ΔTE is proposed, which allows for the consideration of the difference between the CF of the electrical network with or without EIPE throughout its useful life. The result can be positive, negative, or zero. If it is negative, the results represent the saved emissions during the useful life of the device. If it is positive, the results show the total emissions generated during the useful life of the device. If it is zero, the electricity grid does not suffer changes in its emissions with or without a device.

2.4. End-of-Life Stage

When a product reaches the end-of-life, it can be recycled, reused, or disposed of. Methods that consider emissions in this stage are based on PAS 2050 [19] and depend on the final treatment given to the EIPE product or any other electrical product. When an electrical product reaches end of life, it cannot be used to produce energy through combustion because of its properties. Section (d) of Figure 1 shows the options to dispose of EIPE products. If the EIPE product is disposed of like waste in a landfill, it will not generate considerable GHG emissions because it is not composed of organic matter. The remaining disposal methods, reuse and recycling, generate considerable emissions.

2.4.1. Treatment of Emissions Associated with Reuse

Reuse refers to giving new use to some part of the product instead of discarding it. PAS 2050 [19] considers the next expression to simplify the calculation of GHG emissions through Equation (4):

$$\text{GHG emissions} = \frac{a+f}{b} + c + d + e \tag{4}$$

where a is the total life cycle GHG emissions of the product, excluding use-phase emissions; b is the anticipated number of reuse instances for a given product; c refers to emissions arising from an instance of the refurbishment of the product to make it suitable for reuse; d refers to emissions arising from the use stage; e is emissions arising from transport returning the product for reuse; and f is emissions arising from disposal [19].

2.4.2. Treatment of Emissions Associated with Recycling

Recycling transforms matter using energy demand to execute it. PAS 2050 [19] presents different options to assess emissions related with this process, depending on the transformation that suffers the product. One method considers that the recycled material does not maintain the same inherent properties as the virgin material input; for this case, emissions are estimated using Equation (5). The other method considers that recycled material maintains the same inherent properties as the virgin material input; for this case, emissions are estimated using Equation (6). Selection of the method depends on the knowledge of the material and its capacity to be recycled.

$$E = (1 - R_1)E_V + R_1 E_R + (1 - R_2)E_D \tag{5}$$

$$E = (1 - R_2)E_V + R_2 E_R + (1 - R_2)E_D \tag{6}$$

where R_1 is the proportion of recycled material input; R_2 is the proportion of material in the product that is recycled at end-of-life; E_R are emissions and removals arising from recycled material input per unit of material; E_V are emissions and removals arising from virgin material input per unit of material; and E_D are emissions and removals arising from disposal of waste material per unit of material.

As mentioned before, in these stages, the location of the product is important to estimate the CF. When considering the recycling process, data availability of *EF* per material in each location limits the estimation of GHG emissions. Moreover, comparison of GHG emissions is complex because of the differences encountered in the collection and recycling methods of materials according to the technological level applied [31] for different locations. However, different methods to obtain recycling *EF* per material are based on ISO 14067 and PAS 2050 standards, which is why emission factors are directly applied in the proposed methodology.

2.5. Total Life Cycle Carbon Equivalent Emission

LCA stages are consecutive and interlinked and represent a portion of the total CF of the product. Therefore, to obtain the total CF for the EIPE product using a life cycle assessment, CF estimations for stages (a), (b), (c), and (d) are added up to get the total CF (stage (e)).

3. Results

In this section, the multi-pronged approach discussed above to estimate CF in a cradle-to-grave scenario is applied to a Distribution-Static Var Compensator (D-STATCOM) prototype of 30 kvar. The D-STATCOM analyzed is shown in Figure 2.

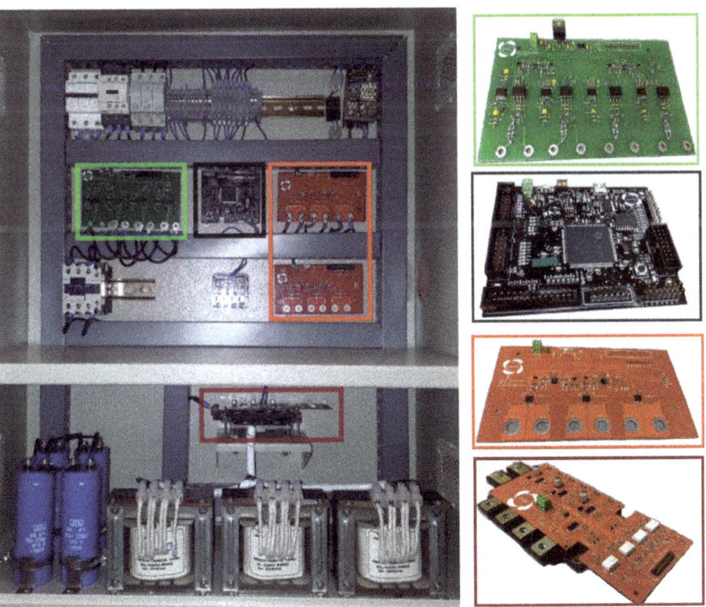

Figure 2. Distribution-Static Var Compensator (D-STATCOM) prototype of 30 kvar.

The D-STATCOM has more than 100 components. Most of the components were analyzed through data sheets provided by the producer of the components. Parts without available data were weighted using a microbalance. Also, to perform the estimation of stage (a), components of material were identified and version 2.2 of the database provided by the association EcoInvent in Zurich (Switzerland) [32] was used to obtain information on CO_2eq emitted during materials extraction. GHG emissions involved in the manufacturing stage were estimated during the development of the D-STATCOM. However, CF estimation does not correspond strictly to an integration of the supply chain of the product due to CF estimations depending on data available or a correlation with similar products.

Emissions related to manufacturing, assembly of the D-STATCOM, and its functional tests are measured as direct emissions with control of the operator in the laboratory.

3.1. CF through Economic-Balance Hybrid LCA

In this section, we use the hybrid LCA method by Deng et al. [23] and Vasan et al. [3], described in Equation (1), which considers two correction factors. $E^{Additive}$ takes into account emissions resulting from relevant industries. These emissions are obtained using the EIOLCA model [24] in which specific economic data on requirements per product are available. E^{RV} estimates the emission contributions from processes for which neither materials nor economic data are publicly available.

3.1.1. Process Data Analysis

The D-STATCOM contains four circuit boards, four electrolytic capacitors, one power transistors module, four inductors, one box support, and more than 100 micro-components and support pieces as shown in Figure 2. The production-related emissions of materials were obtained from measurements of weight, material composition, and data available from the EcoInvent databases [32]. Table 1 lists the main materials found in the D-STATCOM with the highest associated emissions, considering material composition and the energy consumption during the component production (see stage (a) in Figure 1). CO_2eq from the D-STATCOM bulk materials is estimated to be around 0.91735 ton CO_2eq.

Table 1. Embodied CO_2eq in materials.

Material	Amount per Converter (kg)	CO_2eq Intensity (kg CO_2eq/kg)	CO_2eq per Converter (ton)
Aluminum	32.78	8.24	0.270114
Arsenic	0.04	4	0.000145
Brass	0.03	4.39	0.000153
Cadmium	0.00	5.8	0.000002
Carbon	0.10	6.5	0.000648
Ceramic	0.15	5.5	0.000841
Copper	8.63	3.5	0.032747
Epoxy	0.70	9	0.006321
Fiber glass	0.0002	8.1	0.000002
Glass	0.002	0.85	0.000001
Gold	0.02	3000	0.048786
Iron	0.19	1.91	0.000368
Lead	0.00	20	0.000026
Manganese	0.10	2.6	0.000289
Nickel	0.19	12.4	0.002323
Nylon	0.06	6.5	0.000420
Phosphor	0.02	3.39	0.000081
PVC	0.08	2.2	0.000343
PBT	0.08	7.9	0.000647
Silicon	0.29	10	0.002880
Silver	0.20	754	0.150227
Steel	13.85	5.9	0.081720
Tantalum	1.21	260	0.313812
Tin	0.28	13.7	0.003888
Zinc	0.07	9	0.000596
Total	59.07	-	0.917350

Table 2 summarizes the associated assembly process emission data, estimated as energy consumption per hour in the factory (the laboratory where the prototype was developed) located in Medellín (Colombia). This estimation corresponds to the "Components assembly CO_2eq emissions" shown in stage (a) of Figure 2. Other processes involved in the production of pieces are considered in Section 3.1.2. The total life cycle CO_2eq emissions for this section is around 0.00279 ton CO_2eq. However, the chemicals used in bulk materials production, D-STATCOM manufacturing, and functional tests were not taken into consideration in the process-sum analysis.

Table 2. Embodied CO_2eq in manufacturing and assembly processes.

Process	Time (h)	Avg. Energy Consumption (kWh)	CO_2eq Intensity (g CO_2eq/kWh)	CO_2eq per Converter (ton)
Assembly D-STATCOM	1.5	0.22	182	0.000060
Functional tests	1	15	182	0.002730
Total	**2.5**	-	-	0.002790

Finally, total CO_2eq emissions associated with the process section are 0.92014 ton CO_2eq and correspond to the $E^{Process}$ contribution in Equation (1).

3.1.2. Economic Input–Output Correction

The economic value of the electronic chemicals used to manufacture the D-STATCOM is obtained based on the expenses involved in the processes mentioned in Table 2, corresponding to $E^{Additive}$. The estimation considers the amount of electronic chemicals used per device. The value obtained is approximately $5.5 in 2018 per device. This value has to be adjusted to the 2002 US dollar monetary unit in order to use the US 2002 Benchmark model purchaser price from the EIOLCA model. This value corresponds to $4 in 2002.

E^{RV} estimates the contribution of processes that are not included in either process-sum analysis or additive IO. The producer price in 2018 of the D-STATCOM was $1775 in 2018, which was adjusted to the dollar value in 2002, that is $1268. Table 3 shows the values estimated with the EIOLCA model [24]. The results show that $620 account for the process-sum analysis and the additive estimation. Then, RV is approximately $652 in 2002. In this estimation, the fraction accounted for the sum of processes was determined by selecting the related economic sectors of the 2002 model Benchmark purchaser price from the EIOLCA. The model has 247 economic sectors; then, the related sectors of each component manufacture are selected. However, the economic sectors involved in the previous steps must be omitted to estimate the RV. The top 39 selected sectors represent 99.9% of RV shares, but around 7–12 sectors per each component were removed in order to avoid double counting.

Then, it was only considered 82% of $E^{Additive}$ and E^{RV} with a value of $652 in 2002. According to The Bank Group analysis, CO_2 emissions per USD depend on the GDP [33]; this quantity corresponds to 0.321 ton CO_2eq.

Finally, the previous results are added up following Equation (1) to obtain CO_2eq emissions associated with stage (a) in Figure 1. The total CO_2eq emissions in the manufacturing stage are 1.24107 ton CO_2eq.

Table 3. Remaining Value (RV) input–output (IO) correction calculation and embodied CO_2eq per D-STATCOM.

	Value per Converter ($)	Fraction Accounted (%)	Value Accounted for Analysis ($)
Process-sum			
1. Breakers	$155	38	$59
2. Electrolytic capacitor	$262	38	$100
3. Circuits boards	$17	40	$7
4. Transistor	$76	38	$29
5. Bulk materials	$396	45	$178
6. Manufacturing	$357	41	$146
7. Support box	$114	88	$101
Additive IO			
8. Electronic chemicals	$4		$4
Total value accounted in process-sum and addition IO			$620
Producer price in 2002 USD			$1268
Producer price in 2018 USD			$1775
Remaining Value (RV)			$652
Total RV per D-STATCOM (ton CO_2eq)			0.321

3.2. CF Estimation Using Standards After the Product Leaves the Factory

In this section, the final location of the D-STATCOM has to be taken into account. However, there is no control in the stages of distribution, usage, and final disposal. To estimate CF, it is considered that the device will be distributed, used, and disposed of in two different countries: Stockholm (Sweden) and Macau (China). These countries are selected because Sweden is the country with the cleanest energy matrix and China has an energy matrix that generates significant GHG emissions [34,35].

3.2.1. Exporting a New Reactive Power Compensation D-STATCOM

This section shows how to obtain CO_2eq emissions related to stage (b) in Figure 1, which considers the distribution of the device. The D-STATCOM will be sent from Medellín (Colombia) to Buenaventura (Colombia) by road and continue their trips by sea to the final destination (either China or Sweden). GHG emissions were estimated through Equation (2) using a carbon calculator [25]. The calculator allowed using data to simulate in real time the shipping route and corresponding CF estimation. This tool is used because it is based on the guidelines outlined in the GHG Protocol [30] and follows the requirements by European Emissions Trading System (EU-ETS) [36], EN 16258 [37] (based on PAS 2050 [19]), and ISO 14067 [20]. In Equation (2), weight and volume of the D-STATCOM are 68.4 kg and 0.4 m^3, respectively. These values were considered for different types of transport involved in the route to reach the final destination. Emissions factors were also obtained from the carbon calculator [25]. The CFs associated with the shipment of the device to Sweden and China are 0.0176 and 0.0226 ton CO_2eq, respectively. Values used to estimate CF for the distribution stage are described in Table 4.

Table 4. EIPE product distribution stage.

Destination	Distance Road (km)	EF by Road (kg CO_2eq/kgkm)	Distance Sea (km)	EF by Sea (kg CO_2eq/kgm^3)	TE CO_2eq (ton)
Stockholm (Sweden)	471.61	0.0000837	11671.21	0.0031959	0.017620
Macau (China)	471.61	0.0000837	18327.30	0.00271453	0.022600

3.2.2. D-STATCOM in a New Location

D-STATCOM was chosen as the case to be analyzed due to the fact that it is an EIPE device that significantly contributes to the improvement of the efficiency. Figure 3 shows a power system that is composed of (1) power generators in charge of energy generation; (2) transmission lines, distribution lines, and transformers in charge of energy transportation; (3) an electric load that demands the electricity, transforming it in other forms of energy; and, of course, (4) a D-STATCOM in charge of reactive power compensation. When the D-STATCOM is not connected, power generators must deliver active and reactive power flows in order to satisfy the requirements of the electric load. Active and reactive power flows cause losses in transmission lines, distribution lines, and transformers.

Active power is a useful power, while reactive power is an inefficient power and must be compensated to obtain high standards of power quality and efficiency. For this reason, D-STATCOMs are connected in parallel near to the electric load for compensating the reactive power flow and avoiding its circulation through the power system; thus, power generators can supply an equivalent quantity of active power (useful power) to other loads. After the compensation, reactive power bidirectionally flows between the D-STATCOM and the load; however, a small quantity of active power is required to satisfy D-STATCOM internal losses.

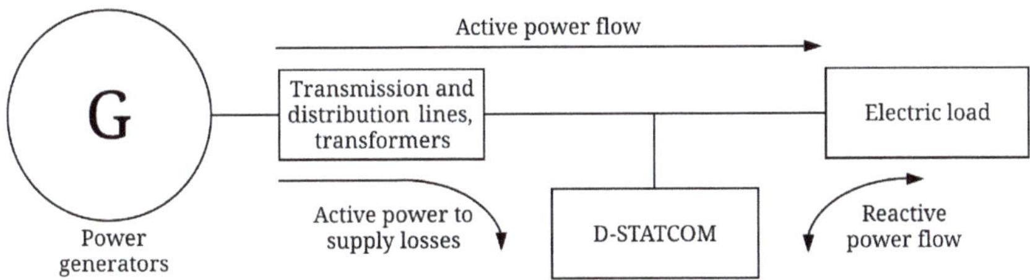

Figure 3. Hypothetical power system with a D-STATCOM.

CF of the usage stage, as depicted in stage (d) in Figure 1, is estimated using Equation (3). The ΔTE, which allows assessing the impact of the incorporation of the device, is determined considering the difference of emissions between a power grid with and without the application of an EIPE product. For this, it was estimated that the D-STATCOM has a useful life of 15 years and its efficiency is 95%, while efficiency of transmission systems is typically 80%. It is estimated that the energy expenditure on the power system without the device is 788.4 MWh and with the device is 236.5 MWh throughout its useful life. The EF used for each of the two countries is taken from the Organization for Economic Cooperation and Development [38]. ΔTE obtained for Sweden is -7.2 ton CO_2eq and for China is -392.4 ton CO_2eq. The negative sign means that the incorporation of the device results in saved emissions in both countries. Results are shown in Table 5.

Table 5. CO_2eq emissions for the usage stage per location.

Location	*EF* 2013 (g CO_2eq/kWh)	*TE* Case without Device (ton CO_2eq)	*TE* Case with Device ton CO_2eq)	ΔTE (ton CO_2eq)
Stockholm (Sweden)	13	10.2	3.0	-7.2
Macau (China)	711	560.6	168.2	-392.4

In Table 5, ΔTE values for Sweden and China vary considerably although the electrical network where they are applied is the same. Therefore, the energy EF per country is what causes the difference in the results. Even though the D-statcom saves the same amount of kW in both countries, the device application in China saves more emissions than in Sweden because the Chinese EF per kW is one of the largest in the world. Consequently, the environmental benefit of our D-statcom is directly proportional to the energetic emission factor.

3.2.3. End-of-Life of the Product

The electronic industry generates scraps which are not degradable. Moreover, reuse of some parts or components is limited because producers prefer to use a new component for its products even if it is possible to utilize a used part. Electronic products are difficult to recycle due to its size. Also, the recycling processes of the unions of electronic components are expensive. Then, it is cheaper to make new electronic components than to try to recycle them. However, metal components and cabling in the electronic devices are recycled because they are considered metallic waste [39].

For the reasons mentioned above, it is considered that only metal parts and cabling of the device are totally recycled. The prototype has 3 metal parts that represent about 80% of the total weight: the support box, craft coils, and cabling. The other 20% are components considered electronic waste with low CO_2 emissions.

Equation (6) was used to estimate emissions related to the recycling process of the D-STATCOM, which is considered in stage (d) in Figure 1. It is supposed that the device will be separated by macro-components and that these will be divided by metals if possible. In addition, it is considered that the parts that are totally metallic are recycled at the end of their useful life. Based on Equation (6), in Sweden, E_R associated to aluminum and steel recycling was estimated by Hillman et al. [40] and E_R associated with copper recycling was approximated by Agency [41]. In China, E_R associated to aluminum and steel recycling was estimated by Reference [26] and E_R associated with copper recycling was estimated by Reference [42]. The results are shown in Table 6.

Table 6. Embodied CO_2eq in recycling materials.

Material	Weight (kg)	CO_2eq Intensity (kg CO_2eq/kg)		CO_2eq per Converter (ton)	
		Sweden	China	Sweden	China
Aluminium	31.8	0.4	1.26	0.01272	0.040068
Copper	8.9	1.5	2.94	0.01338	0.026231
Steel	12.7	0.3	1.02	0.00381	0.012942
Total	-	-	-	0.02991	0.079240

The estimates of the end of life stage for this scenario were simplified because there was not enough data available on the level of technology and the capacity of the factories for each metal in the countries where the product is located. CO_2eq emissions for aluminum, copper, and steel were respectively 0.01272, 0.01338, and 0.00381 ton CO_2eq for Sweden and 0.04007, 0.02623, and 0.01294 for China. Then, total emissions associated with stage (d) in Figure 2 are obtained, adding up emissions generated by the recycling of each metal. The total emissions for this stage are 0.02991 and 0.07924 ton CO_2eq for Sweden and China, respectively.

3.3. Total Life Cycle CF of an EIPE Product

As stated in Section 2.5, the sum of all stages described in Figure 1 results in the CF of the entire life cycle of the product; this corresponds to stage (e). Hence, emissions related to the EIPE manufacturing using an integrated hybrid LCA approach are 1.241 ton CO_2eq (see Tables 1–3). The next stages were estimated based on PAS 2050 and ISO 14607 standards considering Sweden and China as reference countries. The distribution emissions were 0.020 ton CO_2eq in Sweden and 0.024 ton CO_2eq in China. Usage stage emissions consider the difference in emissions generated in an electric system with and without the EIPE device. Saved emissions are 7.2 ton CO_2eq in Sweden and 392.4 ton CO_2eq in China. End-of-life emissions through metal recycling are 0.029 and 0.079 ton CO_2eq in Sweden and China, respectively. Finally, according to Table 7, the total life cycle CO_2 emissions are −5.88 ton CO_2eq in Sweden and −391.04 ton CO_2eq in China. The results show that the incorporation of an EIPE device in a power system allows saving CO_2 emissions in both countries.

Table 7. Embodied CO_2eq in recycling materials.

Country	Manufacturing (ton CO_2eq)	Distribution (ton CO_2eq)	Usage (ton CO_2eq)	End-of-Life (ton CO_2eq)	Total LCA Emissions (ton CO_2eq)
Sweden	1.24107	0.017620	−7.2	0.02991	−5.88
China	1.24107	0.022600	−392.4	0.07924	−391.04

In general terms, emissions are mainly produced in the manufacturing stage; in this stage, most of the emission is generated in the process of raw extraction material. Producers could improve their designs by trying to use alternative materials to aluminum and silver where emissions are concentrated. Additionally, economic input–output correction is necessary due to RV corresponding to nearly 25% of emissions produced in the manufacturing stage. Compared with the manufacturing stage, the distribution and end-of-life stages do not significantly produce CO_2 emissions; however, the highest emissions in an LCA were presented at the usage stage, but when considering the electrical implications of the device on the network, it results in saved emissions. This implies that CF is decreased and, at the same time, contributes to the energy efficiency of the system even though the energy comes from the same generator. The results show that the usage stage has a positive environmental impact which is consistent with a negative CF and contributes to getting a CF overcompensation in the total estimation. Because emissions saved due to the reduction of energy consumption are greater than the sum of the emissions of all the other stages, it is recommended to install D-STATCOMs or EIPE products as long as they contribute to the improvement of energy efficiency. In conclusion, EIPE products generate environmental benefits superior to the impacts generated throughout the life cycle of the product. Therefore, it is considered that EIPE products can help to comply with energy efficiency policies stipulated in international agreements like the Kyoto protocol or COP 21. In particular, EIPE products help to reduce energy, which is in line with energy efficiency resource standard policies or strategic energy management plans. In this regard, governments can implement in their strategic plans the incorporation of reactive power compensation devices at transmission line strategic points to improve their energy efficiency and to thus decrease their CF.

4. Conclusions

The proposed methodology for estimating CF of EIPE products considers a multi-pronged assessment that was achieved using three scopes: (1) direct emissions (owned operations), (2) indirect emissions (energy consumption), and (3) indirect emissions (downstream transportation and distribution). Also, the methodology was designed to be applied to a cradle-to-grave scenario which is composed of an integrated hybrid approach and a standard based approach (ISO 14067 and PAS 2050). The methodology was explained in detail and is presented as a complete methodology to estimate CF during the life cycle of the product. Schematic diagrams to link all stages of the methodology were included to provide a better understanding. EIPE products are designed to contribute to efficiency improvement, so an energy calculation is presented in the usage stage in order to refine CF estimation.

The proposed methodology allows for the estimatation of CF for the distribution stage, considering different scenarios for the transportation. This feature is useful when considering the application of these estimations to propose mitigation actions that seek to reduce GHG emissions. Another highlight of the methodology applied to EIPE products is the estimation of the usage stage, which is not considered in other approaches. This estimation allows for the consideration of environmental implications of the incorporation of an EIPE device in an electrical grid.

The methodology was applied to a 30 kvar D-STATCOM manufactured in Colombia. A comparative analysis was made in two countries with very different energy matrices, Sweden and China. The difference between the possible emissions in the electrical grid with and without the device was estimated. Taking into account the useful life, it was found that CF for both countries is totally compensated, and furthermore, a negative CF (overcompensation) occurs, which is considered as a beneficial environmental impact. It can be concluded that the emissions saved due to the reduction of energy consumption in the usage stage are greater than the sum of the emissions of all the other stages of its life cycle, so a calculation of the energy saved during the operation of EIPE products is absolutely essential for a more realistic CF estimation.

The results exhibit a great difference between the Swedish and Chinese environmental impacts. The ton CO_2eq saved for Sweden was 5.88 while for China was 391.04. The reason for this is that the energy matrix for Sweden is mainly composed of renewable or clean sources while China has a diversified energy matrix (based on coal, oil, solar PV, and wind, among others). Therefore, it is recommended to prioritize the installation of EIPE products in countries where the energy matrix includes fossil fuels to significantly reduce/compensate CF. This feature of an EIPE device would facilitate nations to reduce their CF related to the energy sector, as it is stated in the COP 21 international agreement.

Finally, there is a limitation in the end-of-life stage since the values of the *EF* are not estimated under the same parameters because, in each country, the collection methods and the technological level in the transformation of materials vary according to waste recycling plants. However, emission factors which are estimated based on ISO 14067 and/or PAS 2050 were considered comparable for the proposed methodology. To solve this limitation, it is recommended that the governments be obligated to report the GHG emissions per material emitted by waste recycling plants as is mentioned in the COP 21 agreements.

Author Contributions: Data curation, G.A.Q.-P.; formal analysis, G.A.Q.-P., S.C.V.-A., and N.M.-G.; project administration, S.C.V.-A.; validation, G.A.Q.-P.; writing—original draft, G.A.Q.-P., S.C.V.-A., and N.M.-G.; writing—review and editing, G.A.Q.-P., S.C.V.-A., and N.M.-G.

Funding: This research project is financially supported by the Colombia Scientific Program within the framework of the call Ecosistema Científico (Contract No. FP44842- 218-2018).

Acknowledgments: The authors gratefully acknowledge the financial support provided by the Colombia Scientific Program within the framework of the call Ecosistema Científico (Contract No. FP44842- 218-2018).

Conflicts of Interest: The authors declare no conflict of interest.

Abbreviations

The following abbreviations are used in this manuscript:

OECD	Organization for economic cooperation and development
MDPI	Multidisciplinary Digital Publishing Institute
EIOLCA	Economic Input–Output Life Cycle Assessment
EIPE	Electro-intensive power electronic
DOAJ	Directory of open access journals
LCA	Life cycle assessment
TLA	Three letter acronym
IOA	Input–output based
GHG	Greenhouses gases
LD	Linear dichroism
CF	Carbon footprint
EU	European Union
PA	Process based

References

1. UNFCCC. The Kyoto Protocol to the United Nations Convention on Climate Change. 1997. Available online: http://unfccc.int/kyoto_protocol/items/2830.php (accessed on 27 March 2019).
2. UNFCCC. The Paris Agreement. 2018. Available online: https://unfccc.int/process-and-meetings/the-paris-agreement/the-paris-agreement (accessed on 27 March 2019).
3. Vasan, A.; Pecht, M.; Sood, B. Carbon footprinting of electronic products. *Appl. Energy* **2014**, *136*, 636–648. [CrossRef]

4. Environmental Protection Agency (EPA). Mandatory Reporting of Greenhouse Gases. 2010. Available online: https://www.govinfo.gov/content/pkg/FR-2009-10-30/pdf/E9-23315.pdf (accessed on 29 August 2019)

5. Congress of the United States. Policy Options for Reducing CO_2 Emissions. 2008. Available online: https://www.cbo.gov/sites/default/files/110th-congress-2007-2008/reports/02-12-carbon.pdf (accessed on 28 March 2019).

6. 110th Congress of the USA. Consolidated Appropriations Act. 2008. Available online: https://www.govinfo.gov/content/pkg/BILLS-110hr2764enr/pdf/BILLS-110hr2764enr.pdf (accessed on 28 March 2019).

7. Congress of the United States. Effects of a Carbon Tax on the Economy and the Environment. A Congressional Budget Office (CBO) Study. 2013. Available online: https://www.cbo.gov/publication/44223 (accessed on 28 March 2019).

8. Mapped: Climate Change Laws around the World. 2017. Available online: https://www.carbonbrief.org/mapped-climate-change-laws-around-world (accessed on 15 March 2019).

9. Xiao, M.; Simon, S.; Pregger, T. Energy system transitions in the eastern coastal metropolitan regions of China—The role of regional policy plans. *Energies* **2019**, *12*, 389. [CrossRef]

10. Mittal, S.; Liu, J.Y.; Fujimori, S.; Shukla, P.R. An assessment of near-to-mid-term economic impacts and energy transitions under "2 °C" and "1.5 °C" scenarios for India. *Energies* **2018**, *11*, 2213. [CrossRef]

11. Mostert, C.; Ostrander, B.; Bringezu, S.; Kneiske, T.M. Comparing electrical energy storage technologies regarding their material and carbon footprint. *Energies* **2018**, *11*, 3386. [CrossRef]

12. Siraganyan, K.; Perera, A.T.D.; Scartezzini, J.L.; Mauree, D. Eco-Sim: A parametric tool to evaluate the environmental and economic feasibility of decentralized energy systems. *Energies* **2019**, *12*. [CrossRef]

13. Xie, Y.; Fu, Z.; Xia, D.; Lu, W.; Huang, G.; Wang, H. Integrated planning for regional electric power system management with risk measure and carbon emission constraints: A case study of the Xinjiang Uygur Autonomous Region, China. *Energies* **2019**, *12*. [CrossRef]

14. Inakollu, S.; Morin, R.; Keefe, R. Carbon footprint estimation in fiber optics industry: A case study of OFSS fitel, LLC. *Sustainability* **2017**, *9*, 865. [CrossRef]

15. Minx, J.C.; Wiedmann, T.; Wood, R.; Peters, G.P.; Lenzen, M.; Owen, A.; Scott, K.; Barrett, J.; Hubacek, K.; Baiocchi, G.; et al. Input–output analysis and carbon footprinting: An overview of applications. *Econ. Syst. Res.* **2009**, *21*, 187–216. [CrossRef]

16. Lenzen, M. Errors in conventional and input-output—Based life—cycle inventories. *J. Ind. Ecol.* **2008**, *4*, 127–148. [CrossRef]

17. Wiedmann, T. Editorial: Carbon footprint and input–output analysis—An introduction. *Econ. Syst. Res.* **2009**, *21*, 175–186. [CrossRef]

18. Joshi, S. Product environmental life-cycle assessment using input-output techniques. *J. Ind. Ecol.* **1999**, *3*, 95–120. [CrossRef]

19. British Standard Institute. *Publicly Available Specification (PAS) 2050: Specifications for the Assessment of the Life Cycle Greenhouse Gas Emissions of Goods and Services*; BSI Report PAS 2050; British Standard Institute: London, UK, 2011.

20. ISO. *Greenhouse Gases—Carbon Footprint of Products—Requirements and Guidelines for Quantification and Communication*; ISO 14067; ISO: Geneva, Switzerland, 2013.

21. Krishnan, N.; Boyd, S.; Somani, A.; Raoux, S.; Clark, D.; Dornfeld, D. A hybrid life cycle inventory of nano-scale semiconductor manufacturing. *Environ. Sci. Technol.* **2008**, *42*, 3069–3075. [CrossRef] [PubMed]

22. Nakamura, S.; Murakami, S.; Nakajima, K.; Nagasaka, T. Hybrid input-output approach to metal production and its application to the introduction of lead-free solders. *Environ. Sci. Technol.* **2008**, *42*, 3843–3848. [CrossRef]

23. Deng, L.; Babbittb, C.W.; Williams, E.D. Economic-balance hybrid LCA extended with uncertainty analysis: Case study of a laptop computer. *J. Clean. Prod.* **2011**, *19*, 1198–1206. [CrossRef]

24. Carnegie Mellon University. US 2002 Producer Price Model—Economic Input-Output Life Cycle Assessment—Carnegie Mellon University. Available online: http://www.eiolca.net/Models/USmodels/US07ProducerPrice.html (accessed on 10 March 2019).

25. DHL. Carbon Calculator. Available online: https://www.dhl-carboncalculator.com/#/scenarios (accessed on 26 March 2019).

26. Damgaard, A.; Larsen, A.W.; Christensen, T.H. Recycling of metals: Accounting of greenhouse gases and global warming contributions. *Waste Manag. Res.* **2009**, *27*, 773–780. [CrossRef]

27. Scipioni, A.; Manzardo, A.; Mazzi, A.; Mastrobuono, M. Monitoring the carbon footprint of products: A methodological proposal. *J. Clean. Prod.* **2012**, *36*, 94–101. [CrossRef]
28. Miller, R.E.; Blair, P.D. *Input-Output Analysis: Foundations and Extensions*; Cambridge University Press: Cambridge, UK, 2009.
29. BEA U.S. Department of Commerce. Concepts and Methods of the U.S. Input-Output Accounts. 2009. Available online: https://www.bea.gov/sites/default/files/methodologies/IOmanual_092906.pdf (accessed on 10 February 2019).
30. Greenhouse Gas Protocol. Available online: https://ghgprotocol.org/ (accessed 20 January 2019).
31. Yanjia, W.; Chandler, W. The Chinese nonferrous metals industry—Energy use and CO_2 emissions. *Energy Policy* **2010**, *38*, 6475–6484. [CrossRef]
32. Ecoinvent. Available online: https://www.ecoinvent.org/ (accessed on 24 January 2019).
33. The World Bank. CO_2 Emissions (kg per 2010 US$ of GDP) Data. Available online: https://data.worldbank.org/indicator/EN.ATM.CO2E.KD.GD (accessed on 5 February 2019).
34. Climate Council. 11 Countries Leading the Charge on Renewable Energy. 2019. Available online: https://www.climatecouncil.org.au/11-countries-leading-the-charge-on-renewable-energy/ (accessed on 29 August 2019)
35. Union of Concerned Scientists. Each Country's Share of CO_2 Emissions. 2018. Available online: https://www.ucsusa.org/global-warming/science-and-impacts/science/each-countrys-share-of-co2.html (accessed on 29 August 2019)
36. European Commission. EU Emissions Trading System (EU ETS). 2016. Available online: https://ec.europa.eu/clima/policies/ets_en (accessed 29 March 2019).
37. European Standards. DIN EN 16258. 2013. Available online: https://www.en-standard.eu/din-en-16258-methodology-for-calculation-and-declaration-of-energy-consumption-and-ghg-emissions-of-transport-services-freight-and-passengers/ (accessed 29 March 2019).
38. The Organisation for Economic Co-Operation and Development. Compare Your Country—Climate Change Mitigation Policies. Available online: //www.compareyourcountry.org/climate-policies?lg=en (accessed on 24 October 2018).
39. Tanskanen, P. Management and recycling of electronic waste. *Acta Mater.* **2013**, *61*, 1001–1011. [CrossRef]
40. Hillman, K.; Damgaard, A.; Eriksson, O.; Jonsson, D.; Fluck, L. *Climate Benefits of Material Recycling*; Nordic Council of Ministers: Copenhagen, Denmark, 2015. [CrossRef]
41. Agency, S.E.P. *National Inventory Report Sweden 2018. Greenhouse Gas Emission Inventories 1990–2016. Submitted under the United Nations Framework Convention on Climate Change and the Kyoto Protocol*; Swedish Environmental Protection Agency: Stockholm, Sweden, 2018.
42. Kuckshinrichs, W.; Zapp, P.; Poganietz, W.R. CO_2 emissions of global metal-industries: The case of copper. *Appl. Energy* **2007**, *84*, 842–852. [CrossRef]

Article

Trade Openness and Carbon Leakage: Empirical Evidence from China's Industrial Sector

Bin Fan [1], Yun Zhang [2], Xiuzhen Li [3],* and Xiao Miao [1,4],*

[1] Department of Dermatology, Yueyang Hospital of Integrated Traditional Chinese and Western Medicine, Shanghai University of Traditional Chinese Medicine, Shanghai 200437, China; drfanbin76@163.com

[2] School of Finance, Shanghai Lixin University of Accounting and Finance, Shanghai 201620, China; zhangyunphd@163.com

[3] School of Economics and Trade, Shanghai Lixin University of Accounting and Finance, Shanghai 201620, China

[4] Innovation Research Institute of Traditional Chinese Medicine, Shanghai University of Traditional Chinese Medicine, Shanghai 201203, China

* Correspondence: lixiuzhenlixin@163.com (X.L.); 0000002623@shutcm.edu.cn (X.M.)

Received: 11 February 2019; Accepted: 5 March 2019; Published: 21 March 2019

Abstract: China is a large import and export economy in global terms, and the carbon dioxide emissions and carbon leakage arising from trade have great significance for China's foreign trade and its economy. On the basis of trade data for China's 20 industrial sectors, we first built a panel data model to test the effect of trade on carbon dioxide emissions and the presence of carbon leakage for all industrial sectors. Second, we derived a single-region input–output model for open economies based on the industrial sectors' diversity and carbon dioxide emissions, and performed an empirical test. We estimated the net carbon intensity embodied in export, which is $0.237tCO_2$/ten thousand RMB, to divide all sectors (ACSs) into high-carbon sectors (HCSs) and low-carbon sectors (LCSs). The results show that higher trade openness leads to a reduction in the intensity of CO_2 emissions and gross emissions and that there are obvious structural differences in different sectors with different carbon emission intensity. The coefficient of trade openness for LCSs is -0.073 and is statistically significant at the 1% level, so higher trade openness for LCSs leads to a reduction in the CO_2 emissions intensity. However, the coefficient for HCSs is 0.117 and is statistically significant at the 10% level, indicating that higher trade openness increases the CO_2 emissions' intensity for HCSs. The difference is that higher trade openness in LCSs can help reduce the CO_2 emissions' intensity without the problem of carbon leakage and with the existence of the environmental Kuznets curve (EKC), whereas there is no EKC for HCSs and carbon leakage may happen. We introduced dummy variables and found that a "pollution haven" effect exists in HCSs. The test results in HCSs and LCSs are exactly the opposite of each other, which shows that the carbon leakage of ACSs cannot be determined. The message that can be drawn for policy makers is that China does not need to worry about the adverse impact on the environment of trade opening up and should, in fact, increase the opening up of trade, while becoming acclimatized to environmental regulation of a new trade mode and new standards. This will help amplify the favorable impact of trade opening up on the environment and improve China's international reputation. The policies related to trade should encourage structural adjustment between the sectors via the formulation of differential policies and impose a restraint on sectors that have high levels of CO_2 emissions embodied in export.

Keywords: CO_2 emissions embodied in trade; trade openness; carbon leakage; EKC; industrial sector

---

1. Introduction

The Paris conference on climate change in December 2015 set up the global climate governance mechanism from 2020 to 2030, and China restated that it would peak carbon dioxide emissions around 2030. The China–US Joint Announcement on Climate Change was released in November 2014. It is expected that total CO_2 emissions in China will peak around 2030, and the Chinese government has announced that it will make efforts to reach this peak sooner. Before the Joint Announcement, China's target for reducing CO_2 emissions was a 40–45% decline of CO_2 emissions per unit of GDP by 2020 as compared to 2005. This was a relative goal with no upper limit imposed. However, in this Announcement, China promised that the peak of CO_2 emissions will be reached before 2030, which sets a ceiling on CO_2 emissions associated with China's industrialization and urbanization. China is the world's largest trading nation, and trade openness is closely related to a transition to a low-carbon economy, energy saving, and emissions reduction. Opening up of foreign trade is very likely to cause the transfer of polluting sectors and carbon leakage (Figure 1). This fact has considerable bearing on policy-making related to trade openness and achieving the CO_2 emissions reduction commitment under the context of the new Chinese economic normality, and it even affects the responsibility definition of CO_2 emissions and climate negotiation results.

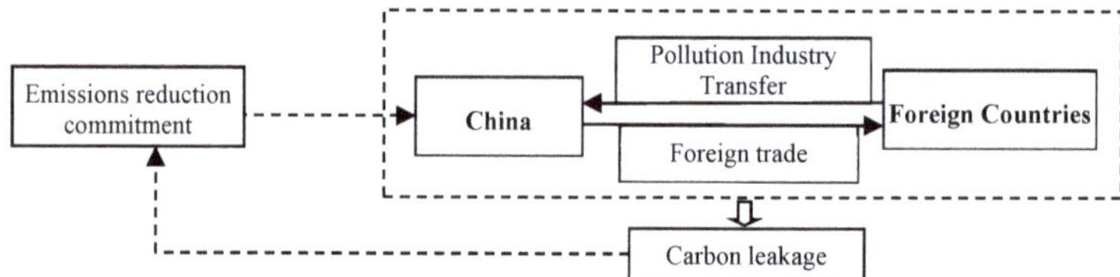

Figure 1. Carbon leakage affects emission reduction commitments.

2. Literature Review

The relationship between trade and environment is an important academic research topic. Since the 1990s, the positive and negative effects of trade opening on the environment have aroused heated discussion, which has brought forth several theories, including the environmental Kuznets curve (EKC), the pollution haven hypothesis, and the race-to-the-bottom corollary. Despite theoretical advancement, the empirical studies tend to come to inconsistent conclusions. Some propose that the pollution haven hypothesis and the factor endowment effect exist side-by-side but play opposite roles [1–3]. The pollution haven hypothesis posits that low-income developing countries have poor environmental regulation intensity and enjoy comparative advantages in pollution-intensive sectors seeking low external costs. The factor endowment effect posits that developed countries enjoy comparative advantages in capital-intensive polluting sectors. Unfortunately, the fact is that a country may feature either high income and capital abundance or low income and low capital, so the two traits always go hand in hand. Thus, the structural effect induced by trade cannot be used for theoretical prediction but serves as a question to be tested empirically [4].

As China is experiencing drastic growth of foreign trade, the connection between trade and CO_2 emissions has become a hot topic. Most of the existing literature deals with either of two topics. One is the CO_2 emissions embodied in China's foreign trade. Shui and Harriss (2006) examined the influence of US–China trade on national and global emissions of carbon dioxide during 1997–2003 and found about 7–14% of China's current CO_2 emissions were a result of producing exports for US consumers [5]. Weber et al. (2008) found that around one-third of Chinese emissions (1700 Mt CO_2) were due to production of exports in 2005, and this proportion rose from 12% (230 Mt) in 1987 and only 21% (760 Mt) as recently as 2002 [6]. Su and Ang (2013) finished empirical studies using five Chinese input–output (IO) tables, 1997, 2000, 2002, 2005, and 2007 and showed that the transitions of China's emissions embodied in imports to those in the exports account for around 4.6–13.3% of total emissions if the competitive imports assumption is used [7]. Wu et al. (2018) used input–output analysis to measure CO_2 emissions (CEs) embodied in Chinese provinces' exports and imports and explore the interprovincial spillover of CEs induced by exports through the domestic supply chains [8]. Besides this, Ahmad and Wyckoff (2003), Yan and Yang (2010), Lin and Sun (2010), Du et al. (2011), Chen et al. (2011), Chen and Yu (2012), and Lin et al. (2014) have done similar research and estimation from a unilateral, bilateral, or multilateral perspective [9–15]. Most studies agree that the CO_2 emissions embodied in China's export are huge. It is thus inferred that China has become a pollution haven for the transfer of CO_2 emissions from developed countries and that carbon leakage does exist due to trading between China and the developed countries.

The second topic is testing for the existence of EKC. Tao et al. (2008) found that there was a long-run cointegrating relationship between the per capita emission of three pollutants and the per capita GDP, and all three pollutants are inverse U-shaped [16]. Song et al. (2013) tested the existence of EKC for thirty provinces and cities in Mainland China through Copeland model, and empirical analysis indicates that EKC does not exist for Liaoning, Anhui, Fujian, Hainan, and Qinghai, while EKC of Shanghai, Guizhou, Tibet, Jilin, and Beijing have reached inflection point [17]. Liu et al. (2018) introduced two subdivisions of trade diversification–export product diversification and export market diversification as proxy variables for economic development in rectification of the EKC, and found the evidence that GDP per capita and export diversification had a robust relationship with ecological footprint and, therefore, the EKC hypothesis holds in Korea, Japan, and China in the long run [18]. Empirical study conclusions diverge greatly. Some researchers have even derived the curves of a non-inverted U-shape. Shen (2006) used Chinese provincial data from 1993 to 2002 to examine the existence of EKC and found an EKC relationship was found in COD (Chemical Oxygen Demand), Arsenic, and Cadmium emissions in China, but SO_2 showed a U-shaped curve and Dust Fall indicates no relationship with income level [19]. Other researchers, such as Jalil and Mahmud (2009), Diao et al. (2009), He and Wang (2012), Liu (2012), Ho et al. (2013), and Yin et al. (2015), have also done relevant empirical research [20–25]. The absolute CO_2 emissions embodied in export depend on the volume of exported goods, and the effect of export volume may play a decisive role in the growth of CO_2 emissions embodied in export. Moreover, the test for EKC is affected by indicator variables and data availability, especially the choice of time periods under investigation.

According to the pollution haven hypothesis and the race-to-the-bottom corollary, there will be a transfer of high-pollution and high-emission industries from developed countries to developing countries; this gives rise to the problem of carbon leakage. If a developed country raises its environmental standards, its production activities will inevitably decrease, while the output of a developing country with lower environmental standards will increase. The net result is an overall increase in global CO_2 emissions. Some representative studies on carbon leakage have come to diverse conclusions. Barker et al. (2007) investigated potential carbon leakage from six EU Member States (MSs) that implemented Environmental Tax Reform (ETRs) unilaterally over the period of 1995–2005 and constructed a counterfactual Reference case, assuming that the six countries did not introduce ETRs, and then alternative scenarios were developed to assess the effects of the ETRs, including effects on CO_2 emissions for the EU25 economies [26]. Most MSs recorded a reduction in CO_2 emissions when comparing the Baseline case to the Reference case, and the results show that carbon leakage is very small and in some cases negative, due to technological spillover effects. Kuik and Hofkes (2010) explored some implications of border adjustment measures in the EU ETS, for sectors that might be subject to carbon leakage and found that border adjustment might reduce the sectoral rate of leakage of the iron and steel industry rather forcefully, but that the reduction would be less for the mineral products sector, including cement, and the reduction of the overall or macro rate of leakage would be modest [27]. Elliott et al. (2014) used a full CGE model with many countries and many goods to measure effects and varied elasticity of substitution and confirmed the analytical model's prediction that negative leakage depends on the ability of consumers to substitute into the untaxed good and the ability of firms to substitute from carbon emissions into labor or capital [28].

BöHringer et al. (2017) showed that the combination of output-based rebating and a consumption tax for emission-intensive and trade exposed goods can be equivalent with border carbon adjustment, and supplementing output-based rebating with consumption tax constituted robust policies to mitigate carbon leakage [29]. Felder and Rutherford (1993), Smith (1998), Paltsev (2001), Aukland et al. (2003), Gerlagh and Kuik (2007), Rosendahl and Strand (2009), Eichner and Pethig (2011), Baylis et al. (2013), and Carbone (2013) have also done relevant research [30–38]. Surprisingly, some researchers argue that carbon leakage can be negative by model or analysis based on assumptions. One country, typically a developed country, will reduce production activities to meet higher environmental standards, whereas a developing country with lower environmental standards will increase its output; in this process, the intensity of CO_2 emissions in developing countries is lessened through the flow of technological innovation and production factors, ultimately leading to a reduction of global CO_2 emissions. Fullerton et al. (2011) used a general equilibrium model to solve for effects of a small increase in carbon tax on leakage, and reported that the taxed sector substitutes away from carbon into capital so it may absorb capital, which shrinks the other sector, causing negative leakage [39]. Winchester and Rausch (2013) investigated leakage in computable general equilibrium models under alternative fossil fuel supply elasticity values and factor mobility assumptions, and found that fossil fuel supply elasticity must be either equal or close to infinity to generate net negative leakage [40].

In summary, existing research has carried out a creative theoretical design and obtained different conclusions, which means that the structural effect induced by trade cannot be used for theoretical prediction but serves as a question to be tested empirically. There have been many results in empirical research. Although the amount of CO_2 emissions embodied in China's trade is huge, it does not indicate that there must be carbon leakage, especially as the results of the EKC test are inconsistent. We believe that, considering these dynamics, CO_2 emissions embodied in trade, EKC, and carbon leakage can be integrated into one framework so that the logistic connections between them can be analyzed. Carbon leakage is the result of transferring high-carbon sectors (HCSs) from developed countries to developing countries and the developed countries needing to import large volumes of goods from the developing countries—the huge level of CO_2 emissions embodied in the export directly lead to carbon leakage. However, this is only a necessary condition, not a sufficient condition. If the developing countries are indeed a pollution haven, there will be no inverted U-shaped EKC; if there is an inverted U-shaped EKC, persistent carbon leakage will not occur. CO_2 emissions embodied in trade and EKC are mostly studied empirically, whereas carbon leakage is estimated using theoretical models. We built our panel data model based on the trade data of China's industrial sectors. Through empirical analysis, the existence of carbon leakage due to China's opening up of trade was proven. Then, the existence of EKC was examined by introducing the linear term and the quadratic term of per capita GDP of the sector. The reliability of the regression was enhanced by adding a control variable, such as the number of employees in the sector. The innovations of this study are summarized as follows: (1) The effect of foreign investments in China on CO_2 emissions shows large variation across the sectors (Lin and Yang, 2013) [41]. Therefore, this article takes China's industrial sector sub-sector as the research object, which is different from most previous studies, and a single-region input–output model (SRIO model) is derived for open economies. The low-carbon sectors (LCSs) and HCSs are divided based on the intensity of CO_2 emissions embodied in net export for each sector. This method fully considers the distinct features of foreign trade, as compared with the classification by CO_2 emissions' intensity for each sector. The research design is also more in line with the characteristics of developing countries and can more effectively test the trade and economic development of developing countries. (2) The cross term of trade openness and the sector's development level are added into the regression equation to test carbon leakage. Moreover, the dummy variable is designed to capture the pollution haven hypothesis. (3) In terms of policy recommendations, we analyzed the policies with 2030 emission reduction commitments latest proposed by the Chinese government, which is also the significance of the research.

3. Empirical Model and SRIO Model Derivation

3.1. Empirical Modeling

Grossman and Krueger (1991) laid the foundation for the analytical framework of the relationship between trade and environment [42]. Antweiler et al. (2001) also proceeded from this framework. We built our panel data model using the sector-based data of China's industrial sectors from 2000 to 2014 [1]. Usually, trade openness of a country can be measured in two ways: One is by measuring the trade barrier directly—the lower the nominal trade barrier, the higher the trade openness; the other is by measuring the trade flow, from which the trade openness is estimated indirectly. The closer the trade flow to the estimate corresponding to completely free trade, the higher the trade openness (Li and Huang, 2006) [43]. Referring to the existing literature, such as Zhao and Ding (2012) and Ma and Li et al. (2012) [44,45], we measured trade openness by the ratio of the sum of imports and exports to the output for the sector. Bao et al. (2003) estimated China's trade openness using five indicators: degree of dependence on foreign trade, effective tariff rate, cost of black market transaction, Dollars' index, and corrected degree of dependence on foreign trade [46]. It is now recognized that only the degree of dependence on foreign trade (namely the ratio of the sum of imports and exports to GDP) is a better reflection of the connections between China's trade openness and economic growth.

Many empirical analyses have been done concerning EKC hypothesis. De Bruyn et al. (1998) were the first to study the dynamics of one country's EKC using time series [47]. In recent years, Dinda and Coondoo (2006) and Soytas and Saria et al. (2009) have argued that analysis based on the cause-and-effect relationship was more fruitful [48,49]. The most common approach is analysis based on panel data (Dinda, 2004) [50], and most studies use per capita GDP as an indicator of economic growth when it comes to the question of how economic growth is related to environmental pollution. The effect of trade openness on CO_2 emissions of different sectors and the existence of EKC were tested by regression analysis. However, this direct effect cannot explain the structural effect of trade openness. That is, the effect of trade openness on China's industrial structure cannot be determined under the constraint of comparative advantage. We introduce the cross term about trade openness proposed by Richard and Piergiuseppe et al. (2012) [51] and test the existence of carbon leakage via trade with China at the industry level. The cross term is the product of trade openness and the sector's development level. The analysis model is constructed as follows:

$$\ln CEM_{i,t} = \alpha_0 + \alpha_1 \ln Y_{i,t} + \alpha_2 (\ln Y_{i,t})^2 + \alpha_3 \ln IDL_{i,t} + \alpha_4 TOI_{i,t} \cdot IDL_{i,t} + \beta X_{i,t} + \varepsilon_{i,t}$$

where subscript i denotes the sector; subscript t denotes time (unit: year); $CEM_{i,t}$ is the CO_2 emissions' intensity in sector i in the t-th year (or total CO_2 emissions); $Y_{i,t}$ is the per capita GDP in sector i in the t-th year; $TOI_{i,t}$ is trade openness in sector i in the t-th year; $TOI_{i,t} \cdot IDL_{i,t}$ is the cross term of trade openness and the sector's development level $IDL_{i,t}$; $X_{i,t}$ is other control variable, e.g., number of employees in the sector $POP_{i,t}$, intensity of research and development $TOR_{i,t}$, and intensity of economic activities $ACT_{i,t}$. $\varepsilon_{i,t}$ is a random error term. To eliminate the influence of heteroscedasticity, the indicator data are expressed in the form of natural logarithm.

3.2. Variable Explanation and Data Source

The data used came from the China Statistical Yearbook, the China Industry Economy Statistical Yearbook, the China Population & Employment Statistics Yearbook, the China Statistical Yearbook on Environment, and the China Energy Statistical Yearbook. Part of the data has been downloaded from the OECD database, DRCnet, and the CEInet Statistics Database [52–56]. The variables are explained as follows:

(1) CO_2 emissions' intensity (or total CO_2 emissions): $CEM_{i,t}$. Two indicators are used to measure China's CO_2 emissions, namely, total CO_2 emissions and CO_2 emissions' intensity. Total CO_2 emissions are a scale indicator that depends on the output scale, output structure, and production conditions of

the sector. It is obtained by calculating the product of energy consumption and the carbon emission coefficient. CO_2 emissions' intensity is a relative indicator that highlights the effect of output structure and production conditions. It is calculated as the CO_2 emissions per unit output of the sector.

(2) Per capita GDP: $Y_{i,t}$. According to the EKC hypothesis, income level is an important control variable that affects pollutant emissions. As mentioned above, per capita GDP reflects how increases of output and income affect scale effect and technology effect of CO_2 emissions. Here we use actual per capita GDP as the variable related to economic growth to test the EKC hypothesis in China's industrial sectors. If the coefficient α_1 of the linear term of per capita GDP is positive and the coefficient α_2 of the quadratic term of per capita GDP is negative, then the relationship between environmental pollution and the income level can be described by an inverted U-shaped curve. Thus, EKC exists.

(3) Trade openness: $TOI_{i,t}$. As in most studies, the ratio of the sum of imports and exports to the output of a sector is calculated as the indicator of trade openness. Trade openness is associated with the structural effect caused by import and export. The trade data of the sectors come from the OECD database. All data from International Standard Industrial Classification (ISIC) are classified into 20 sectors' data compared with the industry classification of input–output (IO) and are then expressed in Renminbi (RMB) using the annual average exchange rate. If the estimated coefficient α_3 of trade openness is positive, it means that higher trade openness aggravates environmental pollution; if α_3 is negative, it means that higher trade openness alleviates environmental pollution and contributes to emission reduction.

(4) Cross term: $TOI_{i,t} \cdot IDL_{i,t}$. The cross term is constructed between trade openness and a sector's development level to examine carbon leakage. If the coefficient α_4 of the cross term is negative, it can be confirmed that there is not a transfer of high-emission sectors from developed countries to developing countries and carbon leakage does not exist. The sector's development level is measured by the added value of the industrial sector $IVA_{i,t}$ and the per capita output $GPC_{i,t}$ of the sector. The data came from the China Statistical Yearbook and the China Industry Economy Statistical Yearbook. Adjustments were made if necessary.

(5) Control variable: $X_{i,t}$. The control variables used are the number of employees in the sector $POP_{i,t}$, the intensity of research and development $TOR_{i,t}$ and the intensity of economic activities $ACT_{i,t}$. To some extent $POP_{i,t}$ influences per capita CO_2 emissions of the sector and is also an important factor influencing per capita income and per capita GDP. This indicator is usually considered when evaluating the damage done by CO_2 emissions and making policies on emission reduction. $TOR_{i,t}$ has an impact on the technology effect of emission reduction. This indicator is expressed as the ratio of research and development expenditures to the output. The data on research and development expenditures come from the China Statistical Yearbook on Science and Technology. A higher $ACT_{i,t}$ usually indicates greater CO_2 emissions of the sector. The intensity of economic activities is expressed in terms of electric power consumption, according to Peng et al. (2013) [57].

3.3. Derivation of SRIO Model for Embodied Emissions Measurement

Based on the overall test of all industries in China's industrial sector, we attempt to conduct a classification test in order to obtain the results of different industries. It is common to divide HCSs and LCSs by the CO_2 emissions' intensity of sectors. Fu and Zhang (2014) directly estimated the CO_2 emissions' intensity as a criterion to divide HCSs and LCSs [58]. Although the cross-sector differences in CO_2 emissions can be directly determined, this method provides no differentiation between CO_2 emissions due to domestic or foreign production activities in the context of international trade effect. Different from their studies, we tried to calculate the CO_2 emissions embodied in imports and exports and net embodied emissions, respectively, to divide the sectors into HCSs and LCSs. Pertinently, this method contains the effect of trade openness on CO_2 emissions.

The most common method for estimating CO_2 emissions embodied in trade is the IO. This model was first proposed by US economist Leontief for national economic accounting. Through constant extension and integration with other economic accounting techniques and optimal control theory, the IO model has found wide application. Based on research by Ahmad and Wyckoff (2003) and Lin and Sun (2010) [9,11], we derived the SRIO model based on open economic operation logic. According to classical IO theory, the total output was calculated:

$$X = AX + Y, \text{ i.e., } X = (I - A)^{-1} \cdot Y$$

where X is the column vector of total output; A is the direct consumption coefficient matrix; Y is final consumption, namely final demand; and $(I - A)^{-1}$ is the Leontief inverse matrix, or the complete demand coefficient matrix.

The above formula represents the operation of a completely closed economy. In contrast, for open economies, parts of the imported goods (including services) are consumed by final users, and the remains become intermediate input, which enters domestic production link. The direct consumption coefficient matrix A for an open economy is $A = A^d + A^{im}$. For the part of the imported goods used as intermediate input, the direct consumption coefficient matrix is $A^{im} = MA$, where M is the diagonal matrix of the proportion of imported goods in the direct consumption coefficient; for the domestic products used as intermediate input, the direct consumption coefficient matrix is $A^d = (I - M)A$. In the diagonal matrix M, element m_{ii} is calculated by $m_{ii} = x_i^{im}/(x_i + x_i^{im} - y_i^{ex})$, where x_i^{im} is imported goods, x_i is total output, and y_i^{ex} is exported goods. Total import X^{im} is divided into two parts based on the flow of production and the consumption link, including the imported goods as intermediate input, $A^{im}X$, and the imported goods consumed domestically, Y^{im}. The intermediate input is transformed into final product, $A^{im}(I - A)^{-1}Y$, after domestic production, and, therefore, the total import is expressed as $X^{im} = A^{im}(I - A)^{-1}Y + Y^{im}$.

The intermediate input for an open economy is from the products of domestic or foreign industrial sectors. However, CO_2 emissions are generated throughout the production chain, ranging from production to transportation, to consumption; these emissions are referred to as the embodied CO_2 emissions. The CO_2 emission coefficient for sector j is divided into the direct emission coefficient e_j^d and the embodied emission coefficient E_j^d; and the embodied emission coefficient $E^d = e^d(I - A)^{-1}$. The goods produced and consumed by an open economy are divided as either imported or exported goods. Based on where the goods are produced and consumed, the embodied CO_2 emissions are divided into four categories (Lin and Sun, 2010; Yan and Zhao, 2012) [11,59]. Y^{ex} indicates the goods that are domestically produced and intended for export. The four categories of embodied CO_2 emissions are represented by I, II, III, and IV, respectively, as shown in Table 1.

Table 1. Four categories of embodied CO_2 emissions.

Classification	Domestically Consumed	Consumed in Foreign Countries
Domestically produced	I	II
Produced in foreign countries	III	IV

Category I is the CO_2 emissions embodied in goods that are domestically produced and consumed: $E^d(Y - Y^{ex})$;

Category II is the CO_2 emissions embodied in goods that are domestically produced and consumed in foreign countries: $E^d Y^{ex}$;

Category III is the CO_2 emissions embodied both in imported goods that are domestically consumed and in the domestically consumed goods transformed from the imported goods as intermediate input in domestic production: $E^d Y^{im} + E^d A^{im}(I - A)^{-1}(Y - Y^{ex})$;

Category IV is the CO_2 emissions embodied in the exported goods that are produced domestically using the imported goods as intermediate input: $E^d A^{im}(I - A)^{-1}Y^{ex}$.

The sum of category II and IV is the CO_2 emissions embodied in the exported goods, and the sum of category III and IV is the CO_2 emissions embodied in the imported goods. The net CO_2 emissions embodied in export Q^{net} are calculated by deducting the CO_2 emissions embodied in the imported goods from the CO_2 emissions embodied in the exported goods:

$$Q^{net} = E^d Y^{ex} - E^d Y^{im} - E^d A^{im} (I - A)^{-1} (Y - Y^{ex})$$

4. Test Based on Industrial Sector Classification

4.1. Empirical Test of China's Industrial Sector as a Whole

We conducted our empirical analysis using Eviews 6.0 and Matlab. EViews is the abbreviation of Econometrics Views, which is developed by economists and mainly used in the field of economics, and MATLAB is a commercial mathematics software produced by MathWorks Company in the United States. Firstly, we performed a data stationary test. An LLC test and a Fisher-ADF test indicated that the data of each indicator were non-stationary panel data under the 1% significance level. The first-order differential data were stationary. Then, the variable combinations were subjected to a co-integration test. Here, we used the Hausman test to determine whether the fixed effects model or the random effects model fit under the 5% significance level. The data of the cross term were subjected to centering to reduce the errors in regression caused by collinearity [60].

The results of empirical analysis using panel data of China's 20 sectors are shown in Table 2. The first four columns (1)–(4) are the results of regression of the intensity of CO_2 emissions in each sector; the last four columns (5)–(8) are the results of regression on total CO_2 emissions. The regression of the control variables in (1) and (5) indicates that the coefficient is statistically significant at the 1% level. After introducing $TOI_{i,t}$ into (2) and (6), the regression fitting degree was improved, indicating the significance of trade openness in influencing CO_2 emissions. Moreover, high trade openness leads to a reduction in both CO_2 emissions' intensity and total CO_2 emissions. As to the effect of the control variable on CO_2 emissions in each sector, the coefficient for intensity, $ACT_{i,t}$, is always positive, hence, the more active the economic activities, the higher the CO_2 emissions' intensity and total CO_2 emissions. The coefficient for the number of employees in the sector $POP_{i,t}$ is positive in the regression of carbon emission intensity and negative in the regression of total CO_2 emissions, which indicates that higher $POP_{i,t}$ will reduce the CO_2 emissions' intensity in the sector but increase total CO_2 emissions. Higher intensity of research and development will reduce the CO_2 emissions of the sector, but, when considering other influencing factors, the coefficient becomes positive and insignificant.

Carbon leakage via international trade was tested. If the regression coefficient of the sector's development level is positive and the regression coefficient of the cross term of trade openness is negative, this means there is no carbon leakage, and China is not a pollution haven for HCSs. According to the regression, the coefficient of the added value of the industrial sector is positive, as might have been expected. Thus, higher added value of the industrial sector increases CO_2 emissions. The coefficient of the cross term between trade openness and the added value of the industrial sector is negative but insignificant, which means that, when either the CO_2 emissions intensity or the total CO_2 emissions is used, carbon leakage result cannot be proven. Testing of the EKC hypothesis by regression indicates that the coefficient of the linear term is negative and that of the quadratic term is positive. This is inconsistent with the EKC hypothesis, and the regression coefficient is not statistically significant. From this we can infer that the variations of CO_2 emissions in China's 20 sectors are characterized by uncertainty. China has greatly diverse sectors, with huge cross-sector differences. Lin (2013) found that the influence of foreign investment on China's CO_2 emissions varies greatly and has very different characteristics in different sectors [61]. So, to further analyze the effect of trade openness on CO_2 emissions in different sectors, we carried out our empirical analysis by dividing the sectors into HCSs and LCSs.

Table 2. Results of regression for all sectors.

Variable	Carbon Emission Intensity in the Sector				Total CO$_2$ Emissions in the Sector			
	(1)	(2)	(3)	(4)	(5)	(6)	(7)	(8)
Constant	4.614 ***	4.704 ***	5.075 ***	4.944 ***	4.614 ***	4.703 ***	5.075 ***	4.942 **
	(16.767)	(17.456)	(20.664)	(2.679)	(16.768)	(17.457)	(20.665)	(2.678)
lnPOP	−0.375 ***	−0.399 ***	−0.666 ***	−0.482 ***	0.624 ***	0.600 ***	0.333 ***	0.517 ***
	(−5.916)	(−6.422)	(−9.777)	(−8.593)	(9.858)	(9.650)	(4.885)	(9.218)
lnTOR	−0.080 ***	−0.045 **	0.002	0.002	−0.080 ***	−0.045 **	0.002	0.002
	(−3.697)	(−1.902)	(0.130)	(0.094)	(−3.697)	(−1.903)	(0.129)	(0.093)
lnACT	0.267 ***	0.228 ***	0.158 ***	0.158 ***	0.267 ***	0.228 ***	0.158 ***	0.158 ***
	(7.124)	(−3.106)	(4.412)	(4.527)	(7.127)	(5.931)	(4.414)	(4.529)
lnTOI		−0.115 ***	−0.027			−0.115 ***	0.158	
		(−3.106)	(−0.788)			(−3.106)	(4.414)	
lnIVA			0.180 ***				0.180 ***	
			(6.570)				(6.570)	
lnTOI·IVA			−0.001				−0.001	
			(−1.069)				(−1.069)	
ln$Y$				−0.105				−0.104
				(−0.362)				(−0.361)
$(\ln Y)^2$				0.011				0.011
				(1.029)				(1.028)
Adjusted R-squared	0.9969	0.9971	0.9977	0.9977	0.9970	0.9972	0.9978	0.9978
Mode Type	Fixed Effect	Fixed Effect	Fixed Effect	Fixed Effect	Fixed Effect	Fixed Effect	Fixed Effect	Fixed Effect

Note: (1) The symbol in parentheses is the *t* value; (2) *, **, and *** indicate either that the data are statistically significant or that the null hypothesis is rejected at the 10%, 5%, and 1% significance levels, respectively.

4.2. Embodied Emissions Estimation and Industrial Sector Classification

China's 2010 Input–Output Table and the China Energy Statistical Yearbook 2011 were used as the data sources. Combining the output and export data of 20 sectors from the DRCnet database, the diagonal matrix of export coefficient $\overline{M}$ was calculated. To reduce the errors in CO$_2$ emissions' calculation that appeared during the conversion into either standard coal or solid–liquid–gas energy, we estimated CO$_2$ emissions using 18 physical quantities of energy consumption and the corresponding emission coefficients in the China Energy Statistical Yearbook 2011. IPCC (2006) provided the formula of CO$_2$ emission coefficient of energy unit identification: $\theta_i = NCV_i \times CC_i \times COF_i \times (44/12)$, where NCV_i is the average low heating value; CC_i is the CO$_2$ emission factor; and COF_i is the carbon oxidation factor. The default value is 1 according to IPCC, and 44 and 12 are the molecular weights of carbon dioxide and carbon, respectively. The CO$_2$ emissions of 20 sectors were estimated to calculate the intensity of direct CO$_2$ emissions, and then the CO$_2$ emissions embodied in the exported goods and the imported goods were calculated for each sector to finally obtain the intensity of CO$_2$ emissions embodied in net export.

The average intensity of CO$_2$ emissions is usually used to divide HCSs and LCSs. The average intensity of net CO$_2$ emissions embodied in export is 0.237tCO$_2$/ten thousand RMB. Using this criterion, seven sectors were classified as HCSs, namely, production and supply of electric power and thermal power, coal mining and dressing, non-metallic mineral products, metal smelting and calendaring, petroleum processing, coking and nuclear fuel processing, the chemical industry, paper making, printing and stationery, and sporting goods. The remaining 13 sectors are LCSs. Fu and Zhang (2015) used the intensity of direct CO$_2$ emissions as the criterion [58], and, as the average intensity in China's sectors is 0.3718 tCO$_2$/ten thousand RMB, classified six sectors as sectors with high intensity of direct CO$_2$ emissions; these are coal mining and dressing, non-metallic mineral products, metal smelting and calendaring, petroleum processing, coking and nuclear fuel processing, the chemical industry, paper making, printing and stationery, and sporting goods.

Although both the gas production and supply industry and the oil and gas exploration industry are also sectors with high intensity of direct CO$_2$ emissions, they are not classified as HCSs. This is probably because the direct CO$_2$ emissions in these two sectors are not caused by trade. We calculated the net CO$_2$ emissions embodied in export, aiming to reveal the relationship between trade openness and CO_2 emissions. The scatter plot showing the relationship between trade openness and CO_2 emissions' intensity was drawn for HCSs and LCSs (Figure 2). It is obvious that the trade openness is positively correlated with the CO_2 emissions' intensity for HCSs, with a mild linear increasing trend; however, for LCSs, the trade openness is negatively correlated with the CO_2 emissions' intensity. Thus, the relationship between trade openness and intensity of carbon emissions varies for different sectors, which shows that the empirical results of industrial industry as a whole are inadequate and need to be classified and tested. Meanwhile, it is proven from another perspective that the effect of trade openness on CO_2 emissions is uncertain for China's industrial sectors on the whole.

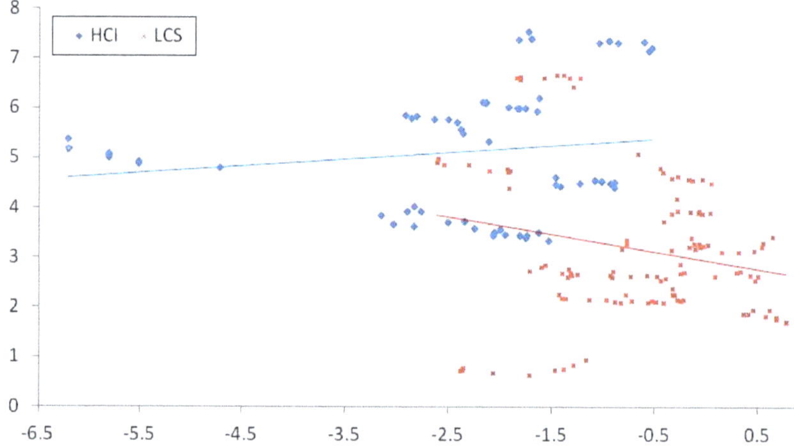

Figure 2. Scatter plot of trade openness vs. CO_2 emissions' intensity for high-carbon sectors (HCSs) and low-carbon sectors (LCSs).

4.3. Empirical Analysis for HCSs and LCSs

In Table 3, columns (1)–(4) and (5)–(8) are the regression results of CO_2 emissions' intensity for 7 HCSs and 13 LCSs. After introducing $TOI_{i,t}$ into the two equations, the fitting degree was improved, indicating the significance of trade openness to CO_2 emissions, which is consistent with the regression result of all sectors (ACSs). For LCSs, higher trade openness leads to a reduction in the CO_2 emissions intensity; however, for HCSs, the coefficient is positive and statistically significant at the 10% level for $TOI_{i,t}$, indicating that higher trade openness increases the CO_2 emissions' intensity for HCSs. The regression results of control variables are basically consistent with the regression for ACSs. The economic activities' intensity varies in the same direction as the CO_2 emissions' intensity for the sector. However, the situation is the opposite for the number of employees in the sector, and the intensity of research and development has no significant impact on the CO_2 emissions' intensity.

Table 3. Regression analysis of CO_2 emissions' intensity in HCSs and LCSs.

Variable	HCSs				LCSs			
	(1)	(2)	(3)	(4)	(5)	(6)	(7)	(8)
Constant	4.155 ***	4.411 ***	6.374 ***	7.646 ***	4.009 ***	4.110 ***	4.245 ***	0.213
	(6.113)	(6.541)	(10.784)	(3.753)	(12.627)	(12.767)	(13.603)	(0.054)
lnPOP	−0.331 **	−0.383 **	−0.870 ***	−0.686 ***	−0.356 ***	−0.372 ***	−0.548 ***	−0.424 ***
	(−2.055)	(−2.405)	(−6.206)	(−5.853)	(−5.360)	(−5.570)	(−6.498)	(−6.405)
lnTOR	−0.061 *	−0.027	0.006	0.018	−0.046	−0.029	0.010	−0.009
	(−1.980)	(−0.787)	(0.228)	(0.758)	(−1.512)	(−0.907)	(0.321)	(−0.303)
lnACT	0.428 ***	0.373 ***	0.203 ***	0.158 ***	0.236 ***	0.213 ***	0.147 ***	0.183 ***
	(5.010)	(4.261)	(2.851)	(4.527)	(5.897)	(5.023)	(3.388)	(4.347)
lnTOI		0.117 *	0.043	0.161 **		−0.073 ***	−0.028	
		(1.975)	(0.814)	(2.301)		(−1.55)	(−0.624)	
lnIVA			0.265 ***				0.134 ***	
			(6.497)				(3.530)	
ln$TOI{\cdot}IVA$			0.001				−0.031 **	
			(0.095)				(−2.250)	
ln$Y$				−0.284				0.565 *
				(−1.004)				(1.908)
(ln$Y$)^2				0.021 *				−0.017 **
				(1.909)				(−3.702)
Adjusted R-squared	0.9922	0.9955	0.9975	0.7605	0.9903	0.9963	0.9967	0.3207
Mode Type	Fixed Effect	Fixed Effect	Fixed Effect	Random Effect	Fixed Effect	Fixed Effect	Fixed Effect	Random Effect

Note: (1) The symbol in parentheses is the t value; (2) *, **, and *** indicate either that the data are statistically significant or that the null hypothesis is rejected at the 10%, 5%, and 1% significance levels, respectively.

As to the testing of carbon leakage, the regression coefficient is positive for the added value of both types of sectors, which is consistent with ACSs. This means that the higher added value of the industrial sectors causes the increase in the CO_2 emissions' intensity. The coefficient of the cross term between trade openness and added value of LCSs is negative and statistically significant at the 5% level, indicating the absence of carbon leakage. However, the coefficient of the cross term is positive and statistically insignificant for HCSs, indicating the possible existence of carbon leakage (albeit uncertain). The results again demonstrate the above uncertainty of carbon leakage. The coefficient is positive for the linear term of per capita income and negative for the quadratic term in the regression for LCSs—significant at the 10% level—so the EKC hypothesis is confirmed for LCSs. However, for HCSs, the coefficient of the linear term is negative, and that of the quadratic term is positive, which does not agree with the EKC hypothesis. Moreover, the coefficient of the linear term is statistically insignificant. The testing of EKC hypothesis for the two types of sectors also proves the uncertainty of the existence of the EKC. In other words, higher trade openness leads to a reduction in CO_2 emissions' intensity for both LCSs and HCSs, but the existence of carbon leakage and EKC is uncertain for both LCSs and HCSs, which provides valuable information for making decisions on trade openness and industrial policies with the purpose of energy saving and emission reduction.

5. Robustness Test

5.1. Cross Term Between Trade Openness and Per Capita Output

A robustness test was performed subsequently. The sector's development level was measured by per capita output $GPC_{i,t}$ of the sector, and then we tested the existence of carbon leakage for ACSs, HCSs, and LCSs using the cross term $TOI_{i,t} \cdot GPC_{i,t}$ between trade openness and per capita output.

The regression results are shown in Table 4. Due to limited space, the coefficients of constant term and control variables are not reported. We can see that the regression coefficient is positive and statistically significant for per capita output of the sector. Hence, higher per capita output increases the CO_2 emissions' intensity. The coefficient of the cross term between trade openness and per capita output is negative and statistically insignificant for ACSs, demonstrating the uncertainty of carbon leakage. The coefficient of the cross term between trade openness and per capita output is positive and statistically insignificant for HCSs, indicating the possible existence of carbon leakage in HCSs (albeit uncertain). The coefficient of the cross term is negative and statistically significant for LCSs, indicating non-existence of carbon leakage for LCSs. These regression results are the same as that of the cross term between trade openness and added value, which proves good robustness.

Table 4. Regression analysis of cross term between trade openness and per capita output.

Variable	ACSs	HCSs	LCSs
$\ln TOI$	−0.008 **	0.094 **	−0.023
	(−0.237)	(1.963)	(−0.503)
$\ln GPC$	0.187 ***	0.278 ***	0.141 ***
	(6.685)	(7.362)	(3.478)
$\ln TOI{\cdot}GPC$	−0.001	0.029	−0.038 *
	(−1.379)	(0.170)	(−1.812)
Adjusted R-squared	0.9977	0.9980	0.9967
Mode Type	Fixed Effect	Fixed Effect	Fixed Effect

Note: (1) The symbol in parentheses is the *t* value; (2) *, **, and *** indicate either that the data are statistically significant or that the null hypothesis is rejected at the 10%, 5%, and 1% significance levels, respectively.

5.2. Testing of the Pollution Haven Hypothesis

There is no carbon leakage in LCSs, but there is probably carbon leakage in HCSs. For ACSs, carbon leakage is more likely to be absent. To further test this conclusion, we tried to use the method by Halkos (2003) and Peng et al. (2013) by introducing the dummy variables [57,62]. As a result of higher trade openness, the countries (regions) or sectors implementing lower environmental regulations tend to be pollution havens for HCSs. Therefore, the CO_2 emissions' intensity can reflect the intensity of implementation of the environmental regulations. Pollution havens will be those countries with lower intensity of implementation of environmental regulations.

$$\ln CEM_{i,t} = \lambda Z_{i,t} + \phi_1 \ln TOI_{i,t} + \phi_2 \ln TOI_{i,t} \cdot CEM_{i,t} DUM_{i,t} + \mu_{i,t}$$

where $CEM_{i,t}$ is total CO_2 emissions (or emission intensity) of the sector; $Z_{i,t}$ are the variables, except for $\ln TOI_{i,t}$; the cross term is the product of the dummy variable $CEM_{i,t} DUM_{i,t}$ of the CO_2 emission intensity and trade openness. For HCSs, the value of the dummy variable is 1; otherwise, it is 0. Thus, for LCSs and HCSs, the effect of trade openness on CO_2 emissions is ϕ_1 and $\phi_1 + \phi_2$, respectively. If $\phi_1 \prec 0$, $\phi_1 + \phi_2 \succ 0$, then higher trade openness leads to environmental improvement for sectors implementing higher environmental regulations and having lower CO_2 emissions' intensity, but it is not conducive to environmental improvement for sectors implementing lower environmental regulations and having higher CO_2 emissions' intensity. According to our results, $\phi_1 = -0.155$, $\phi_1 + \phi_2 = 0.030$. Thus, higher trade openness leads to reduction in CO_2 emissions' intensity for LCSs, but increases CO_2 emissions' intensity for HCSs. The pollution haven hypothesis is confirmed for China's HCSs, which are the destination for the transfer of higher-carbon sectors from the developed countries.

6. Conclusions and Policy Suggestions

We carried out an empirical analysis of the effect of trade openness on CO_2 emissions and tested the existence of carbon leakage for all and classified industrial sectors in China, by building a panel data model and deriving a single-region input–output model for the classification of high-carbon sectors (HCSs) and low-carbon sectors (LCSs). Firstly, the empirical test results of the industry as a whole and the classification industry in China's industrial sectors have confirmed that higher trade openness leads to reduction of both CO_2 emissions' intensity and gross CO_2 emissions in the industrial sectors. So, China's policy makers do not need to worry too much about the adverse impact of trade openness on the environment. Instead, the environmental standards should be raised to further benefit the process of environmental improvement in an open economy. Some countries have discussed new standards and a new trade mode with greater emphasis on environmental cooperation, which represents both a challenge and an opportunity. As China is faced with an urgent need for energy saving and environmental protection, raising the environmental standards is not only a pathway to achieving economic transition, but also a means of strengthening the long-term competitiveness of China's exports. This will also reduce the pressure on China, as the world's largest CO_2 emitter, in climate negotiation.

Secondly, analysis based on industrial sector classification indicates dramatic cross-sector difference in CO_2 emissions' intensity. Higher trade openness can help reduce the CO_2 emissions' intensity. For China's industrial sectors on the whole, no carbon leakage exists, whereas EKC does. For HCSs, there is no EKC, but carbon leakage probably occurs. The pollution haven hypothesis was confirmed by introducing the dummy variable. However, the divergent conclusions reached for LCSs and HCSs point to the uncertainty of carbon leakage. China should encourage structural adjustment through relevant policies during green low-carbon development, as well as the transition and upgrading of exports. Since net CO_2 emissions embodied in export tend to vary from one sector to another, differential policies with structural features should be adopted to perform structural adjustment, and government can place tighter restraint on HCSs to appropriate to achieve low carbon development of industrial structure. This is in accord with the economic transition taking place under the new normality of the Chinese economy.

Thirdly, the regression analysis of the control variables also provides some clues. The Chinese government has announced that China's CO_2 emissions will peak around 2030 and efforts will be made to reach this peak sooner. Reducing the CO_2 emissions embodied in export is favorable to acquire the CO_2 emission space for Chinese economic development. From the empirical results, China's industrial sector has an optimized space for improving the structure of human capital, as the reduction effect of China's human resources is not obvious and higher population density will enhance the response of environmental policies to the income level [63]. As a country with a large population, China needs to further optimize its human resource structure and play its role in reducing emissions. In addition, the impact of R&D investment in China's industrial sector on reducing carbon intensity has not yet emerged clearly. Therefore, enhancing the emission reduction techniques is an important pathway to optimizing not only energy efficiency and the carbon production rate, which are a qualitative index, but also optimizing the number of high- and low-carbon industries, which is a quantitative index.

Author Contributions: Conceptualization, B.F., Y.Z., X.L. and X.M.; Methodology, X.L. and X.M.; Software, Y.Z. and X.L.; Validation, B.F., X.L. and X.M.; Formal Analysis, B.F.; Investigation, B.F. and Y.Z.; Resources, B.F.; Data Curation, B.F. and Y.Z.; Writing-Original Draft Preparation, B.F. and Y.Z.; Writing-Review & Editing, X.L. and X.M.; Visualization, B.F.; Supervision, X.L. and X.M.; Project Administration, X.L. and X.M.; Funding Acquisition, Y.Z., X.L. and X.M.

Funding: This research was funded by the *Shanghai Pujiang Program* (No.16PJC065) and the *"Shuguang Program"* (No.16SG51) supported by Shanghai Education Development Foundation and Shanghai Municipal Education Commission, and China Postdoctoral Science Foundation (NO. 2017M611424), and Shanghai Philosophy and Social Science Youth Project Research (No. 2014EJB001), and Humanities and social sciences youth research project by Ministry of Education of China (NO. 17YJC790081) and the Shanghai High-level University Development Project.

Conflicts of Interest: The authors declare no conflict of interest. The founding sponsors had no role in the design of the study; in the collection, analyses, or interpretation of data; in the writing of the manuscript, and in the decision to publish the results.

References

1.	Antweiler, W.; Copeland, B.R.; Taylor, M.S. Is free trade good for the environment? *Am. Econ. Rev.* **2001**, *91*, 877–908. [CrossRef]

2.	Cole, M.A.; Elliott, R.J.R. Determining the trade-environment composition effect: The role of capital, labor and environmental regulations. *J. Environ. Econ. Manag.* **2003**, *46*, 363–383. [CrossRef]

3.	Managi, S.; Hibiki, A.; Tsurumi, T. Does trade openness improve environmental quality? *J. Env. Econ. Manag.* **2009**, *3*, 346–363. [CrossRef]

4. Peng, S.; Bao, Q. Economic growth and environmental pollution: An empirical test for the environmental Kuznets curve hypothesis in China. *Res. Financ. Econ. Issues* **2006**, *8*, 3–17.
5. Shui, B.; Harriss, R.C. The role of CO_2 embodiment in US–China trade. *Energy Policy* **2006**, *34*, 4063–4068. [CrossRef]
6. Weber, C.L.; Peters, G.P.; Guan, D.; Hubacek, K. The contribution of Chinese exports to climate change. *Energy Policy* **2008**, *9*, 3572–3577. [CrossRef]
7. Su, B.; Ang, B.W. Input-output analysis of CO_2 emissions embodied in trade: Competitive versus non-competitive imports. *Energy Policy* **2013**, *56*, 83–87. [CrossRef]
8. Wu, S.M.; Wu, Y.R.; Lei, Y.L.; Li, S.H.; Li, L. Chinese provinces' CO_2 emissions embodied in imports and exports. *Earth's Future* **2018**, *6*, 867–881. [CrossRef]
9. Ahmad, N.; Wyckoff, A. *Carbon Dioxide Emissions Embodied in International Trade of Goods*; Technology and Industry Working Papers; OECD Science: Paris, France, 2003.
10. Yan, Y.F.; Yang, L.K. China's foreign trade and climate change: A case study of CO_2 emissions. *Energy Policy* **2010**, *1*, 350–356.
11. Lin, B.; Sun, C. Evaluating carbon dioxide emissions in international trade of China. *Energy Policy* **2010**, *1*, 613–621. [CrossRef]
12. Du, H.; Guo, J.; Mao, G.; Smith, A.M.; Wang, X.; Wang, Y. CO_2 emissions embodied in China–US trade: Input–output analysis based on the emergy/dollar ratio. *Energy Policy* **2011**, *39*, 5980–5987. [CrossRef]
13. Chen, Z.M.; Chen, G.Q. Embodied carbon dioxide emission at supra-national scale: A coalition analysis for G7, BRIC, and the rest of the world. *Energy Policy* **2011**, *39*, 2899–2909. [CrossRef]
14. Chen, L.; Yu, S. The co-integration analysis of carbon emission and service trade: From China's empirical analysis. *Adv. Inf. Sci. Ser. Sci.* **2012**, *9*. [CrossRef]
15. Lin, J.; Pan, D.; Davis, S.J.; Zhang, Q.; He, K.; He, K.; Wang, C.; Streets, D.G.; Wuebbles, D.J.; Guan, D. China's international trade and air pollution in the United States. *Proc. Natl. Acad. Sci. USA* **2014**, *5*, 1736–1741. [CrossRef]
16. Tao, S.; Zheng, T.; Lianjun, T. An empirical test of the environmental Kuznets curve in China: A panel cointegration approach. *China Econ. Rev.* **2008**, *3*, 381–392.
17. Song, M.L.; Zhang, W.; Wang, S.H. Inflection point of environmental Kuznets curve in mainland China. *Energy Policy* **2013**, *57*, 14–20. [CrossRef]
18. Liu, H.; Kim, H.; Liang, S.; Kwon, O.-S. Export diversification and ecological footprint: A comparative study on EKC theory among Korea, Japan, and China. *Sustainability* **2018**, *10*, 1–12. [CrossRef]
19. Shen, J. A simultaneous estimation of environmental Kuznets curve: Evidence from China. *China Econ. Rev.* **2006**, *4*, 383–394. [CrossRef]
20. Jalil, A.; Mahmud, S.F. Environment Kuznets curve for CO_2 emissions: A cointegration analysis for China. *Energy Policy* **2009**, *37*, 5167–5172. [CrossRef]
21. Diao, X.D.; Zeng, S.X.; Tam, C.M.; Tam, V.W. EKC analysis for studying economic growth and environmental quality: A case study in China. *J. Clean. Prod.* **2009**, *17*, 541–548. [CrossRef]
22. He, J.; Wang, H. Economic structure, development policy and environmental quality: An empirical analysis of environmental Kuznets curves with Chinese municipal data. *Ecol. Econ.* **2012**, *76*, 49–59. [CrossRef]
23. Liu, L. Environmental poverty, a decomposed environmental Kuznets curve, and alternatives: Sustainability lessons from China. *Ecol. Econ.* **2012**, *73*, 86–92. [CrossRef]
24. Ho, J.W.; Lee, A.M.; Macfarlane, D.J.; Fong, D.Y.; Leung, S.; Cerin, E. An empirical study of the environmental Kuznets curve in Sichuan province, China. *Environ. Pollut.* **2013**, *2*, 204–209.
25. Yin, J.; Zheng, M.; Chen, J. The effects of environmental regulation and technical progress on CO_2 Kuznets curve: An evidence from China. *Energy Policy* **2015**, *77*, 97–108. [CrossRef]
26. Barker, T.; Junankar, S.; Pollitt, H.; Summerton, P. Carbon leakage from unilateral environmental tax reforms in Europe, 1995–2005. *Energy Policy* **2007**, *12*, 6281–6292. [CrossRef]
27. Kuik, O.; Hofkes, M. Border adjustment for European emissions trading: Competitiveness and carbon leakage. *Energy Policy* **2010**, *38*, 1741–1748. [CrossRef]
28. Elliott, J.; Fullerton, D. Can a unilateral carbon tax reduce emissions elsewhere? *Res. Energy Econ.* **2014**, *36*, 6–21. [CrossRef]
29. Böhringer, C.; Rosendahl, K.E.; Storrøsten, H.B. Robust policies to mitigate carbon leakage. *J. Public Econ.* **2017**, *149*, 35–46. [CrossRef]

30. Felder, S.; Rutherford, T.F. Unilateral CO_2 reductions and carbon leakage: The consequences of international trade in oil and basic materials. *J. Environ. Econ. Manag.* **1993**, *25*, 162–176. [CrossRef]

31. Smith, C. Carbon leakage: An empirical assessment using a global econometric model. *Int. Compet. Environ. Policies* **1998**, 143–169.

32. Paltsev, S.V. The Kyoto Protocol: Regional and sectoral contributions to the carbon leakage. *Energy J.* **2001**, *22*, 53–79. [CrossRef]

33. Aukland, L.; Costa, P.M.; Brown, S. A conceptual framework and its application for addressing leakage: The case of avoided deforestation. *Clim. Policy.* **2003**, *2*, 123–136. [CrossRef]

34. Gerlagh, R.; Kuik, O. Carbon leakage with international technology spillovers. *SSRN Electr. J.* FEEM Working paper No. 33. **2007**. [CrossRef]

35. Rosendahl, K.E.; Strand, J. Carbon Leakage from the Clean Development Mechanism. Discussion Papers. 2009. Available online: https://www.researchgate.net/publication/4645542 (accessed on 1 March 2019).

36. Eichner, T.; Pethig, R. Carbon leakage, the green paradox, and perfect future markets. *Int. Econ.Rev.* **2011**, *52*, 767–805. [CrossRef]

37. Baylis, K.; Fullerton, D.; Karney, D.H. Leakage, welfare, and cost-effectiveness of carbon policy. *Am. Econ. Rev.* **2013**, *103*, 332–337. [CrossRef]

38. Carbone, J.C. Linking numerical and analytical models of carbon leakage. *Am. Econ. Rev.* **2013**, *103*, 326–331. [CrossRef]

39. Fullerton, D.; Karney, D.; Baylis, K. Negative leakage. *Natl. Bur. Econ. Res. Work. Papers* **2011**, *1*, 51–73.

40. Winchester, N.; Rausch, S. A numerical investigation of the potential for negative emissions leakage. *Am. Econ. Rev.* **2013**, *3*, 320–325. [CrossRef]

41. Lin, J.; Yang, L.K. FDI, export and carbon dioxide emissions: A comparative study with east. *J. Int. Trade.* **2013**, *10*, 129–137.

42. Grossman, G.M.; Krueger, A.B. *Environmental Impacts of the North American Free Trade Agreement*; NBER Working Paper No. 3914; National Bureau of Economic Research: Cambridge, MA, USA, 1991; p. 9.

43. Li, K.; Huang, J.L. Openness to trade in China's manufacturing sector. *J. World Econ.* **2006**, *8*, 11–22.

44. Zhao, J.W.; Ding, L.T. Trade opening, external shocks and Inflation: Based on the analysis of nonlinear models STR. *J. World Econ.* **2012**, *9*, 61–83.

45. Ma, Y.; Li, J.; Yu, G.S. Trade openness, economic growth and restructuring of labor-intensive industries. *J. Int. Trade.* **2012**, *9*, 96–107.

46. Bao, Q.; Xu, H.L.; Lai, M.Y. Trade opening and economy development: Theory and empirical from China. *World Econ.* **2003**, *2*, 10–18.

47. De Bruyn, S.M.; van den Bergh, J.C.J.M.; Opschoor, J.B. Economic growth and emissions: Reconsidering the empirical basis of environmental Kuznets Curves. *Ecol. Econ.* **1998**, *25*, 161–175. [CrossRef]

48. Dinda, S.; Coondoo, D. Income and emission: A panel-data based cointegration analysis. *Ecol. Econ.* **2006**, *57*, 167–181. [CrossRef]

49. Soytas, U.; Sari, R. Energy consumption, economy growth and carbon emissions: Challenges faced by an EU candidate member. *Ecol. Econ.* **2009**, *6*, 1667–1675. [CrossRef]

50. Dinda, S. Environmental Kuznets curve hypothesis: A survey. *Ecol. Econ.* **2004**, *49*, 431–455. [CrossRef]

51. Richard, K.; Piergiuseppe, F. International trade and carbon emissions. *Eur. J. Dev.t Res.* **2012**, *24*, 509–529.

52. National Bureau of Statistics, People's Republic of China. *China Statistical Yearbook*; China Statistics Press: Beijing, China, 2002–2013.

53. National Bureau of Statistics, People's Republic of China. *China Industry Economy Statistical Yearbook*; China Statistics Press: Beijing, China, 2002–2013.

54. National Bureau of Statistics, People's Republic of China. *China Population & Employment Statistics Yearbook*; China Statistics Press: Beijing, China, 2002–2013.

55. National Bureau of Statistics, People's Republic of China. *China Statistical Yearbook on Environment*; China Statistics Press: Beijing, China, 2002–2013.

56. National Bureau of Statistics, People's Republic of China. *China Energy Statistical Yearbook*; China Statistics Press: Beijing, China, 2002–2013.

57. Peng, S.J.; Zhang, W.C.; Cao, Y. Does composition effect of trade opening aggravate environment pollution in China? An empirical analysis based on panel data of prefecture cities. *J. Int. Trade.* **2013**, *8*, 119–132.

58. Fu, J.Y.; Zhang, C.J. International trade, carbon leakage and CO_2 emissions of manufacturing industry. *Chin. J. Popul. Res. Environ.* **2015**, *13*, 139–145.

59. Yan, Y.F.; Zhao, Z.X. CO_2 Emissions embodied in China's international trade: A perspective of allocating international responsibilities. *J. Int. Trade* **2012**, *1*, 131–142.

60. Aiken, L.S.; West, S.G.; Reno, R.R. *Multiple Regression: Testing and Interpreting Interactions*; Sage: Thousand Oaks, CA, USA, 1991.

61. Lin, B.Q.; Jiang, Z.J. Forecasting and influencing factors on Chinese carbon dioxide environmental Kuznets curve. *Manag. World* **2009**, *4*, 27–36.

62. Halkos, G.E. Environmental Kuznets curve for sulfur: Evidence using GMM estimation and random coefficient panel data models. *Environ. Dev. Econ.* **2003**, *4*, 581–601. [CrossRef]

63. Copeland, B.R.; Taylor, M.S. North-South trade and the environment. *Q. J. Econ.* **1994**, *109*, 755–787. [CrossRef]

Concept Paper

A Common Risk Classification Concept for Safety Related Gas Leaks and Fugitive Emissions?

Torgrim Log [1,2,3,*] **and Wegar Bjerkeli Pedersen** [3,4]

1. Department of Fire Safety and HSE Eng., Fire Disaster Research Group, Western Norway University of Applied Sciences, 5545 Haugesund, Norway
2. Equinor Kårstø, 5565 Kårstø, Norway
3. Equinor Hammerfest, 9601 Hammerfest, Norway; wpe@equinor.com
4. Department of Engineering and Safety, The Arctic University of Norway, 9019 Tromsø, Norway
* Correspondence: torgrim.log@hvl.no; Tel.: +47-900-50001

Received: 9 September 2019; Accepted: 22 October 2019; Published: 24 October 2019

Abstract: Gas leaks in the oil and gas industry represent a safety risk as they, if ignited, may result in severe fires and/or explosions. Unignited, they have environmental impacts. This is particularly the case for methane leaks due to a significant Global Warming Potential (GWP). Since gas leak rates may span several orders of magnitude, that is, from leaks associated with potential major accidents to fugitive emissions on the order of 10^{-6} kg/s, it has been difficult to organize the leaks in an all-inclusive leak categorization model. The motivation for the present study was to develop a simple logarithmic table based on an existing consequence matrix for safety related incidents extended to include non-safety related fugitive emissions. An evaluation sheet was also developed as a guide for immediate risk evaluations when new leaks are identified. The leak rate table and evaluation guide were tested in the field at five land-based oil and gas facilities during Optical Gas Inspection (OGI) campaigns. It is demonstrated how the suggested concept can be used for presenting and analysing detected leaks to assist in Leak Detection and Repair (LDAR) programs. The novel categorization table was proven valuable in prioritizing repair of "super-emitter" components rather than the numerous minor fugitive emissions detected by OGI cameras, which contribute little to the accumulated emissions. The study was limited to five land based oil and gas facilities in Norway. However, as the results regarding leak rate distribution and "super-emitter" contributions mirror studies from other regions, the methodology should be generally applicable. To emphasize environmental impact, it is suggested to include leaking gas GWP in future research on the categorization model, that is, not base prioritization solely on leak rates. Research on OGI campaign frequency is recommended since frequent coarse campaigns may give an improved cost benefit ratio.

Keywords: fugitive emissions; hydrocarbon leaks; optical Gas Imaging (OGI); leak detection and repair (LDAR)

---

1. Introduction

Natural gas is currently the third largest source of energy, covering about 20% of the world's primary energy demand. Unlike other fossil fuels, the global demand of natural gas is expected to increase over the next few decades. In the International Environment Agency (IEA) Sustainable Development Scenario in the World Energy Outlook 2018, designed to be fully aligned with the Paris Agreement goal of keeping the global average temperature rise below 2 °C, natural gas is expected to supply one quarter of the world's primary energy demand by 2040 [1]. The increasing demand is driven by the development in Asia, governmental policies of carbon-taxing and coal-to-gas switch in industries and buildings. The increasing supply is primarily met by the US shale gas revolution.

Methane is the most prominent gas in natural gas mixtures. During processing, the natural gas is dried, CO_2 and traces of inorganic materials—for example, sulphur and mercury—are removed as required by the customer specifications. Wet gas and rich gas are typically separated from the oil before the gas is treated in a gas processing facility. The wet gas or NGL (Natural Gas Liquids), contains a mixture of heavier gases (ethane, propane, butanes and naphtha), which may be further separated to produce clean gas components or gas mixtures of varying properties. Rich gas has a lower content of ethane and other heavier hydrocarbons. Dry gas consists mainly of methane with some ethane and only minor fractions of heavier hydrocarbons. In some facilities, gas containing only methane and some ethane is compressed and cooled to LNG (liquefied natural gas), while in other facilities the gas is used to produce other products, for example, methanol.

Natural gas and in particular its primary gas component methane, represents the lowest emitting fossil fuel when combusted. Natural gas therefore has many environmental advantages over coal and oil. The flexibility provided by the natural gas when converted to LNG also makes the transition to less CO_2 release from the energy consumers easier. It may therefore be concluded that natural gas will play a major role in the global energy transition towards a low-carbon future.

In all engineered systems, there is unfortunately a loss of energy. In for example, the electrical power grid system, 5% of the transmitted power is typically lost [2]. Processing and transporting natural gas, as well as compression and cooling of natural gas to produce LNG, require operations at high pressures. Transport pipelines may operate at pressures above 100 bar while production of LNG prior to cooling typically involves pressures in the range of 20–60 bar. Leaks therefore occur both in the processing and transport of gas to the markets. Unfortunately, methane is a potent greenhouse gas (GHG) with a global warming potential (GWP) of 28–36 times CO_2 in a 100-year time frame. Having both a shorter atmospheric lifetime and higher energy absorption potential than CO_2, methane's short-term global GWP is estimated to 84–87 [3,4]. The way the greenhouse gas emissions are assessed is not straight forward, as many factors are part of the picture [5,6]. It is, however, clear that uncontrolled methane leaks and emissions along the natural gas value chain may drastically reduce the climate benefits of natural gas over other fossil fuels, especially in the short term.

When studying fugitive emissions from the UK high-pressure pipeline transport system, that is, up to 85 bar pressure, Boothroyd et al. [7] detected both soil and air emissions. They concluded that the loss to the air accumulated to as much as 627 tonnes CH_4/km/yr (241–1123 tonnes CH_4/km/yr interquartile range). The estimated natural gas loss to the soil was 62.6 kt/year. For one particular pipeline, the average distance between leaks indicated that nearly all pipe welding joints were leaking. By investigating the Bangladesh Titas Gas distribution network by means of soap screening and a Gasurveyor 500 series instrument (Gas Measurement Instruments Ltd, Renfrew, Scotland, UK), Mandal and Morshed [8] found several leak sources. Scrubber dump valves and pressure relief valves were identified as the most severe leak sources with respectively average leak rates of 217 L/min and 438 L/min. Average leak rates of respectively 4.0, 8.0 and 1.6 L/min were found for insulation points, tube fittings and connectors. These examples indicate that the methane losses may be significant and may thereby reduce the environmental benefit of natural gas as a fuel compared to more carbon rich fuels.

According to Miller and Michalak [9], the current emission measurement and estimation methodologies are associated with high degrees of uncertainty. Results from top-down (e.g., remote atmospheric measurements) and bottom-up (e.g., direct component measurements) studies show significant differences and much research on this topic is initiated by environmental agencies and oil and gas operator consortiums. According to the IEA, the methane emissions must be lower than 3% for natural gas to represent a cleaner energy provider in the short-term than coal [10]. The general consensus is, however, that the global methane losses along the value chain are lower than 3%. Recent studies estimate that the methane emission from US oil and gas supply chain is close to 2.3% of the gross US gas production [11].

National environmental protection agencies (EPAs) are increasingly concerned about the accumulated leak rates in the oil and gas industry. In several countries, for example, Norway and USA, the EPAs want to quantify the leak rates as a basis for issuing environmental taxes. There are indications of methane leak rates being higher than the respective EPAs estimates [12,13]. Loss of methane gas to the atmosphere must be prevented as these losses counteract an improvement in GWP relative to carbon rich fuels.

Leak rates and greenhouse gas (GHG) footprint of both the Norwegian LNG and pipeline gas exported to Europe have been found to be well below recently reported EU averages [14]. The estimated CH_4 emissions were 0.01–0.04% of the production rate. But there are leaks and the Norwegian Environmental Agency (NEA) requires Differential Absorption Lidar (DIAL) recordings, which is also frequently used to record methane gas emissions from industry areas [15] and landfills [16]. In some special cases, the DIAL technique can also be used to record leaks from for example, large process units and storage tanks. However, to search for leaks on a component level, close-up methods are required.

It is well known internationally that a minor fraction of the emission sources dominates the accumulated methane emissions [3,17]. Rather than searching for numerous marginal small leaks on the order of 1×10^{-6} kg/s, to start looking for the "super-emitters" may be worthwhile. If the average period a few "super-emitters" are active is half a year, that is, before detected in a yearly OGI campaign, they may dominate the accumulated leak rate. Searching more frequently for "super-emitters and not devote so much time to the marginal leaks, could be very beneficial with respect to environmental impacts, safety and plant economy.

Methods that can be used near the leak sources may reveal exactly where the leaks occur and which leaks contribute most to the accumulated emissions, making it easier to prioritize repair of the "super-emitters" [3,17]. On the component level, since such "super-emitters" may emit orders of magnitude more gas than minor fugitive emissions, one may question whether a high accuracy is necessary, that is, over-engineering may be an issue [3]. Since the "super-emitters" contribute so much to the accumulated leak rates, it is important to locate and repair these.

In the present study, we have developed and tested a concept for improving the focus on the leaks contributing most to the accumulated emissions. An incident consequence matrix used by the Petroleum Safety Authority of Norway, as well as oil and gas companies operating on- and offshore in Norwegian territories, has been extended and tested as a tool also for categorizing fugitive emissions. It was also tested for communicating the importance of addressing "super-emitters" with less focus on a high number of minor leaks significantly smaller than these "super-emitters." A leak risk evaluation guide is presented and a discussion regarding too high focus on quantifying all detected leaks, rather than focusing on the major contributors, is included. Some comments are included regarding safety risk versus fugitive emissions and possible benefits of a common gas leak categorization system. The novelty of the paper is the concept presented and the way it can be implemented by operators and managers.

2. Gas Leak Detection in Processing Facilities

2.1. Fire Safety Related Properties of Natural Gas

Natural gas typically consists of 90–95 vol% methane (CH_4), 3–6 vol% ethane (C_2H_6), with some minor amounts of longer chain alkanes, for example, propane (C_3H_8), butane (C_4H_{10}) and so forth, as well as varying amounts of N_2, CO_2 and H_2S, which is generally removed when processing the gas. The molar mass of methane and ethane are 16.04 g/mol and 30.07 g/mol, respectively. A representative molar mass for natural gas may typically be 19 g/mol, while dry air has molar mass 28.97 g/mol. At 20 °C, the density of natural gas is about 0.79 kg/m^3 while the density of air is 1.21 kg/m^3. Unless very cold due to the expansion from a high-pressure source, that is, the Joule Thompson effect or resulting from for example, an evaporating LNG leak, a gas leak will be lighter than the ambient air.

Natural gas is flammable, with a heat of combustion of about 50 kJ/g. It has a minimum ignition energy (MIE) of 0.25 mJ, that is, it is easily ignited when mixed with air. The lower flammability limit in air, which is also denoted as the lower explosion limit (LEL), is 5 vol% and the upper flammability limit of 15 vol% [18].

Due to the high pressures associated with natural gas processing and pipeline transport, even small leak openings may give substantial leak rates. Ignited large leaks have given many major accidents worldwide. Early detection of gas leaks is therefore given high priority to prevent fires and/or explosions.

2.2. Fixed Gas Detectors for Safety Related Leak Detection

Gas processing plants, compression facilities for gas transport and LNG plants are equipped with fixed gas detectors to detect safety related leaks. These gas detectors usually come in two principle categories, that is, point detectors and line-of-sight detectors. There are many different types of point gas detectors. Previously, combustion-based gas detectors were used. These detectors relied on recording the oxidation of gas on a heated catalyst and had quite long response times. Today, infra-red (IR) light absorption detectors dominate, at least in Norwegian on- and offshore facilities. These are based on recording the IR absorption of the C-H bonds in the gas at about 3.4 μm wavelength. Readings from point detectors are usually reported in % LEL. For new equipment, the detection limit for high alarm (H) is typically 10% LEL with high-high alarm (HH) at 20% LEL. Line of sight detectors work similarly, however with a sender and a receiver, where any absorption of wavelengths corresponding to the C-H bond IR signature between these, is recorded. Line of sight detectors typically report the values in LELm (LEL meter), with set values for H alarm and HH alarm typically at 1 LELm and 2 LELm, respectively. At H alarm, the plant operators are warned and non-essential electrical equipment is automatically disengaged. Even stronger actions are taken at HH alarm or when more than one gas detector H alarm is activated.

2.3. Investigation of Gas Release Incidents

In a number of cases in the world-wide oil and gas industry, large gas releases have been ignited and thereby caused major accidents. Containment is therefore generally accepted as the most important barrier against severe incidents. Diluting gas leaks to below the LEL is also a possible safety measure. In other industries, such as the coal mining, where methane is continually released, ventilation is critical for ensuring safe conditions [19].

Major oil and gas companies relentlessly investigate situations that have resulted in safety related gas leaks, regardless of the release being ignited or not. Investigation of such leak incidents is indeed required by the Norwegian regulations, stating that "Situations that occur frequently or that have great actual or potential consequences, shall be investigated" [20]. The motivation for investigating gas leaks is to learn from the incidents and enable the companies to prevent future accidents and near misses.

Since loss of containment is a severe risk, most companies have developed criteria for leak severity categorization represented by the leak rate for lasting leaks and accumulated leak for brief leak bursts. A representative general incident and near miss consequence matrix is presented in Figure 1, where also other incidents such as for example, personnel injury, uncontrolled discharge/emissions to the environment and loss of reputation are included. In Norway, such matrixes have been used for more than three decades, with only minor adjustments. For flammable oil and gas leaks, the consequence matrix is organized from the most severe to the less severe leak rates in the following order: leak rate > 10 kg/s → Red 1; 10 kg/s > leak rate > 1 kg/s → Red 2; 1 kg/s > leak rate > 0.1 kg/s → Yellow 3; leak rate < 0.1 kg/s → Green 4; leak rate << 0.1 kg/s → Green 5. It is interesting to notice that the leak rate categories are organized logarithmically based of the leak rates, both for actual leaks and for potential consequences under slightly altered circumstances.

Degree of serious-ness	Injury		Work related illness (WRI)		Uncontrolled discharge/emissions		Oil-/gas/flammable liquids leakages		Fire/explosion		Failure in safety/security functions and barriers		Reputation	
	Actual	Potent.	Actual	Potent.	Actual	Potent.	Actual	Potent.	Actual	Potent.	Actual	Potent.	Actual	Potent.
1	Fatality		Work related illness that result in death		Single spill with long term effect on the environment. Release to air > yearly expected emission of component		>10 kg/s or brief leakages >100 kg		Whole facility/plant exposed		Threaten whole facility/plant		Great international negativ exposure in mass media and among organisations	
2	Serious lost time injury/serious injury		Serious work related illness		Single spill with medium term effect on the environment. Release to air > monthly expected emission of component		1–10 kg/s or brief leakages >10 kg		Large part of facility/plant exposed		Threaten large part of facility/plant		Medium international negative exposure in mass media and among organisations	
3	Other lost time injury or injury involving substitute work		Work related illness that results in short-term absence or restricted/ alternative work		Single spill with short term effect on the environment. Release to air > weekly expected emission of component		0.1–1 kg/s or brief leakages >1 kg		Parts of facility/plant exposed		Threaten parts of facility/plant		National negative exposure in mass media, from authorities on national level	
4	Medical treatment injury		Work related illness that results in treatment from authorised health care personnel		Single spill with minor effect on the environment. Release to air < weekly expected emission of component		<0.1 kg/s		Local area of facility/plant exposed		Threaten local area		Local/regional negative exposure in mass media, from authorities and customers	
5	First aid injury		Other work related illnesses		Single spill or release to air with negligible effect on the environment		<<0.1 kg/s (significantly less than 0.1 kg/s)		Negligible risk for facility/plant		Negligible risk for facility/plant		Limited to a few persons or a single customer	

Figure 1. Representative consequence matrix for incidents and near misses in the Norwegian oil and gas industry.

2.4. Previous Detection of Fugitive Emissions from Production Facilities

In recent years, campaigns for detecting methane leaks at land-based hydrocarbon processing facilities in Norway were mostly based on the Differential Absorption Lidar (DIAL) technique, since that method was considered as best available technology (BAT). Leak detection programs were mandatory and enforced by the NEA due to the methane greenhouse gas potential. Leak rates recorded by third-parties were used to issue taxes. The measurements were typically performed by a DIAL truck visiting the site [15,16] to record methane gas in the atmosphere upwind and downwind of the facility. Usually, the truck had to be positioned within the "hot plant," that is, where there should be no ignition sources. As a non-Ex safe unit, there was a need for work permits for this operation. The truck was at each site for 2–3 weeks and the recordings could be severely affected by surrounding nature, for example, mire habitats releasing methane at changing rates when for example, heated by sun radiation. Only in a few cases, the DIAL technique could reveal which equipment was indeed leaking. Besides these issues, the DIAL technology can get a fair representation of the total fugitive emissions, however, generally without providing details about any individual "super-emitters."

For single leak identification, close-up methods are required. Until recently, the preferred solution at Norwegian facilities was sniffing with handheld gas detectors held 10 cm from potential leak sources and the readings were either in ppm (parts per million) or in %LEL. The measurements at 10 cm distance were done on a regular basis, for example, weekly, as detection at this distance was recognized as important for detecting safety related leaks. Measurement campaigns using 1 cm distance were done less frequently, for example, once a year, to detect the smallest fugitive emissions.

2.5. Detecting Minor Gas Leaks and Seepages

Unless directly exposing a point gas detector, a natural gas seepage of for example, 5 cm^3/s (3.5×10^{-6} kg/s, 125 kg/y) will not result in a H alarm of neither point gas detectors nor line-of-sight gas detectors. The released gas will be too diluted to result in alarm activation. Such small leak rates in an open area or in an indoor ventilated area, do not represent a safety issue. However, when undetected by the fixed gas detectors, these leaks may release gas for several years. When accumulating the contribution from numerous non-safety related leaks, it is apparent that they contribute to a significant environmental impact, as well as a loss of valuable products.

Handheld infrared (IR) based optical gas imaging (OGI) cameras have now been introduced for leak detection. Ravikumar et al. [21] recently investigated whether the OGI technology could represent an effective method for methane leak detection. This was done based on the U.S. EPA proposed regulations requiring use of OGI passive IR technologies in LDAR programs. It was concluded that 80% of the emissions could be detected at 10 m distance. This work was followed up with a study where an OGI camera was used for blind tests at mock-ups resembling flange leaks [17]. At 6 m imaging distance, a 50% likelihood detection limit was about 20 g CH_4/h (0.0056 g/s), corresponding to a volumetric rate of 7.0 cm^3/s. The 90% detection likelihood limit followed a power-law relationship with distance.

There are several benefits when being able to walk into the field and visually see gas leaks. One can see exactly which point the leak originates as well as getting an estimate of the leak rate by the gas plume character. In a recent OGI study, several methane leaks of about 1 cm^3/s (0.0008 g/s) were identified in an LNG plant [22].

2.6. Leak Rate Quantification

Leak rates may be estimated by a variety of methods. Test campaigns have been performed at Total's Lacq Platform in France using gas spectral imaging systems, that is, multispectral Long-Wavelength InfraRed (LWIR) band (7–9 μm) camera, mobile hyperspectral cameras in the LWIR band (7.7–12 μm) and Light Detection and Ranging (LIDAR) multi gas system. Teams from France, Spain, USA and Norway were invited to assess remote-sensing systems for methane leak quantification [23]. The testing showed that methane leaks in the range from 0.7 g/s to 140 g/s could be visualized and quantified in real time using mobile Telops Hyper-Cam, while also confirming the performance of several remote sensing technologies. Open path laser systems have been used [24] for gas leak detection at rates of about 1 L/min. Low-cost, off-the-shelf, metal oxide-based methane (CH_4) gas sensors have also been tested for monitoring methane leaks. They gave quite reliable data in laboratory conditions when properly correcting for the influence of air temperature and relative humidity [25], as well as in the field for recording ambient concentrations [26]. Dyakowska and Pęgielska [27] tested trained operators using the EN 15446 and the Hi Flow sampler technique and found that the Hi Flow sampler gave the best results. By carefully following the measurement procedures, the best operator was able to quantify the leak rates to within 3.5% in blind tests.

For an LDAR program, it may however not be necessary to identify leak rates to within this accuracy, that is, when the main goal is to prioritize leaks for repair to reduce emissions. As identified leak rates are of varying orders of magnitude, a less refined mesh may be sufficient for identifying the "super-emitters," which, according to Brandt et al. [3] should be focused.

In the present study, OGI cameras were used for identifying leaks. For quick estimates of leak rate, a very simple guide was developed to support the operators in doing soap bubble tests using spray (CRC Leak Finder Spray). The guide, as shown in Figure 2, presents the volume of a sphere as a function of diameter to assist in volume rate estimates of the bubbles generated by leaks. The operators were instructed to count to for example, 10 s (1001, 1002, 1003, …), record the dimension of the individual bubbles or bubble groups, using a ruler, note the total bubble volume, correct for the shape of the bubbles (e.g., half sphere) and divide by the bubble development time. This gives a fair leak rate estimate in cm^3/s. Blind tests were also conducted in a lab, where 90% of the estimates were within ±30% of the correct value. In the field it may be expected that volume estimates are slightly less accurate. Leaks that generate bubbles so fast that the operators were unable to quantify these leaks were simply identified as "super-emitters." Such leaks could then, if needed and considered safe, be quantified by for example, Hi Flow sampling.

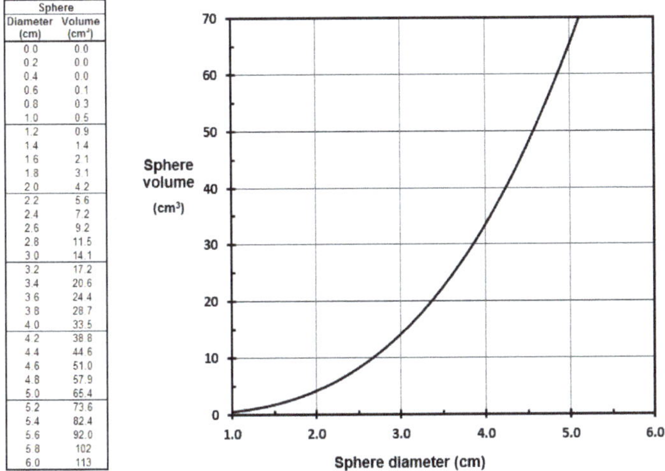

Figure 2. Guide for estimating volume rate of minor gas leaks.

3. Proposed Leak Categorization Concept and Risk Evaluation Guide

3.1. Leak Categorization Concept

All gas leaks represent environmental challenges, while in outdoor conditions, the smallest leaks do not represent a safety risk, that is, fire or explosion risk. Since all leaks are relevant to the environment, it would be beneficial to have a common consequence classification concept, that is, ranging from more than 10 kg/s to less than for example, 10^{-6} kg/s. This span would include release rates associated with potential major accidents and the smallest detectable fugitive emissions. Assuming natural gas, that is, with at a density not that much short of 1 kg/m^3, a leak rate of 10^{-6} kg/s corresponds to a volume rate of about 1 cm^3/s at ambient conditions. In open areas, this leak rate obviously does not represent a significant safety risk. A leak rate three orders of magnitude larger (10^{-3} kg/s, that is, about 1.0 L/s, corresponding to a heat release rate of 50 kW if ignited) may, however, be considered a safety related leak, at least when it comes to potential personnel exposure. To be somewhat conservative, one may define all leaks >10^{-4} kg/s as safety related leaks when considering for example, burns risk for field operators.

With leak rates spanning several orders of magnitude, the logarithmic based safety related incident categorization matrix, which is familiar to managers and operators in Norway, represents an interesting concept. By extending it to smaller leak rates, fugitive emissions may then be included. It may be discussed where the cut between the safety related leaks and purely environmental risk should be drawn. In the suggested categorization table, a new category "Green 6," that is, with leak rates in the range 1×10^{-4}–1×10^{-3} kg/s, is introduced and the cut between the safety related risk and purely environmental risk is proposed at 1×10^{-4} kg/s, as indicated in Figure 3.

	Safety related leaks						Fugitive emissions		
	Red	Red	Yellow	Green	Green	Green			
Level:	1	2	3	4	5	6	7	8	9
Mass rate:	>10 kg/s	<10 kg/s	<1.0 kg/s	<0.1 kg/s	<10^{-2} kg/s	<10^{-3} kg/s	<10^{-4} kg/s	<10^{-5} kg/s	<10^{-6} kg/s
Vol. rate:	>10 m^3/s	<10 m^3/s	<1.0 m^3/s	<0.1 m^3/s	<10 L/s	<1.0 L/s	<0.1 L/s	<10 cm^3/s	<1.0 cm^3/s
HRR	>500 MW	<500 MW	<50 MW	<5.0 MW	<0.50 MW	<50 kW	<5.0 kW	<0.5 kW	<50 W
Acc. mass:	-	-	-	-	-	<3.6 kg/h	<360 g/h	<36 g/h	<3.6 g/h
	-	-	-	-	-	<32 tons/y	<3.2 tons/y	<320 kg/y	<32 kg/y

Figure 3. Suggested leak rate categories including safety related leaks and fugitive emissions.

The heat release rates of the potentially ignited gas leaks, assuming a heat of combustion of about 50 kJ/g [18] and 100% combustion efficiency, are also included in Figure 3. This was done to increase the understanding of the dimensions of a fire if a leak should become ignited. The last lines of Figure 3 include leak rates in units of hours and years, as that is relevant for evaluating and reporting fugitive emissions for environmental impact.

It should be noted that whether a hydrocarbon gas leak may represent a fire and explosion hazard is very dependent on the ventilation conditions. In unventilated enclosures even minor leaks, when undetected, may result in hazardous concentrations in hours, days or weeks depending on the size of the enclosure. It is in the present work assumed that such leaks will be discovered by fixed point gas or line-of-sight detectors as the concentration reaches the H alarm level, for example, 10% LEL or 1 LELm, respectively.

3.2. Risk Evaluation Guide

In order to repair a leak of for example, 5 cm^3/s, depressurization of a process module may be necessary. This would normally include flaring, which also releases hydrocarbons due to partly incomplete combustion. Depressurization to repair a leak and pressurization again after the repair and start-up of the involved module, may also result in new leaks. It is therefore necessary to evaluate whether the processes involved when repairing a leak could result in any associated negative consequences.

Leak rates may not be constant but may develop with time. Under certain circumstances, the leak rates may develop very fast. The risk associated with a leak should therefore not be assessed only based on the observed leak rate, as such. In a high-pressure system, a minor leak of for example, 1 cm^3/s originating from a corrosion attacked pipe should get much more attention than a similar leak from for example, a valve stem cage gasket. An immediate depressurization would be a proper action when a situation involving a sudden loss of containment may be anticipated. The risk evaluation and proper actions regarding an identified leak should therefore include issues relevant to potential immediate or long-term leak rate development, negative consequences if a repair is undertaken and so forth. A guide for instant risk evaluation was therefore developed and tested in the present work, as presented in Figure 4.

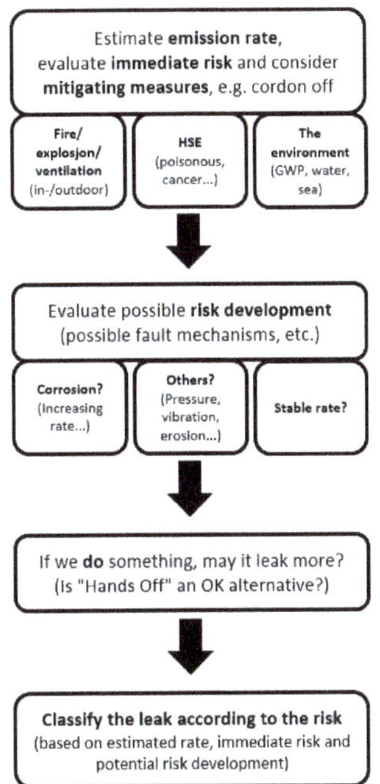

Figure 4. Sheet for simple minor leaks risk evaluation for prioritizing leak repairs.

4. Experience with the Leak Categorization Concept and the Risk Evaluation Guide

4.1. Leak Detection Campaign

The extended safety related leak rate concept was tested during leak detection campaigns at five onshore production facilities in Norway, that is, an oil refinery, a gas processing plant, a compressor plant, a methanol production plant and an oil terminal. None of these facilities had previously been screened by OGI campaigns. The screening was done by a third-party (The Sniffers, Balen, Belgium) using OGI cameras (FLIR GFx320, FLIR, Wilsonville, OR, USA) for identifying leaks and a HI Flow Sampler (Bacharach, Inc., New Kensington, PA, USA) technique for measuring leak rates. It should be noted that only about 40% of the identified leaks were quantified. This was partly due to for example, inaccessibility, weather conditions or general time constraints. Five leaks were immediately reported as potential safety related to the plant operators who took the proper actions for handling these leaks, either the actions were to cordon off, to depressurize and repair or to stay "hands-off" for a period based on a risk evaluation as presented in Figure 4.

The facilities varied substantially in size, from the large refinery and gas plant to the medium size compressor plant and the methanol production plant and to the minor size oil terminal. All the facilities had been regularly surveyed by plant operators using hand-held sniffers. No extra sniffing campaigns were done prior to the OGI campaigns. The number of leaks identified was quite consistent with the size of the facilities, that is, 124, 73, 29, 14 and 11 respectively. The number of quantified leaks, arranged in accordance with the suggested categorization system, is presented in Figure 5.

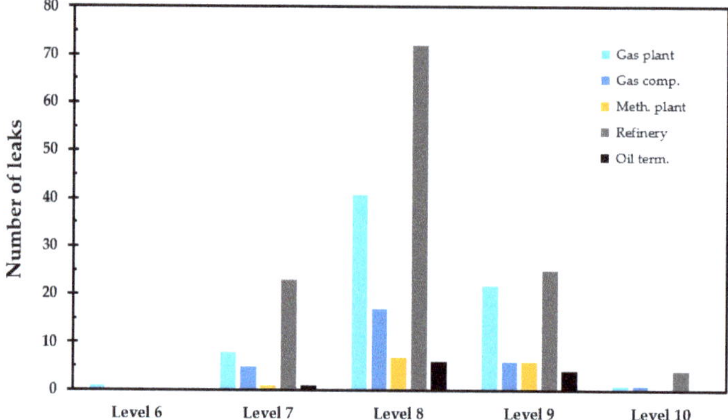

Figure 5. The number of leaks for each facility arranged per leak level.

The total accumulated leak rate from these 251 leaks was 0.73 g/s (2.6 kg/h). If unattended for one year, this would result in loss of 49 tons to the atmosphere. It is seen from Figure 5 that leaks in Level 8 dominated the total number of leaks identified at all the facilities. However, when studying the leak rates in each category, as presented in Figure 6, a conspicuous single Level 6 leak at the gas production plant stands out as a major contributor. It is also seen that the leaks in Level 7 contribute much to the accumulated leak rate.

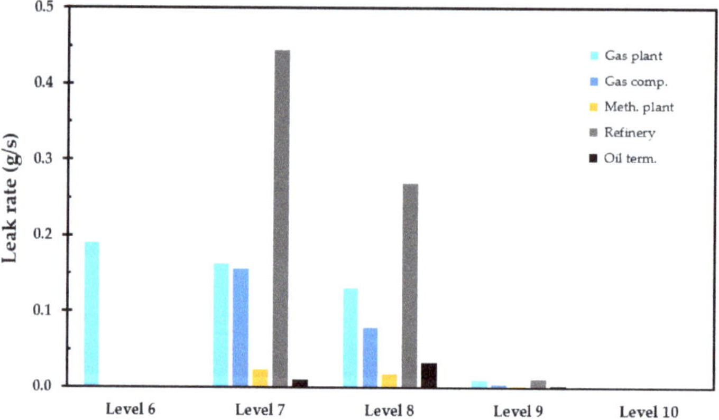

Figure 6. Leak rate for each facility arranged per leak level.

The distribution of quantified leaks from all the facilities combined, is presented in Figure 7. Though the Level 8 leaks dominate in numbers in Figure 7a, the leaks in Level 7 contributed most to the emission of hydrocarbons, as seen in Figure 7b, while the single leak in Level 6 contributed 12% to the accumulated leak rate. For both the gas processing plant and the gas compression facility, the leaking gas was mostly methane.

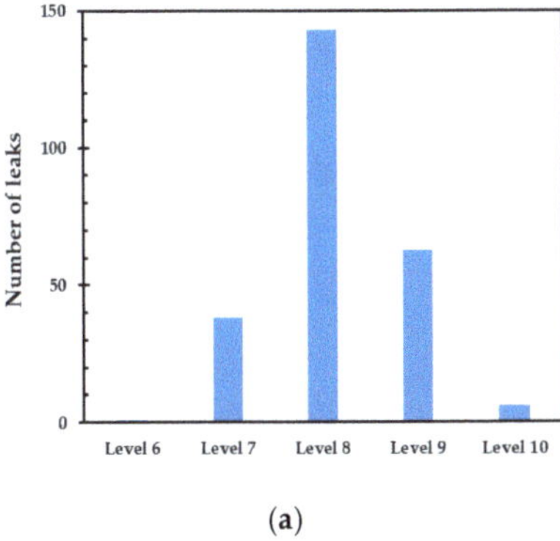 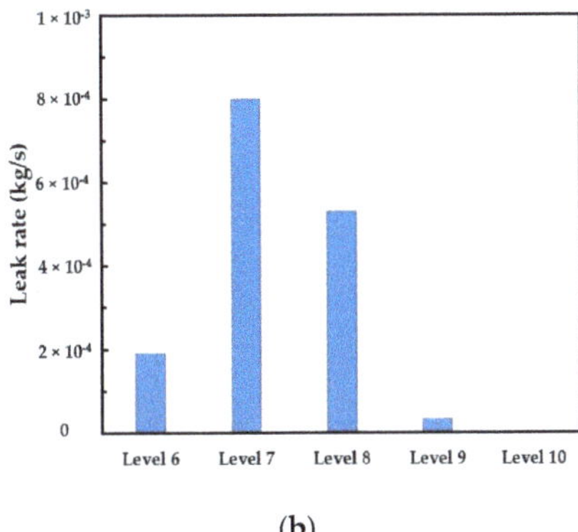

(a)   (b)

Figure 7. Number of leaks (**a**) and leak rate distribution (**b**) arranged according to the suggested categorization system.

The leak rates, arranged from the largest to the smallest leak, are presented in Figure 8a and the accumulated leak contribution is presented in Figure 8b. The largest of these 251 leaks. accounting for 12% of the total release rate, may then qualify for the label "super-emitter." The 21 largest leaks, that is, 8% of the leaks, accounted for 50% of the total release rate. The combined Level 6 and Level 7 leaks, that is, 39 leaks (15% of the leaks), accounted for 63% of the total emissions, as indicated in Figure 8b.

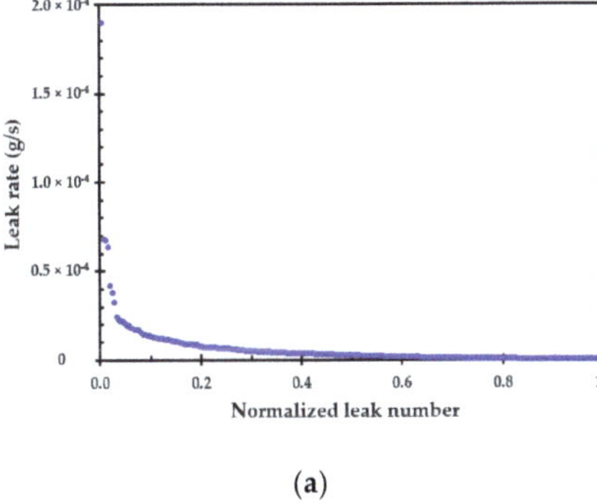 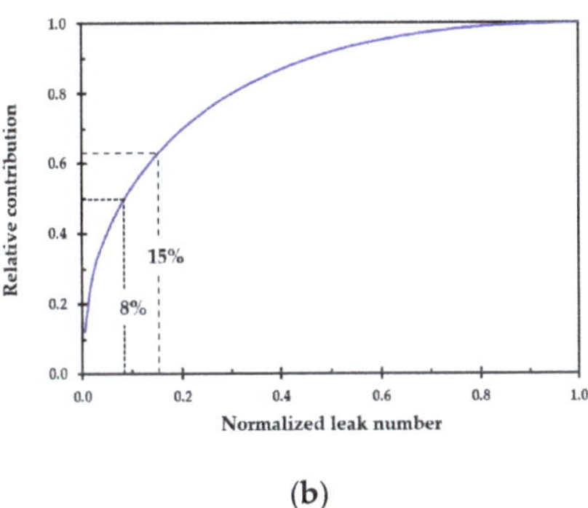

(a)   (b)

Figure 8. Leak rate (**a**) and accumulated leak rate (**b**) for the 251 quantified leaks arranged according to decreasing leak rate.

To be able to study the leak distribution for each plant in the same diagram, the leaks were arranged based on the leak rates, from large to small leaks. The leak number was normalized by the total number of leaks for each facility, as shown in Figure 9. For three of the facilities, that is, the gas compressor facility, the methanol plant and the oil terminal, a quite small number of leaks were identified and quantified. The results may, however, still indicate that there are some differences between the results from these facilities. The leaks detected for the methanol plant generally seem to be smaller than the leaks detected from the two other facilities. There also seem to be similar differences between the gas processing plant and the refinery. Except for 2–3 possible "super-emitters," the former generally exhibits lower leak rates than the latter. Since the same third-party surveyors did the leak search and leak quantification, this could possibly indicate some differences between the studied facilities.

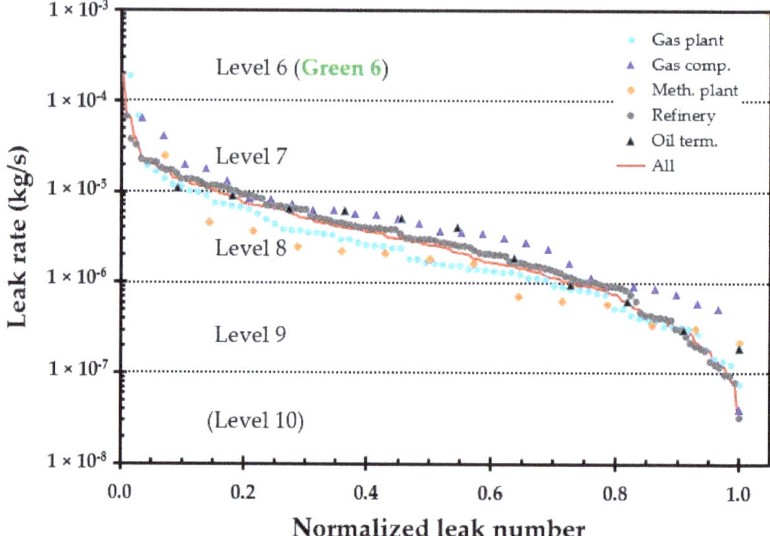

Figure 9. Leak rate for each individual quantified leak for the facilities, arranged per normalized number of leaks at each facility. The suggested leak categorization levels are indicated on the figure. The notion "All" refers to all the 251 quantified leaks, normalized accordingly.

It is generally known that the gas processing plant for several years has had a quite detailed fugitive emission leak detection program and has followed up their fugitive emissions in site maps and markings in the field. Historically, this facility had a higher number of known fugitive emission (FE) leaks on their FE map. The explanation for this could either be that this facility was leaking more than the other facilities or that it had a better leak detection program. This was also the case for the methanol production facility. An OGI campaign in more windy conditions could, however, mask the minor leaks and therefore skew the results in a diagram such as presented in Figure 9 towards higher emission rates since only the higher leak rates could then be detected. To draw any conclusion on this issue based on the OGI campaigns would therefore require information about the weather conditions during the OGI campaigns.

Very few of the detected leaks were in Level 10. This may to a large extent be explained by the difficulty to detect leaks at this level, that is, $<10^{-7}$ kg/s (<0.1 cm^3/s). Their contribution to the accumulated leak rate is also very small. Leaks in Level 9 are also generally quite difficult to detect. They do, as well, only to a minor extent contribute to the accumulated leak rate, as seen in Figure 7b. Since leaks in Level 9 and Level 10 are both difficult to detect and are observed to contribute little to the accumulated leak rate, it is suggested to combine these in the smallest practical level of leaks rates, that is, $<10^{-6}$ kg/s. In accordance to Figure 3, all leaks $<10^{-6}$ kg/s are therefore suggested to be labelled as Level 9 leaks.

4.2. Some Comments on the Obtained Results

Rather than consuming much time for quantifying leaks in Level 9 (and Level 10), it may be worthwhile focusing more on the leaks that contribute most regarding environmental impact and safety related risk. Repair of leaks with rates in Level 7, Level 6 and so forth, should therefore be prioritized. When presenting the results from the leak detection campaign in figures like Figures 7 and 8, it became quite clear that repairing the "super-emitters" should be focused. It therefore turned out to be an efficient way of communicating the OGI campaign results.

4.3. Leak Rate Diagram Guide

For a few leaks, it was decided to monitor the possible development in leak rate over time using the bubble technique (CRC Leak Finder spray) and the diagram presented in Figure 2. This was based on an uncertainty whether the leak could gradually develop, that is, needed attention as suggested in the risk evaluation guide presented in Figure 4. It was then needed to give some information about the expected accuracy of such monitoring. It was, however, apparent that Figures 3 and 4 represented an aid in such risk evaluation processes.

4.4. Benefits of OGI Campaigns Versus Regular Sniffing Campaigns

The previous aspiration (sniffing) detection campaigns at the respective facilities were generally done at areas of the plant with easy access for the operators. Equipment associated with potentially leaking components had been given attention, while for example, possible leaks surfacing from cladding distant from a leaking component were not investigated. An important benefit of having an OGI camera survey was the possibility of detecting previously undetected leaks at places with limited or no access or where leaks were not expected.

Several of the detected leaks were at locations where quantification could not be done without erecting scaffolding. This limited the possibility for leak quantification by the Hi Flow sampler technique. However, from both an environmental and a technical safety point of view, detecting the gas leaks that required immediate action was important. Since some of these leaks could represent a hazard to the Hi Flow sampler operators, any work to quantify these leaks was off course deemed inappropriate. It may very well be that one or more of these leaks were larger than the accumulated rate found for the 251 leaks quantified.

4.5. Leak Rate Risk Evaluations in the Field

A plant operator always followed the third-party OGI inspector and the Hi Flow Sampler operator. The plant operators at the investigated facilities were well trained and knew the gases and liquids, as well as the process conditions (pressures and temperatures), in different modules at the facilities. Until now, the plant operators have, however, had little experience with the risks involved with leaks other than on an overarching level, for example, combustible gas or liquid products, liquids flashing when exposed to atmospheric pressure or self-ignition of hot oils when exposed to air, to name a few risks. When suggesting that plant operators should evaluate the risk related to a fugitive emission, for example, 5 cm^3/s leak rate, they could not give any description of the associated risk beyond general terms, for example, it may ignite and cause a fire or an explosion. To relate the risk to something familiar seemed farfetched.

The presented risk evaluation guide, that is, as shown in Figure 4, was intended to assist in understanding the fire and explosion risk associated with a hydrocarbon gas leak. In Norway, most plant operators are familiar with the heat release rates of for example, a camping stove (2–4 kW) and that a burning candlelight represents about 50–80 W. Using such familiar examples represented knowledge anchors for understanding that an ignited release rate in for example, Level 8 is between these two familiar heat release rates and that a release in for example, Level 5 is far more. Some operators even knew that a fully involved passenger car fire [28] would typically be representative for the heat release rates associated with level Green 4. Adding the row of estimated heat release rate (HRR) of an ignited leak in the leak category table (Figure 3) clearly helped them to comprehend the risk associated with a potentially ignited hydrocarbon gas leak.

5. Discussion

In the present study, a simple guide for estimating volumetric leak rate for minor gas leaks was presented in Figure 2. The guide worked well when tested by plant operators. The results from such quantifications were intended for further risk evaluation, for example, monitoring development in the leak rate and so forth.

The consequence matrix for incidents and accidents in the Norwegian oil and gas industry was used as a template for categorizing safety related gas leaks and fugitive emissions. In line with the incident and accident consequence matrix, the suggested leak categories were arranged based on the leak rate in kg/s, in orders of magnitude in leak rate, descending to $<1 \times 10^{-6}$ kg/s. The suggested leak categories were used to analyse and present the results from third party OGI campaigns at five land-based facilities in Norway. The facilities comprised an oil refinery, a gas processing plant, a gas compression plant, a methanol production plant and an oil terminal. A total of 630 leaks were identified by the OGI campaigns. About 40% of these leaks, that is, 251 leaks, could be quantified by the HI Flow sampler technique. The number of quantified leaks at the mentioned facilities was respectively 124, 73, 29, 14 and 11 with an accumulated leak rate of 0.73 g/s (2.6 kg/h).

The reasons for 60% of the leaks identified by the OGI campaigns not being quantified were associated with limited access to the leaking components and weather conditions. It should be noted that five leaks were not quantified as it was considered unsafe for the HI Flow Sampler operator to approach these leaks, which were cordoned off for subsequent leak repair.

Leaks in Level 8 (10^{-5} kg/s, 10^{-6} kg/s] dominated the number of quantified leaks. The single leak in Level 6 (10^{-3} kg/s, 10^{-4} kg/s], that is, the largest leak quantified, contributed 12% to the accumulated leak rate. The leaks in Level 7 (10^{-4} kg/s, 10^{-5} kg/s] together with the one leak in Level 6, contributed to 64% of the accumulated leak rate. The 21 largest leaks, that is, the 8% of the leaks, accounted for 50% of the accumulated leak rate.

Leaks not quantified were probably associated with leak rates similar to those quantified. The exceptions were the five leaks that could not be quantified due to safety reasons. These may have been much larger than the largest quantified leak. Some of these may therefore have been in Level 5. If so, they would have dominated the accumulated leak rate. These leaks had not been detected by the regular sniffing campaigns at the facilities. Nor were they large enough or close enough, to be detected by the 10 % (or 20 %) LEL H-alarm on point gas detectors or the 1 LELm H-alarm on the line of sight gas detectors. Without an OGI campaign at the facilities, these leaks could have continued leaking for years, with negative impacts on the environment and safety, as well as representing loss of valuable products. Getting these large leaks repaired was probably more important than all the rest of the leaks identified both regarding safety as well as environmental impact.

In the present study, it may be concluded that the OGI technology is a great tool for field use, as also concluded by Ravikumar et al. [21]. The best available OGI cameras are, however, quite expensive. When looking for "super-emitters," less expensive OGI technologies may be "good enough." Less costly cameras may also allow for using such equipment in organized training sessions and for demonstrating how the gases spread after larger or smaller releases [29,30].

The present study is limited to presenting the leaks according to their mass loss rates. Since the hydrocarbon gases have different GWP, the leak rates in kg/s cannot be directly transformed to environmental impact. This could, however, in principle be done as each quantified leak originates from an equipment with known approximate gas composition. Rather than using resources for getting these details sorted for all the 251 leaks, one could introduce the approximate gas composition as a criterion for prioritizing repairs. A suggestion could be to multiply the leak rates with representative GWPs versus CO_2 for the leaking gas mixture and prioritize according the respective leak's GWP. This would lead to a discussion of what factor to use [5] but would ensure that methane rich leaks got increased attention due to the high environmental impact of this gas.

The experience with the suggested classification system for fugitive emissions was quite good. It helped in getting the attention away from the large number of minor leak sources identified and focus on leak rates. This resulted in focus on the "super-emitters" contribution to the accumulated emissions. Acknowledging that a limited number of leaks accounted for a large proportion of the accumulated leaks may help addressing the "super-emitters" more frequently in the future.

To directly view a leak in the OGI camera or subsequently on a recorded video, reveals much information about the leak. Since the plant operators know the leaking hydrocarbon gas from a given equipment, including, for example, the approximate average mole mass, operational pressure and temperature, the way the gas plume behaves can give much information even without any leak rate quantification. By occasional leak quantification by for example, soap screening (Leak Finder Spray) as a calibration of the visual information, the plant operators might be able to classify most leaks in the correct level in the suggested categorization table. Or, if the leak is large, classify it as a "super-emitter" for immediate attention. It is suggested that this is further investigated since and if this works, the focus can be shifted to early identify any "super-emitters" and thereby significantly reduce the accumulated leak rates, that is, reduce the environmental impacts and loss of valuable products.

The international focus on greenhouse gas emissions from the oil and gas industry is expected to be intensified to limit the global warming. This is especially important for production, transport as well as end consumers of products that may result in methane emissions [31–34]. The present study contributes to reduce the emissions at land based mid-stream facilities, where it is shown that identifying and repairing "super-emitters" would give a high benefit cost ratio. It seems quite clear to the authors that there must be good economy in identifying and repairing the "super-emitters."

6. Conclusions

The suggested system based on orders of magnitude in leak rates, from safety related leaks to the smallest fugitive emissions detectable by OGI camera, worked quite well. The simple risk guide sheet also worked well in assisting the operators during the risk evaluation process. The suggested categorization table was used for presenting and analysing detected leaks to help prioritizing repair of "super-emitters," which contributed most to the accumulated leak rates. A change in LDAR programs to coarser and more frequent OGI campaigns to earlier detect any new "super-emitter" is suggested. This would also improve the search for safety related leaks as well as prevent loss of valuable products.

Author Contributions: Conceptualization, T.L. and W.B.P.; methodology, T.L. and W.B.P.; field operations, T.L.; literature review, T.L. and W.B.P.; resources, T.L. and W.B.P.; data curation, T.L. and W.B.P.; writing—original draft preparation, T.L.; writing—review and editing, T.L. and W.B.P.; project administration, T.L.; funding acquisition, W.B.P.

Funding: The work related to organizing and presenting the study received no external funding.

Acknowledgments: The cooperation with Nick Mertens, Toon Van Nooten and Arno Tuerlinckx, The Sniffers Inc., Belgium, is much appreciated.

Conflicts of Interest: The authors declare no conflict of interest.

References

1. IEA. *World Energy Outlook*; International Energy Agency: Paris, France, 2018; Available online: www.iea.org/weo2018/ (accessed on 15 May 2019).
2. IEA. Energy Efficiency Indicators: Fundamentals on Statistics. 2014. Available online: https://www.iea.org/statistics/ (accessed on 22 January 2019).
3. Brandt, A.R.; Heath, G.A.; Cooley, D. Methane Leaks from Natural Gas Systems Follow Extreme Distributions. *Environ. Sci. Technol.* **2016**, *50*, 12512–12520. [CrossRef] [PubMed]
4. Stocker, T.F.; Qin, D.; Plattner, G.-K.; Alexander, L.V.; Allen, S.K.; Bindoff, N.L.; Bréon, F.-M.; Church, J.A.; Cubasch, U.; Emori, S.; et al. Technical Summary. In *Climate Change 2013: The Physical Science Basis*; Contribution of Working Group I to the Fifth Assessment Report of the Intergovernmental Panel on Climate Change; Stocker, T.F., Qin, D., Plattner, G.-K., Tignor, M., Allen, S.K., Boschung, J., Nauels, A., Xia, Y., Bex, V., Midgley, P.M., Eds.; Cambridge University Press: Cambridge, UK; New York, NY, USA, 2013; 222p, ISBN 978-92-9169-138-8.
5. Balcombe, P.; Speirs, J.F.; Brandon, N.P.; Hawkes, A.D. Methane emissions: Choosing the right climate metric and time horizon. *Environ. Sci. Process. Impacts* **2018**, *20*, 1323–1339. [CrossRef] [PubMed]

6.   Crow, D.J.G.; Balcombe, P.; Brandon, N.; Hawkes, A.D. Assessing the impact of future greenhouse gas emissions from natural gas production. *Sci. Total Environ.* **2019**, *668*, 1242–1258. [CrossRef] [PubMed]

7.   Boothroyd, I.M.; Almond, S.; Worrall, F.; Davies, R.-K.; Davies, R.J. Assessing fugitive emissions of CH4 from high-pressure gas pipelines in the UK. *Sci. Total Environ.* **2018**, *631–632*, 1638–1648. [CrossRef] [PubMed]

8.   Mandal, P.C.; Morshed, S.M. Localization of fugitive methane emission from natural gas distribution network of Titas Gas. *Pol. J. Chem. Technol.* **2017**, *19*, 127–131. [CrossRef]

9.   Miller, S.M.; Michalak, A.M. Constraining sector-specific CO2 and CH4 emissions in the US. *Atmos. Chem. Phys.* **2017**, *17*, 3963–3985. [CrossRef]

10.   IEA. *World Energy Outlook*; International Energy Agency: Paris, France, 2017; Available online: www.iea.org/weo2017/ (accessed on 15 May 2019).

11.   Alvarez, R.A.; Zavala-Araiza, D.; Lyon, D.R.; Allen, D.T.; Barkley, Z.R.; Brandt, A.R.; Davis, K.J.; Herndon, S.C.; Jacob, D.J.; Karion, A.; et al. Assessment of methane emissions from the U.S. oil and gas supply chain. *Science* **2018**, *361*, 186–188. [CrossRef]

12.   Brandt, A.R.; Heath, G.A.; Kort, E.A.; O'Sullivan, F.; Petron, G.; Jordaan, S.M.; Tans, P.; Wilcox, J.; Gopstein, A.M.; Arent, D.; et al. Methane Leaks from North American Natural Gas Systems. *Science* **2014**, *343*, 733–735. [CrossRef]

13.   Harriss, R.; Alvarez, R.A.; Lyon, D.; Zavala-Araiza, D.; Nelson, D.; Hamburg, S.P. Using Multi-Scale Measurements to Improve Methane Emission Estimates from Oil and Gas Operations in the Barnett Shale Region, Texas. *Environ. Sci. Technol.* **2015**, *49*, 7524–7526. [CrossRef]

14.   Borda, E.S.; Korre, A.; Nie, Z.; Durucan, S. Comparative assessment of life cycle GHG emissions from European natural gas supply chains. In Proceedings of the 14th International Conference on Greenhouse Gas Control Technologies, GHGT-14, Melbourne, Australia, 21–25 October 2018.

15.   Robinson, R.; Gardiner, T.; Innocenti, F.; Woods, P.; Coleman, M. Infrared differential absorption Lidar (DIAL) measurements of hydrocarbon emissions. *J. Environ. Monit.* **2011**, *13*, 2213–2220. [CrossRef]

16.   Innocenti, F.; Robinson, R.; Gardiner, T.; Finlayson, A.; Connor, A. Differential Absorption Lidar (DIAL) Measurements of Landfill Methane Emissions. *Remote. Sens.* **2017**, *9*, 953. [CrossRef]

17.   Ravikumar, A.P.; Wang, J.; McGuire, M.; Bell, C.S.; Zimmerle, D.; Brandt, A.R. "Good versus Good Enough?" Empirical Tests of Methane Leak Detection Sensitivity of a Commercial Infrared Camera. *Environ. Sci. Technol.* **2018**, *52*, 2368–2374. [CrossRef] [PubMed]

18.   Drysdale, D. *An Introduction to Fire Dynamics*, 2nd ed.; John Wiley: New York, NY, USA, 1999; 476p, ISBN 0-471-97291-6.

19.   Tutak, M.; Brodny, J. Analysis of the Impact of Auxiliary Ventilation Equipment on the Distribution and Concentration of Methane in the Tailgate. *Energies* **2018**, *11*, 3076. [CrossRef]

20.   PSA. *Regulations Relating to Management and the Duty to Provide Information in the Petroleum Activities and at Certain Onshore Facilities (The Management Regulations)*; Petroleum Safety Authority of Norway: Stavanger, Norway, 2017; 15p.

21.   Ravikumar, A.P.; Wang, J.; Brandt, A.R. Are Optical Gas Imaging Technologies Effective For Methane Leak Detection? *Environ. Sci. Technol.* **2017**, *51*, 718–724. [CrossRef]

22.   Log, T.; Pedersen, W.B.; Moumets, H. Optical Gas Imaging (OGI) as a Moderator for Interdisciplinary Cooperation, Reduced Emissions and Increased Safety. *Energies* **2019**, *12*, 1454. [CrossRef]

23.   Watremez, X.; Marblé, A.; Baron, T.; Marcarian, X.; Dubucq, D.; Donnat, L.; Cazes, L.; Foucher, P.Y.; Danno, R.; Elie, D.; et al. Remote Sensing Technologies for Detecting, Visualizing and Quantifying Gas Leaks, Society of Petroleum Engineers. In Proceedings of the SPE International Conference and Exhibition on Health, Safety, Security, Environment, and Social Responsibility, SPE-190496-MS, Abu Dhabi, UAE, 16–18 April 2018. [CrossRef]

24.   Gibson, G.; van Well, B.; Hodgkinson, J.; Pride, R.; Strzoda, R.; Murray, S.; Bishton, S.; Padgett, M. Imaging of methane gas using a scanning, open-path laser system. *New J. Phys.* **2006**, *8*, 26. [CrossRef]

25.   van den Bossche, M.; Rose, N.T.; De Wekker, S.F.J. Potential of a low-cost gas sensor for atmospheric methane monitoring. *Sens. Actuators B* **2017**, *238*, 501–509. [CrossRef]

26.   Eugster, W.; Kling, G.W. Performance of a low-cost methane sensor for ambient concentration measurements in preliminary studies. *Atmos. Meas. Tech.* **2012**, *5*, 1925–1934. [CrossRef]

27. Dyakowska, E.; Pęgielska, M. Comparison of the accuracy of two methods of methane fugitive emissions measurements—One according to EN 15446 standard and the other using the Hi Flow Sampler device—GERG (The European Gas Research Group) project results. *Nafta Gaz* **2016**, *72*, 660–665. [CrossRef]

28. Park, Y.; Ryu, J.; Ryou, H.S. Experimental Study on the Fire-Spreading Characteristics and Heat Release Rates of Burning Vehicles Using a Large-Scale Calorimeter. *Energies* **2019**, *12*, 1465. [CrossRef]

29. Metallinou, M.M. Liquefied Natural Gas as a New Hazard; Learning Processes in Norwegian Fire Brigades. *Safety* **2019**, *5*, 11. [CrossRef]

30. Log, T.; Moi, A.L. Ethanol and Methanol Burn Risks in the Home Environment. *Int. J. Environ. Res. Public Health* **2018**, *15*, 2379. [CrossRef] [PubMed]

31. Dalianis, G.; Nanaki, E.; Xydis, G.; Zervas, E. New Aspects to Greenhouse Gas Mitigation Policies for Low Carbon Cities. *Energies* **2016**, *9*, 128. [CrossRef]

32. Lin, H.-C.; Chen, G.-B.; Wu, F.-H.; Li, H.-Y.; Chao, Y.-C. An Experimental and Numerical Study on Supported Ultra-Lean Methane Combustion. *Energies* **2019**, *12*, 2168. [CrossRef]

33. Few, S.; Gambhir, A.; Napp, T.; Hawkes, A.; Mangeon, S.; Bernie, D.; Lowe, J. The Impact of Shale Gas on the Cost and Feasibility of Meeting Climate Targets—A Global Energy System Model Analysis and an Exploration of Uncertainties. *Energies* **2017**, *10*, 158. [CrossRef]

34. Balcombe, P.; Anderson, K.; Speirs, J.F.; Brandon, N.P.; Hawkes, A. The Natural Gas Supply Chain: The Importance of Methane and Carbon Dioxide Emissions. *ACS Sust. Chem. Eng.* **2017**, *5*, 3–20. [CrossRef]

Article

CO_2 Efficiency Break Points for Processes Associated to Wood and Coal Transport and Heating

Robert Baťa, Jan Fuka *, Petra Lešáková and Jana Heckenbergerová

Institute of Administrative and Social Sciences, Faculty of Economics and Administration,
University of Pardubice, Studentská 84, 532 10 Pardubice, Czech Republic; robert.bata@upce.cz (R.B.);
petra.lesakova@upce.cz (P.L.); jana.heckenbergerova@upce.cz (J.H.)
* Correspondence: jan.fuka@upce.cz

Received: 27 May 2019; Accepted: 8 October 2019; Published: 12 October 2019

Abstract: This paper aims to deal with CO_2 emissions in energy production process in an original way, based on calculations of total specific CO_2 emissions, depending on the type of fuel and the transport distance. This paper has ambition to set a break point from where it is not worthwhile to use wood as an energy carrier as the alternative to coal. The reason for our study is the social urgency of selected problem. For example, in the area of public sector decision-making, wood heating is promoted regardless of the availability within the reasonable distance. From the current state of the research, it is also clear that none of the studies compare coal and biomass fuel transportation from the point of view of CO_2 production. For this purpose, an original methodology has been proposed. It is based on a modified life cycle assessment (LCA), supplemented with a system of equations. The proposed methodology has a generalizable nature, and therefore, it can be applied to different regions. However, calculation inputs and modelling are based on specific site data. Based on the presented numerical analysis, the key finding is the break point for associated processes at a distance of 1779.64 km, since when that it is better to burn brown coal than wood in terms of total CO_2 emissions. We can conclude that, in some cases, it is more efficient to use coal instead of wood as fuel in terms of CO_2 emissions, particularly in regard to transport distance and type of transport.

Keywords: biomass; efficiency; heating system; renewable energy; decision making process; transport; LCA; break point

---

1. Introduction

In past decades, the issues of nature's conservation, sustainability, energy intensity, and greenhouse gas emissions reduction have been not only problems for practitioners and scholars, but it also, increasingly, are becoming political problems. An example might be the upward discussions on this subject, which in some cases present hugely different opinions of the regional but also national political leaders. There is the question of how much importance is attached to individual opinion streams. When searching for the objective attitude, it is important to rely on quantifiable data and high-quality research. Even though the aforementioned areas have been relatively well researched, some gaps still might be identified. The political decision-making process and related presentation of key theses and strategies might be ideologically burdened. The correctness and relevance of the proposed policies should not be relativized, particularly because of the gravity of this issue. In this case, a science based on empirical and quantifiable findings is the only fair and verifiable tool. The challenge of this article is to look at the selected issues using the optics of quantifiable and measurable variables. In the context of the urgency of the open topic, the total specific CO_2 emissions from the transport and burning processes of coal and wood, depending on the transport distances, were selected as indicators of environmental burden and energy intensities. How effective is the replacement of fossil fuels with biofuels (wood in our case) in relation to transport distance for the reduction of CO_2 emissions? This question will be answered in the following text. Many authors have dealt with the issue of CO_2 emissions, transport, and solid fuels. Therefore, we would like to mention the most important work that has been published in this area. Describing current state of the art will help us to create a broader theoretical basis for our practical part, and at the same time to find a research gap that has not been explored so far.

The introductory part of this work will be opened by a global perspective of the problem, from the point of view of the international authorities, because energy efficiency is directly related to climate change issues, primarily in terms of searching for new, more efficient, and sustainable technologies. The United Nations (UN) initiated the establishment of the International Panel on Climate Change (IPCC). This UN body, among other agenda, has been publishing reports on climate change, in particular, the current Fifth Assessment Report (AR5). The report takes into account the impact of human activities on climate change. Factors that can amplify the effects of adaptation and mitigation can be considered good-quality public administration, such as the use of green technologies or a sustainable way of life [1]. Key documents with global impact are the United Nations Framework Convention on Climate Change and the Kyoto Protocol and the Paris Agreement. The United Nations Framework Convention on Climate Change (the convention) has been the initial platform for international climate negotiations from 1992. Its objective is to stabilize the concentration of greenhouse gases in the atmosphere in order to prevent dangerous changes in the climate system [2]. The 1998 Kyoto Protocol (UN) obliges the countries involved to reduce their greenhouse gas emissions. The Czech Republic signed and ratified this document in 1998 and 2001 [3]. In 2015, the Paris Agreement defining a long-term perspective on climate protection was adopted by the stakeholders of the Convention. EU countries agreed to reduce greenhouse gas emissions by at least 40% by 2030 compared to 1990 and they ratified the agreement in October 2017 [4]. At European Union level, a few documents with significant impact on energy and efficiency needed to be mentioned. Firstly, the Covenant of Mayors for Climate and Energy whose participants declare to act according to Paris Agreement [5]. Currently, there are 11 signatories to this initiative from the total of approximately 6250 municipalities and towns in the Czech Republic [6]. Directive 2010/31/EU of the European Parliament and of the Council on the energy performance building is a document from 2010, according to which buildings account for 40% of total energy consumption in the Union [7]. Directive 2012/27/EU of the European Parliament and of the Council on energy efficiency, amending Directives 2009/125/EU and 2010/30/EU and repealing Directives 2004/8/EC and 2006/32/EU from 2012 states that the energy efficiency is a tool to fight the dependence on energy imports, energy shortages, economic difficulties, and climate change [8]. The Czech Republic as a member state of the European Union, and therefore, coordinates and harmonizes the priorities and objectives of its policies. An important state body is the Government Council for Sustainable Development of the Czech Republic, which is administered by the Office of the Government of the Czech Republic. The importance of this office is underlined by the fact that the chairman is the Prime Minister, and other significant members include the Minister of the Environment, the Minister of Finance, the Minister of Industry and Trade, the Minister of Labour and Social Affairs [9]. It consists of nine Committees, such as the Sustainable Energy Committee, which deals with possibilities of implementing international sustainable development documents and other energy documents into Czech environment [10]. The Advisory and Working Body, whose activities are provided by the Ministry of Industry and Trade, is the Government Council for energy and raw materials strategy of the Czech Republic [11]. In case of legislation, it is necessary to mention Act number 165/2012 Coll., on supported energy sources and on amendments to certain acts that regulates the field of renewable sources, secondary energy sources, and high-efficiency combined heat and power generation, including adjustments of stakeholders' behaviour (state administration, natural persons, and legal entities) [12]. The state energy policy from 2004 (updated in 2010 and 2015) acknowledges clearly formulated priorities and strategic objectives of the Czech Republic for future decades; e.g., principles of sustainable development [13,14]. The Ministry of the Environment of the Czech Republic, in cooperation with other institutions, created The Strategy on Adaptation to Climate Change in the Czech Republic, which was approved by the Government of the Czech Republic in 2015. The document assesses the impacts of climate change [15]. In 2010, the Government of the Czech Republic approved The Strategic Framework for Sustainable Development of the Czech Republic, defining the concepts of sustainable development in the Czech Republic, defining basic principles, objectives, priorities, and economic, social, and environmental indicators [16]. In 2017, this document was followed by the strategic framework called The Czech Republic 2030. Sustainable development should be measured, according to the document, by improving quality of life of individuals and society, taking into account the legacy to future generations [17].

2. Literature Review

There are many authors studying impact of fossil fuel combustion emissions on human health and the air quality; therefore, the aforementioned issues are reflected in both academic and scientific circles [18–20]. There are no doubts that coal use results in environmental degradation and causes negative health consequences [20,21]. According to [22], where the human health and ecotoxicological impacts of electricity production from wood to coal fuel were compared, "Improvements in power plant efficiency, silviculture management, and reduced transport distance have the potential to reduce the respiratory effects of bioenergy systems." In terms of emissions, studies mainly deal with pollutant emissions produced by the combustion of solid fuels by households or industrial activities [23–26]. Other types of studies deal with the reduction of emissions of CO_2, NO_x, and other pollutants in the atmosphere caused by burning fossil fuels [27,28]. As [29] claims, bioenergy represents a sustainable greenhouse gas (GHG) reduction option. In their work, they raise concerns about the climate change impacts of bioenergy and uncertainties within the bioenergy supply chains, and that evaluation methods generate large variations in emission profiles. However, as [29–31] claim, biomass in terms of forest residues is supposed to have large global availability and might achieve large GHG emissions savings. In the study of Thakur et al. [32] it is concluded that "Forest residues can provide an almost carbon neutral energy source that has lower GHG emissions than fossil fuels and requires very little energy for processing and growth compared to what is produced." However, those authors deal with an issue of chipping options when preparing forest residues for power plant processing. In their work, the transportation distance seems to be crucial in terms of reduction of energy consumption and emissions. The general view has been that carbon emitted from biomass combustion is assumed to be low-level or carbon neutral—as supported by IPCC [33]. The explanation is that amount of CO_2 released from biomass fuel (e.g., wood as the major biomass resource) combustion equals the amount of CO_2 trees absorb during their growth. However, this opinion about biomass carbon neutrality has been questioned in recent years. Scientists have been arguing that biomass energy produces emissions, and therefore, is unlike other renewables. Sedjo and Cherubini at al. explain in their works, that sustainable foresting may be carbon neutral. However, in the short term, using wood biomass energy can generate increases in atmospheric carbon. "The issue arising from the violation of the temporal boundary is the waiting time needed to achieve carbon neutrality" [34,35]. Nian and Johnson state, that in some cases, biomass fuels can be far more carbon positive than fossil fuels [36,37].

Even the Environmental Protection Agency (EPA) is considering regulations against biomass energy's carbon emissions [34].

However, it should not be overlooked that the carbon contained in wood was absorbed from the atmosphere, unlike that of fossil fuels, a relatively short time ago, usually in tens or hundreds of years, when it was grown. By returning it back to the atmosphere, it is not possible to change the balance of its concentration if the area of forest land is maintained, where the growth of new tree species absorbs this carbon dioxide again, thus it is sustainable. Since this study assumes this kind of sustainability, we consider CO_2 emissions from biomass to be CO_2-neutral. Let us add that, of course, extensive clearing or burning of forest stands does not represent sustainability.

However, at the same time European Commission, and Joint Research Centre on Directorate Energy, Transport, and Climate states, that the European Union's CO_2 emissions increased again, by 1.3%, in 2015 [38]. This was caused by an increase of 4.6% in natural gas consumption, mainly utilized in power generation and space heating, and by an increase of 4% in diesel consumption in transport [38]. Additional emissions from the combustion of gas could be offset by the employment of local biomass sources, especially wood as fuel, but additional emissions due to transport are not obvious for this solution. The International Energy Agency compares the CO_2 emissions of fossil fuels in a broad way, but does not compare them with neutral biofuels [39]. In terms of biomass trade, there are several studies that have been investigating trade in biomass for energy purposes. In study [40], an initial overview of the global status of the production and biomass trade for energy is presented. Proskurina, Junginger, and Heinino [41] investigated emerging energy biomass trade streams, in other words, biomass producing and consuming countries regarding liquid and solid biofuels (including roundwood).

Nevertheless, environmental risks caused by coal energy includes, besides direct emissions from the coal burning process, indirect emissions, such as coal mining emissions and transportation emissions. The carbon emission factor of the coal-to-energy chain is calculated based on the life-cycle assessment that is provided in [42]. Based on his conclusions, CO_2 is the most direct GHG emission and mainly results from coal combustion, which accounts for 93.8% of the total GHG emissions. The remaining 6.2% of total GHG indirect emissions are from the energy consumption in the mining, transportation, and washing processes. Different authors have introduced their LCA (life cycle assessment) studies on the carbon emissions of coal-fired electricity generation in different countries; for example, in the UK [43], Japan [44], Canada [18], and Germany [45]. Only very few works focus on coal and biomass together; for example, Morrison and Golde [46] analysed the environmental impacts associated with producing electricity from wood pellets and coal. According to their conclusions, "Utilizing wood pellets in lieu of coal results in a GWP reduction of 90–92% per kWh of electricity generation." One of the most current papers of Sterman, Siegel, and Rooney-Varga [47] solves a very similar problem, whether replacing coal with wood lowers CO_2 emissions. However, it does not include the transport effect, which can change the whole emission reduction effect, since the wood is significantly less dense, so its transportation with a growing distance may be significantly less efficient than the transportation of coal. The work of Zhang et al. [18] proves that there is huge advantage for electricity-generating companies to substitute biomass fuel for coal (reducing emissions by 91% and 78% relative to a coal and natural gas combined cycle). A study on the Chinese environment investigated a life-cycle comparison of the energy, environmental, and economic impacts of coal versus wood pellets for generating heat. In that work, the authors presented the conclusion on wood pellets system significantly reduces various emissions in comparison to coal [18]. In the following chapters, the data collection, evaluation, and processing; the methods used; the methods of calculations; and the modelling, will be described. Subsequently, on the basis of the proposed methodology and modelling results, the overall environmental burden (the indicator was CO_2) monitored will be described. Then, the results will be discussed and conclusions and recommendations will be provided. In the last chapter, the results of the theoretical and practical part of the research will be summarized; we will try to formulate the weaknesses, and the possibilities of further research. The aim of this paper is to examine how the total volume of CO_2 produced in connection with the transport and combustion of selected solid fuels develops. Therefore, the following working hypothesis can be defined: the break point at which the total CO_2 emissions from the coal and wood transport and combustion process equals is less than 1500 km. From the above, a possible range of methods follows.

Based on the provided facts in the analyses of LCA, transportation of solid fuel is one of the biggest indirect sources of GHG emissions. It is also clear that none of those studies compares coal and biomass fuels' transportations from the point of view of CO_2 production. In addition, the LCA method is quite problematic where it comes to its practical application. Among the most common LCA issues in practice are, first, the high demands on time and input data; second, the uncertainty regarding the content of some particles of the analysed LCA chain; and third, that in practice, it is not necessary to analyse the complete LCA chain, which may be debatable with respect to the abovementioned point. That last statement is supported by several works [48–51]. That logic brings us, for practical purposes, unambiguously to a deeper analysis of the selected part of the LCA chain. The analysis is presented in this paper for two parts of the LCA, the first of which is represented by the transport process and its output streams, and the second is represented by the energy utilization of the conveyed fuel and its output streams. The latter is represented by the total calculated CO_2 emissions. The first reason is that transport is an integral part of both processes in terms of emissions associated with the use of these fuels, and therefore, forms one whole with the combustion process. Solid fuels cannot be used energetically without being transported to the final consumer. Alternatively, other means of transport may also be used; other results would be obtained when using an alternative mode of transport (train, ship, etc.). The second reason is that it was possible to obtain unambiguous data for both partial processes, and in that connection, to obtain clear outputs with respect to the nature of the phenomenon under investigation. The overall view would probably be changed by taking into account other parts of the LCA chain. However, this article, based on the above-mentioned arguments, prefers accurate and practically useful calculations of the defined part of the LCA. The third reason is that during the solution, it became clear that the model was easily transferable to other conditions. An example may be the fact that the originally intended application of the model in the Czech environment has been limited to a certain extent, due to the relatively short transport distances (see chapter 3). While under conditions of longer transport distances, the model clearly demonstrates the expectedly significant increase in inefficiency for shipping, including the so-called break point: the point from which coal is more efficient than wood in terms of CO_2 emissions (see chapter 3). Nevertheless, our study emphasizes issue of CO_2 emissions that are generated by transportation of coal and wood, depending on transport distance. In other words, our study solves the question of coal and wood transportation effectiveness in terms of CO_2 production by the break point determination.

3. Materials and Methods

After detailed literature research, which deals with CO_2 emissions from wood combustion and transportation, we have come up with following findings.

1. Studies dealt with the partial aspects of CO_2 production, without dependence on the transport of individual raw materials. CO_2 emissions are comparable in both processes under review and represent at the same time energy consumption.
2. Previous studies were concerned with setting specific LCA values for the ratio of direct and indirect emissions in specific regions. In this study, we deal with general but significant connections of the transport of solid fuels and direct emissions from their combustion.
3. The output of this study is the determination of the unique relationship between the selected, and some of the most widely used types of solid fuels, and the transport distance and total specific CO_2 emissions from these processes.

The main output of the work is the analysis of CO_2 emissions' development from transport depending on the type of solid fuel considered, while direct emissions from the combustion of these solid fuels are taken into account also. In the analysis, a break point was defined. In other words, a break point where is the maximum meaningful distance for timber transport. This point clearly quantifies the distance in which the total CO_2 production from transport and wood burning process equals total CO_2 production from transport and coal burning process. The analysis was divided into two sub-processes; namely, the determination of the CO_2 emissions from the combustion of the considered fuels and CO_2 emissions from transport.

3.1. Fuel

For auxiliary calculations of different fuels, the energy contents, volumes, and CO_2 emissions of selected fossil fuels' combustions need to be determined. Those are average values of efficiency and corresponding emissions for commonly utilized types of boilers and stoves fired by coal and wood. The energy content parameters of the combustion process for coal were determined from two sources.

In general, the parameters for coal combustion may vary; however, in this work, brown coal was analysed. It may vary in parameters such as ash content, sulphur, water, and so on, but CO_2 emissions should be comparable regardless of the specific source. The value for the CO_2 emissions produced is, therefore, calculated from two independent sources for different types of lignite.

The first source was the coal-fired process from Umberto's software tool library, which predicts a 70% efficiency of the burning stove. The second one was the information on the website of the Czech coal producer called Severočeské doly, a.s. According to Umberto library, 80.80 kg of coal produces 1,000,000 kJ of heat. This corresponds to 12.376 MJ/kg by taking into account the 70% efficiency of the heating system.

The primary energy input is calculated according to the formula:

$$En_p = En/\eta \tag{1}$$

where En_p represents primary energy consumption, En represents energy obtained by burning a certain volume of coal in kg, and η represents efficiency of the process.

The CO_2 emissions from the combustion of 1 MJ of primary energy, e_p, is evaluated as

$$e_p = e_m \times \eta \tag{2}$$

where e_m represents specific emissions and η represents efficiency of the process.

For the assessment of transport effects, it is important to take into account the amount of energy that can be transported, in each particular form of fuel, in one shipment. This requires data on the density and energy contents of the transported fuels. For coal, this data is directly available [52].

For wood, it is necessary to employ the following formula:

$$\rho_w = M_n \times v_{ff} \tag{3}$$

where ρ_w represents density of 1 m^3 of cut and chopped loose wood, M_n is the average mass of wood calculated from a combination of different tree species in 1 m^3, and v_{ff} represents the percentage volume of clean wood in free-flowing 1 m^3 of cut and chipped wood.

The following option is considered for wood. For calculating itself, it is necessary to set additional values, especially the energy contained in the wood E_p. The precise type of wood, its proportion, or wood moisture is unknown both in short-term and in long-term; therefore, the value of E_p was determined in accordance with the average calorific value of the wood and in accordance with expected moisture content based on [53]. CO_2 emissions from the biomass burning process itself are herein considered as neutral, so they cannot be counted as total additional CO_2 emissions.

The value of the wood mass, m_w, corresponding to this volume, is calculated for the bulk space meters according to the relationship:

$$m_w = V_w \times \rho_w \tag{4}$$

where the mass of wood is in kg and the volume of wood V_w is in m^3. Analogically m_c and m_b for coal are evaluated from volume V_c of coal respectively.

The determination of the calorific value for wood is a little bit problematic in Central European conditions. A relatively wide range of different tree species can be used, which differ both in calorific value and in density. Another problem is that the calorific value of the same wood type varies according to the volume of moisture contained. The information portal TZB info, which focuses on the issue of energy in the long term, provides data from which these typical calorific values can be derived [54], and subsequently, adjusted for real conditions. When determining the average energy content of firewood, items that are not meaningfully usable from an economic, environmental or physical point of view have been omitted. In Table 1 there are expected values of wood with 40–50% moisture contained, which is not possible to burn in common types of boilers, and wood with 0% moisture contained, that does not exist in normal conditions.

Table 1. Wood types and characteristics.

Wood Type	Moisture [%]	Calorific Value [MJ/kg]
Deciduous wood (oak, beech 50/50%)	15	14.605
Coniferous wood (spruce, pine 50/50%)	15	15.584
Logs (spruce, pine 50/50%)	20	14.28
Logs (spruce, pine 50/50%)	30	12.18
Wood chips (oak, beech, spruce, pine each 25%)	10	16.4
Wood chips (oak, beech, spruce, pine each 25%)	20	14.28
Wood chips (oak, beech, spruce, pine)	30	12.18

Source: own processing according to [54].

Let us assume in this study that wood is cultivated in the Czech Republic and it is commonly dried for a year in roofed areas; exceptionally wet wood stays there for 2 years. Let us, therefore, assume that the wood has a total calorific value of 14.215 MJ/kg, which is the average calculated from the values given in Table 1. The selected wood types represent the most commonly used species in the Czech Republic. The final value is, therefore, average, and at the same time corresponds to a water content of 20%.

The total amount of energy included in one load was calculated based on this calculation of energy content from values shown in Table 1 as follows:

$$E_{w1l} = E_w \times m_w \tag{5}$$

where E_w is the energy included in 1 m^3 of wood. Alternatively, from energies E_c in coal, the total amount of energy included in one load E_{c1l} coal was evaluated.

For brown coal, it was also necessary to determine the corresponding CO_2 emissions from the combustion process. Due to the neutrality of biomass emissions, CO_2 emissions for wood were set to zero ($S_w = 0$). The emissions for coal are determined according to the formula:

$$S_b = E_{b1l} \times e_p \tag{6}$$

3.2. Transportation

In addition to emissions from energy consumed by transport, the associated processes, such as transport, are examined in this case also. As the means of transport were considered truck, train and ship. Data for a train is specific to the conditions of the Czech Republic, as it is based on the structure of the electric energy mix and is, therefore, not transferable. However, the data for shipping are transferable, since the propulsion of river cargo ships is similar and most of the ships in Europe that use diesel fuel.

When considering a truck with an average transport capacity of 8–10 m^3 (in our study, the Tatra 815 S3 was selected with parameters according to Table 2. This type of truck is commonly utilized not only in the Czech Republic, but also in other countries. It has an average transport capacity of 8–10 m^3.), the increase in total transport emissions of fossil fuel and renewable energy in the form of wood is caused due to the increasing consumption of fossil fuel consumed by transporting the wood to its customer. For analysis, a truck with the parameters listed in Table 2 was selected.

Table 2. Technical specifications of selected vehicle.

Type of Vehicle	T-815 S3 6 × 6
Curb weight	11,300 kg
Payload	10,700 kg
Total vehicle weight	22,000 kg
Engine type	T-3–929 –11
Engine displacement	15,825 cm^3
Highest engine power	280/2200 kW/Nm
Basic fuel consumption	32.5/63l/100 km
Volume of the hull	8 m^3

Source: [55–57].

As the mass of transported fuels differs considerably, it is obvious that this effect will be reflected in the fuels' consumption during transport. Therefore, it was necessary to establish a function that describes the relationship between the weight of the transported cargo and the fuel consumption. Let us assume for simplicity that the consumption depends on the additional mass linearly. The dependency function for selected Tatra 815 S3 vehicle is in the form:

$$y = 32.5 + (x/350.8197) \tag{7}$$

where y represents the total consumption in l/100 km and x represents the weight of the load in kg.

Based on knowledge of fuel consumption in l/100 km, CO_2 emissions can also be determined. For CO_2 emissions produced by transport, Ekoblog.cz reports the calculation of the specific emissions of diesel fuels [58]: "Specific emissions of CO_2 per kilometre driven by diesel combustion = 10,084/3.7584 × Specific consumption (l/100 km)/100 = Specific consumption × 26.83 (g CO_2/km)".

Transport-related emissions from transport by truck as the second component of the total CO_2 emissions $S_{t(truck)}$, can thus be evaluated based on formula:

$$S_{t(truck)} = q_t \times y \tag{8}$$

where q_t is the coefficient of diesel fuel CO_2 emission (26.83 in this case) and y represents the consumption in litres per 100 km. Last but not least, the total number of trips to transport the same amount of energy, Nt, should be evaluated. This number can be determined as the ratio of the amount of energy transported per carriage relative to the brown coal according to the formula:

$$Nt_w = E_{c1l}/E_{w1l} \tag{9}$$

where Nt_w represents the number of trips needed to transport the same amount of energy in the form of wood, which corresponds to one fully loaded car with a coal. The value Nt_c is less than always 1, because it is given by relation $Nt_c = E_{c1l}/E_{c1l}$. Consumers will primarily demand a certain amount of energy, not fuel; however, when a specific fuel volume is demanded in the order, estimation of the amount of energy is required.

The data are available in a different form for train and ship transport. According to the original study, which was conducted in 2004 under the conditions of the Czech Republic [59].

The following values were researched for rail transport. It is true that the study may not be perceived to be completely up-to-date, but given the developments in the field of automobile transport, significant changes in energy consumption cannot be expected, since the measures being considered, especially recovery, have not yet been implemented. In addition, these are average values.

Since similar studies are rare in this field and the data structure exactly matches the purpose of the research, we consider these data to be sufficient.

Shipping transports according to Table 3. 410,000,000 t of cargo over a distance of 1 km, consumes 128,000 GJ. It uses 0.312 MJ for the transport of one ton of cargo per km. Emissions were calculated analogously.

Table 3. Outputs, energy consumption and number of ton-kilometers (tkm) per 1 TJ of energy consumed by transport type.

Outputs, Corrected Energy Consumption, and Number of Ton-Kilometers (tkm) per 1 TJ of Energy Consumed in the Czech Republic in 2004.			
Type of transport	Transport volume (10^6 tkm)	Energy consumption (TJ)	Number of tkm/TJ
Truck transport	46,010	58,116	791,693
Railway motor transport	1690	2272	743,908
Railway electric transport	13,040	2761	4,723,200
Shipping	410	128	3,203,125

Source: [59].

Railway motor transport transports 1,690,000,000 tons of cargo per km and consumes 2,272,000 GJ. It therefore consumes 2,272,000 GJ/1,690,000,000 tkm for the transport of one ton of cargo per km, which is 1.344 MJ.

Therefore, we need 1.344 MJ of diesel to transport one ton of cargo over a distance of 1 km using diesel railways.

In case of electric railway transport, we calculated the outputs according to the same formula. Railway electric transport transports 13,040,000,000 t of cargo per 1 km and consumes 2,761,000 GJ. It therefore consumes 2,761,000 GJ/13,040,000,000 tkm for the transport of one ton of cargo per km, which is 0.212 MJ.

Therefore, we need 0.212 MJ of electricity to transport one ton of cargo over a distance of 1 km using an electrified railway.

The cargo volume according the Table 3 for electric and motor railway transport is a ratio of 88.53/11.47%. Therefore, the resulting emissions calculated for rail transport were combined from those two items, which together constitute the weighting for the weighted average calculation.

Specific CO_2 emissions can be calculated in a similar way as for truck by substituting into the Formula (there).

Under the conditions of the current electric energy mix, according to a decree of the Ministry of Trade and Industry [60], 1 kWh (3600 kJ) of electricity is produced with 1.17 kg of CO_2 in case of the Czech Republic.

Total emissions from river transport (shipping) are also calculated by substituting for Equation (8). The auxiliary calculation that had to be performed here is to convert the tonne/kilometre data according to Table 3 and to specific CO_2 emissions according to the formula.

$$S_{t(ship)} = (E_{tkm}/E_d) \times 2.683 \tag{10}$$

where E_{tkm} represents the energy consumption for transport of 1 t cargo for distance 1 km by ship, E_d represents the amount of energy included in 1 L of diesel fuel, and the number 2.683 is the weight of CO_2 emitted by combustion of 1 L of diesel in kg.

Total rail transport emissions can be also calculated from ton-kilometers, but according to Formula (8) CO_2 emissions can only be calculated for that part of rail transport which is realized using diesel locomotives. Under the conditions of the Czech Republic, this represents 11.47% of freight rail transport, while the remaining 88.53% is transport on electrified lines. To calculate CO_2 emissions, it is therefore, necessary to use data from the Ministry of Industry and Trade, which states an emission coefficient of 1.17 CO_2 per kJ [60].

Emissions for freight transport on electrified lines were calculated by multiplying the amount of energy per ton transported over a kilometer by an electrified railway by the amount of emissions determined by the Ministry of Industry and Trade. The formula used was:

$$S_{t(train-e)} = ((E_{tkm} \times Ef)/3600) * 1000 \tag{11}$$

where E_{tkm} represents amount of energy for transport of 1 ton cargo over distance 1 km on an electrified line, and Ef is the emission coefficient according the directive 425/2004 Coll. of the Ministry of Industry and Trade ČR. The 3600 is the conversion factor from kWh to J, and multiplication by 1000 was necessary to express the result in grams.

The following three relationships are original and key results of proposed methodology. These functions define the relationship between CO_2 emissions and monitored process parts. The total amount of wood CO_2 emissions, y_w, for transport and combustion together is given by:

$$y_w = (x \times S_{t(L=0)} + x \times S_{t(L=w)}) \times Nt_w \tag{12}$$

$$y_c = ((x \times S_{t(L=0)} + x \times S_{t(L=c)}) \times Nt_c) + S_c \tag{13}$$

Equations (12) and (13) characterize the total CO_2 emission of wood and the total CO_2 emission of brown coal y_c. Analogously2, CO_2 emissions for other modes of transport are calculated.

The whole methodology can be illustrated as depicted in Figure 1. The first arrow symbolises transportation and the second one corresponds to the combustion process. The variable S_w is not considered in the Equation (11), because of its zero value. But for the completeness and better clarity of proposed procedure, it is shown in Figure 1 and distinguished by the grey text colour. This model can then be further extended to other transport types.

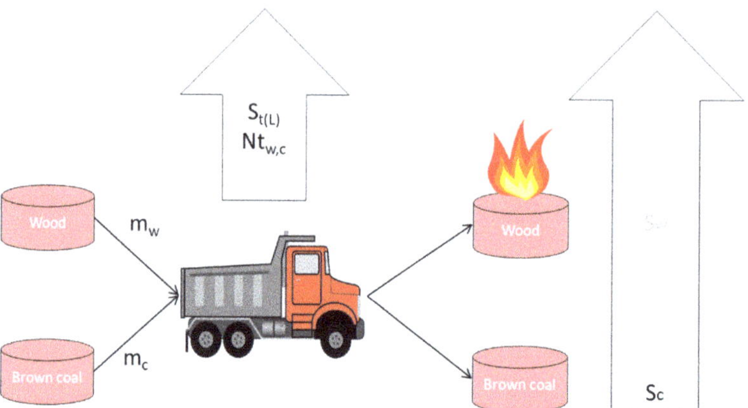

Figure 1. Illustration of process of the total CO_2 emission evaluation for transport and combustion together. Source: own processing.

4. Results

As we recognized the knowledge gap during the literature review, the previous chapter proposed an original methodology for calculating inefficiency in terms of CO_2 emission and transport distance.

This inefficiency relates to the nature of the material being transported. In the following paragraphs, the case study with specific empirical values (valid for the Czech Republic) is provided and the break point for wood is revealed.

4.1. Calculations

In the following chapters are the results of calculations based on formulae defined in the methodology section, first, for the CO_2 from the combustion of fuels used for heating, and second for fuels and energy types used for transport of fuels used for heating.

The following formulas show the process of the evaluation of the specific CO_2 emissions per unit of energy (MJ) when coal is assumed as the fuel. The emission data obtained from the Umberto software environment library [61] and the Bílina mine website were used for the calculation [62].

In line with the proposed methodology, the primary energy consumption, En_p, was determined at first by substitution into Equation (1).

$$En_p = En/\eta = 12.376/0.7 = 17.68 \text{ MJ/kg} \tag{14}$$

The result, giving the calorific value of the coal itself, without taking into account the efficiency of the equipment, is equal to 17.68 MJ/kg. This is in correspondence with calorific value for Ledvice coal, that is equal to 17.6 MJ/kg, as listed on the site of the mining company [62].

For this coal, Biom suggests emissions of 102.9 g CO_2/MJ for combusted brown coal with a humidity of 39.5% [63]. The Umberto software-based calculation gave 151.27g CO_2/MJ, which is 70% relative to the efficiency of the combustion plant. By substituting into Equation (2), we calculated the emission value for the fuel itself:

$$e_{p2} = e_m \times \eta = 151.27 \times 0.7 = 105.889 \text{ g } CO_2/\text{MJ} \tag{15}$$

For the evaluation of total CO_2 emissions from coal, the average value of these two values was used:

$$e_p = (102.9 + 105.889)/2 = 104.3945 \text{ g } CO_2/\text{MJ} \tag{16}$$

Substituting the empirical data from the above-mentioned sources [61,62], quite similar values were obtained. From those partial results, which are presented in the Equations (15) and (16), the average value was calculated. This is a substantial sub-result, since it has been found that the specific CO_2 emissions converted to energy in MJ do not differ significantly for different coal types. This fact makes it possible to assume that the resulting value can be utilized with little error for other types.

Wood as a Fuel no Number if There Is only One Subsection

The average energy value of firewood used most often in the Czech Republic for the following calculations was set in methodology chapter as 14.215 MJ/kg. In case of applying this procedure to another area with a different composition of used tree species, it is possible to update the calorific value and recalculate the results for any conditions.

Within the following relationship (Equation (17)), the weight of 1 m^3 of pure wood mass was calculated after processing to a length of 33 cm. The value ρ_w indicates the specific value of the wood processed to 33 cm and loose to the body of the transport device. The weight was calculated for this blend of wood as 582 kg/m^3 of pure wood; the expected volume of bulk timber was 0.41 m^3 of pure wood volume per 1m^3 of the loading area. A value of 0.41 m^3 is the diameter for a non-irrigated field of 33 cm, which is the most commonly used one [64].

$$\rho_w = M_n \times v_{ff} = 582 \times 0.41 = 238.62 \text{ kg/}m^3 \tag{17}$$

In the case of fuel transport, it is necessary to work with data that can be compared for further actions. Therefore, Table 4 provides a summary and supplementation of the data and weights of the fuels compared to their volume.

Table 4. Bulk density of the fuel.

Type of Fuel	Bulk Density of Fuel ρ_w [kg/m^3]
Wood	238.62
Brown coal	1100–1500

Source: own processing from [52].

Another important limiting factor is the mode of transport. The reason for this is the possibility of using different types of transport, and therefore, different specific CO_2 emissions can be expected, due to variable energy consumption per kilometre and variable transport capacity.

In Sections 4.2 and 4.3, the amount of energy contained in the transported solid fuels is described with full utilization of the capacity of the means of transport considered.

It is also necessary to take into account maximum weight of the load to be transported with respect to the weighed means of transport. The reason for this is to verify the possibility of using the full transport volume potential for the considered fuel and type of vehicle. The article considered that Tatra 815 S3 truck has two important limits: the weight limit of 10,700 kg and the volume limit of 8 m^3 of transported cargo.

4.2. Calculation for Wood

Based on the aforementioned calculations and data in Table 4, the bulk density of wood was taken to be 238.62 kg/m^3. The weight of transported wood was evaluated by substituting into Equation (4):

$$m_w = V_w \times \rho_w = 8 \times 238.62 = 1909 \text{ kg} \tag{18}$$

A fully loaded truck brings 8 m^3 of wood that weighs 1909 kg. Even if there the maximal load, the weight capacity of 10,700 kg will not be exceeded. Therefore, wood transport is limited by the volume of the hull.

To express the total CO_2 emissions from the monitored processes, it is necessary to express the total transported volumes of energy. This is necessary in order to determine the number of trips corresponding to the transport of the same amount of energy in the form of different types of fuels (wood and coal).

Therefore, it would be possible to transport 1.9 t of wood in one trip. Substituting into Equation (5) we got the energy content of that mass of wood:

$$E_{w1l} = E_w \times m_w = 14.23 \text{ MJ/kg} \times 1909 \text{ kg} = 27{,}165 \text{ MJ} \tag{19}$$

The resulting value represents the amount of energy in wood that can be transported by fully loaded transport vehicle.

4.3. Calculation for Brown Coal

Analogous calculations were done and results for brown coal were obtained. Those values were already calculated for bulk material.

Substituting into Equation (4):

$$m_c = 8 \times 1300 = 10{,}400 \text{ kg} \tag{20}$$

8 m^3 of brown coal would weigh 10,400 kg. Even of there were maximum load capacity, the weight limit would not be exceeded.

Substituting into Equation (5):

$$E_{cll} = 17.6 \text{ MJ/kg} \times 10{,}400 \text{ kg} = 183{,}040 \text{ MJ} \tag{21}$$

The resulting value of E_{cll} represents the amount of energy that can be transported in the case of a fully loaded transport vehicle and brown coal. Again, unlike wood, it was necessary to calculate the amount of CO_2 emissions generated by burning this brown coal.

Substituting into Equation (6):

$$Sc = 183{,}040 \text{ MJ} \times 104.3945 \text{ g} = 19.108 \text{ t } CO_2 \tag{22}$$

The corresponding CO_2 emissions contained in 8 m^3 of brown coal are 19.108 t of CO_2.

The results in Sections 4.2 and 4.3 present the first part of both combined processes to monitor CO_2 emissions. The next section opens the second part of the combined process calculations, focusing on the emissions from the transport process.

4.4. Calculation for Transport

The Tatra 815 S3 with an average transport capacity of 8–10 m^3, has fuel consumption for any load weight within the load interval shown in Figure 2. That dependency is given by Equation (7).

Expected consumption will differ because the weight will vary considerably in the cases considered. The basic fuel consumption of 32.5/63 L/100 km represents the range for an empty and fully loaded truck. Consumption, however, is given above all by the weight of the load, not its size (considering the air resistance does not make sense to include). The graph of Tatra 815's diesel fuel consumption is shown in Figure 2.

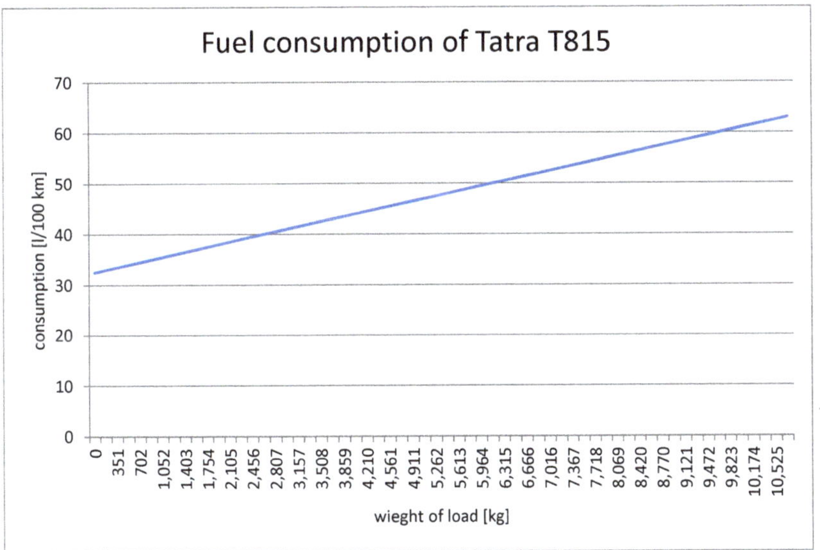

Figure 2. The result of the function describing the consumption of Tatra T815 S3, depending on the weight of the load. Source: own processing.

By substituting into Equation (7), consumption for the transport of the material concerned was obtained. Then, with the help of determined functional dependency, the fuel consumption of the selected vehicle was able to be calculated for any weight of the load being carried.

The results for the distribution of individual types of fuels are given in Table 5. The table also summarizes the results for the combustion process of transported fuels, which is the summary of the previous parts of the calculations. The values in Table 5 were converted to 8 m^3, representing the fully loaded Tatra 815 S3. The first column shows the weight of a fully loaded vehicle with a particular solid fuel; the second column shows the CO_2 emissions of selected solid fuels; and the third column summarizes the fuel consumption per 100 km of the right fuels selected by the vehicle.

Table 5. Summary table.

Fuel	Weight of Load Carried by 1 Ride ($m_{w,c}$) [kg]	CO$_2$ Emissions from Combustion of Fuel by Weight According to Column 2 (S_c ($S_{w\,=\,0}$)) [t]	Corresponding Consumption of Fuel with the Load According to the Column 2 (y) [L/100 km]
Without load	0	0	32.5
Wood	1909	0	37.94
Brown coal	10,400	19.108	62.14

Source: own processing.

Specific consumption and specific CO_2 emissions are included in Table 6. Because CO_2 emissions were the chosen identifier of the reviewed (analysed) processes, the emissions of CO_2 per 1 km were determined (Table 6). The results were calculated based on [58] by substituting the values of the third column of Table 5 as a variable x in Equation (7). The results were further converted to g/km and presented in Table 6. The reason is to compare the emissions related to transport of each type of solid fuel (mentioned in 1st column of Table 5). In general, is also possible to use this data for comparisons with other types of trucks or even different transport means.

Table 6. Specific consumption of l/100 km and specific CO$_2$ emissions.

Specific Consumption of L/100 km (y)	Specific CO$_2$ Emissions (S_t) [g/km]
32.5	871.975
37.94	1017.9302
50.74	1361.3542
62.14	1667.2162

Source: own processing.

Table 6, therefore, summarizes the emissions per kilogram of a given load. The order of rows is identical as in Table 5. The value of 871.975 g/km, contained in the second column of Table 6, corresponds to emissions per kilometre without a load; the value of 1017.9302 g/km corresponds to emissions per kilometre of a fully wood-loaded truck; the following values apply accordingly for driving a fully loaded car carrying brown coal. Since energy is being demanded in the form of energy volume, it is then necessary to express how much energy is transported in a fully loaded truck. That could be expressed by how many times the path must be taken to transport the same amount of energy relative to the chosen fuel variant. That is shown in Table 7, where $Nt_{(w,c)}$ are calculated according to Equation (9).

The first column of Table 7 describes the solid fuel types; the second column contains the amount of energy contained in the solid fuel per trip considering the type of transport. The third column shows the recalculated number of trips to be performed in order to carry the same amount of energy in all cases. Trips are recalculated according to the volume of solid fuel transported. The initial values and the results after substitution into Equation (5) are given in the second column. As a reference, the value for brown coal was used.

Table 7. Transported energy and number of rides for selected fuels.

Fuel	Transported Energy on a Fully Loaded T815 (E_{w1l}) [MJ], Train and Ship	Number of Trips ($Nt_{(w,c)}$)
Wood (w)	27.165	6.738
Brown coal (c_c)	183.040	1
Brown coal (c_t)	183.040	x
Brown coal (c_s)	183.040	x

Source: own processing.

The number of trips were left in decimal form, because research's subject was the specific emissions of CO_2 for the whole process. They are average values, not specific numbers of trips and specific deliveries. These values apply when carriers use their transmission capacities optimally. The actual condition is likely to exhibit worse parameters, since the used vehicle is not always loaded at 100%.

As can be seen from the data in Table 7, the smallest amount of transported energy is represented by transport of wood. For transporting the same amount of energy (that falls on one fully loaded carriage of brown coal) an average of 6.738 rides is needed.

Break Point Determination

The key finding of this article was the determination of CO_2 efficiency break point for different kinds of transport that can be evaluated directly from Figure 3.

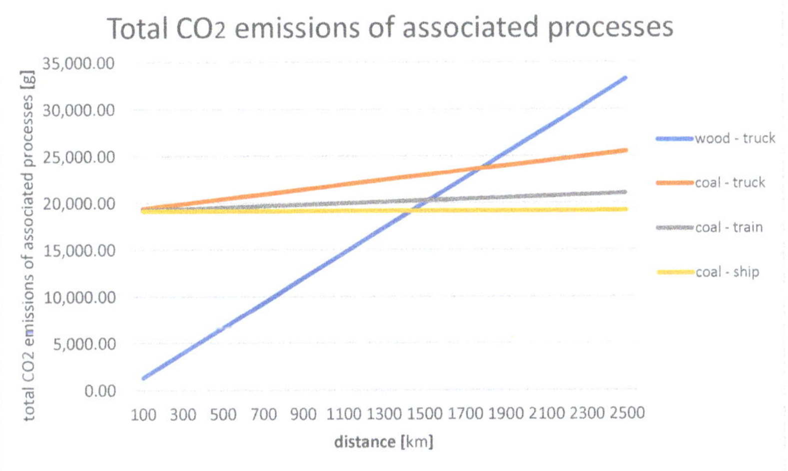

Figure 3. Total CO_2 emissions of associated processes of selected fuels and related break points. Source: own processing.

The evaluation of the CO_2-efficiency break point for associated processes has not been explored so far, so it was confirmed as a research gap. This was already demonstrated in the introduction, where a detailed overview of current state of the art and review of previous works with similar topic can be found. There are many studies with comparable topics, but neither of them examined the overall fossil CO_2 emissions by combining several processes within the supply chain as we have proposed. Studies [29–31] addressed biofuel reduction of CO_2 emissions; however, they did not address the problem of combining those factors. Contribution [32] dealt with the combined process of biomass combustion and transport; nevertheless, in terms of total energy involving the woodworking process. Environmental break points were not established for the efficiency of these processes, as we proposed, and a comparison with the combined process for fossil fuel was not included. Many other sources have been found, e.g., [38–42], addressing either the combination of transport and biofuels, or the determination of fossil fuel emissions. However, they examine the problem from a different point of view, with a different range and context. In studies [43–45], authors applied the LCA methodology, but none of them combined the LCA methodology together with the associated processes involving transport and combustion, and the comparison of biofuels with fossil fuels. Contributions [46,47] dealt with emissions from biomass and coal combustion together, but only for electricity generation. As was explained in detail earlier, all of those studies addressed selected issues from different points of view and with different scopes and contexts. Moreover, none of them had the ambition to set a specific breakpoint for the combination of transport and post-combustion of fossil fuels and biofuels in the CO_2 emissions context.

From a theoretical point of view, it has been shown that such points actually exist. Based on the solution of the system of defined equations, its value was calculated for the selected conditions: ca. 1780 km for transport by truck, ca. 1500 km by train, and ca. 1400 km by ship. As can be seen from Figure 3, according to various fuel parameters, break points for different fuel and transport combinations can be also determined. In the case of Figure 3, it shows a combination of wood and brown coal truck, train, and ship transport. On a practical level, a break point can be used to assess the effectiveness, sustainability, and environmental impact of any associated process. This coupled process is bounded by fuel transport and combustion.

From our results, it is obvious that, for normal distances in the order of kilometres or tens of kilometres, the difference in total specific CO_2 emissions is not significant. However, the overall emission curve for the wood transport process is considerably larger due to its lower weight curves than those for brown coal. This points to the fact that the combined process of burning wood and its transport is prone to a rapid increase in total CO_2 emissions over longer distances. As the model neglects CO_2 emissions associated with logging, the results for wood will be even more unfavourable than the ones show in Figure 2.

For distances of 1780, 1500, and 1400 km for trucks, trains, and ships, respectively (points where the blue curve crosses the others), it is better to burn brown coal than wood in terms of total CO_2 emissions.

5. Discussion and Conclusions

At the beginning of the discussion and before stating final conclusion, it is necessary to be aware of following facts, which underline the originality of the approach of researched in this paper, and at the same time, illustrate its benefits.

First, the aim of this paper was to deal with CO_2 emissions in energy production process based on calculation of the total specific CO_2 emissions of a combined process and setting the break point according to chosen factors. The outcomes are evident from the graphical interpretation of the results in Figure 3. Key factors were the type of fuel and the transport distance. The break point, where it is not worthwhile to use wood as an energy carrier as the alternative to coal, has been found. As can be seen in the search section, such an approach has not yet been applied in this context. Also, the dependence of specific CO_2 emissions from transport on the type of transported solid fuel was also revealed. Results were achieved within the original methodology, applied within the LCA framework. The general novelty of this paper is, with reference to the literature search, in the original view of analysing CO_2 emissions in the case of associated processes and with reference to the political level. Novel benefits are outlined in the following paragraphs. An example could be the discussed fuel transport in terms of energy volume or the interconnection of transport, cargo carried, and specific CO_2 emissions.

Second, the proposed methodology opens two issues to be discussed. First issue concerns its practical use. The second issue is about its general use of the methodology. In case of the practical use, it should be taken into account that, although the process and the associate input parameters were designed for the Central European environment (namely, the Czech Republic), the calculations facilitate its use in other regions of the world. This is mainly due to the conditions in Central Europe.

Original results show an unusually large transport distance for which the total specific CO_2 emissions are lower for one fuel (the calculated break-point).

Limitations of Study

While working on this study, we came across limitations that we consider necessary to state in order to broaden the perspective of the researched issue. What matters is the locality, because the choice of locality and its specifics determines other input parameters. Those are mainly the type of wood, type of transport, and transport distances. In case of the locality we chose, it is the environment of the Czech Republic, which in many aspects can be extended to the Central European region. This implies the specifics for the selected type of transport, also the means of transport, as well as certain above-mentioned parameters in case of rail transport and related CO_2 emission factors. Selected types of wood and their parameters are connected with the choice of the locality, which of course significantly contribute to the final results. At the same time, the selected location determines the transport distances common to the region. It should be noted, however, that when developing the methodology, the emphasis was placed on the possibility of its universality in terms of transferability to other environments and related, different input parameters. We believe that due to careful construction and a detailed description of the methodology, non-local application should be feasible without major obstacles.

Practical Use of the Methodology

Since the aim was to determine a theoretical break-point, it was necessary to think about the practical application of this result. In Europe, the usual fuel transport distance when using trucks is within range of tens to hundreds of kilometres. The break-points calculated in this study were 1779.64, 1500, and 1400 km, showing the fact that in Europe, transport efficiency is not a major problem in terms of specific CO_2 emissions. On the other hand, this distance limits the usability of the breakpoints itself and the circumstances resulting from it, even though the model shows the increasing inefficiency expressed by specific CO_2 emissions from traffic for any distance. But there are regions in the world where truck transport is realized for hundreds to thousands of kilometres. An example might be South America or some regions in Asia, where, due to insufficient rail networks trucks are used for long-distance transport. Thus, even though the original intention was to concentrate on the practical application of the model in Central Europe, the calculations in practice appear to be inappropriate for this region; but possibly worth considering for other regions. Because we did not consider it essential to change the input parameters typical of Central Europe (namely, the Czech Republic) for the purposes of this article, we used the Tatra 815 S3, which is a truck typical of the Czech Republic. In addition, parameters typical for coal and wood extracted in the Czech Republic were also used. However, the strength of this model lies in its easy generalization by the alteration of several input parameters to match the specific conditions of particular regions. For application in other regions, it is necessary to perform a parameter correction for a particular means of transport used. That could be related to another mode of transport as well. As far as trucks are concerned, the consumption and transport capacity of a particular vehicle needs to be known. For more accurate calculations, it is also appropriate to take into account the weight and average calorific value for the usual composition by species of burned wood, in regions where there is a completely different composition of trees (e.g., tropical). For coal, it is recommended to verify the parameters, especially in terms of its weight per unit volume and calorific value of a specific type of coal. For example, in South America, the practical application of this model gains it merit. For Europe's conditions, the meaning of the results can be seen not directly in determining the breakpoint itself, but in determining the rate of increasing inefficiency of wood use as a fuel with growing distance (in terms of total specific CO_2 emissions).

General Use of the Methodology

At the theoretical level, it has been shown that CO_2 is a suitable indicator for conjoined processes' analyses, even though the use of CO_2 for such applications has been underestimated. This was underlined by the works in the literature review; e.g., [18,46,47,65] compared to the results of this article.

Generally speaking, the intent that we opened at the beginning of the study was to deal with transport and combustion as a combined process. Parts of this process cannot be separated in practice, and we believe that it is not possible to deal only with emissions issues in the combustion process as a determinant that will represent the induced environmental burden. Calculations of total specific CO_2 emissions from the combustion and transport of wood can be considered only as a partial result. On this basis, it was possible to establish a functional relationship between the distance and overall emissions. This functional relationship was expressed by three equations and it was considered the basis of the original methodology. The break point, which was the final outcome of this work, was then, the final result of the above-mentioned equations.

Overall Conclusions and Discussion

The proposed methodology can also be evaluated in terms of overall conclusions, in terms of some partial results, and finally, in terms of the possibilities of generalization. The structure of the two parts of the work provide both a relatively complex view of, and some partial conclusions for the process of determination of total specific-emissions of CO_2 from the combustion of solid fuels and the emissions from the transportation of those fuels. The part which deals with the determination of CO_2 emissions from the combustion processes of selected solid fuels shows one of multiple possible views on the relatively accurate determination of some parameters of the combustion process's outputs. The values for wood were based on knowledge of the average composition of harvested trees corresponding to the conditions of Central Europe. The portability of the model to other regions is, of course, also possible, as already mentioned. However, a possible correction of parameters for the composition of local tree species is appropriate. Values for coal can usually be found on suppliers' or mining companies' websites. For these fuels, it is also possible to recommend validation and eventual value correction when applying the model to other regions. The energy content of the coal can shift the proposed breakpoint either to a greater distance, or alternatively, shorten the distance.

Increasing the amount of energy in wood will move the breakpoint closer to the beginning of the graph. Reducing the amount of energy will move the breakpoint to greater distances. For transport parameters, a simple model was designed to determine the change in consumption in relation to the freight load. In practice, the course of the function is not entirely linear. Most likely, it will be exponential, as the rolling resistance of the tires will also appear. As the straight line is not used at the speeds considered therein, such influences are not very significant; therefore, the linear function $y = 32.5 + (x/350.8196721)$ was constructed as is depicted on Figure 2. It can be said that the analysis has demonstrated the sensitivity of the timber supply process to transport when the process is considered in terms of CO_2 emissions. The results show that the total specific emissions of CO_2 from the comparison processes equalize only at the distance of 1779.64 km. With the reference to practical utilization, this problem should be seen in the fact that, already, at a distance of 355 km, the amount of CO_2 corresponds to 1/5 of the emissions that would be produced by heating with brown coal. Such a quantity of emissions largely limits the positive effects resulting from the neutrality of CO_2 emissions from the energy use of biomass. Besides the practical application, it raises questions about the results of the policies aiming to meet the requirement of the Paris Climate Conference. The final summary of this transmits this simple message: We have shown that there is a point that represents a turning point in which it is more efficient to burn coal than wood. Efficiency refers to the total CO_2 emissions of the combined process. Therefore, it can be stated that wood is not always a more environmentally friendly alternative, because, as we have shown, under certain circumstances it is more environmentally friendly to use coal.

Author Contributions: Conceptualization, R.B. and J.F.; Methodology, R.B. and J.F.; Software, R.B.; Validation, R.B. and J.F.; Formal Analysis, R.B. and J.F.; Investigation, R.B., J.F. and P.L.; Resources, R.B., J.F. and P.L.; Data Curation, R.B., J.F. and J.H..; Writing-Original Draft Preparation, R.B. and J.F.; Writing-Review & Editing, R.B. and J.F.; Visualization, R.B. and J.F.; Supervision, R.B. and J.F.; Project Administration, R.B., J.F., P.L. and J.H.; Funding Acquisition, R.B., J.F., P.L. and J.H.

Funding: This article was supported by the grant No. SGS_2019_19 of the Student Grant Competition.

Conflicts of Interest: The authors declare no conflict of interest.

References

1. IPCC. Climate Change 2014, Synthesis Report, Summary for Policymakers. 2014. Available online: 1url.cz/Ft9K6 (accessed on 7 October 2017).
2. United Nations. United Nations Framework Convention on Climate Change. 1992. Available online: 1url.cz/Ct9rt (accessed on 7 October 2017).
3. United Nations. Kyoto Protocol to the United Nations Framework Convention on Climate Change. Available online: https://unfccc.int/resource/docs/convkp/kpeng.pdf (accessed on 7 October 2017).
4. European Commission. Covenant of Mayors for Climate & Energy. Available online: http://www.covenantofmayors.eu/IMG/pdf/covenantofmayors_text_en.pdf (accessed on 7 October 2017).
5. United Nations. Paris Agreement. Available online: 1url.cz/3t9rr (accessed on 7 October 2017).
6. Ministry of the Environment of the Czech Republic. Covenant of Mayors for Climate and Energy. Available online: https://www.mzp.cz/cz/pakt_starostu_a_primatoru (accessed on 7 October 2017).
7. Recast, E.P.B.D. Directive 2010/31/EU of the European Parliament and of the Council of 19 May 2010 on the Energy Performance of Buildings. *Off. J. Eur. Union* **2010**, *18*, 2010.
8. European Parliament. Directive 2012/27/EU of the European Parliament and of the Council of 25 October 2012 on energy efficiency, amending Directives 2009/125/EC and 2010/30/EU and repealing Directives 2004/8/EC and 2006/32/EC. *Off. J. Eur. Union* **2012**, *315*, 1–56.
9. Government of the Czech Republic. Government Council for Sustainable Development. Available online: https://www.vlada.cz/cz/ppov/radavlady-pro-udrzitelny-rozvoj--120432/ (accessed on 30 December 2017).
10. Government of the Czech Republic. Government Council for Sustainable Development. Available online: https://www.vlada.cz/cz/ppov/udrzitelny-rozvoj/vybory-rvur/vybor-proududitelnou-energetiku-130368/ (accessed on 30 December 2017).
11. Ministry of Industry and Trade. Government Council for Energy and Raw Materials Strategy of the Czech Republic. 2017. Available online: https://www.mpo.cz/dokument147240.html (accessed on 30 January 2018).
12. Chamber of Deputies of the Czech Republic. *Act No. 165/2012 Coll. On Supported Energy Sources and on Amendments to Certain Acts*; Chamber of Deputies of the Czech Republic: Prague, Czech Republic, 2012.
13. Ministry of Industry and Trade. State Energy Policy. Available online: https://mpo.cz/dokument5903.html (accessed on 30 December 2017).
14. Ministry of Industry and Trade. State Energy Policy. Available online: https://www.mpo.cz/dokument158059.html (accessed on 30 December 2017).
15. Ministry of the Environment of the Czech Republic. Strategy on Adaptation to Climate Change in the Czech Republic. Available online: https://www.mzp.cz/C125750E003B698B/en/strategy_adaptation_climate_change/$FILE/OEOK_Adaptation_strategy_20171003.pdf (accessed on 30 December 2017).
16. Ministry of Industry and Trade. Strategic Framework for Sustainable Development of the Czech Republic. Available online: https://www.mpo.cz/assets/dokumenty/41073/45840/553766/priloha001.pdf (accessed on 20 February 2018).
17. Office of the Government of the Czech Republic, Department for Sustainable Development. Strategic framework Czech Republic 2030. Available online: https://www.cr2030.cz/wp-content/uploads/Strategický_rámec_ČR2030_komplet.zip (accessed on 20 August 2018).
18. Zhang, Y.; Mc Kenzie, J.; Cormier, D.; Lyng, R.; Mabee, W.; Ogino, A.; MacLean, H.L. Life cycle emissions and cost of producing electricity from coal, natural gas, and wood pellets in Ontario, Canada. *Environ. Sci. Technol.* **2009**, *44*, 538–544. [CrossRef] [PubMed]
19. Butt, E.W.; Rap, A.; Schmidt, A.; Scott, C.E.; Pringle, K.J.; Reddington, C.L.; Richards, N.A.D.; Woodhouse, M.T.; Ramirez-Villegas, J.; Yang, H.; et al. The impact of residential combustion emissions on atmospheric aerosol, human health, and climate. *Atmos. Chem. Phys.* **2016**, *16*, 873–905. [CrossRef]
20. Mumford, J.L.; He, X.Z.; Chapman, R.S.; Harris, D.B.; Li, X.M.; Xian, Y.L.; Jiang, W.Z.; Xu, C.W.; Chuang, J.C. Lung cancer and indoor air pollution in Xuan Wei, China. *Science* **1987**, *235*, 217–221. [CrossRef]
21. Finkelman, R.B.; Tian, L. The health impacts of coal use in China. *Int. Geol. Rev.* **2018**, *60*, 579–589. [CrossRef]
22. Weldu, Y.W.; Assefa, G.; Jolliet, O. Life cycle human health and ecotoxicological impacts assessment of electricity production from wood biomass compared to coal fuel. *Appl. Energy* **2017**, *187*, 564–574. [CrossRef]

23. Miehe, R.; Scheumann, R.; Jones, C.M.; Kammen, D.M.; Finkbeiner, M. Regional carbon footprints of households: A German case study. *Environ. Dev. Sustain.* **2016**, *18*, 577–591. [CrossRef]
24. Chafe, Z.; Brauer, M.; Héroux, M.E.; Klimont, Z.; Lanki, T.; Salonen, R.O.; Smith, K.R. Residential Heating with Wood and Coal: Health Impacts and Policy Options in Europe and North America. 2015. Available online: http://www.euro.who.int/__data/assets/pdf_file/0009/271836/ResidentialHeatingWoodCoalHealthImpcts.pdf (accessed on 20 August 2018).
25. Liu, Z.; Guan, D.; Wei, W.; Davis, S.J.; Ciais, P.; Bai, J.; Peng, S.; Zhang, Q.; Hubacek, K.; Marland, G.; et al. Reduced carbon emission estimates from fossil fuel combustion and cement production in China. *Nature* **2015**, *524*, 335–338. [CrossRef]
26. Junninen, H.; Mønster, J.; Rey, M.; Cancelinha, J.; Douglas, K.; Duane, M.; Borowiak, A. Quantifying the impact of residential heating on the urban air quality in a typical European coal combustion region. *Environ. Sci. Technol.* **2009**, *43*, 7964–7970. [CrossRef]
27. Walmsley, M.R.; Walmsley, T.G.; Matthews, L.; Atkins, M.J.; Neale, J.R.; Kamp, P.J. Pinch analysis techniques for carbon emissions reduction in the New Zealand industrial process heat sector. *Chem. Eng. Trans.* **2015**, *45*, 1087–1091. [CrossRef]
28. Nussbaumer, T. Combustion and co-combustion of biomass: Fundamentals, technologies, and primary measures for emission reduction. *Energy Fuel* **2003**, *17*, 1510–1521. [CrossRef]
29. Röder, M.; Whittaker, C.; Thornley, P. How certain are greenhouse gas reductions from bioenergy? Life cycle assessment and uncertainty analysis of wood pellet-to-electricity supply chains from forest residues. *Biomass Bioenergy* **2015**, *79*, 50–63. [CrossRef]
30. AEBIOM European Biomass Association; Network, B.B.; US Industrial Pellet Association. *Forest Sustainability and Carbon Balance of EU Importation of North American Forest Biomass for Bioenergy Production*; BC Bioenergy Network: Vancouver, BC, Canada, 2013.
31. Colnes, A.; Doshi, K.; Emick, H.; Evans, A.; Perschel, R.; Robards, T.; Saah, D.; Sherman, A. *Biomass Supply and Carbon Accounting for Southeastern Forests*; Biomass Energy Resource Center, Forest Guild, Spatial Informatics Group: Burlington, NJ, USA, 2012.
32. Thakur, A.; Canter, C.E.; Kumar, A. Life-cycle energy and emission analysis of power generation from forest biomass. *Appl. Energy* **2014**, *128*, 246–253. [CrossRef]
33. IPCC. Agriculture, Forestry and Other Land Use. Intergovernmental Panel on Climate Change, IPCC Guidelines for National Greenhouse Gas Inventories. 2006. Available online: http://www.ipcc-nggip.iges.or.jp/public/2006gl/vol4.html (accessed on 10 August 2019).
34. Sedjo, R.A. Comparative life cycle assessments: Carbon neutrality and wood biomass energy. *Resour. Future DP* **2013**, 11–13. [CrossRef]
35. Cherubini, F.; Peters, G.P.; Berntsen, T.; Strømman, A.H.; Hertwich, E. CO_2 emissions from biomass combustion for bioenergy: Atmospheric decay and contribution to global warming. *Glob. Chang. Biol. Bioenergy* **2011**, *3*, 413–426. [CrossRef]
36. Nian, V. The carbon neutrality of electricity generation from woody biomass and coal, a critical comparative evaluation. *Appl. Energy* **2016**, *179*, 1069–1080. [CrossRef]
37. Johnson, E. Goodbye to carbon neutral: Getting biomass footprints right. *Environ. Impact Assess. Rev.* **2009**, *29*, 165–168. [CrossRef]
38. Olivier, J.G.J.; Janssens-Maenhout, G.; Muntean, M.; Peters, J.A.H.W. *Trends in Global CO_2 Emissions*; 2016 Report; Netherlands Environmental Assessment Agency: Hague, The Netherlands, 2016.
39. International Energy Agency. *CO_2 Emissions from Fuel Combustion*; IEA: Paris, France, 2017.
40. Heinimö, J.; Junginger, M. Production and trading of biomass for energy—An overview of the global status. *Biomass Bioenergy* **2009**, *33*, 1310–1320. [CrossRef]
41. Proskurina, S.; Junginger, M.; Heinimö, J.; Tekinel, B.; Vakkilainen, E. Global biomass trade for energy—Part 2: Production and trade streams of wood pellets, liquid biofuels, charcoal, industrial roundwood and emerging energy biomass. *Biofuels Bioprod. Biorefin.* **2019**, *13*, 371–387. [CrossRef]
42. Yu, S.; Wei, Y.M.; Guo, H.; Ding, L. Carbon emission coefficient measurement of the coal-to-power energy chain in China. *Appl. Energy* **2014**, *114*, 290–300. [CrossRef]
43. Odeh, N.A.; Cockerill, T.T. Life cycle analysis of UK coal fired power plants. *Energy Convers. Manag.* **2008**, *49*, 212–220. [CrossRef]

44. Hondo, H. Life cycle GHG emission analysis of power generation systems: Japanese case. *Energy* **2005**, *30*, 2042–2056. [CrossRef]
45. Schreiber, A.; Zapp, P.; Kuckshinrichs, W. Environmental assessment of German electricity generation from coal-fired power plants with amine-based carbon capture. *Int. J. Life Cycle Assess.* **2009**, *14*, 547–559. [CrossRef]
46. Morrison, B.; Golden, J.S. Life cycle assessment of co-firing coal and wood pellets in the Southeastern United States. *J. Clean. Prod.* **2017**, *150*, 188–196. [CrossRef]
47. Sterman, J.; Siegel, L.; Rooney-Varga, J. Does replacing coal with wood lower CO_2 emissions? Dynamic lifecycle analysis of wood bioenergy. *Environ. Res. Lett.* **2018**, *13*. [CrossRef]
48. Millet, D.; Bistagnino, L.; Lanzavecchia, C.; Camous, R.; Poldma, T. Does the potential of the use of LCA match the design team needs? *J. Clean. Prod.* **2007**, 335–346. [CrossRef]
49. Tarantini, M.; Loprieno, A.D.; Cucchi, E.; Frenquellucci, F. Life Cycle Assessment of waste management systems in Italian industrial areas: Case study of 1st Macrolotto of Prato. *Energy* **2009**, 613–622. [CrossRef]
50. Ng, C.Y.; Chuah, K.B. Evaluation of Eco design alternatives by integrating AHP and TOPSIS methodology under a fuzzy environment. *Int. J. Manag. Sci. Eng. Manag.* **2012**, *7*, 43–52. [CrossRef]
51. Omar, W.M.S.; Doh, J.H.; Panuwatwanich, K. Variations in embodied energy and carbon emission intensities of construction materials. *Environ. Impact Assess.* **2014**, *49*, 31–48. [CrossRef]
52. Periodická Tabulka. Hustota Pevných Látek. Available online: http://www.prvky.com/hustota.html (accessed on 20 February 2018).
53. TZB Info. Calorific Value and Measurement Units of Firewood. 2016. Available online: http://www.tzb-info.cz/tabulky-a-vypocty/12-vyhrevnosti-a-merne-jednotky-palivoveho-dreva (accessed on 23 August 2018).
54. TZB Info. Calorific Value of Fuels. 2017. Available online: http://vytapeni.tzb-info.cz/tabulky-a-vypocty/11-vyhrevnosti-paliv (accessed on 29 August 2018).
55. Tatra. Technical Data of Individual Versions. Available online: http://tatratech.wz.cz/prospekty/t815-2/260s43.html (accessed on 25 August 2018).
56. Švik, P. Nákladní Doprava. Available online: http://www.petrsvik.cz/autodoprava.htm (accessed on 29 August 2018).
57. Tatra. Technical Data. Available online: http://tatratech.wz.cz/prospekty/t815/t815s3.html (accessed on 13 December 2017).
58. Ekoblog.cz. Technologie a Životní Prostředí. 2017. Available online: http://www.ekoblog.cz/?q=emise (accessed on 20 February 2018).
59. Zeman, J. Specific Energy Intensity of Individual Modes of Transport in the Czech Republic. (In Czech: Měrná Energetická Náročnost Jednotlivých Druhů Dopravy v ČR). Ekolist. Available online: https://ekolist.cz/cz/publicistika/nazory-a-komentare/merna-energeticka-narocnost-jednotlivych-druhu-dopravy-v-cr (accessed on 28 August 2019).
60. Doležel, J. Calculation of Carbon Dioxide (CO_2) Savings (in Czech: Výpočet Úspor Emisí Oxidu Uhličitého (CO_2)) Ministry of Industry and Trade ČR. Available online: https://www.mpo.cz/dokument6794.html (accessed on 28 August 2019).
61. UMBERTO Software. Ver. 5.5. Hamburg 2009. Advanced Tool for Material and Energy Flow Management. For Windows XP or Higher. Available online: https://www.ifu.com/umberto/ (accessed on 28 August 2019).
62. Doly Bílina. Bílinské Uhlí z SD a.s. Doly Bílina—Ledvícké Uhlí. Available online: http://www.bilinske-uhli.cz (accessed on 7 September 2017).
63. Biom. Alternativní Energetické Zdroje a Měrné Emise CO_2. Available online: https://biom.cz/cz/odborne-clanky/alternativni-energeticke-zdroje-a-merne-emise-co2 (accessed on 12 January 2018).
64. Dřevoprodukt. Measurement of Firewood. Available online: http://www.drevoprodukt.cz/upload/mereni_paliv_drivi.pdf (accessed on 12 January 2018).
65. Wang, C.; Chang, Y.; Zhang, L.; Pang, M.; Hao, Y. A life-cycle comparison of the energy, environmental and economic impacts of coal versus wood pellets for generating heat in China. *Energy* **2017**, 374–384. [CrossRef]

Article

The Efficiency of the Sustainable Development Policy for Energy Consumption under Environmental Law in Thailand: Adapting the SEM-VARIMAX Model

Pruethsan Sutthichaimethee * and Sthianrapab Naluang

School of Law, Assumption University, 592/3 Ramkhamhaeng 24, Hua Mak, Bangkok 10240, Thailand
* Correspondence: pruethsan.s@chula.ac.th; Tel.: +66-639-645-195

Received: 25 June 2019; Accepted: 9 August 2019; Published: 12 August 2019

Abstract: This research aims to predict the efficiency of the Sustainable Development Policy for Energy Consumption under Environmental Law in Thailand for the next 17 years (2020–2036) and analyze the relationships among causal factors by applying a structural equation modeling/vector autoregressive model with exogenous variables (SEM-VARIMAX Model). This model is effective for analyzing relationships among causal factors and optimizing future forecasting. It can be applied to contexts in different sectors, which distinguishes it from other previous models. Furthermore, this model ensures the absence of heteroskedasticity, multicollinearity, and autocorrelation. In fact, it meets all the standards of goodness of fit. Therefore, it is suitable for use as a tool for decision-making and planning long-term national strategies. With the implementation of the Sustainable Development Policy for Energy Consumption under Environmental Law $(S.D.EL)$, the forecast results derived from the SEM-VARIMAX Model indicate a continuously high change in energy consumption from 2020 to 2036the change exceeds the rate determined by the government. In addition, energy consumption is predicted to have an increased growth rate of up to 185.66% (2036/2020), which is about 397.08 ktoe (2036). The change is primarily influenced by a causal relationship that contains latent variables, namely, the economic factor $(ECON)$, social factor $(SOCI)$, and environmental factor $(ENVI)$. The performance of the SEM-VARIMAX Model was tested, and the model produced a mean absolute percentage error $(MAPE)$ of 1.06% and a root-mean-square error $(RMSE)$ of 1.19%. A comparison of these results with those of other models, including the multiple linear regression model (MLR), back-propagation neural network (BP model), grey model, artificial neural natural model (ANN model), and the autoregressive integrated moving average model (ARIMA model), indicates that the SEM-VARIMAX model fits and is appropriate for long-term national policy formulation in various contexts in Thailand. This study's results further indicate the low efficiency of Sustainable Development Policy for Energy Consumption under Environmental Law in Thailand. The predicted result for energy consumption in 2036 is greater than the government-established goal for consumption of no greater than 251.05 ktoe.

Keywords: environmental law; latent variables; structural equation modelling; sustainable development policy; energy consumption; vector autoregressive model

---

1. Introduction

Sustainable development policy has been given increasingly serious attention around the world. It is used side by side to define national strategies of various countries for different time scales; short-term, medium-term and long-term [1–3]. In the area of Environmental Law specifically, it is part of the driving mechanisms to run such a policy for economic, social and environmental sustainability [4–6]. In order to make national development more sustainable, mutual coordination between the national management policy and legislation is required, especially integrating and incorporating environmental law in order to achieve long-run sustainability [6–8].

For Thailand, the main goal of sustainable development policy is to play a core role in creating sustainability, as stated in the constitution of the Kingdom of Thailand B.E. 2560, with the say of government policy under Article 72. In addition, the government is given the role of managing the environment under Article 57, and protecting it under Article 58. Furthermore, this version of the constitution provides a new provision to guarantee the rights of the people and community toward the environment under Article 43, as well as grant the right to charge the government or government agencies with the responsibility for protecting the environment under Article 41. While the National Environmental Quality Promotion and Preservation Act (Version 2) B.E. 2561 [9] comes with a significant focus on the formulation of environmental protection policies, as follows: (1) promoting the participation of people and NGOs in protecting the environment, particularly Articles 6 to 8; (2) establishing an Environmental Committee under Articles 12 to 21; (3) establishing a Pollution Control Committee under Articles 52 to 54 as the main organization to determine pollution control policy; (4) establishing the Environment Fund under Articles 22 to 31; (5) overseeing the environmental quality management short-term plans of 5 and 20 years under Articles 35 to 41, which is deemed significant, especially the Sustainable Development Policy for Energy Consumption under Environmental Law; (6) establishing environmental standards under Articles 32 to 34; (7) establishing environmental protection zones under Articles 42 to 45; (8) establishing pollution control zones under Articles 59 to 63; (9) assessing environmental impact under Articles 46 to 51/7 and 101/1 to 101/2; and (10) determining the civil responsibility of polluters under Articles 96 to 111.

In fact, from 1995 to 2018, Thailand has tremendously improved its economic development, when Gross Domestic Product (GDP) of Thailand improves with an increasing growth rate [10]. The Thai government has continuously established policies to increase in national revenues. The main revenue-generated bases the government has attended to are the continuation of export activities to major trading partners, while increasing the diversification of exported goods with good quality and strong marketability. This is done together with establishing various measures to broadening market shares [9,10]. In addition, it can be observed that the government has adopted certain strategies by allowing others countries engaging in local investments in various industrial projects. Besides, the government allows joint investments together with other foreign countries within the main industries of Thailand, as well as promotes Thailand as a strategically important production base. These strategies are from both proactive and receptive approaches, including tax exemption for foreigners to land production bases in Thailand, and the promotion of international tourism, for instance [11]. Besides, the government seeks to promote and implement social policies at the same time. This has resulted in development and an increased growth rate. In general, the government has played a significant role in formulating different policies, such as the promotion in Employment Opportunities, Health and Illness, Social Security, Consumer Protection, as well as monitoring and follow-up programs [12,13]. However, with robust development in economic and social development, it has simultaneously led to the environmental change as well. By noting from the past (1995) until today (2018), the greenhouse gas rate has increased continuously, especially the increment of CO_2 emission from the energy-based sector. This cause of energy consumption tends to rise continuously in all sectors. The most sectors are the electronic sector, transportation sector and industrial sector, generating greenhouse gas up to 90.05 percent (2018) [13,14]. From the above discussion, it can be noticed that Thailand has succeeded with economic and social policy, yet environmental policy is not much given serious attention in development; resulting in the reduction of carrying capacity in the ecosystem. One of major reasons is that the inefficiency and weak enforcement of Environmental Law [9]. Moreover, there is still a lack of tools in the implementation of the Sustainable Development Policy in Energy Consumption under Environmental Law in Thailand to drive the nation towards the sustainability.

Establishing a sustainable development policy for Thailand is considered to be an important strategy for driving the country toward sustainability. This requires a powerful tool for ensuring that the outcomes of national policies and plans have the highest possible efficiency and effectiveness over short- and long-term periods. Meanwhile, the government can aim to mitigate or solve problems, particularly during the formulation process of the Sustainable Development Policy in Energy Consumption under Environmental Law. This is also seen as a necessity for national development under the national strategic plan because of its effects on economic, social, and environmental dimensions, which constitute part of a holistic approach to developing the nation. Any country that is able to strategize such an approach and turn it into reality will benefit by attaining sustainability in both the short term and long term. In the long run, there is a high possibility that problems and hurdles will occur within and outside the nation, and these challenges are usually difficult to control or even monitor. Thus, strategic planning must evolve from strong knowledge, capacities, and resources, since the output of this action will determine the future of the nation. To date, the implementation of Thailand's Sustainable Development Policy in Energy Consumption under Environmental Law is still weak and poorly planned. Moreover, there is still no single tool to facilitate a solution to this matter, which affects economic, social, and environmental systems. Therefore, determining the relationship among factors by developing a causal model that integrates economic, social, and environmental aspects, as well as enforcing the environmental law, has become crucial. In addition, studying the relevant research (discussed in the literature reviews section) reveals a gap that no other studies have focused on when proposing models for different contexts in various sectors. In fact, the previous studies have applied the same research methodologies, leading to insufficient analysis in different contexts and sectors. Therefore, the paper has understood the gap and problem, and starts introducing the SEM-VARIMAX model as a tool for national policy formulation and in all short-term, medium and long-term future planning.

2. Literature Review

Upon reviewing the relevant literature from available resources, many streamlined studies have highlighted the evolving concept of sustainable development, which has made significant progress in different areas worldwide. Zhou et al. [15] investigated the evolution of sustainable development-related politics and laws in China, and they found that ecological civilization tends to broadly tackle problems, focus on public participation, as well as fill the gap in environmental legislation. In their review of the historical experience of successful development in the Su-style furniture industry in the Ming Dynasty using a diamond model, Fan and Feng [16] found that style, material, skill, and government contributions, as well as consumer demand, had significant roles in gaining competitive advantages during that period. In fact, Boyd et al. [17] reviewed 10 Clean Development Mechanism (CDM) projects according to their sustainability privileges. The study later illustrated that sustainable development concerns have been marginalized in some countries. Joseph [18] has observed that, most Malaysian local authorities' personnel do not understand the concept of sustainable development and sustainability reporting. In Bangladesh, Bahauddin [19] revisited the environmental protection history and other relevant interests. This visit has made a new initiative possible by paving core best practices. Strengthening and restructuring key environmental organizations are of few guidelines that must be done. As of understanding the concept of sustainable development, Rivera [20] studied Sustainable Development Goals (SDGs), and investigated whether any work has been done between science and policy. The study pointed out the failure to fulfill some set criterion. Ali et al. [21] have investigated the connection between environmental degradation and economic growth in Pakistan through a test of Environmental Kuznets Curve along with Autoregressive Distributed Lag (ARDL) model. There comes the result of which inverted U-shaped relationship exists between those two spaces, implying the positive impact of population density on per capita carbon emission. In addition, the rise of energy consumption tends to degrade the environmental aspect. However, the role of multi-stakeholder partnerships dealing with climate change and sustainable development for developing countries was examined by Pinkse and Kolk [22]. Upon analysis, it revealed the participation of all parties in the creation of linkage between issues. To Choi and Ng [23], they attempted to understand consumers' responses on two sustainability concepts in terms of environmental and economic aspects, along with price. The study explained a positive consumer behavior on two-sustainability-focused companies, while there was no in favor of low price reaction when the consumers are aware of the firm with poor environmental sustainability. Amesheva [24] touched upon the environmental impact on development and social inequality along with recent legislative measures. As of a result, it revealed the need of reformation due to governance challenge. In the meanwhile, Bakari [25] shad some highlights on the challenges of sustainable development implementation in term of global governance, confirming less impact on the whole global governance system. While the need of economy and environment was found by Martin [26] with the affirmation of the assessment of environmental and welfare policies. The main aim of all the relevant studies in the review was to address the concept of sustainability in economic, social, and environmental aspects.

However, the UN Secretariat [27] reported the possible continuation of Urban development, and this aspect can be further improved by adopting the New Urban Agenda at the United Nations Conference on Housing and Sustainable Urban Development (Habitat III). Khalifa and Connelly [28] has found advantages to decision makers after the introduction of sustainable development indicators along with an index appropriate as compared to the current method of locally calculated Human Development Indices (HDI). In addition, Wuelser et al. [29] proposed an analytical framework assisting research on sustainable development by using theoretical conceptions and in-depth analysis by paving the setting of joint learning in policy making, shared visions and knowledge creation which in line with sustainable development's objectives. While Mueller et al. [30] discussed four different standards (ISO 14001, SA 8000, FSC and FLA), and revealed basic conditions for stakeholders to uphold, like CSR in supply chains, for instance. Based on other existing studies, Zhang et al. [31] evaluated the overall robustness of Ecological Footprint (EF) for decision-making on sustainability, while seeking ways to improve the EF. The new three methods were proposed, a correction factor for bio-capacity measurement, three-dimensional ecological footprint model and modified carbon footprint measurement. To add on, Wang et al. [32] introduced the ecological carrying capacity intensity (EC Intensity) according to the revised version of three-dimensional ecological footprint (3DEF) model. The findings of the study disclosed that EC Intensity has raised slowly with stronger capacity for regional development. Singh and Debnath [33] did a study to comprehend the Clean Development Mechanism (CDM). The finding gave them the fact that sustainable development is reachable if there is an emphasis on strategic goals and mission.

On the other hand, Giddings et al. [34] integrated environment, economy, and society to sustainable development and this required more than technical changes and a shift in human worldview. With the investigation of Sapukotanage et al. [35] on the sustainable practices of the manufacturing firms in a developing nation in South Asia, there was evidence of such sustainable practices leading towards sustainable performance. Sutthichaimethee [36] predicted the sustainable development policy implementation in the sanitary and service sectors of Thailand by 2045 with the result of potential growth of Thai economy system by 25.76% along with changes. Gradually, the Greenhouse gas emissions are found to increase by 49.65%. To Greaker et al. [37], they established a benchmark for climate policy at a national level. The greenhouse gas mitigation projects at certain cost and acquisitions of emission permits were part of the benchmark, as discussed in the study. It is also worth noting that many studies have shown the significance of why the concept comes into existence. Cetindamar and Husoy [38] understood why companies act environmentally responsible and that came with more than one reason, while ethical and economic reasons are found to be among the reasons. Bedore [39] further investigated the impact of new Canadian legislation, Federal Sustainable Development Act in 2008 on sustainable development. The result of this investigation pointed out the improvement of Canadian sustainable development planning systems due to this Act. Lee et al. [40] looked at Korea's official development assistance (ODA) projects in Sri Lanka as the basis of identifying policy issues on the sustainable development of developing countries, and the observation has shown the improvement made by the projects in term of environmental policy enhancement, public awareness, increased communication and cooperation between participating countries and follow-up management. While Aguilera-Caracuel et al. [41] sought to see the impact of institutional distance between the home and the host country, and the headquarters' financial performance on the environmental standardization decision among multinational companies. The finding showed that when an environmental institutional distance is high, it would slow down the standardization of environmental practices, while high-profit headquarters are ready to take part. Pires, Fidelis and Ramos [42] measured and compared local sustainable development based on common indicators by seeing through some constraints and achievements. As of their finding, it revealed that the communication, limited political support, and application of such indicators are the main issues, and these limit indicators' capacity towards sustainable development.

Giannetti, Demetrio, Bonilla, Agostinho and Almeida [43] later diagnosed an environmental energy of Brazil compared to Russia, India, China, South Africa and United States. The study concluded what actions may be put in place; reducing total energy use in developed economies and decreasing exportation of indigenous resources in developing economies. Wysokinska [44] analyzed the impact of eight UN Millennium Development Goals implementation, drawing a further implication that triggers the fight against poverty, hunger, disease, and environmental destruction, rather than mitigate the risk of climate change, global hunger, and the economic fallout. Panzaru and Dragomir [45] firmly stood with high importance of managers' involvement in predicting economic growth for sustainable and economic development. Byrch et al. [46] have found the participant maps in promoting business, and accommodating economic growth and development, as the key player in the sustainable development after they revisited the meaning of sustainable development held by New Zealand. In addition, Casey and Galor [47] examined the carbon emissions in terms of their effect of lower fertility. Regardless of its complexity, population policies were found to be part of the approach to tackling global climate change. Also, Ramakrishnan et al. [48] have established an environmental model for economic growth by integrating sustainability principles. The model produced a number of outcomes, showing the energy demand would decrease when the regional agricultural share rises.

However, another exploration on the engagement of sustainable development with legislation gives a better understanding of how such development can be enforced. Ladan [49] tried to establish a significant nexus between the SDGs, human rights and climate change. This study has concluded that national law must come into play in order to archive the above objective. Craig et al. [50] sought to study the flexibility and stability in governance. They came into a conclusion of which an attention to process and procedure along with increased use of substantive standards would improve and better the substantive flexibility level to operate with legitimacy and fairness. Whereas, Wang [51] reviewed ongoing debates pertaining to environmental regulation in developing countries and other aspects. During this exploration, China has been found to face environmental problems, yet China has made a serious long-term campaign to confront these issues. Furthermore, Bartel and Barclay [52] have applied Motivational Posture Theory to examine motivational attitudes on relevant areas, including government, environmental problems, environmental laws and regulations and farm management behaviors in the context of Australian agriculture and environmental regulation. Here, the compliance was found and supported both government and regulations. At the same arena, Kim and Mackey [53] exhibited the international environmental law and found it to be a complex network of treaties and institutions. Huber [54] visited the recurrent political challenge for environmental policymakers, and has found the matter of regulatory cost and change-resistant legal and institutional policy arrangements becomes the main challenge. Alongside, Tecklin et al. [55] explored the environmental policymaking process while examining the character and impact of the environmental governance. Here, the study was evident of the strongly market-enabling quality for the governance instead of the market-regulating one. In addition, Zeben [56] managed to introduce additional criteria for competence allocation, and this further expanded its application in the regulatory process, be it norm setting, implementation or enforcement.

If other studies are put further into the discussion, Bodansky [57] clarified the nature of the legitimacy challenge on environmental problem with the claim of decision-making deficit for both individual states and international institutions. With his study of the European Union, it demonstrated that the magnitude of legitimacy positively depends on the strength of the institution. Heinzerling [58] encouraged the lawyers furthering their efforts to attain proper laws and institutions that can reduce the effect of the polluting state. In China, Chang and Wang [59] began to tap on climate change, including most environmental governance system. However, the results showed that pollution discharge permit system is built upon insufficient resources, leading to differing standards for different places in China. Nonetheless, Periconi and Jokajtys [60] pointed out the importance of modern environmental laws in New York restricting on certain harmful practices for the environment, and those laws shall be continued for the current applications. Latham, Schwartz and Appel [61] investigated the intersection of tort and environmental law, and later found that such intersection should be narrowed in order to harmonize both statutory and common law. To this extent, Wood [62] addressed the failure of environmental law in the United States as all juristic agencies allowed so. In order to reduce such failure, the study suggested that all government institutions shall be held accountable for their discretions. While Gibson [63] put 10 basic design principles as part of environmental assessment consideration in Canada, triggering a new trend of global attention for the future version of environmental assessment. In particular, Fast and Fitzpatrick [64] explored the Environmental Rights Act of the Government of Manitoba in Bill 20, indicating the importance of the Environmental Bill of Rights in the legislation, and it must be placed in the on-going efforts to restructure the provincial environmental protection system. De Moerloose [65] has further compiled papers for the "2016 Law and Development Conference: From the Global South Perspectives." This compilation exhibited the disconnection between law and development, and that leads to further action on reconnecting law with development.

Nevertheless, Tania [66] has reviewed the trade–sustainable development debate in the view of Rio+20 and its relevant green economic policy. Here, the market access barriers for least developed country (LDC) is turned out to be the main concern for developed countries towards the sustainable development. Whereas Chepaitis and Panagakis [67] engaged legal philosophy in bridging individual capacity and environmental degradation, and this justified the return of greenhouse gases in the atmosphere with the absence of individual responsibility. Miao [68] analyzed the situations of the right to justice in environmental matters in China from a legal perspective. The study's findings have shown three main focuses in order to protect such right; engaging, effectiveness, and efficiency. To a broader aspect, Pourhashemi et al. [69] examined the international treaties and the United Nations Framework on Climate Change Convention in particular, as well as to evaluate the existing forms of legal and operational protection in relation with climate change. From this study, they have found many issues, and a failure to protect the rights of refugees and immigrants comes before hand. By tackling the above issues, it could actually result in efficient management of this crisis and stop the possible chaos across the globe. While Ruhl [70] investigated the context and policy dynamics of climate change and its trends while exploring normative and structural impacts on how environmental law fits in. The study has illustrated three main areas that environmental law plays: pollution control and ecological conservation, climate change mitigation, and its adaptation.

The structural equation model is a forecasting method commonly used in studies in a variety of contexts and for various objectives. Moreno et al. [71] applied a Structural Equation Model (SEM) and Confirmatory Factor Analysis (CFA) to understand the nature of classroom conflict in schools in Spain. Boccia and Sarnacchiaro [72] examined consumer attitudes pertaining to companies' corporate social responsibility initiatives by applying a structural equation model. Baumgartner and Homburg [73] assessed the applications of structural equation modeling in marketing and consumer research in three aspects, examining problematic issues and suggesting ways to improve them. Furthermore, Mai et al. [74] analyzed latent variables by comparing exploratory structural equation modeling

(ESEM) with structural equation modeling (SEM) and manifest regression analysis (MRA). In their study, ESEM was determined to provide the least biased estimation of regression coefficient. Ryu and Mehta [75] examined multilevel factorial invariance in n-level structural equation modeling (nSEM) by optimizing a multigroup multilevel confirmatory factor analysis. Lei and Lomax [76] examined structural equation modeling under nonnormality conditions using two different estimation methods. Significantly, the study showed no effect of estimation methods and nonnormality conditions on the standards errors of parameter estimates. Cugnata et al. [77] used Bayesian Networks (BN) to investigate factors regarding overall customer satisfaction to determine appropriate actions to improve customer satisfaction. Nylund et al. [78] simulated a study on the performance of latent class analysis (LCA), factor mixture model (FMA) and growth mixture model (GMM) to identify the number of classes in different sample sizes. In Japan, Saito et al. [79] estimated the effects of daily CO_2 exchange on environmental variables by using a path analysis, which showed soil temperature having a significant impact on ecosystem CO_2 exchange throughout the year. Yang and Yuan [80] proposed ridge generalized least squares (RGLS) as part of a structural equation modeling procedure for the development of formulas. Here, RGLS were found beneficial for enhancing parameter estimate efficiency.

A number of studies in various countries have attempted to optimize different forecasting models. In China, Chang et al. [81] deployed a fuzzy-based grey modeling (GM) procedure in the estimation of sulfur dioxide emissions. The study showed the effectiveness of the model and the forecasting indicated a decline in such emissions. Wang et al. [82] predicted air temperature by introducing a new integrated model, the Variational Mode Decomposition-Autoregressive Integrated Moving Average (VMD-ARIMA), which was found to be effective in providing accurate temperature forecasting. Ma et al. [83] predicted provincial vehicle ownership utilizing the Gompertz model, estimating a rapid growth in vehicle ownership in each province by 2050. Zhao et al. [84] used a giant information history simulation to estimate the value-at-risk (VaR) of oil prices, analyzing how various VaR factors from online news sources can most accurately measure crude oil VaR. Xiong et al. [85] incorporated a novel linear time-varying grey model (1,N) to predict haze while comparing it with the original GM model, finding that the novel linear time-varying GM model outperformed the original model. In New Zealand, Zhao et al. [86] explored the connection between household energy use and residential building costs by using time series methods, the exponential smoothing method, the autoregressive integrated moving average (ARIMA) model and the artificial neutral networks (ANNs) model. In the study, the ANNs model was proven to be the most accurate for cost forecasting. In the U.S., Barari and Kundu [87] revisited the role of the U.S. Federal Reserve in triggering the recent housing crisis by using a VAR model and found that federal funds rate did not lead to house price increases. Looking at the European Union, Tucki et al. [88] proposed a new method to investigate the development of the electromobility sector in Poland and the EU states. Their study concluded that Poland and the EU states require new approaches in terms of energy management and vehicle operation management. In Africa, Ahmed et al. [89] applied the ANNs model to forecast GRACE data of African watersheds and found that the model provided the most accurate forecast. Ramsauer et al. [90] adapted a Factor-Augmented Vector Autoregression Model (FAVAR) with an extension of a Kalman Filter for Factors to measure the impact of monetary policy in a case study.

From the review of the relevant literature, it was found that the research patterns and methodologies varied in terms of the research process and the statistics used to create forecasting models. This study is distinct from others; it addresses a gap in the research while also developing the research process and pattern and applying advanced statistics. In addition to this, the authors conducted other studies to support the forecasting models with different indicators. Those studies are titled "The efficiency of long-term forecasting model on final energy consumption in Thailand's petroleum industries sector: enriching the LT-ARIMAXS Model under a sustainability policy" [91] and "A relational analysis model of the causal factors influencing CO_2 in Thailand's industrial sector under a sustainability policy adapting the VARIMAX-ECM Model" [92]. The mentioned studies used a stationary process, while adapting the concept of a co-integration and error correction mechanism in order to analyze the real impact of the indicators on the dependent variable. This research extends and develops the above studies in depth using a model with improved accuracy by optimizing the advantages of those studies. From those advantages, this study analyzes direct and indirect effects, and it adjusts each latent variable toward equilibrium. The output can be used as a tool to formulate future national policies and plans. Furthermore, this study aims to create new knowledge and act as a guide for research and education. Importantly, this study features a model that is applicable to various sectors and contexts, and it was developed with the aim of producing efficient and effective study outcomes. As mentioned earlier, this study applied advanced statistics in a proper context to make the model applicable to different sectors. As a whole, this construct is called "Structural Equation Modeling-Vector Autoregressive with Exogenous Variables Model" (SEM-VARIMAX Model). The research also utilizes Linear Structural Relations (LISREL) [93] software along with Econometric Views (EVIEWS) [94,95]. The above-mentioned model is assessed in term of its Model Validity and Best Modelling, as well as "Best Linear Unbiased Estimated (BLUE) assessment. This is to ensure that there will be no issues of Heteroskedasticity, Multicollinearity, and Autocorrelation. Once the complete model is obtained, it is then deployed to analyze the future trend together with qualitative analysis. The Model Validity Test is done through Triangulation. The research flow is explained below, as shown in Figure 1.

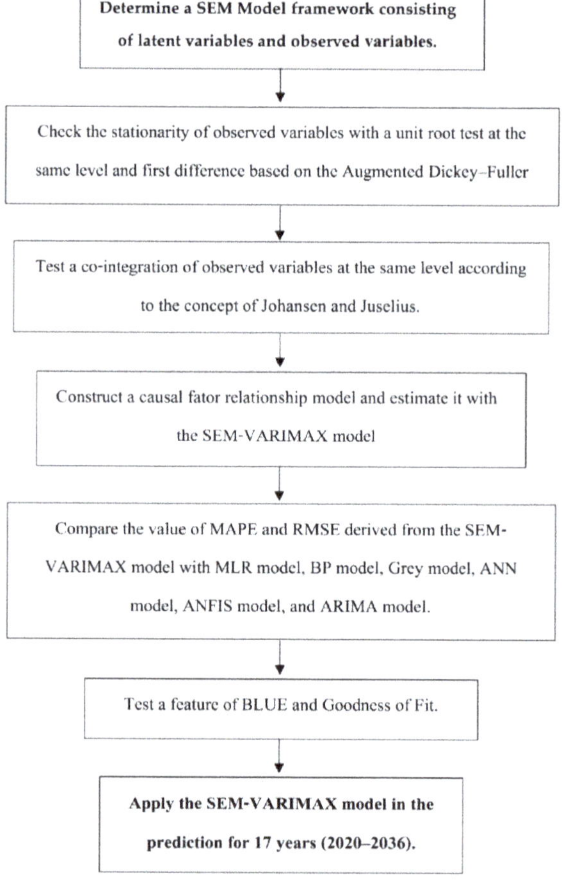

Figure 1. The flowchart of the structural equation modeling/vector autoregressive model with exogenous variables (SEM-VARIMAX) model.

1. Determine a variable framework based on the SEM-VARIMAX model, which contains both latent variables of Sustainable development policy under environmental law (*S.D.EL*), economic (*ECON*), social (*SOCI*), and environmental (*ENVI*), and observed variables of Sustainable development policy under environmental law indicators, Error Correction Mechanism (ECM_{t-1}) and energy consumption (*EC*). As of the economic indicators, there are per capita GDP (*GDP*), urbanization rate (*UR*), industrial structure (*IS*), net exports ($E - M$), indirect foreign investment (*IF*), foreign tourists (*FT*). The social indicators are employment (*EM*), health and illness (*HI*), social security (*SS*), consumer protection (*CP*). The environmental indicators include Carbon Dioxide Emissions (CO_2) and energy intensity (*EI*).
2. Examine and check the stationarity of observed variables with a unit root test based on the concept of the Augmented Dickey–Fuller [96].
3. Test the co-integration of observed variables at the same level [97–99].
4. Construct a causal factor relationship model and estimate its relationship with the SEM-VARIMAX model along with other test of BLUE characteristic and its Goodness of Fit [36,100].
5. Compare the effectiveness of the SEM-VARIMAX model with other models, including with MLR model, BP model, Grey model, ANN model, ANFIS model, and ARIMA model. through a performance measure of *MAPE* and *RMSE* [101,102].
6. Forecast the future Sustainable Development policy for energy consumption under Environmental Law with the use of a sample indicator, which is energy consumption (*EC*), by deploying the SEM-VARIMAX model for the year of 2020 to 2036, totaling 17 years. The flowchart of the SEM-VARIMAX model is shown below.

3. Materials and Methods

The SEM-VARIMAX was developed through the application of advanced statistics, consisting of the causal factor relationship called structural equation modeling (SEM), and the estimation of such a relationship with the vector autoregressive model. The model was structured to be the best model, hereafter referred to as the SEM-VARIMAX model. Furthermore, the model is characterized with the best linear unbiased estimate, and is not spurious. The details are provided below.

3.1. SEM-VARIMAX Model

Structural equation modeling (SEM) is a second-generation model, which can analyze the relationships between multiple levels of SEM. This is inclusive of completely analyzing relationships in the inner model (structure model) and outer model (measurement model). This feature differs from first generation modeling, such as regression analysis, ANOVA, and MANOVA, which are used to analyze a single subject at a time and so may take longer for the path model. Even though the outcome is no different, of which the study findings is parallel with this result, it is not a numerical value of regression coefficients, statistical values t (t-test) and other indicators. This is because these values are commonly different as their method is different; however, they still have similar values [72,73,103].

SEM produces models that indicate the relationships between variables, and it is explained below [104].

Given that $X = \{ X_1, X_2, \dots, X_H \}$ represent the observed value of the exogenous latent variable.

Given $\xi = \{ \xi_1, \xi_2, \dots, \xi_H \}$ is the exogenous latent variable (we may call the latent variable the score or component).

Given that $Y = \{ Y_1, Y_2, \dots, y_K \}$ represent the observed value of the endogenous latent variable.

Given $\eta = \{ \eta_1, \eta_2, \dots, \eta_K \}$ is the endogenous latent variable.

The coefficient π_j is the multiple regression coefficient.

The reflective relationship has been developed to positively connect observer variables (MV) and latent variables (LV), and it is a loading, and/or the regression coefficient must be a positive value. Yet, it is allowable to have a negative value, but such value tells us some issues with data. For instance, it indicates an incompatibility in the scale of measurement, and that the mean or variance does not reflect the real meaning of the data.

Therefore, the solution is to change the estimation method by using the Vector Autoregressive model at p: VAR(p) as follows.

There are 2 series of time series, which are Y_t and Z_t. They can be written in the VAR(p) model as shown below [105,106].

$$Y_t = a_{10} + a_{11,1}Y_{t-1} + a_{12,1}Z_{t-1} + a_{11,2}Y_{t-2} + a_{12,2}Z_{t-2} + \ldots + a_{11,p}Y_{t-p} + a_{12,p}Z_{t-p} + u_{1t} \tag{1}$$

$$Z_t = a_{20} + a_{21,1}Y_{t-1} + a_{22,}Z_{t-2} + a_{21,2}Y_{t-2} + a_{22,2}Z_{t-2} + \ldots + a_{21,p}Y_{t-p} + a_{22,p}Z_{t-p} + u_{2t} \tag{2}$$

If we have n series of time series, including $X_{1t}, X_{2t}, \ldots, X_{nt}$, we will write that time series in the VAR(p) model as illustrated below.

$$X_t = A_0 + A_1 X_{t-1} + A_2 X_{t-2} + \ldots + A_p X_{t-p} + u_t \tag{3}$$

$$\text{where } X_t = \begin{bmatrix} X_{1t} \\ X_{2t} \\ \vdots \\ X_{nt} \end{bmatrix}_{n \times 1}, A_0 = \begin{bmatrix} a_{01} \\ a_{02} \\ \vdots \\ a_{0n} \end{bmatrix}_{n \times 1}, A_i = \begin{bmatrix} a_{11,i} & \cdots & a_{1n,i} \\ a_{21,i} & \cdots & a_{2n,i} \\ \vdots & \vdots & \vdots \\ a_{n1,i} & \cdots & a_{nn,i} \end{bmatrix}_{n \times n}, i = 1,\ldots,P, \text{ and } u_t = \begin{bmatrix} u_{1t} \\ \vdots \\ u_{nt} \end{bmatrix}_{n \times 1}.$$

As for measuring the mean and variance of the VAR(p) model, the same method can be used as that of the VAR(1) model. When observing the VAR(p) model, the value of the parameter is many, that is constant in the number of n. In addition, the parameters as the coefficient value of $X_{t-1}, X_{t-2}, \ldots, X_{t-p}$ are $n^2 + n^2 + \ldots + n^2 = pn^2$. Hence, all parameters of the VAR model are $n + pn^2$. Here, it indicates that the greater the number of time series is by 1 unit or sequence of VAR is bigger by 1 unit, the parameters will also be greater at the same time. Thus, any time series used in the VAR model should be an impactful time series, that can explain each other's effect.

However, constructing a model as the best model requires a BLUE feature. In the actual context, there should be exogenous variables in the modelling. This simplifies that the model should have white noise and be free from a spurious in which heteroskedasticity, multicollinearity and autocorrelation are eliminated. The authors, therefore, developed a new model called the SEM-VARIMAX model, which effectively incorporates various exogenous variables in different contexts or various sectors. The details of the SEM-VARIMAX model are explained as follows [107].

$$Y_t = \beta_{10} - \beta_{12}Z_t + \gamma_{11}Y_{t-1} + \gamma_{12}Z_{t-1} + \varepsilon_{yt} \tag{4}$$

$$Z_t = \beta_{20} - \beta_{21}Y_t + \gamma_{21}Y_{t-1} + \gamma_{22}Z_{t-1} + \varepsilon_{yt} \tag{5}$$

$$BX_t = \Gamma_0 + \Gamma_1 X_{t-1} + \varepsilon_{yt} \tag{6}$$

$$\text{where } B = \begin{bmatrix} 1 & \beta_{12} \\ \beta_{21} & 1 \end{bmatrix}, X_t = \begin{bmatrix} Y_t \\ Z_t \end{bmatrix}, \Gamma_0 = \begin{bmatrix} \beta_{10} \\ \beta_{20} \end{bmatrix}, \Gamma_1 = \begin{bmatrix} \gamma_{11} & \gamma_{12} \\ \gamma_{21} & \gamma_{22} \end{bmatrix}, \varepsilon_t = \begin{bmatrix} \varepsilon_{yt} \\ \varepsilon_{zt} \end{bmatrix}.$$

When considering Equation (4), it indicates that ε_{yt} will affect Y_t, while Y_t will affect Z_t when considering Equation (5) (or briefly written as $\varepsilon_{yt} \to Y_t \to Z_t$). Hence, we can say that $Cov\left(Z_t, \varepsilon_{yt}\right) \neq 0$ or time series of Z_t and ε_{yt} are related, indicating the assumption of CLRM is incorrect. Therefore, the parameter estimation in Equation (4) will be a biased estimator. Even if the sample is large, it still finds that the probability of the estimator with the least squares method with not be as the actual value (inconsistent estimator). Equation (5) will also give the same result as above. However, when the SEM-VARIMAX model is transformed into a deformed model, or VAR(1) model by multiplying B^{-1} throughout Equation (6), Equation (7) will be as follows [94,108,109].

$$X_t = A_0 + A_1 X_{t-1} + u_t \tag{7}$$

where $A_0 = B^{-1}\Gamma_0 = \begin{bmatrix} a_{01} \\ a_{02} \end{bmatrix}$, $A_1 = B^{-1}\Gamma_1 = \begin{bmatrix} a_{11} & a_{12} \\ a_{21} & a_{22} \end{bmatrix}$ and $u_t = B^{-1}\varepsilon_t = \begin{bmatrix} u_{1t} \\ u_{2t} \end{bmatrix}$, or rewritten as Equations (8) and (9) like below.

$$Y_t = a_{10} + a_{11}y_{t-1} + a_{12}Z_{t-1} + u_{1t} \tag{8}$$

$$Z_t = a_{20} + a_{21}y_{t-1} + a_{22}Z_{t-1} + u_{2t} \tag{9}$$

We can see that the VAR(1) model will not cause any problems like what occurred in the SEM-VARIMAX(1) model. Besides, we can estimate the parameters in the VAR(1) model with the least squares method. By observing the SEM-VARIMAX(1) model and the VAR(1) model, it was found that

- The parameters in the VAR(1) model are actually caused by the parameters in the SEM-VARIMAX(1) model, or the parameters of both models are related.
- The number of parameters in the VAR(1) model was 9, namely the $a_{10}, a_{11}, a_{12}, a_{20}, a_{21}, a_{22}$, parameters of $Var(u_{1t})$, the parameters of $Var(u_{2t})$ and the parameter of $Cov(u_{1t}, u_{2t})$.
- The number of parameters in the SEM-VARIMAX(1) model was 10, namely the β_{10}, β_{11}, β_{20}, $\beta_{21}, \gamma_{11}, \gamma_{12}, \gamma_{21}, \gamma_{22}$, parameters of $Var(\varepsilon_{1t})$ and the parameters of $Var(\varepsilon_{2t})$.

It can be observed that the number of parameters in the VAR(1) model is less than the SEM-VARIMAX(1) model's. Even though all the parameters can be estimated in the VAR(1) model, we still cannot use the relationship between the parameters of both models to find the parameter estimator in the SEM-VARIMAX(1) model [91,92].

However, if we can place some limitations in the SEM-VARIMAX(1) model, then it would cause a reduction in the number of parameters to 9, allowing us to use the parameter estimator in the VAR model(1) in discovering the parameter estimator of the SEM-VARIMAX(1).

As for the SEM-VARIMAX(p) model, it can run up to sequence (p) as determined in the study, aiming at benefiting future applications.

3.2. Measurement of the Forecasting Performance

In this research, tested the performance of the SEM-VARIMAX model by comparing it with other exiting models, like the MLR, BP, Grey, ANN, ANFIS, and ARIMA models. In this comparison, we use the *MAPE* and *RMSE* values to examine the forecasting accuracy in each model. The calculation equations are shown as follows [101–103]:

$$MAPE = \frac{1}{n}\sum_{i=1}^{n}\left|\frac{\hat{y}_i - y_i}{y_i}\right| \tag{10}$$

$$RMSE = \sqrt{\frac{1}{n}\sum_{i=1}^{n}(\hat{y}_i - y_i)^2} \tag{11}$$

4. Empirical Analysis

4.1. Screening of Influencing Factors for Model Input

In this paper, the structure equation modeling framework was determined. Four factors were modelled as latent variables as follows: Sustainable development policy under environmental law

$(S.D.EL)$, economic $(ECON)$, social $(SOCI)$, and environmental $(ENVI)$, while the observed variables comprised of 13 indicators inclusive of energy consumption (EC). The economic indicators were per capita GDP (GDP), urbanization rate (UR), industrial structure (IS), net exports $(E-M)$, indirect foreign investment (IF), and foreign tourists (FT). The social indicators are employment (EM), health and illness (HI), social security (SS), consumer protection (CP). The environmental indicators comprised Carbon dioxide emissions (CO_2) and energy intensity (EI). This research analyzed the influence of the relationship of the causal factors with the SEM-VARIMAX model. All the causal factors used in the model must be stationary at the same level only. Here, the natural logarithm of every variable is taken, so that linear data can be obtained and tested for stationary. The value obtained from this process is then compared to MacKinnon critical value at level I (0) based on the Dickey-Fuller theory. In this paper, all variables were found to be non-stationary at level I (0). Therefore, those variables were carried forward to perform stationary tests at the first difference I (1), as illustrated in Table 1.

Table 1. Stationary at first difference I (1).

Stationary		MacKinnon Critical Value		
Variables	**Tau Test**	**1%**	**5%**	**10%**
$\Delta\ln(EC)$	-5.69 ***	-4.15	-3.20	-2.50
$\Delta\ln(GDP)$	-5.21 ***	-4.15	-3.20	-2.50
$\Delta\ln(UR)$	-5.16 ***	-4.15	-3.20	-2.50
$\Delta\ln(IS)$	-4.65 ***	-4.15	-3.20	-2.50
$\Delta\ln(E-M)$	-5.05 ***	-4.15	-3.20	-2.50
$\Delta\ln(IF)$	-4.50 ***	-4.15	-3.20	-2.50
$\Delta\ln(FT)$	-4.32 ***	-4.15	-3.20	-2.50
$\Delta\ln(EM)$	-4.29 ***	-4.15	-3.20	-2.50
$\Delta\ln(HI)$	-4.68 ***	-4.15	-3.20	-2.50
$\Delta\ln(SS)$	-4.91 ***	-4.15	-3.20	-2.50
$\Delta\ln(CP)$	-5.01 ***	-4.15	-3.20	-2.50
$\Delta\ln(CO_2)$	-5.85 ***	-4.15	-3.20	-2.50
$\Delta\ln(EI)$	-5.34 ***	-4.15	-3.20	-2.50

Notes: EC is energy consumption, GDP is per capita GDP, UR is urbanization rate, IS is industrial structure, $E-M$ is the net exports, IF is indirect foreign investment, FT is foreign tourists, EM is employment, HI is health and illness, SS is social security, CP is consumer protection, CO_2 is carbon dioxide emissions, EI is energy intensity. *** denotes a significance, $\alpha = 0.01$, compared to the Tau test with the MacKinnon critical value, Δ is the first difference, and ln is the natural logarithm.

Table 1 shows that all factors were stationary at the first difference or stationery at Level I (1). When calculating the Tau test of every causal factor, the values were found to be greater than the MacKinnon critical value, which indicates all causal variables were stationary at a significance level of 1%, 5%, and 10%. When all causal factors were stationary at the same level, we used them for the co-integration test proposed by Johansen and Juselius, as shown in Table 2.

Table 2. Co-integration test by Johansen and Juselius.

Variables	Co-Integration Test		Mackinnon Critical Value	
$\Delta\ln(EC), \Delta\ln(GDP), \Delta\ln(UR), \Delta\ln(IS),$ $\Delta\ln(E-M), \Delta\ln(IF), \Delta\ln(FT), \Delta\ln(EM), \Delta\ln(HI),$ $\Delta\ln(SS), \Delta\ln(CP), \Delta\ln(CO_2), \Delta\ln(EI)$	Trace statistic test	Max-Eigen statistic test	1%	5%
	215.75 ***	121.01 ***	15.25	11.75

Notes: *** denotes significance $\alpha = 0.01$.

4.2. Analysis of Co-Integration

According to Table 2, the co-integration test result based on Johansen and Juselius shows that all causal factors, which were stationary at first difference in the SEM-VARIMAX model, were co-integrated at the significance level of 1% and 5% because the Trace statistic test value (215.75) and the Maximum Eigen statistic test value (121.01) were greater than the MacKinnon critical values at significance levels of 1% and 5%, respectively. Therefore, all variables could be used in analyzing the impact of causal factors using the SEM-VARIMAX model, as shown in Figure 2 and Table 3.

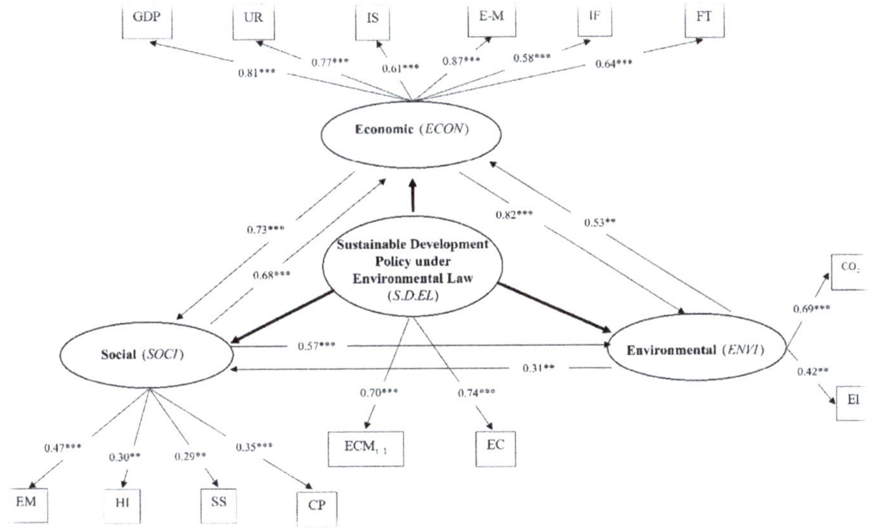

Figure 2. The casual relationship in the SEM-VARIMAX model.

Table 3. Results of the relationship of size analysis of the structural equation modeling/vector autoregressive model with exogenous variables (SEM-VARIMAX) model.

Dependent Variables	Type of Effect	Independent Variables			
		Economic (ECON)	Social (SOCI)	Environmental (ENVI)	Error Correction Mechanism (ECM_{t-1})
Economic (ECON)	DE	-	0.68 ***	0.53 **	−0.39 ***
	IE	-	0.05 ***	0.02 **	-
Social (SOCI)	DE	0.73 ***	-	0.31 **	−0.26 ***
	IE	-	-	0.09 **	-
Environmental (ENVI)	DE	0.82 ***	0.57 ***	-	−0.05 ***
	IE	0.12 ***	0.09 ***	-	-

Note: In the above, *** denotes significance $\alpha = 0.01$, ** denotes significance $\alpha = 0.05$, χ^2/df is 1.19, RMSEA is 0.05, RMR is 0.003, GFI is 0.95, AGFI is 0.90, R-squared is 0.94, the F-statistic is 225.05 (probability is 0.00), the ARCH test is 22.85 (probability is 0.1), the LM test is 1.35 (probability is 0.10), DE is the direct effect, and IE is the indirect effect.

4.3. Formation of Analysis Modeling with the SEM-VARIMAX Model

The SEM-VARIMAX model is a model that consists of short-term and long-term causal relationships, which show the impact of the latent variables, with the analysis explained as follows.

Figure 2 shows the impact of the causal factor relationship in the SEM-VARIMAX model determined by Sustainable development policy under environmental law (S.D.EL), where the latent variables are: economic (ECON), social (SOCI), and environmental (ENVI); the observed variables consist of energy consumption (EC), per capita GDP (GDP), urbanization rate (UR), industrial structure (IS), net exports (E − M), indirect foreign investment (IF), foreign tourists (FT), employment (EM), health and illness (HI), social security (SS), consumer protection (CP), Carbon dioxide emissions (CO_2), energy intensity (EI), and error correction mechanism (ECM_{t-1}). The study findings reveal which factors had direct and indirect effects, as can be seen in Table 3.

Table 3 illustrates the parameters of the SEM-VARIMAX model at the statistically significant level of 1% and 5%. With the analyzed findings, the SEM-VARIMAX model features with the goodness of fit standards, where the value of RMSEA and RMR is not far from 0 (zero), while the GFI and AGFI values approach 1. Furthermore, the BLUE testing indicates that the SEM-VARIMAX model has a BLUE feature, indicating that the model is not spurious yet it is reliable. This is due to the absence of heteroskedasticity, multicollinearity, and autocorrelation. In contrast, the F-test matters at the significance level of 1%. Besides, the SEM-VARIMAX model explains a lot about the model featured under Sustainable Development Policy with Environmental Law ($S.D.EL$). In detail, the economic factor ($ECON$) has a direct impact on the environmental factor ($ENVI$) amounting to 82% at a significance level of 1%, the economic factor ($ECON$) has a direct impact on the social factor ($SOCI$) of 73% at a significance level of 1%, the social factor ($SOCI$) has a direct impact on the environmental factor ($ENVI$) totaling 57% at a significance level of 1%, the environmental factor ($ENVI$) has a direct impact on the social factor ($SOCI$) at 31% at a significance level of 5%, and the environmental factor ($ENVI$) has a direct effect on the economic factor ($ECON$) of 53% at a significance level of 5%.

In the case of ECM_{t-1}, this has a direct effect on the economic factor ($ECON$), where the parameter value is −0.39 at a significance level of 1%, suggesting that the economic factor ($ECON$) has the ability to adjust toward the equilibrium at 39%. For the same case of ECM_{t-1}, this has a direct effect on social factor ($SOCI$), where the parameter value is −0.26 at a significance level of 1%, telling us that the social factor ($SOCI$) has the same stated ability of 26%, as is the same for ECM_{t-1}, which has a direct effect on the environmental factor ($ENVI$), where the parameter value is −0.05 at a significance level of 1%, showing that the environmental factor ($ENVI$) has the same ability at 5%.

As for this SEM-VARIMAX mode, it has been measured for performance monitoring of the forecasting model in comparison with other models, including the MLR, BP, Grey, ANN, ANFIS, and ARIMA models by using the *MAPE* and *RMSE*, as illustrated below.

Table 4 explains the SEM-VARIMAX model in terms of *MAPE* and *RMSE*, and they are found to be lower than any other existing model at 1.06% and 1.19%, respectively. If considering the performance monitoring result of the forecasting model for other models, the following was found. For the ARIMA model, its *MAPE* and *RMSE* were 6.29% and 3.41%, respectively; the ANFIS model generated *MAPE* and *RMSE* with a value of 6.42% and 6.89%, respectively; the ANN model generated *MAPE* and *RMSE* with a value of 8.65% and 10.15%, respectively; the Grey model generated *MAPE* and *RMSE* with a value of 12.11% and 14.48%, respectively; the BP model generated *MAPE* and *RMSE* with a value of 13.50% and 16.87%, respectively; and the MLR model generated *MAPE* and *RMSE* with a value of 20.06% and 22.91%, respectively.

Table 4. Performance monitoring of the forecasting models.

Forecasting Model	*MAPE* (%)	*RMSE* (%)
MLR model	20.06	22.91
BP model	13.50	16.87
Grey model	12.11	14.48
ANN model	8.65	10.15
ANFIS model	6.42	6.89
ARIMA model	6.29	3.41
SEM-VARIMAX model	1.06	1.19

Therefore, the above calculations show that the SEM-VARIMAX model is particularly suitable for future forecasting, especially long-term forecasting to support in strategy and effective planning.

4.4. The Forecasting Model and the Efficiency of the Sustainable Development Policy for Energy Consumption under Environmental Law in Thailand based on the SEM-VARIMAX model

For forecasting purposes, the SEM-VARIMAX model was applied to predict energy consumption for the next 17 years (2020–2036) so as to gauge the efficiency of the Sustainable Development Policy in Energy Consumption under Environmental Law in Thailand based on the national strategy set to support policy formulation of Thailand in the future (from the present to 2036), as illustrated in Figure 3.

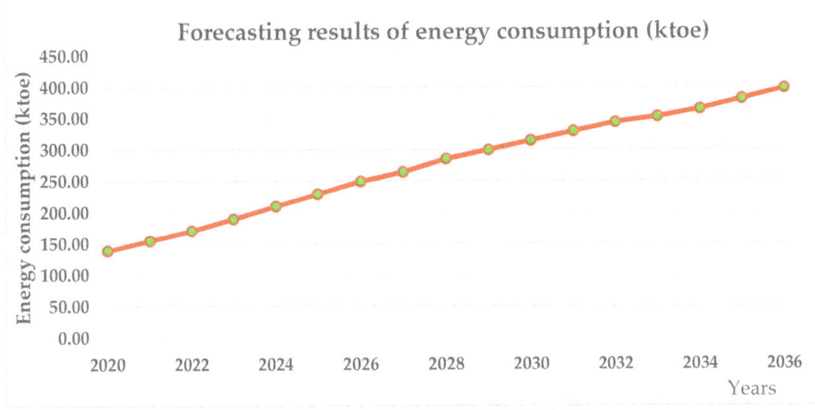

Figure 3. The forecasting results of energy consumption from 2020 to 2036 in Thailand.

Figure 3 shows that energy consumption from 2020 to 2036 under the Sustainable Development Policy for Energy Consumption under Environmental Law in Thailand will continuously increase from 2020 to 2036 with an increased growth rate from 185.66 (2036/2020) to 397.08 ktoe by 2536.

5. Conclusions and Discussion

This research was developed from relevant theories and advanced statistics to develop the SEM-VARIMAX model. This model differs from previous ones because it attempts to close several gaps and generate a functional model for the present, leading to research outcomes that are inclusive of special features relevant to different sectors and contexts. The software used in this research was LISREL incorporating EVIEWS, which are widely regarded to be the best choices for application. This research was conducted thoroughly and carefully while considering factors that influence policy implementation. This effort differentiates this research from other studies. In addition, this forecasting model determines real relationships between relevant causal and influential factors and the efficiency of Sustainable Development Policy for Energy Consumption under Environmental Law in Thailand (*S.D.EL*). The model comprises three latent variables: economic (*ECON*), social (*SOCI*), and environmental (*ENVI*) factors, while the observed variables are energy consumption (*EC*), per capita GDP (*GDP*), urbanization rate (*UR*), industrial structure (*IS*), net exports ($E - M$), indirect foreign investment (*IF*), foreign tourists (*FT*), employment (*EM*), health and illness (*HI*), social security (*SS*), consumer protection (*CP*), Carbon dioxide emissions (CO_2) and energy intensity (*EI*). Each factor has undergone various processes to ensure its significance. To the greatest extent possible, this research also eliminated potential issues that may lead to spurious results, a problem that is faced in the development of any model. This elimination ensures the absence of heteroskedasticity, multicollinearity, and autocorrelation [103,110]. In this research, we identified the relevant factors and their impact direction. Most importantly, the approach used was able to adjust each latent variable toward equilibrium, which renders it significant for studying the change and balance resulting from the Sustainable Development Policy for Energy Consumption under Environmental Law in Thailand. The key measurement indicator is the error correction mechanism. In addition, the SEM-VARIMAX model passed the qualification for the BLUE feature and thus met the goodness-of-fit requirements. This research also examined the performance of the model in terms of *MAPE* and *RMSE* values and compared them with those of existing models, including the MLR, BP, Gray, ANN, ANFIS, and ARIMA models. It was found that the SEM-VARIMAX model had the lowest *MAPE* and *RMSE*, followed by the ARIMA, ANFIS, ANN Grey, BP, and MLR models, ordered in descending performance.

The SEM-VARIMAX model was used to predict energy consumption for the period of 2020–2036, and energy consumption was found to have an increased growth rate from 185.66 (2036/2020) to 397.08 ktoe by 2036. This rate is obviously higher than the government's set target, which is 251.05 ktoe by 2036. This further reflects the inefficiency of the Sustainable Development Policy for Energy Consumption under Environmental Law in Thailand; the findings predict the disruption of the government's plan for carrying capacity under the environmental law currently enforced in Thailand. In addition, this research found that the adjustment to the environmental equilibrium in Thailand can be measured by the parameter of the error correction mechanism. The value of the parameter reflected a low adjustment rate of only 5% in the environmental aspect, while the economic side had an adjustment rate of 39%. For the social aspect, the adjustment rate was 26% of total capacity. This finding presents clear evidence of the inefficiency of the environmental law in Thailand. In fact, this law has not been fully updated and modernized for the current context.

Recommendations for the future application of this research include the selection of appropriate statistics and research procedures that fit efficient long-term forecasting. This forecasting requires the best model and a white-noise-type model. The findings of this research explicitly reveal that the country will be at disadvantage if the policy is implemented according to the past or present practice or according to ordinary least squares or only the ARIMA model is used in the research. Quality research must focus on the forecasting task, with an emphasis on forecasting quality and validity. This is to avoid any possible damage that may arise.

The limitation of this research is that Thailand's policy planning does not consider causal factors or their impact. This is evident from Thailand's attempt to improve the economy and society through the implementation of various measures. However, the environmental law in Thailand does not fit the current situation or context, which constantly change, leading to a failure to support a green environment. In addition, some factors were found to be inconsistent with the model because of the intervening factor of fuel price control and government interference in certain sectors, resulting in the imbalance between demand and supply. This further causes instability in these factors at certain times. However, long-term forecasting is always a challenging task that requires detail and consistency. The SEM-VARIMAX model can be applied to different contexts and sectors, but it has to be carefully utilized because of its in-depth analysis, complexity, and advanced statistics used in the modeling process. If this model is properly applied, it will have great potential to provide extensive knowledge in the future.

The policy recommendations of Thailand derived from this research all concern the environmental law of Thailand. Although Thailand currently has the National Environmental Quality Promotion and Preservation Act (Version 2) B.E. 2561, it is still too weak to achieve the efficiency of sustainable development policy under environmental law. This is due to the failure to achieve the determined target along with some weaknesses, that require correction. In order to improve the environmental law, the following is suggested.

1. Increase community participation in the management and preservation of natural resources and the environment. For instance, there is the requirement of community representatives, state and public representatives who are elected or nominated to be part of the National Environment Board (Category 1).

2. Revisit the direction of the National Environmental Development Plan locally, provincially and nationally.

3. Adopt the concept and philosophy of Thai traditions about the environment, as well as universal environmental law concepts, such as environmental justice being integrated into the constitution, so that the law and ordinances become supportive.

4. There are currently many environmental laws under different wings of administrations. Therefore, this has to be organized and systemized.

5. At present, the environmental case is under the judicial process of both the civil court and administrative court. Hence, all environmental cases shall be dealt with only by the administrative court, as the cases are wholly a matter of environmental justice, relating to the benefits of individuals, society and public interest. This settlement requires special expertise, which differs from civil cases. This further requires a revision of the law to switch the jurisdiction.

6. Mobilize scientific experts about the environment in various fields to help making legal decisions, conducting research, and developing environmental knowledge, as well as keeping environmental laws up to date.

7. Revise the processes and penalties from the polluter pays principle (PPP) with serious implementation and clarity.

Author Contributions: P.S. and S.N. were involved in the data collection and preprocessing phase, model constructing, empirical research, results analysis and discussion, and manuscript preparation. Both authors have approved the submitted manuscript.

Funding: This research received no external funding.

Acknowledgments: This research is supported by the School of Law, Assumption University.

Conflicts of Interest: The authors declare no conflict of interest.

References

1. Savaresi, A. The Paris Agreement: An Early Assessment. *Environ. Policy Law* **2016**, *46*, 14–18.
2. Laina, E. Sustainable Development in Operation. *Environ. Policy Law* **2016**, *46*, 47–49.
3. Sustainable Development Goals. Available online: http://www.solarimpulse.com/sustainable-development-goals? (accessed on 30 June 2019).
4. Krapp, R. Sustainable Development in the Second Committee. *Environ. Policy Law* **2016**, *46*, 10–13.
5. Uddin, M.K. Climate Change and Global Environmental Politics: North-South Divide. *Environ. Policy Law* **2017**, *46*, 106–114. [CrossRef]
6. Moore, P.; Pereira, E.S.; Duggin, G. Developing Environmental Law for All Citizens. *Environ. Policy Law* **2015**, *45*, 88–98.
7. Global Goals and the Environment: Progress and prospects. Available online: https://www.iisd.org/library/global-goals-and-environment-progress-and-prospects (accessed on 30 June 2019).
8. Savaresi, A. Developments in Environmental Law. *Environ. Policy Law* **2012**, *42*, 365–369.
9. Office of Natural Resources and Environmental Policy and Planning. Available online: http://www.onep.go.th (accessed on 1 April 2019).
10. Office of the National Economic and Social Development Board (NESDB). Available online: http://www.nesdb.go.th/nesdb_en/more_news.php?cid=154&filename=index (accessed on 1 April 2019).
11. National Statistic Office Ministry of Information and Communication Technology. Available online: http://web.nso.go.th/index.htm (accessed on 1 April 2019).
12. Department of Alternative Energy Development and Efficiency. Available online: http://www.dede.go.th/ewtadmin/ewt/dede_web/ewt_news.php?nid=47140 (accessed on 2 April 2019).
13. Thailand Greenhouse Gas Management Organization (Public Organization). Available online: http://www.tgo.or.th/2015/thai/content.php?s1=7&s2=16&sub3=sub3 (accessed on 2 April 2019).
14. Achawangkul, Y. Thailand's Alternative Energy Development Plan. Available online: http://www.unescap.org/sites/default/files/MoE%20_%20AE%20policies.pdf (accessed on 3 April 2019).
15. Zhou, K.; Zhang, H.; Baum, J.; Chen, W. The Evolution of Policy and Law for Sustainable Development in China. *Front. Law China* **2014**, *9*, 389–402.

16. Fan, K.-K.; Feng, T.-T. Discussion on Sustainable Development Strategies of the Traditional Handicraft Industry Based on Su-Style Furniture in the Ming Dynasty. *Sustainability* **2019**, *11*, 8. [CrossRef]

17. Boyd, E.; Hultman, N.; Roberts, J.T.; Corbera, E.; Cole, J.; Bozmoski, A.; Ebeling, J.; Tippman, R.; Mann, P.; Brown, K.; et al. Reforming the CDM for sustainable development: Lessons learned and policy futures. *Environ. Sci. Policy* **2009**, *12*, 280–831. [CrossRef]

18. Joseph, C. Understanding sustainable development concept in Malaysia. *Soc. Responsib. J.* **2013**, *9*, 441–453. [CrossRef]

19. Bahauddin, K.M. Environmental system management and governance needs in a developing country. *Environ. Syst. Decis.* **2014**, *34*, 342–357. [CrossRef]

20. Rivera, M. Political Criteria for Sustainable Development Goal (SDG) Selection and the Role of the Urban Dimension. *Sustainability* **2013**, *5*, 5034–5051. [CrossRef]

21. Ali, S.; Bibi, M.; Rabbi, F. A New Economic Dimension to the Environmental Kuznets Curve: Estimation of Environmental Efficiency in Case of Pakistan. *Asian Econ. Financ. Rev.* **2014**, *4*, 68–79.

22. Pinkse, J.; Kolk, A. Addressing the Climate Change—Sustainable Development Nexus: The Role of Multistakeholder Partnerships. *Bus. Soc.* **2012**, *51*, 176–210. [CrossRef]

23. Choi, S.; Ng, A. Environmental and Economic Dimensions of Sustainability and Price Effects on Consumer Responses. *J. Bus. Ethics* **2011**, *104*, 269–282. [CrossRef]

24. Amesheva, I. Environmental Degradation and Economic Development in China: An Interrelated Governance Challenge. *Law Dev. Rev.* **2017**, *10*, 425–450. [CrossRef]

25. Bakari, M.E.K. Sustainable Development in a Global Context: A Success or a Nuisance? *New Glob. Stud.* **2015**, *9*, 27–56. [CrossRef]

26. Martin, E.J. Economic rights, sustainable development, and environmental management. *Public Adm. Manag.* **2011**, *16*, 121–144.

27. United Nations Secretariat. Urbanization and sustainable development in Asia and the Pacific: Linkages and policy implications. Available online: https://www.unescap.org/commission/73/document/E73_16E.pdf (accessed on 1 April 2019).

28. Khalifa, M.A.; Connelly, S. Monitoring and guiding development in rural Egypt: Local sustainable development indicators and local Human Development Indices. *Environ. Dev. Sustain.* **2009**, *11*, 1175–1196. [CrossRef]

29. Wuelser, G.; Pohl, C.; Hadorn, G.H. Structuring complexity for tailoring research contributions to sustainable development: A framework. *Sustain. Sci.* **2012**, *7*, 81–93. [CrossRef]

30. Mueller, M.; Dos Santos, V.G.; Seuring, S. The Contribution of Environmental and Social Standards towards Ensuring Legitimacy in Supply Chain Governance. *J. Bus. Ethics* **2009**, *89*, 509–523. [CrossRef]

31. Zhang, L.; Dzakpasu, M.; Chen, R.; Wang, X.C. Validity and utility of ecological footprint accounting: A state-of-the-art review. *Sustain. Cities Soc.* **2017**, *32*, 411–416. [CrossRef]

32. Wang, Y.; Jiang, Y.; Zheng, Y.; Wang, H. Assessing the Ecological Carrying Capacity Based on Revised Three-Dimensional Ecological Footprint Model in Inner Mongolia, China. *Sustainability* **2019**, *11*, 2002. [CrossRef]

33. Singh, R.; Debnath, R.M. Modeling sustainable development: India's strategy for the future. *World J. Sci. Technol. Sustain. Dev.* **2012**, *9*, 120–135. [CrossRef]

34. Giddings, B.; Hopwood, B.; O'Brien, G. Environment, economy and society: Fitting them together into sustainable development. *Sustain. Dev.* **2002**, *10*, 187–196. [CrossRef]

35. Sapukotanage, S.; Warnakulasuriya, B.N.F.; Yapa, S.T.W.S. Outcomes of Sustainable Practices: A Triple Bottom Line Approach to Evaluating Sustainable Performance of Manufacturing Firms in a Developing Nation in South Asia. *Int. Bus. Res.* **2018**, *11*, 89–104. [CrossRef]

36. Sutthichaimethee, P. Forecasting Economic, Social and Environmental Growth in the Sanitary and Service Sector Based on Thailand's Sustainable Development Policy. *J. Ecol. Eng.* **2018**, *19*, 205–210. [CrossRef]

37. Greaker, M.; Stoknes, P.E.; Alfsen, K.H.; Ericson, T. A Kantian approach to sustainable development indicators for climate change. *Ecol. Econ.* **2013**, *91*, 10–18. [CrossRef]

38. Cetindamar, D.; Husoy, K. Corporate Social Responsibility Practices and Environmentally Responsible Behavior: The Case of the United Nations Global Compact. *J. Bus. Ethics* **2007**, *76*, 163–176. [CrossRef]

39. Bedore, J. An Evaluation of Canada's Environmental Sustainability Planning System and the Federal Sustainable Development. Act. Master's Thesis, Queen's University, Kingston, ON, Canada, 2008.

40. Lee, D.; Park, H.; Park, S.K. Policy Issues for Contributing ODA to Sustainable Development in Developing Countries: An Analysis of Korea's ODA and Sri Lankan Practices. *Asian Perspect.* **2018**, *42*, 623–646. [CrossRef]

41. Aguilera-Caracuel, J.; Aragon-Correa, J.A.; Hurtado-Torres, N.E.; Rugman, A.M. The Effects of Institutional Distance and Headquarters' Financial Performance on the Generation of Environmental Standards in Multinational Companies. *J. Bus. Ethics* **2012**, *105*, 461–474. [CrossRef]

42. Pires, S.M.; Fidélis, T.; Ramos, T.B. Measuring and comparing local sustainable development through common indicators: Constraints and achievements in practice. *Cities* **2014**, *39*, 1–9. [CrossRef]

43. Giannetti, B.F.; Demétrio, J.F.C.; Bonilla, S.H.; Agostinho, F.; Almeidan, C.M.V.B. Emergy diagnosis and reflections towards Brazilian sustainable development. *Energy Policy* **2013**, *63*, 1002–1012. [CrossRef]

44. Wysokińska, Z. Millenium Development Goals/UN and Sustainable Development Goals/UN as Instruments for Realising Sustainable Development Concept in the Global Economy. *Comp. Econ. Res.* **2017**, *20*, 101–118. [CrossRef]

45. Pânzaru, S.; Dragomir, C. The Considerations of the Sustainable Development and Eco-Development in National and Zonal Context. *Rev. Int. Comp. Manag.* **2012**, *13*, 823–831.

46. Byrch, C.; Kearins, K.; Milne, M.; Morgan, R. Sustainable "what"? A cognitive approach to understanding sustainable development. *Qual. Res. Account. Manag.* **2007**, *4*, 26–52. [CrossRef]

47. Casey, G.; Galor, O. Population Growth and Carbon Emissions. Available online: https://www.brown.edu/academics/economics/sites/brown.edu.academics.economics/files/uploads/2016-8_paper_0.pdf (accessed on 1 April 2019).

48. Ramakrishnan, S.; Hishan, S.S.; Nabi, A.A.; Arshad, Z.; Kanjanapathy, M.; Zaman, K.; Khan, F. An interactive environmental model for economic growth: Evidence from a panel of countries. *Environ. Sci. Pollut. Res.* **2016**, *23*, 14567–14579. [CrossRef]

49. Ladan, M.T. Achieving Sustainable Development Goals through Effective Domestic Laws and Policies on Environment and Climate Change. *Environ. Policy Law* **2018**, *48*, 42–63. [CrossRef]

50. Craig, K.R.; Garmestani, A.S.; Allen, C.R.; Arnold, C.A.; Birgé, H.; DeCaro, D.A.; Fremier, A.K.; Gosnell, H.; Schlager, E. Balancing stability and flexibility in adaptive governance: An analysis of tools available in U.S. environmental law. *Ecol. Soc.* **2017**, *22*, 3. [CrossRef]

51. Wang, A.L. The Search for Sustainable Legitimacy: Environmental Law and Bureaucracy in China. *Harv. Environ. Law Rev.* **2013**, *37*, 365–440. [CrossRef]

52. Bartel, R.; Barclay, E. Motivational postures and compliance with environmental law in Australian agriculture. *J. Rural Stud.* **2011**, *27*, 153–170. [CrossRef]

53. Kim, R.E.; Mackey, B. International environmental law as a complex adaptive system. *Int. Environ. Agreem.* **2014**, *14*, 5–23. [CrossRef]

54. Huber, B.R. Transition Policy in Environmental Law. *Harv. Environ. Law Rev.* **2011**, *35*, 91–130.

55. Tecklin, D.; Bauer, C.; Prieto, M. Making environmental law for the market: The emergence, character, and implications of Chile's environmental regime. *Environ. Politics* **2011**, *20*, 879–898. [CrossRef]

56. Zeben, J.V. Subsidiarity in European Environmental Law: A Competence Allocation Approach. *Harv. Environ. Law Rev.* **2014**, *38*, 415–464.

57. Bodansky, D. The Legitimacy of International Governance: A Coming Challenge for International Environmental Law? *Am. J. Int. Law* **1999**, *93*, 596–624. [CrossRef]

58. Heinzerling, L. New directions in environmental law: A climate of possibility. *Harv. Environ. Law Rev.* **2011**, *35*, 263–273.

59. Chang, Y.-C.; Wang, N. Environmental regulations and emissions trading in China. *Energy Policy* **2010**, *38*, 3356–3364. [CrossRef]

60. Periconi, J.J.; Jokajtys, M.R. Shining Some Light Back on the Dark Ages: New York State's Early Environmental Law and Its Implications for Today's Environmental Insurance Coverage Disputes. *Environ. Claims J.* **2014**, *26*, 287–300. [CrossRef]

61. Latham, M.; Schwartz, V.E.; Appel, C.E. The intersection of tort and environmental law: Where the twains should meet and depart. *Law Rev.* **2011**, *80*, 737–773.

62. Wood, M.C. "You Can't Negotiate with a Beetle": Environmental Law for a New Ecological Age. *Nat. Resour. J.* **2010**, *50*, 167–210.

63. Gibson, R.B. In full retreat: The Canadian government's new environmental assessment law undoes decades of progress. *Impact Assess. Proj. Apprais.* **2012**, *30*, 179–188. [CrossRef]

64. Fast, H.; Fitzpatrick, P. Modernizing environmental protection in Manitoba: The environmental bill of rights as one component of environmental reform. *J. Environ. Law Pract.* **2017**, *30*, 295–320.

65. De Moerloose, S. Law and Development as a Field of Study: Connecting Law with Development. *Law Dev. Rev.* **2017**, *10*, 179–186. [CrossRef]

66. Tania, S.J. Is There a Linkage Between Sustainable Development and Market Access of LDCs? *Law Dev. Rev.* **2013**, *6*, 143–223. [CrossRef]

67. Chepaitis, D.J.; Panagakis, A. Individualism Submerged: Climate Change and the Perils of an Engineered Environment. *UCLA J. Environ. Law Policy* **2010**, *28*, 291–342.

68. Miao, H. Sustainable Development through the Right to Access to Justice in Environmental Matters in China. *Sustainability* **2019**, *11*, 900.

69. Pourhashemi, S.A.; Khoshmaneshzadeh, B.; Soltanieh, M.; Hermidasbavand, D. Analyzing the individual and social rights condition of climate refugees from the international environmental law perspective. *Int. J. Environ. Sci. Technol.* **2012**, *9*, 57–67. [CrossRef]

70. Ruhl, J.B. Climate Change Adaptation and the Structural Transformation of Environmental Law. *Environ. Law* **2009**, *40*, 363–431.

71. Moreno, E.M.O.; de Luna, E.B.; Gómez, M.D.C.O.; López, E.J. Structural Equations Model (SEM) of a questionnaire on the evaluation of intercultural secondary education classrooms. *Suma Psicol.* **2014**, *21*, 107–115. [CrossRef]

72. Boccia, F.; Sarnacchiaro, P. Structural Equation Model for the Evaluation of Social Initiatives on Customer Behaviour. *Procedia Econ. Financ.* **2014**, *17*, 211–220. [CrossRef]

73. Baumgartnar, H.; Homburg, C. Applications of Structural equation modeling in marketing and consumer research: A review. *Int. J. Res. Mark.* **1996**, *13*, 139–161. [CrossRef]

74. Mai, Y.; Zhang, Z.; Wen, Z. Comparing Exploratory Structural Equation Modeling and Existing Approaches for Multiple Regression with Latent Variables. *Struct. Equ. Model. Multidiscip. J.* **2018**, *25*, 737–749. [CrossRef]

75. Ryu, E.; Mehta, P. Multilevel Factorial Invariance in n-Level Structural Equation Modeling (NSEM). *Struct. Equ. Model. Multidiscip. J.* **2017**, *24*, 936–959. [CrossRef]

76. Lei, M.; Lomax, R.G. The Effect of Varying Degrees of Nonnormality in Structural Equation Modeling. *Struct. Equ. Model. Multidiscip. J.* **2005**, *12*, 1–27. [CrossRef]

77. Cugnata, F.; Kenett, R.; Salini, S. Bayesian Network Applications to Customer Surveys and InfoQ. *Procedia Econ. Financ.* **2014**, *17*, 3–9. [CrossRef]

78. Nylund, K.L.; Asparouhov, T.; Muthen, B.O. Deciding on the Number of classes in Latent Class Analysis and Growth Mixture Modeling: A Monte Carlo Simulation Study. *Struct. Equ. Model. Multidiscip. J.* **2007**, *14*, 534–569. [CrossRef]

79. Saito, M.; Kato, T.; Tang, Y. Temperature controls ecosystem CO_2 exchange of an alpine meadow on the northeastern Tibetan Plateau. *Glob. Chang. Biol.* **2009**, *15*, 221–228. [CrossRef]

80. Yang, M.; Yuan, K.H. Optimizing Ridge Generalized Least Squares for Structural Equation Modeling. *Struct. Equ. Model. Multidiscip. J.* **2019**, *26*, 24–38. [CrossRef]

81. Chang, C.J.; Li, G.; Zhang, S.Q.; Yu, K.P. Employing a Fuzzy-Based Grey Modeling Procedure to Forecast China's Sulfur Dioxide Emissions. *Int. J. Environ. Res. Public Health* **2019**, *16*, 2504. [CrossRef]

82. Wang, H.; Huang, J.; Zhou, H.; Zhao, L.; Yuan, Y. An Integrated Variational Mode Decomposition and ARIMA Model to Forecast Air Temperature. *Sustainability* **2019**, *11*, 4018. [CrossRef]

83. Ma, L.; Wu, M.; Tian, X.; Zheng, G.; Du, Q.; Wu, T. China's Provincial Vehicle Ownership Forecast and Analysis of the Causes Influencing the Trend. *Sustainability* **2019**, *11*, 3928. [CrossRef]

84. Zhao, L.T.; Liu, L.N.; Wang, Z.J.; He, L.Y. Forecasting Oil Price Volatility in the Era of Big Data: A Text Mining for VaR Approach. *Sustainability* **2019**, *11*, 3892. [CrossRef]

85. Xiong, P.; Shi, J.; Pei, L.; Ding, S. A Novel Linear Time-Varying GM(1,N) Model for Forecasting Haze: A Case Study of Beijing, China. *Sustainability* **2019**, *11*, 3832. [CrossRef]

86. Zhao, L.; Liu, Z.; Mbachu, J. Energy Management through Cost Forecasting for Residential Buildings in New Zealand. *Energies* **2019**, *12*, 2888. [CrossRef]

87. Barari, M.; Kundu, S. The Role of the Federal Reserve in the U.S. Housing Crisis: A VAR Analysis with Endogenous Structural Breaks. *J. Risk Financ. Manag.* **2019**, *12*, 125. [CrossRef]

88. Tucki, K.; Orynycz, O.; Świć, A.; Wojtanek, M.M. The Development of Electromobility in Poland and EU States as a Tool for Management of CO_2 Emissions. *Energies* **2019**, *12*, 2942. [CrossRef]

89. Ahmed, M.; Sultan, M.; Elbayoumi, T.; Tissot, P. Forecasting GRACE Data over the African Watersheds Using Artificial Neural Networks. *Remote Sens.* **2019**, *11*, 1769. [CrossRef]
90. Ramsauer, F.; Min, A.; Lingauer, M. Estimation of FAVAR Models for Incomplete Data with a Kalman Filter for Factors with Observable Components. *Econometrics* **2019**, *7*, 31. [CrossRef]
91. Sutthichaimethee, P.; Kubaha, K. The Efficiency of Long-Term Forecasting Model on Final Energy Consumption in Thailand's Petroleum Industries Sector: Enriching the LT-ARIMAXS Model under a Sustainability Policy. *Energies* **2018**, *11*, 2063. [CrossRef]
92. Sutthichaimethee, P.; Kubaha, K. A Relational Analysis Model of the Causal Factors Influencing CO_2 in Thailand's Industrial Sector under a Sustainability Policy Adapting the VARIMAX-ECM Model. *Energies* **2018**, *11*, 1704. [CrossRef]
93. Jöreskog, K.G.; Sörbom, D. *New Features in LISREL 8*; Chicago Scientific Software International: Chicago, IL, USA, 1993.
94. Hunter, M.D. State Space Modeling in an Open Source, Modular. *Struct. Equ. Model. Multidiscip. J.* **2018**, *25*, 304–324. [CrossRef]
95. Sutthichaimethee, P.; Dockthaisong, B. A Relationship of Causal Factors in the Economic, Social, and Environmental Aspects Affecting the Implementation of Sustainability Policy in Thailand: Enriching the Path Analysis Based on a GMM Model. *Resources* **2018**, *7*, 87. [CrossRef]
96. Dickey, D.A.; Fuller, W.A. Likelihood ratio statistics for autoregressive time series with a unit root. *Econometrica* **1981**, *49*, 1057–1072. [CrossRef]
97. Johansen, S.; Juselius, K. Maximum likelihood estimation and inference on cointegration with applications to the demand for money. *Oxf. Bull. Econ. Stat.* **1990**, *52*, 169–210. [CrossRef]
98. Johansen, S. *Likelihood-Based Inference in Cointegrated Vector Autoregressive Models*; Oxford University Press: New York, NY, USA, 1995.
99. MacKinnon, J. *Critical Values for Cointegration Test in Long-Run Economic Relationships*; Engle, R., Granger, C., Eds.; Oxford University Press: Oxford, UK, 1991.
100. Wall, M.M.; Li, R. Comparison of multiple regression to two latent variable techniques for estimation and prediction. *Stat. Med.* **2003**, *22*, 3671–3685. [CrossRef]
101. Enders, W. *Applied Econometrics Time Series*; Wiley Series in Probability and Statistics; University of Alabama: Tuscaloosa, AL, USA, 2010.
102. Harvey, A.C. *Forecasting, Structural Time Series Models and the Kalman Filter*; Cambridge University Press: Cambridge, UK, 1989.
103. Lim, S.; Melville, N.P. Robustness of Structural Equation Modeling to Distributional Misspecification: Empirical Evidence & Research Guidelines. Available online: http://ssrn.com/abstract=1375251 (accessed on 31 August 2019).
104. Sutthichaimethee, P.; Ariyasajjakorn, D. Forecast of Carbon Dioxide Emissions from Energy Consumption in Industry Sectors in Thailand. *Environ. Clim. Technol.* **2018**, *22*, 107–117. [CrossRef]
105. Sutthichaimethee, P.; Ariyasajjakorn, D. The revised input-output table to determine total energy content and total greenhouse gas emission factors in Thailand. *J. Ecol. Eng.* **2017**, *18*, 166–170. [CrossRef]
106. Sutthichaimethee, P. Varimax Model to Forecast the Emission of Carbon Dioxide from Energy Consumption in Rubber and Petroleum Industries Sectors in Thailand. *J. Ecol. Eng.* **2017**, *18*, 112–117. [CrossRef]
107. Sutthichaimethee, P.; Ariyasajjakorn, D. Forecasting Model of GHG Emission in Manufacturing Sectors of Thailand. *J. Ecol. Eng.* **2017**, *18*, 18–24. [CrossRef]
108. Sutthichaimethee, P. Modeling Environmental Impact of Machinery Sectors to Promote Sustainable Development of Thailand. *J. Ecol. Eng.* **2016**, *17*, 18–25. [CrossRef]
109. Sutthichaimethee, P. Model of Environmental Problems Priority Arising from the Use of Environmental and Natural Resources in Machinery Sectors of Thailand. *Environ. Clim. Technol.* **2016**, *17*, 18–29. [CrossRef]
110. Grewal, R.; Cote, J.A.; Baumgartner, H. Multicollinearity and Measurement Error in Structural Equation Models: Implications for Theory Testing. *Mark. Sci.* **2004**, *23*, 519–529. [CrossRef]

Article

Optimization in the Stripping Process of CO_2 Gas Using Mixed Amines

Pao Chi Chen * and Yan-Lin Lai

Department of Chemical and Materials Engineering, Lunghwa University of Science and Technology, Taoyuan 333, Taiwan; g1062141003@gm.lhu.edu.tw
* Correspondence: chenpc@mail2000.com.tw

Received: 30 April 2019; Accepted: 5 June 2019; Published: 10 June 2019

Abstract: The aim of this work was to explore the effects of variables on the heat of regeneration, the stripping efficiency, the stripping rate, the steam generation rate, and the stripping factor. The Taguchi method was used for the experimental design. The process variables were the CO_2 loading (A), the reboiler temperature (B), the solvent flow rate (C), and the concentration of the solvent (monoethanolamine (MEA) + 2-amino-2-methyl-1-propanol (AMP)) (D), which each had three levels. The stripping efficiency (E), stripping rate ($\dot{m}_{CO_2}$), stripping factor (β), and heat of regeneration (Q) were determined by the mass and energy balances under a steady-state condition. Using signal/noise (S/N) analysis, the sequence of importance of the parameters and the optimum conditions were obtained, and the optimum operating conditions were further validated. The results showed that E was in the range of 20.98–55.69%; $\dot{m}_{CO_2}$ was in the range of 5.57×10^{-5}–4.03×10^{-4} kg/s, and Q was in the range of 5.52–18.94 GJ/t. In addition, the S/N ratio analysis showed that the parameter sequence of importance as a whole was A > B > D > C, while the optimum conditions were A3B3C1D1, A3B3C3D2, and A3B2C2D2, for E, $\dot{m}_{CO_2}$, and Q, respectively. Verifications were also performed and were found to satisfy the optimum conditions. Finally, the correlation equations that were obtained were discussed and an operating policy was discovered.

Keywords: heat of regeneration; stripping rate; stripping factor; mixed solvent; Taguchi method

---

1. Introduction

In order to reduce CO_2 gas emissions, many solutions have been proposed for several significant industries, such as coal-fired power plants, petroleum industries, steel industries, and cement industries. In this regard, a number of technologies have been applied, such as post-combustion, pre-combustion, oxyfuel combustion, and chemical looping. Among these technologies, an absorption-desorption process for post-combustion has been widely used [1]. A number of solvents have been adopted for the capture of CO_2 [2–4]. In these solvents, amines are most extensively used in chemical absorption to capture CO_2 [5–7]. As the chemical structure of an amine has at least one OH and amine group, the OH group can reduce the vapor pressure of the amine, and because an amine has alkaline properties, it can absorb acidic gases. Among the various amines, monoethanolamine (MEA) is used most extensively, because of its high solution absorbability, high alkalinity, high reaction rate, regenerability, and low cost. However, some drawbacks have been observed, such as high solvent regeneration energy, corrosion, and degradation.

Capturing CO_2 increases the cost of electricity production by 70%, while the energy required for the regeneration process is estimated to be in the range of 15–30% of a power plant's output [8,9]. Therefore, effectively reducing the cost of electric power has become the key to the success or failure of carbon capture and sequestration (CCS). Achieving minimum heat for regeneration has become a significant challenge, especially for the improvement of the stripper structure, the discovery of new solvents, and improvements in operating conditions [10–20]. In the stripping process, rich solvents are heated in the stripper to allow for the release of CO_2 from the scrubbed solutions. The stripping vapor, involving water vapor, CO_2, and small amounts of solvents, is regenerated in the reboiler, and rises from the reboiler through the column to the top of the stripper. The stripping vapor counter-current contacts a rich-loading feed stream, which absorbs energy from the stripping steam for CO_2 desorption. The remaining vapor is condensed at the top of the column in the overhead condenser. The vapor and liquid contact system are shown in Figure 1, indicating that the system is complex.

The heat of the solvent regeneration in the stripper of the CO_2 capture process can be described as follows [13,21]:

$$Q = Q_{sen} + Q_{abs} + Q_{vap} \tag{1}$$

where Q_{sen} is the sensitive heat, Q_{vap} is the heat of evaporation, and Q_{abs} is the heat of absorption. In general, the three terms are evaluated separately when the relevant thermodynamic data are available.

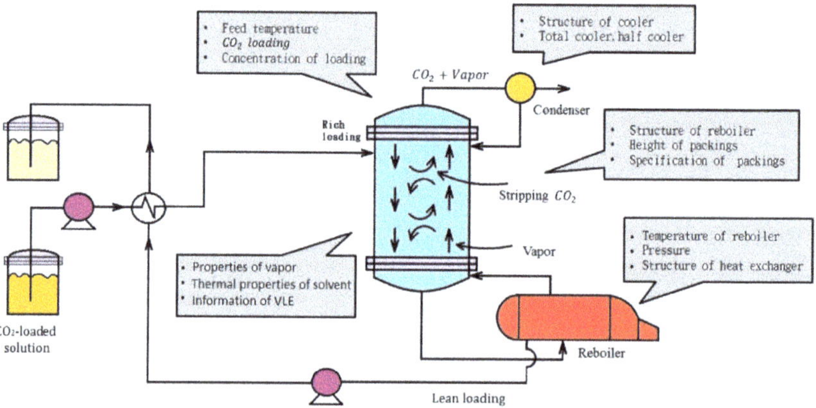

Figure 1. Vapor and liquid contact in the stripper.

The cost of regeneration is influenced by the adopted solvent, the stripping equipment structure, the operating temperature, and the steam cost [8,9,22,23]. In order to minimize the heat of regeneration, many studies have focused on more efficient solvents that show a lower heat of absorption [14,24,25]. These studies showed that blended solvents have been widely studied in the capture of CO_2 gas [6,7,10], while new solvents have also been actively tested [11,14,20,26]. As a result, many scholars have studied blended amines and effective solvents, with the goal of improving regeneration energy [14–20]. Choi et al. [6] found that MEA + 2-amino-2-methyl-1-propanol (AMP) has a relatively high economic benefit. Some studies have indicated that if the gas stripping column uses a split-flow, the energy saved could be at least 20%. The effect of the stripper configuration on the heat duty was also reported by Rochelle [23], who found that the reduction in energy requirement was 5–20%. Additionally, rich loading and lean loading each have a relative effect on heat duty. The findings showed that heat duty in regeneration is relatively higher when rich loading is low, whereas, a lean loading of 0.1–0.2 has a minimum heat duty [21]. Li and Keener [1] reported that many studies focus on reducing sensible heat and the heat of evaporation. In order to understand the contribution of individual heat duties, an investigation of the heat mechanism of regeneration energy can also be explored if the thermodynamic data are available. Table 1 shows the heat of regeneration data under various conditions. The reported levels for the heat of regeneration are in the range of 2.1–11.25 GJ/t, depending on the operating system. However, there are no available empirical equations that could be used to predict the heat of regeneration at the given conditions.

In some processes, the focus on the development of blended amines and new solvents for obtaining a low heat of absorption could be misleading, because the focus on solvents with a low heat of absorption is not reasonable without considering the overall process, as pointed out by Oexmann and Kather [21]. They found that at a low heat of absorption, the heat of the solvent regeneration at a low pressure leads to an increase in power for compression, as well as a lower quality steam. On the other hand, at a high heat of absorption, solvents show an increase in stripper operating pressure, and a reboiler temperature that leads to less water vapor at the stripper head. Thus, less heat must be provided in the reboiler. They also stated that in the evaluation of a solvent, it is necessary to consider the interdependence of the three terms that contribute to the heat of regeneration and their connection to the process parameters. This could be accomplished by balancing the materials and energy in the stripper. However, the regeneration energies are related to the type of solvent, reboiler temperature, solvent flow rate, loading, and structure of the stripper. Therefore, understanding the parameter significance and optimization conditions in order to reduce regeneration energy needs was required. This could be done by using the Taguchi experimental design [4,5].

In this work, the process variables included the concentration of the solvents, the flow rate, the CO_2-loading, and the reboiler temperature. In order to better understand the effects of the process variables on the outcome data, a framework was designed. The stripping efficiency, heat of regeneration, stripping rate, and steam generation rate were calculated using materials and an energy balance model created with the aid of thermal data [2,15]. Finally, the data were used to make a regression to obtain empirical equations, which were then discussed further.

Table 1. Heats of regenerations and operating conditions at various systems. AMP—2-amino-2-methyl-1-propanol; SG—sodium glycinate; MEA—monoethanolamine; AMPD—2-amino-2-methyl-propane-1,3-diol; DETA—Diethylenetriamine; PZ—Piperazine; PZEA—(Piperazinyl-1)-2-ethylamine; TETA—Triethylenetetramine.

Solvents	Conditions	Heat of Regeneration (GJ/t-CO_2)	References
SG	Loading = 0.11–0.52 T = 100–120 °C 3–6 M SG	3.68–10.75	[4]
AMP + PZ	Loading = 0.46 L/G = 2.9 18 wt.% AMP + 17.5 wt.% PZ	3.4–4.4	[7]
DETA	Loading = 1.2–1.4 Solvent flow rate = 3–12 m^3/m^2-h 2–3 M DETA	2.61–4.96	[13]
ACOR100 (MEA + TETA +AMPD + PZEA)	Solvent flow rate = 0.4–0.8 L/min Partial pressure of CO_2 = 54 mbar	2.7–3.9	[24]
KoSol-4	Loading = 0.8 L/G = 1.4–3.1 kg/Sm^3 P = 0.35–0.8 kg/cm^2	3.0–4.1	[14]
SG	Loading = 0.13–0.50 15–40 wt.% L/G = 2–10 L/m^3	5.3–8.5	[26]
AMP	Loading = 0.55 30 wt.% AMP	2.1	[27]
MEA	Loading = 0.25–0.49 30 wt.% MEA	3.3–6.4	[28]
PZ	Loading 0.4 T = 120–150 °C	2.93–3.43	[29]
Ammonia	Loading = 0.0525–0.01236 7–14% Ammonia	11.25	[30]
MEA MEA + ionic liquid + water	30 wt.% MEA 30 wt.% + 40 wt.% + 30 wt.% T = 103 °C	8.19 5.14	[31]
SG MEA	Simulation 30 wt.% T = 40–120 °C Thermodynamic calculation	5.7 4.7	[32]
MEA + AMP	MEA:AMP = 2:1/1:1/1:2 T = 103–110 °C	2.5–5.0	[10]

2. Experimental Features

2.1. Experimental Design

The experimental design had four parameters, namely: the concentration of blended amine (MEA + AMP), the feed rate, the CO_2 loading, and the reboiler temperature. Each parameter had three levels. Originally, MEA and AMP (30 wt.% AMP in total amine) were mixed together; then, the blended amines were poured into a known amount water to prepare the desired amine concentrations. Theoretically, 3^4 or 81 experiments were therefore needed. Using the Taguchi experimental design, the orthogonal array $L_9(3^4)$ showed nine experiments, thereby reducing the number of experiments needed and the research cost by over 80% [4,5]. The blended amine concentrations were 4 kmol/m^3, 5 kmol/m^3, and 6 kmol/m^3; the feed rates were 3×10^{-4}, 6×10^{-4}, and 9×10^{-4} m^3/s; the CO_2 loadings were 0.3, 0.4, and 0.5 kmol-CO_2/kmol-amine; and the reboiler temperatures were 100, 110, and 120 °C, respectively. The MEA/AMP weight fraction ratios obtained in here were 0.1908/0.0818, 0.2388/0.1023, and 0.2869/0.1229 for 4 kmol/m^3, 5 kmol/m^3, and 6 kmol/m^3, respectively. Table 2 shows the factors and levels in this work, while Table 3 presents the combination of experiments in the orthogonal array. The rich loading for the feed solution was obtained by early experimental preparation. Figure 2 illustrates the framework of the research project. The steps involved the input variables, the Taguchi experimental design, outcome data, data analysis, and verification.

Once the measured data were obtained, the experimental data could be evaluated, and the optimum condition and importance of the parameters could be determined with a signal/noise (S/N) ratio using the Taguchi analysis. The S/N ratio is calculated as follows:

$$\left(\tfrac{S}{N}\right)_{SB} = -10 \times \log\left(\tfrac{1}{n} \sum_{i=1}^{n} z_i^2\right) \qquad \text{(smaller is better)} \qquad (1)$$

$$\left(\tfrac{S}{N}\right)_{LB} = -10 \times \log\left(\tfrac{1}{n} \sum_{i=1}^{n} \tfrac{1}{z_i^2}\right) \qquad \text{(larger is better)} \qquad (2)$$

$$(2)$$

where, n is the amount of data, i is the amount of ith data, and z_i is the experimental data, as determined in this work.

Table 2. Factors and levels used in this study.

Factor	1	2	3
CO_2-loading (A) mol-CO_2/mol-amine	0.3	0.4	0.5
Reboiler temperature (B)/°C	100	110	120
Feed rate (C)/m^3/s	$3(10^{-4})$	$6(10^{-4})$	$9(10^{-4})$
Concentration of solvent (D)/kmol/m^3	4	5	6

Table 3. Orthogonal arrays for experimental design.

No.	A (kmol-CO_2/kmol-amine)	B (°C)	C (m^3/s)	D (kmol/m^3)
1	1	1	1	1
2	1	2	2	2
3	1	3	3	3
4	2	1	2	3
5	2	2	3	1
6	2	3	1	2
7	3	1	3	2
8	3	2	1	3
9	3	3	2	1

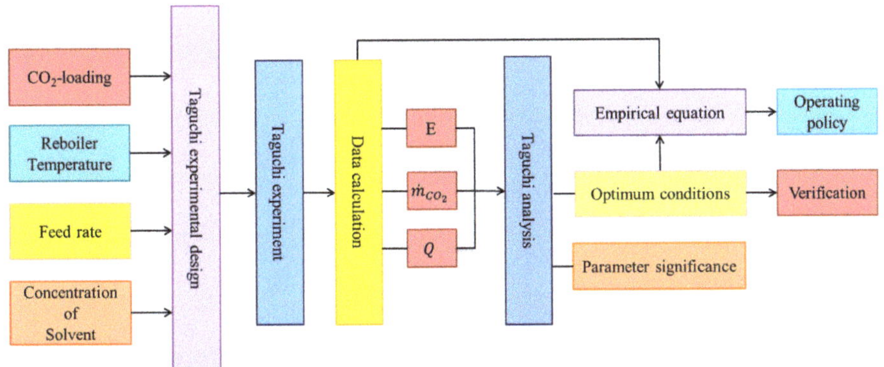

Figure 2. A framework between the parameters and outcome data.

2.2. Experimental Device and Operating Procedure

The stripping system is shown in Figure 3. It included a packed column, a reboiler, a condenser, a heat exchanger, and a heating system. The diameter of the column was 50 mm. It was filled with an 8 × 8 mm θ-ring. The height of the packed column was 800 mm, and the height of the condenser was 500 mm. In addition, a reboiler (12 L in volume) was heated by silicone oil using heating tubes. In order to adjust the pressure in the column, a pressure back valve was adopted as a suitable value at the top and bottom of the column, as shown.

In order to effectively explore the effect of the process variables on the performance of a stripper, a continuous process was adopted. First, the temperature indicators, cooling water circulator, and oil-bath power supply were switched on and adjusted to a preset temperature. Second, the prepared rich loading solution was poured into the reboiler until it flooded. When the oil-bath temperature reached the set temperature, the oil-bath pump power supply was turned on and an oil-bath inlet valve was used to adjust the flow. Third, the inlet temperature of the cooling water was set to the desired condition. Then, the reboiler was regulated to the desired experimental temperature. The experiment began when the temperature of the cooling water and that of the reboiler vapor temperatures reached the set temperatures. The rich loading in the storage tank flowed through a heat exchanger and into a packed bed, and contacted the vapor rising from the bottom of the column. The lean loading was withdrawn at the bottom of the reboiler and went through the heat exchanger, releasing the heat to the rich loading as the input solvent. The lean loading was withdrawn every 30 min for sample examination using a TOC (Total Organic Carbon) meter (Tekmar-Dohrmann Phoenix 800). The measurement procedure was performed according to the operation manual provided by Tekmar-Dohrmann Co. However, we need to prepare standard solutions (KHP and Na_2CO_3) before measurement. The analyzer needs to be calibrated every two-weeks. At this time, it requires being replaced with ultra-pure water. In addition, two antioxidant solutions also need to be replaced every month. One is a 20 wt.% H_2PO_4 solution and the other is one liter of 5 wt.% of H_2PO_4 solution dissolved with 10 g of $Na_2S_2O_8$ solids. In addition, the flow meter was calibrated using a measuring cylinder; then, it had a micro adjustment to a desired value. During the experiment, all of the temperature points indicated in Figure 3 had to be recorded, including the bed temperature (T01-T05). A combination of Taguchi experimental designs was used to effectively explore the effect of multiple variables on the outcome data.

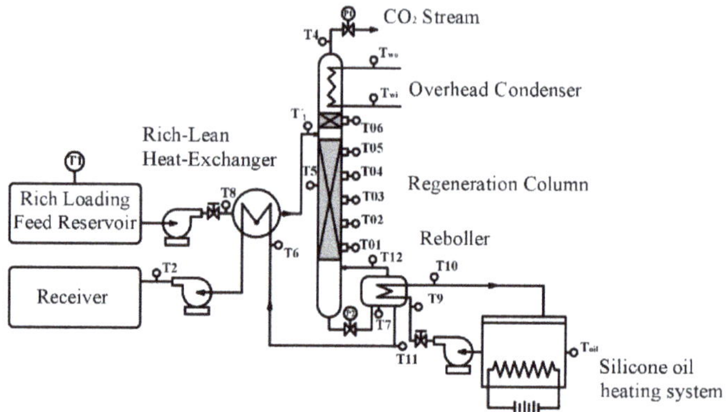

Figure 3. A stripping process used for both modes in this work.

T1:	Inlet Temperature of Rich Loading	T7:	Reboiler Temperature	T12:	Steam Temperature
T2:	Temperature at outlet of heat exchanger	T8:	Inlet temperature of rich loading	T_{wi}:	Inlet temperature of cooling water
T4:	Out temperature at the top of cooler	T9:	Oil temperature at the inlet of reboiler (T_{in})	T_{wo}:	Outlet temperature of cooling water
T5:	Temperature of bed	T10:	Oil temperature at the outlet of reboiler (T_{out})	P1:	Pressure valve at the top of tower
T6:	Liquid temperature at the inlet of heat exchanger	T11:	Outlet temperature the reboiler	P2:	Pressure valve at the bottom of tower

2.3. Determination of Experimental Data

In order to obtain the experimental data, it was necessary to determine the materials and energy balances in the stripper, as shown in Figure 4. The data involving the stripping efficiency, stripping rate, heat of regeneration, and steam generation rate were calculated separately.

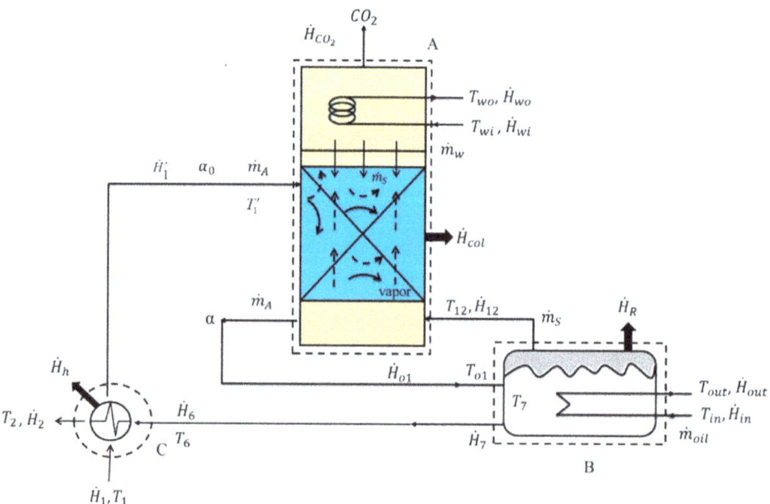

Figure 4. Materials and energy balances for the stripper.

2.3.1. Stripping Efficiency

The stripping efficiency is defined as the following:

$$E = \frac{\alpha_0 - \alpha}{\alpha_0} \times 100\% \tag{3}$$

where α_0 (mol-CO_2/mol-solvent) and α (mol-CO_2/mol-solvent) are the rich loading and lean loading, respectively.

2.3.2. Stripping Rate of CO_2

The stripping rate of CO_2 is defined as follows:

$$\dot{m}_{CO_2} = \dot{n}_A(\alpha_0 - \alpha)M_{CO_2} \tag{4}$$

where $\dot{n}_A$ (mol/s) is the molar rate of the solvent, and M_{CO_2} is the molecular weight of carbon dioxide.

2.3.3. Enthalpy Balance for Heat of Regeneration

From Figure 4, this procedure can be divided into three subsystems (i.e., A, B, and C), as indicated in the figure. Herein, an enthalpy balance can be made at the steady state for the three subsystems, as shown below:

Subsystem A:

$$\dot{H}_{12} + \dot{H}_1' - \dot{H}_{01} - \dot{H}_{col} + \dot{H}_{wi} - \dot{H}_{wo} - \dot{H}_{CO_2} = 0 \tag{5}$$

Subsystem B:

$$\dot{H}_{in} - \dot{H}_{out} + \dot{H}_{01} - \dot{H}_7 - \dot{H}_{12} - \dot{H}_R = 0 \tag{6}$$

Subsystem C:

$$\dot{H}_6 + \dot{H}_1 - \dot{H}_2 - \dot{H}_1' - \dot{H}_h = 0 \tag{7}$$

From the three equations above, and assuming $\dot{H}_6 \approx \dot{H}_7$, it can be rewritten as the following equations:

$$\dot{H}_{CO_2} = \dot{H}_{reb} - \dot{H}_{loss} - (\dot{H}_2 - \dot{H}_1) - (\dot{H}_{wo} - \dot{H}_{wi}) \tag{8}$$

where

$$\dot{H}_{reb} = (\dot{H}_{in} - \dot{H}_{out}) = \dot{m}_{oil}\overline{C}_{p,oil}(T_{in} - T_{out}) \tag{9}$$

and

$$\dot{H}_{loss} = \dot{H}_R + \dot{H}_{col} + \dot{H}_h \tag{10}$$

and

$$\dot{H}_2 - \dot{H}_1 = \dot{m}_A\overline{C}_{pA}(T_2 - T_1) \tag{11}$$

and

$$\dot{H}_{wo} - \dot{H}_{wi} = \dot{m}_w\overline{C}_{pw}(T_{wo} - T_{wi}) \tag{12}$$

As $\dot{H}_{CO_2} = \dot{m}_{CO_2}Q$, Equation (8) becomes the following:

$$Q = \frac{\dot{H}_{reb} - \dot{H}_{loss} - (\dot{H}_2 - \dot{H}_1) - (\dot{H}_{wo} - \dot{H}_{wi})}{\dot{m}_{CO_2}} \tag{13}$$

where, $\dot{m}_{oil}$ is the mass flow rate of oil, $\overline{C}_{p,oil}$ is the mean heat capacity of oil, and T_{in} and T_{out} are the input and output temperatures of the oil, respectively. In addition, $\dot{H}_{loss}$ is the heat loss, including $\dot{H}_{col}$ in the column, $\dot{H}_h$ in the heat exchanger, and $\dot{H}_R$ in the reboiler. Therefore, the heat losses $\dot{H}_{col}$, $\dot{H}_R$, and $\dot{H}_h$, which can easily be calculated separately, can be evaluated from the sensitive heat, while the heat transfer coefficient can be estimated by free convection [33]. Equation (13) states that the heat of regeneration can be determined when the enthalpy of the reboiler, the enthalpy of the heat loss, the enthalpy change between the inlet and outlet in the whole system, and the enthalpy change of the cooler are given.

2.3.4. Steam Flow Rate

In order to determine the bulk steam flow rate, materials and energy balances are required when the temperature of the stripper is fixed (Figure 4). Considering Subsystem B, the enthalpy balance at a steady state becomes the following:

$$\dot{H}_{12} = \dot{H}_{in} - \dot{H}_{out} + \dot{H}_{01} - \dot{H}_7 - \dot{H}_R \tag{14}$$

Equation (14) can be integrated into the following equation:

$$\dot{m}_s = \frac{\dot{m}_{oil}\overline{C}_{P,oil}(T_9 - T_{10}) - \dot{m}_A\overline{C}_{PA}(T_7 - T_{01}) - \dot{H}_R}{\Delta H^{vap}} \tag{15}$$

From the measured data and thermodynamic data [34], the steam flow rate can be evaluated.

3. Results and Discussion

3.1. Steady State Operation

The change in temperature at individual points, as indicated in Figure 3, was observed and recorded during the operation. Several important points, such as T1, T2, T6, T7, and T12, were recorded for No. 1, as shown in Figure 5. It was found that the temperatures remained constant when the operating time was greater than 80 min. In addition, the distribution of the lean loading was found, as shown in Figure 6, which also remained constant after 80 min, which was coincident with Figure 5. In addition, the temperature distribution in the packed bed was recorded, as shown in Figure 7. The distributions approached a steady-state operation after 80 min, except for T05, as it was near the location of the solvent input and was prone to perturbation during operation. On the other hand, the effect of the solvent input on other points (T01 to T04) were weak, because they are far from the solvent input. Because of this, it could be said that the system changed to a steady state operation after 80 min. Under a steady state condition, the outcome data could be evaluated using Equations (3), (4), (13), and (15), as shown in Table 4, for the Taguchi experiments (No. 1–No. 9). In addition, the verification data of the optimum conditions are listed in this table, including No. 10, 11, and 12, which were discussed further. It was found that the data ranges for the Taguchi experiments were 20.96–55.69%, 5.57×10^{-5}–4.03×10^{-4} kg/s, 5.52–18.94 GJ/t, 0.0580–0.203 kg-CO_2/kg-steam, and 8.38×10^{-4}–30.63×10^{-4} kg/s for E, $\dot{m}_{CO_2}$, Q, β, and $\dot{m}_s$, respectively. Except for No. 1 and 3, the values of Q fell between the data (2.1–11.5 GJ/t), as presented in Table 1. However, some data, such as that of no. 11 and 12, were comparable with the data [7,14,24,29] listed in Table 1 for the systems of MEA, ACOR100, Kosol-4, and AMP + PZ. However, the heat of the generation data shown here was higher than that reported by Aroonwilas and Veawab [10]. A possible reason was the difference in heating systems; the former was an oil bath system and the latter was a steam heating system. This could be seen in Equations (9) and (13). In order to reduce the heat of regeneration, it was necessary to choose a lower heat capacity oil that could reduce the $\dot{H}_{reb}$, as shown in Equation (13).

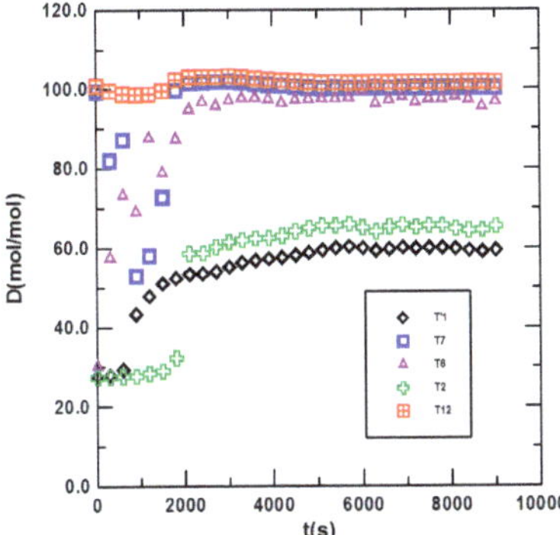

Figure 5. Variations of temperature during operation.

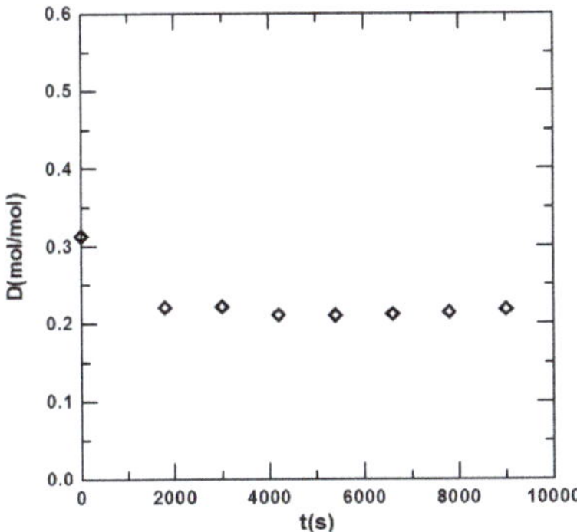

Figure 6. A plot of lean loading versus time.

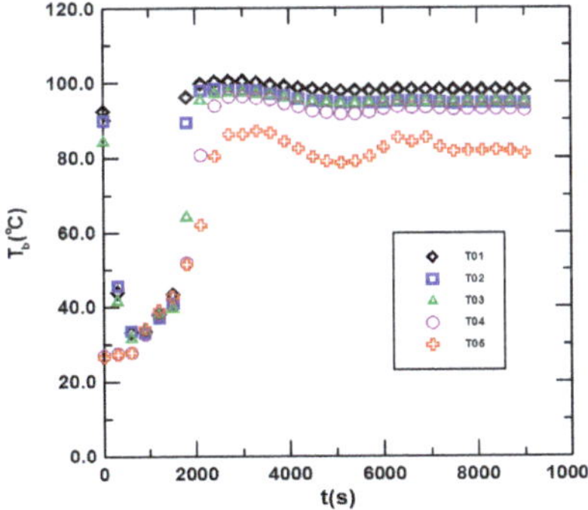

Figure 7. Various temperature distributions in the packed bed.

Table 4. Measured data obtained in this work.

No.	α_0 $\left(\frac{\text{kmol–CO}_2}{\text{kmol–amine}}\right)$	α $\left(\frac{\text{kmol–CO}_2}{\text{kmol–amine}}\right)$	t_R (°C)	C_A (kmol/m^3)	Q_A (10^4) (m^3/s)	Q (GJ/t)	$\dot{m}_{CO_2}$ (10^{-5} kg/s)	$\dot{m}_s$ (10^{-4} kg/s)	β $\left(\frac{\text{kg–CO}_2}{\text{kg–steam}}\right)$	E (%)
1	0.31	0.22	100	4	3	15.69	5.57	8.82	0.063	30.43
2	0.31	0.24	110	5	6	7.29	10.04	8.38	0.120	22.45
3	0.30	0.22	120	6	9	18.94	16.71	30.63	0.058	20.98
4	0.41	0.30	100	6	6	10.05	16.71	12.34	0.144	23.64
5	0.41	0.25	110	4	9	7.52	28.91	14.52	0.199	40.79
6	0.39	0.13	120	5	3	6.75	15.77	9.06	0.174	54.78
7	0.51	0.34	100	5	9	7.69	40.26	20.67	0.195	35.73
8	0.50	0.25	110	6	3	7.08	21.59	10.62	0.203	49.63
9	0.49	0.18	120	4	6	5.52	32.04	17.21	0.186	55.70
10	0.55	0.21	120	4	3	6.89	17.41	7.10	0.245	58.18
11	0.47	0.26	120	5	9	3.97	46.15	19.24	0.236	44.54
12	0.52	0.31	110	5	6	4.44	29.61	13.14	0.225	39.20

3.2. Taguchi Analysis

3.2.1. S/N Ratio for E

The stripping efficiency for the nine data points was analyzed according to the S/N ratio, and the results are shown in Table 5. The parameters, in order of importance, were A > D > C > B, while the optimum condition was A3B3C1D1. Similarly, the same analysis for $\dot{m}_{CO_2}$ and Q were obtained and are listed in Table 6. It was found that the effects of parameters A and D were significant, while those of B and C were minor. The verification of the optimum conditions for the three runs could thus be carried out further.

Table 5. Signal to noise (S/N) ratio analysis, as indicated in Equation (2), for efficiency (E).

Level	A	B	C	D
1	27.71	29.40	32.78	32.26
2	31.49	31.05	29.80	30.95
3	33.30	32.04	29.90	29.27
Delta	5.59	2.64	2.98	2.99
Rank	1	4	3	2

Table 6. S/N ratio analysis to obtain optimum condition and parameter significance.

S/N	Optimum Condition	Parameter Importance
E (Equation (2))	A3B3C1D1	A > D > C > B
$\dot{m}_{CO_2}$ (Equation (2))	A3B3C3D2	A > C > B > D
Q (Equation (1))	A3B2C2D2	A > D > B > C

3.2.2. Confirmation of the Optimum Conditions

The procedure of confirmation was similar to that reported in Section 2.2. The results of the confirmation tests for the three optimum conditions are listed in Table 7. The words printed in red are the optimum values obtained here compared with the Taguchi experimental values, as shown in Table 4. The results indicated that the Taguchi experimental design for this study was reliable. In order to verify the optimum data, the Taguchi data together with the three optimum sets of data were all adopted for regression.

Table 7. Confirmation of optimum conditions.

No	Optimum Condition	E (%)	$\dot{m}_{CO_2}$ (10^5) (kg/s)	Q (GJ/t)
No. 10	A3B3C1D1	(58.18)	17.41	6.89
No. 11	A3B3C3D2	44.54	(46.15)	3.97
No. 12	A3B2C2D2	39.20	29.61	(4.44)

3.3. Empirical Equations

In order to obtain the empirical equations for E, $\dot{m}_{CO_2}$, Q, and $\dot{m}_s$, the experimental data were correlated with suitable parameters, that is, $\Phi = f(\alpha_0, t_R, C_A, Q_A)$. For example, the stripping efficiency was correlated with the CO_2 loading, the reboiler temperature, the solvent concentration, and the flow rate. A total of twelve data sets (listed in Table 3) were used, and the results are as follows:

$$E = 1.94 \times 10^5 \exp\left(-\frac{3702.14}{T_R(K)}\right)\alpha_0^{0.68}[C_A(\text{kmol}/\text{m}^3)]^{-1.39}[Q_A(\text{m}^3/s)]^{-0.33} \tag{16}$$

The root means relative error, based on the measured values for Equation (16), was 4.97%. However, the R^2 obtained in here was 0.72. Figure 8 shows the confidence of the regression. It was found that most data, including the data for the optimum condition, were within a ±20% margin of error, showing that the regression was good. From Arrhenius' law, the temperature dependence of the correlation, such as Equation (16), was determined by the activation energy and temperature level of the correlation, as shown in this equation [35]. The activation energy obtained was found to be 30.78 kJ/mol. The same regression procedure was performed in the others' correlation equations. The results for the stripping rate, steam flow rate, and heat of regeneration are shown below:

$$\dot{m}_{CO_2} = 2.08 \times 10^7 \exp\left(-\frac{1542.05}{T_R(K)}\right)\alpha_0^{2.03}[C_A(\text{kmol}/\text{m}^3)]^{-0.23}[Q_A(\text{m}^3/s)]^{0.68} \tag{17}$$

$$\dot{m}_s = 1.44 \times 10^8 \exp\left(-\frac{2985.14}{T_R(K)}\right)\alpha_0^{-0.097}[C_A(\text{kmol}/\text{m}^3)]^{0.19}[Q_A(\text{m}^3/s)]^{0.78} \tag{18}$$

and

$$Q = 0.10 \exp\left(\frac{895.75}{T_R(K)}\right)\alpha_0^{-1.45}\alpha^{-0.0083}[C_A(\text{kmol}/\text{m}^3)]^{0.35}[Q_A(\text{m}^3/s)]^{-0.011}. \tag{19}$$

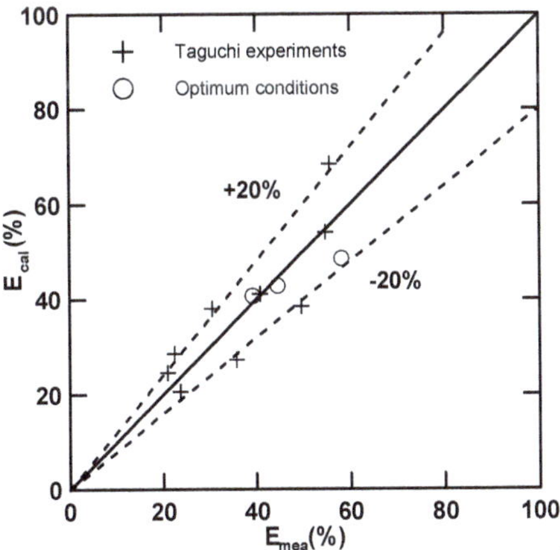

Figure 8. A plot of E_{cal} versus E_{mea}, showing the data distribution.

It was found that most data, including the data for the optimum conditions, were within a ±20% margin of error for the stripping rate, steam flow rate, and heat of regeneration, as shown in Figures 9–11, showing that the regressions were good. However, Figure 11 shows that some data were scattered beyond the ±20% margin of error. The regression errors for the four correlations are listed in Table 8, and were in the range of 4.79–7.91. The activation energy for the four equations is also listed in this table, in which the range was −7.45–30.78 kJ/mol. The negative value for Q means that Q decreased with the increase in T. Correlations with high activation energies are temperature-sensitive; correlations with low activation energies are very temperature-insensitive [35]. Therefore, E was more temperature-sensitive compared with other outcome data. In addition, the exponents of each of the equation is also listed in this table for comparison. The results showed that the effect of α_0 on $\dot{m}_{CO_2}$ was obvious as compared with others, while the effect of C_A on E was the most significant. Finally, the effect of Q_A on $\dot{m}_s$ was more significant, compared with the others' correlations. In addition, the exponents and activation energy for E and Q were in contract. They need to be discussed in later works.

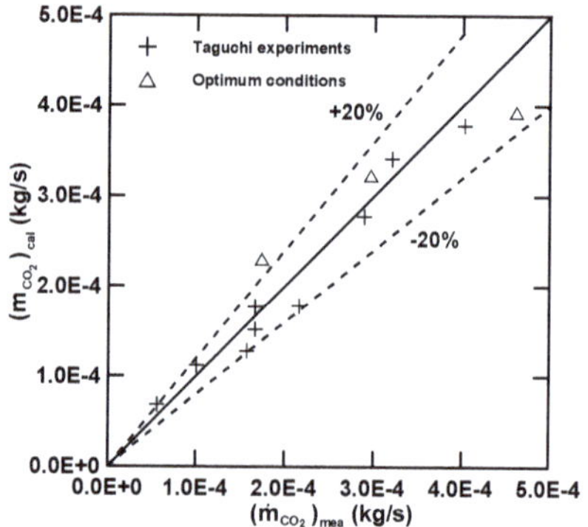

Figure 9. A plot of the calculated stripping rate versus the measured stripping rate.

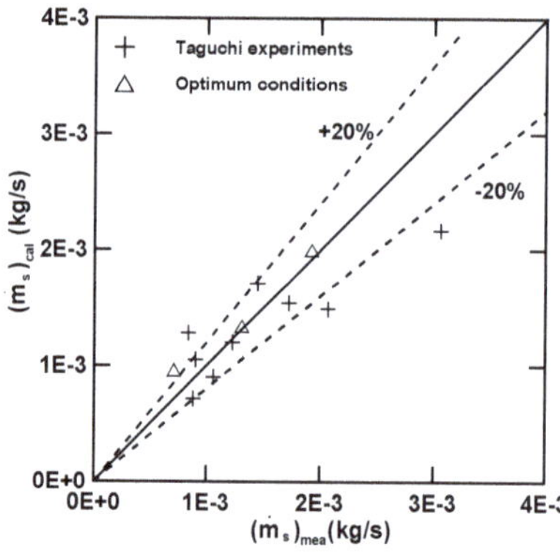

Figure 10. A plot of calculated steam generation rate versus measured values.

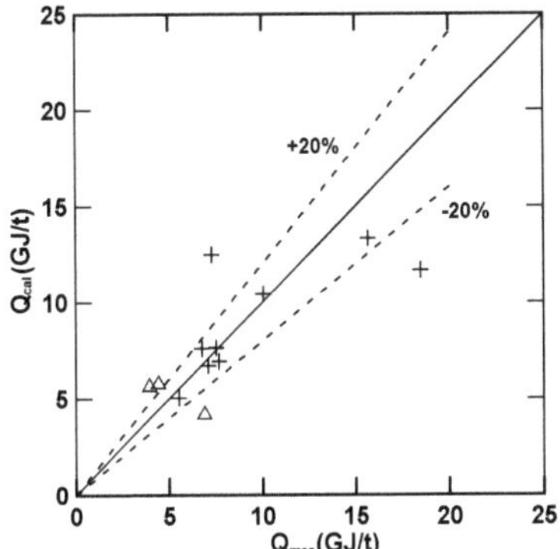

Figure 11. A plot of calculated and measured data for Q, showing regression error.

Table 8. The root mean relative error, activation energy, and exponent for Equations (16)–(19).

Outcome Data	Root Mean Relative Error (%)	Activation Energy (kJ/mol)	Exponent for Each Equation		
			α_0	C_A	Q_A
Equation (16) (E)	4.97 ($R^2 = 0.72$)	30.78	0.68	−1.39	−0.33
Equation (17) ($\dot{m}_{CO_2}$)	7.00 ($R^2 = 0.92$)	12.82	2.03	−0.23	0.68
Equation (18) ($\dot{m}_s$)	5.71 ($R^2 = 0.71$)	24.81	−0.097	0.19	0.78
Equation (19) (Q)	7.91 ($R^2 = 0.56$)	−7.45	−1.45	0.35	−0.011

3.4. Heat of Regeneration and Stripping Factor

Alternatively, the heat of regeneration was used to correlate to the stripping factor, which was defined as the ratio of the stripping rate to the steam flow rate, that is, $\beta = \dot{m}_{CO_2}/\dot{m}_s$. Here, the β value can act as the performance indicator of the stripper. The higher the β value, the better the stripper. Thus, the regression result became the following:

$$Q = 1.62\beta^{-0.84} \tag{20}$$

The R-square value was 0.8036. Figure 12 shows a plot of Q versus β. It was found that the trend of the Q value decreased with the increase in β, which could be calculated using Equations (17) and (18) when α, Q_A, C_A, and T_R were given. The increase in β indicated that, with the same amount of steam, more CO_2 could be desorbed.

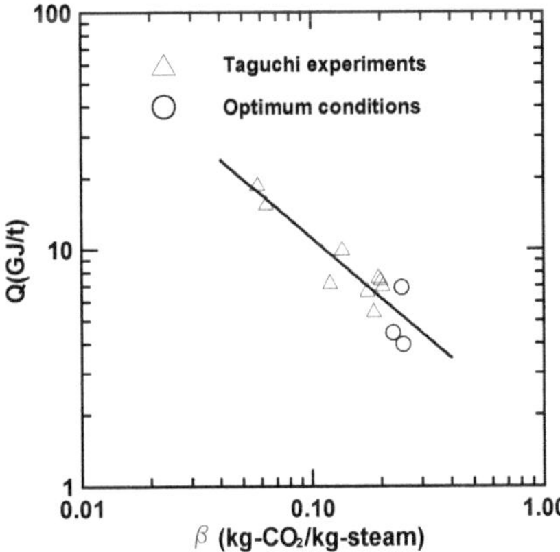

Figure 12. A plot of Q against β, showing the effect of the stripping factor on the heat of regeneration.

3.5. Policy in Operation

Figure 13 is a plot of E versus Q. The plot shows that the trend of E decreased with the increase in Q. Therefore, if Q was set to 3, the corresponding stripping efficiency would be about 80%. From Equation (16), it was found that α_0 and t_R needed to be higher, while C_A and Q_A needed to be lower. On the other hand, when ignoring the effects of the lean CO_2 loading and liquid flow rate shown in Equation (19) for Q, it was found that α_0 and t_R needed to be higher, while C_A needed to be lower. The trend was similar to that obtained from Equation (16) for E. Because of these findings, α_0 and t_R needed to be controlled at higher values, while lower values for C_A and Q_A required reducing the Q value and increasing the E value.

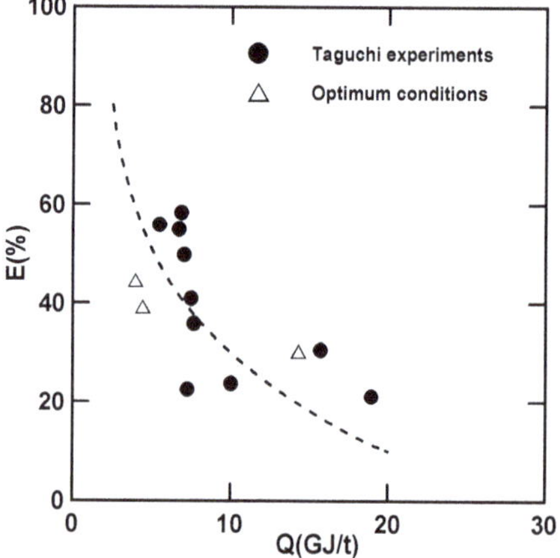

Figure 13. A plot of E versus Q showing the variation of E with Q.

4. Conclusions

This study successfully used a continuous packed-bed stripper with an MEA + AMP solvent containing CO_2 for the study of the heat of regeneration using the Taguchi experimental design. Using mass and energy balances, the stripping efficiency, stripping rate, steam flow rate, stripping factor, and heat of regeneration could be determined under a steady-state condition. Quantitatively, the effects of the variables on E, $\dot{m}_{CO_2}$, β, and Q were explained by the empirical equations obtained in this study. The definition of the stripping factor (β) was used to describe the performance of the stripper. It was found that the heat of regeneration could be correlated with β. The heat of regeneration decreased with the increase in β, showing that more CO_2 could be desorbed. The Taguchi S/N ratio analysis found that the parameter importance sequence was A > D > B > C. Additionally, the optimum conditions for E, $\dot{m}_{CO_2}$, and Q were all verified, thereby showing confidence in the experimental design. In order to obtain a higher E and a lower Q, lower values of Q_A and C_A were required, while higher values of α_0 and t_R were needed.

Author Contributions: conceptualization, P.C.C.; data curation, Y.-L.L.; formal analysis, Y.-L.L.; methodology, P.C.C.; writing (review and editing), P.C.C.

Funding: The authors acknowledge the financial support of the MOST in Taiwan ROC (MOST-106-2221-E-262-015).

Acknowledgments: The authors acknowledge the administration support by Lunghwa University of Science and Technology.

Conflicts of Interest: The authors declare no conflict of interest.

Abbreviation

AMP	2-amino-2-methyl-1-propanol
AMPD	2-amino-2-methyl-propane-1,3-diol
DETA	Diethylenetriamine
MEA	Monoethanolamine
PZ	Piperazine
PZEA	(Piperazinyl-1)-2-ethylamine
SG	Sodium glycinate
S/N	Signal/Noise
TETA	Triethylenetetramine

Nomenclature

C	kmol/m^3	concentration
C_A	kmol/m^3	concentration of amine
C_{Pw}	kJ/kg·K	hat capacity of water
C_{PA}	kJ/kg·K	hat capacity of amine
$C_{p,oil}$	kJ/kg·K	hat capacity of oil
E	%	stripping efficiency
$\dot{H}_R$	kJ/s	reboiler heat loss
$\dot{H}_{col}$	kJ/s	column heat loss
$\dot{H}_{cold}$	kJ/s	heat removed by cooler
H_h	kJ/s	heat exchanger heat loss
H^{vap}	kJ/kg	heat of evaporation
$\dot{H}_{loss}$	kJ/s	total heat loss
$\dot{H}_{reb}$	kJ/s	heat flow rate in reboiler
$\dot{H}_1$	kJ/s	enthalpy at inlet
$\dot{H}_2$	kJ/s	enthalpy at outlet
H_{abs}	kJ/kg	heat of absorption
$\dot{m}_{CO_2}$	kg/s	stripping rate
$\dot{m}_{oil}$	kg/s	oil flow rate
$\dot{m}_s$	kg/s	steam generation rate
$\dot{m}_w$	kg/s	cooling water flow rate
n	-	number of data points
$\dot{n}_A$	kmol/s	amine flow rate
Q	W	total heat flow
Q_A	m^3/s	liquid volumetric flow rate
Q_{abs}	kW	heat flow of absorption
Q_{sen}	kW	sensitive heat flow
Q_{vap}	W	heat flow of evaporation
T_{in}	K	temperature at inlet of reboiler
T_{out}	K	temperature at outlet of reboiler
T_{o1}	K	temperature at the bottom of column
T_R	K	reboiler temperature
T_s	K	steam temperature
T_{wi}	K	inlet temperatureof the cooler
T_{wo}	K	outlet temperatureof the cooler
T_1	K	temperature at inlet of storage tank
t_R	°C	reboiler temperature
z_i	-	value of ith data

Greek Letters

α	kmol-CO$_2$/kmol-amine	rich loading
α_0	kmol-CO$_2$/kmol-amine	lean loading
β	kg-CO$_2$/kg-steam	stripping factor

References

1. Li, T.; Keener, T.C. A review: Desorption of CO_2 From Rich Solutions in Chemical Absorption Processes. *Int. J. Greenh. Gas Control* **2016**, *51*, 290–304. [CrossRef]
2. Aronu, U.E.; Svendsen, H.F.; Hoff, K.A.; Juliussen, O. Solvent Selection for Carbon Dioxide Absorption. *Energy Procedia* **2009**, *1*, 1051–1057. [CrossRef]
3. Adeosun, A.; Goetheer, N.E.H.; Abu-Zahra, M.R.M. Absorption of CO_2 by Amine Blends Solution: An Experimental Evaluation. *Int. J. Eng. Sci.* **2013**, *3*, 12–23.
4. Chen, P.C.; Lin, S.Z. Optimization in the Absorption and Desorption of CO_2 Using Sodium Glycinate Solution. *Appl. Sci.* **2018**, *8*, 2041. [CrossRef]
5. Chen, P.C.; Luo, Y.X.; Cai, P.W. Capture of Carbon Dioxide Using Monoethanolamine in a Bubble-Column Scrubber. *Chem. Eng. Technol.* **2015**, *38*, 274–282. [CrossRef]
6. Choi, W.J.; Seo, J.B.; Seo, S.-Y.; Jung, J.H.; Oh, K.J. Removal Characteristics of CO_2 Using Aqueous MEA/AMP Solutions in the Absorption and Regeneration Process. *J. Environ. Sci.* **2009**, *21*, 907–913. [CrossRef]
7. Dash, S.K.; Sammanta, N.; Bandyopadhyay, S.S. Simulation and Parametric Study of Post Combustion CO_2 Capture Process Using (AMP+PZ) Blended Solvent. *Int. J. Greenh. Gas Control* **2014**, *21*, 130–139. [CrossRef]
8. Leung, D.Y.C.; Caramanna, G.; Maroto-Valer, M.M. An Overview of Current Status of Carbon Dioxide Capture and Storage Technologies. *Renew. Sustain. Energy Rev.* **2014**, *39*, 426–443. [CrossRef]
9. Ho, M.T.; Allinson, G.W.; Wiley, D.E. Comparison of MEA Capture Cost for Low CO_2 Emissions Sources in Australlia. *Int. J. Greenh. Gas Control* **2011**, *5*, 49–60. [CrossRef]
10. Aroonwilas, A.; Veawab, A. Integration of CO_2 Capture Unit Using Blended MEA-AMP Solution into Coal-Fired Power Plants. *Energy Procedia* **2009**, *1*, 4315–4321. [CrossRef]
11. Kim, J.H.; Lee, J.H.; Lee, I.Y.; Jang, K.R.; Shim, J.G. Performance Evaluation of Newly Developed Absorbents for CO_2 Capture. *Energy Procedia* **2011**, *4*, 81–84. [CrossRef]
12. Han, K.; Ahn, C.K.; Lee, M.S. Performance of an Ammonia-Based CO_2 Capture Pilot Facility in Iron and Steel Industry. *Int. J. Greenh. Gas Control* **2014**, *27*, 239–246. [CrossRef]
13. Zhang, X.; Fu, K.; Liang, Z.; Yang, Z.; Rongwong, W.; Na, Y. Experimental Studies of Regeneration Heat Duty for CO_2 Desorption from Aqueous DETA Solution in a Randomly Packed Column. *Energy Procedia* **2014**, *63*, 1497–1503. [CrossRef]
14. Kwak, N.S.; Lee, J.H.; Lee, I.Y.; Jang, K.R.; Shim, J.G. A Study of New Absorbent for Post-Combustion CO_2 Capture Test Bed. *J. TIChE* **2014**, *45*, 2459–2556. [CrossRef]
15. Xiao, J.; Li, C.C.; Li, M.H. Kinetics of Absorption of Carbon Dioxide into Aqueous Solutions of 2-amino-2-methyl-1-propanol+Monoethanolamin. *Chem. Eng. Sci.* **2000**, *55*, 161–175. [CrossRef]
16. Kumar, P.S.; Hogendoorn, J.A.; Versteeg, G.F.; Feron, P.H.M. Kinetics of the Reaction of CO_2 With Aqueous Potassium Salt of Taurine and Glycine. *Aiche J.* **2003**, *49*, 203–213. [CrossRef]
17. Singh, P.; Versteeg, G.F. Structure and Activity Relationships for CO_2 Regeneration from Aqueous Amine-Based Absorbents. *Process Saf. Environ. Prot.* **2008**, *86*, 347–359. [CrossRef]
18. Idem, R.; Wilson, M.; Tontiwachwuthikul, P.; Chakma, A.; Veawab, A.; Aroonwilas, A.; Gelowitz, D. Pilot Plant Studies of the CO_2 Capture Performance of Aqueous MEA and Mixed MEA/MDEA Solvents at the University of Regina CO_2 Capture Technology Development Plant and Boundary Dam CO_2 Capture Demonstration Plant. *Ind. Eng. Chem. Res.* **2006**, *45*, 2414–2420. [CrossRef]
19. Weiland, R.H.; Hatcher, N.A. Post-Combustion CO_2 Capture with Amino-Acid Salts. In Proceedings of the GPA Europe Meeting, Lisbon, Portugal, 22–24 September 2010.
20. Lin, P.H.; Shan, D.; Wong, H. Carbon Dioxide Capture and Regeneration with Amine/Alcohol/Water. *Int. J. Greenh. Gas Control* **2014**, *26*, 69–75. [CrossRef]
21. Oexmann, J.; Kather, A. Minimising the Regeneration Heat Duty of Post-Combustion CO_2 Capture by Wet Chemical Absorption: The Misguided Focus on Low Heat of Absorption Solvents. *Int. J. Greenh. Gas Control* **2010**, *4*, 36–43. [CrossRef]
22. Veawab, V.; Aroonwilas, A. Energy Requirement for Solvent Regeneration in CO_2 Capture Plants. In Proceedings of the 9th International CO_2 Capture Network, Copenhagen, Denmark, 16 June 2006.
23. Rochelle, G.T. Innovative Stripper Configurations to Reduce the Energy Cost of CO_2 Capture. In Proceedings of the Second Annual Carbon Sequestration Conference, Alexandria, VA, USA, 5–8 May 2003.

24. Esmaeili, H.; Roozbehani, B. Pilot-Scale Experiments for Post-Combustion CO_2 Capture from Gas Fired Power Plants with a Novel Solvent. *Int. J. Greenh. Gas Control* **2014**, *30*, 212–215. [CrossRef]

25. Yu, C.H.; Huang, C.H.; Tan, C.S. A Review of CO_2 Capture by Absorption and Adsorption. *Aerosol Air Qual. Res.* **2012**, *12*, 745–769. [CrossRef]

26. Rabensteiner, M.; Kinger, G.; Koller, M.; Gronald, G.; Unterberger, S.; Hochenauer, C. Investigation of the Suitability of Aqueous Sodium Glycinate as a Solvent for Post Combustion Carbon Dioxide Capture on the Basis of Pilot Studies and Screening Methods. *Int. J. Greenh. Gas Control* **2014**, *29*, 1–3. [CrossRef]

27. Ali Khan, A.; Halder, G.N.; Saha, A.K. Carbon Dioxide Capture Characteristics from Flue Gas Using Aqueous 2-amino-2-methyl-1-propanal (AMP) and Monoethanolamine (MEA) Solutions in Packed Bed Absorption and Regeneration Columns. *Int. J. Greenh. Gas Control* **2015**, *32*, 15–23. [CrossRef]

28. Badea, A.A.; Dinca, C.F. CO_2 Capture from Post-Combustion Gas by Employing MEA Absorption Process—Experimental Investigation for Pilot Studies. *U.P.B. Sci. Bull. Ser. D* **2012**, *74*, 21–32.

29. Van Wagener, D.H.; Rochelle, G.T. Stripper Configuration for CO_2 Cpature by Aqueous Monoethhanolamine and Piperazine. *Energy Procedia* **2011**, *4*, 1323–1330. [CrossRef]

30. Yeh, J.T.; Resnik, K.P.; Rygle, K.; Pennline, H.W. Semi-Batch Absorption and Regeneration Studies for CO_2 Capture by Aqueous Ammonia. *Fuel Process. Technol.* **2005**, *86*, 1533–1546. [CrossRef]

31. Yang, J.; Yu, X.; Yan, J.; Tu, S.T. CO_2 Capture Using Solution Mixed with Ionic Lliquid. *Ind. Eng. Chem. Res* **2014**, *53*, 2790–2799. [CrossRef]

32. Song, H.J.; Lee, S.; Park, K.; Lee, J.; Chand Spah, D.; Park, J.W.; Filburn, T.P. Simplified Estimation of Regeneration Energy of 30 wt% Sodium Glycinate Solution for Carbon Dioxide Absorption. *Ind. Eng. Chem. Res.* **2008**, *47*, 9925–9930. [CrossRef]

33. Geankoplis, C.J. *Transport Processes and Unit Operations*; Allyn and Bacon, Inc.: Boston, MA, USA, 1983; pp. 244–250, ISBN 0-205-07788-9.

34. Smith, J.M.; van Ness, H.C.; Abbott, M.M. *Introduction to Chemical Engineering Thermodynamics*; McGraw Hill: New York, NY, USA, 2006; pp. 116–125, ISBN 007-008304-5.

35. Levenspiel, O. *Chemical Reaction Engineering*; Jon Wiley & Sons, Inc.: Hoboken, NJ, USA, 1999; pp. 28–29, ISBN 0-471-25424-X.

Article

An Empirical Study on Low Emission Taxiing Path Optimization of Aircrafts on Airport Surfaces from the Perspective of Reducing Carbon Emissions

Nan Li [1], Yu Sun [1], Jian Yu [2,*], Jian-Cheng Li [3], Hong-fei Zhang [4] and Sangbing Tsai [5,*]

[1] College of Air Traffic Management, Civil Aviation University of China, Tianjin 300300, China; nanli@cauc.edu.cn (N.L.); sunyu8787@gmail.com (Y.S.)
[2] China Civil Aviation Environment and Sustainable Development Research Center, Tianjin 300300, China
[3] School of Software of Dalian University of Technology, Dalian 116620, China; Camouf@mail.dlut.edu.cn
[4] Operation and Control Center, Shandong Airlines, Jinan 250014, China; zhanghf5@sda.cn
[5] Zhongshan Institute, University of Electronic Science and Technology of China, Guangdong 528400, China
* Correspondence: yjian@cauc.edu.cn (J.Y.); sangbing@zsc.edu.cn (S.T.)

Received: 8 April 2019; Accepted: 28 April 2019; Published: 30 April 2019

Abstract: Aircraft emissions are the main cause of airport air pollution. One of the keys to achieving airport energy conservation and emission reduction is to optimize aircraft taxiing paths. The traditional optimization method based on the shortest taxi time is to model the aircraft under the assumption of uniform speed taxiing. Although it is easy to solve, it does not take into account the change of the velocity profile when the aircraft turns. In view of this, this paper comprehensively considered the aircraft's taxiing distance, the number of large steering times and collision avoidance in the taxi, and established a path optimization model for aircraft taxiing at airport surface with the shortest total taxi time as the target. The genetic algorithm was used to solve the model. The experimental results show that the total fuel consumption and emissions of the aircraft are reduced by 35% and 46%, respectively, before optimization, and the taxi time is greatly reduced, which effectively avoids the taxiing conflict and reduces the pollutant emissions during the taxiing phase. Compared with traditional optimization methods that do not consider turning factors, energy saving and emission reduction effects are more significant. The proposed method is faster than other complex algorithms considering multiple factors, and has higher practical application value. It is expected to be applied in the more accurate airport surface real-time running trajectory optimization in the future. Future research will increase the actual interference factors of the airport, comprehensively analyze the actual situation of the airport's inbound and outbound flights, dynamically adjust the taxiing path of the aircraft and maintain the real-time performance of the system, and further optimize the algorithm to improve the performance of the algorithm.

Keywords: carbon emission; reduction; carbon emission; taxiing on airport surfaces; pollutants emissions; taxi time; taxi path optimization

---

1. Introduction

With the rapid growth of traffic flow in the civil aviation transportation industry, the operation of major airports in China has become increasingly congested, and the operational safety and efficiency problems of airport surfaces have become increasingly serious. At the same time, the amount of pollutants emitted by flight operations has gradually increased, bringing immeasurable pressure to the future atmospheric environment. In order to develop air transport sustainably, the realization of a safe, efficient, and environmentally friendly new aircraft operation method has become an important means to maintain the sustainable development of air transport. In this context, the use of scientific methods and efficient path planning can reduce aircraft taxiing time, avoid aircraft taxiing conflicts, improve the overall operation efficiency of the airport, reduce aircraft fuel consumption, and promote the construction of green airports.

Many national and international scholars have investigated taxi path optimization and pollutant emissions of aircrafts on airport surfaces. In 2013, Landry et al. [1] used complex network theory to dynamically detect and resolve conflicts encountered on taxiways and runways, improve the operational efficiency of the airport surface, and ensure the safety of the aircraft. However, the simulation of the model is complex and cannot meet real-time. In 2014, Ravizza [2] studied aircraft path planning in airport surface considering time and fuel consumption, and introduced a sequence diagram-based algorithm to solve this problem. This method increases the authenticity of the model and more accurately estimates the aircraft taxi time. Finally, the purpose of reducing taxi time and reducing fuel consumption was achieved. In 2015, to solve the comprehensive optimization problem of combining runway scheduling and ground motion problems, Weiszer et al. [3] used the multi-objective optimization method. The proposed evolutionary algorithm is based on improved congestion distance, taking into account delay costs and fuel prices. Noticeably, price range uncertainties are defined by preference. In 2016, Ortega Alba [4] summarized current state-of-the-art energy use behaviors and airport trends, and analyzed major energy sources and consumers as well as the application of energy and energy efficiency measures. The airport energy indicators have been established. In 2016, Li and Lv [5,6] analyzed the monitoring data of Hongqiao Airport, and used SVM (Support Vector Machine) to classify and traject the taxiing aircraft in airport surface. The application of data mining technology to the prediction of airport surface taxiing time, taxiing hotspots, and the determination of conflict areas were studied. In 2016, Przemysław, P. [7] studied models and computer software tools for implementing dynamic taxi route selection modules, which can use conflict points to describe airport congestion. The proposed taxi route selection system is integrated with the previously developed system to enable it to be implemented in the routing module in A-SMGCS (Advanced-Surface Movement Guidance and Control System). In 2017, a study by Li and Zhang established fuel consumption and emission calculation models of single-engine taxi, tow-outs, and APU (Auxiliary Power Unit) electric taxi [8]. In this study, the authors analyzed the types of pollutants emitted by aircraft engines, introduced the calculation model of fuel consumption and emissions under the standard landing and take-off cycle (LTO) and aircraft all-engine taxi mode. The fuel consumption and emission correction model was used to correct changes in fuel consumption and emission coefficients caused by external temperature, air pressure, and humidity during the aircraft taxiing stage. Taking Shanghai Hongqiao Airport as an example, pollutant emissions of various aircraft types under different taxiing modes were calculated. The results showed that single-engine taxi and APU electric taxi can reduce HC, CO, and NOx (hydrocarbon, Carbondioxide, and nitrogen oxides) emissions during the aircraft taxiing phase. The tow-outs taxi has little effect on NOx emissions, but can significantly reduce HC and CO emissions. In 2017, Chen et al. [9] considered that the route and schedule generated by taxi time prediction may not be flexible under the uncertain factors such as changing weather conditions, operating scenarios and pilot behavior, and based on multi-objective fuzzy rules. This uncertainty is quantified based on historical aircraft taxi data. In 2018, Zhu et al. [10] used the Shanghai Hongqiao Airport scene monitoring data to study the number of different departure aircraft in the apron and taxiing system under the condition of runway capacity limitation, flight departure rate, departure taxi time, and runway utilization. A departure aircraft taxi time prediction model was established to develop a reasonable launch control strategy for the departing aircraft during the peak hour of the airport without reducing the runway efficiency; ultimately, it reduces aircraft taxiing time and reduces aircraft fuel consumption and pollutant emissions [11–13].

Most traditional taxi path optimization studies have been based on the shortest taxi path, or model optimization based on the shortest taxi time—assuming aircrafts have equal speed taxiing conditions [14–17]. The constructed path optimization model is simple and easy to solve. However, limitations exist in the large gap between the model and the actual taxiing situation of the aircraft.

In fact, due to the presence of the turning section of the airport surface, the speed profile of the aircraft changes and the deceleration during a large turning will cause additional taxiing time. The exact class algorithm has many applications in recent path planning studies because it can fully consider various airport surface limiting factors and find the optimal solution [18–22]. However, there are many variables in this kind of algorithm model; the algorithm is complex, and the calculation amount is huge. It is generally difficult to obtain the global optimal solution in an acceptable time [23–26]. Therefore, this paper comprehensively considers the factors such as the taxiing distance of the aircraft, the large number of turns in the taxiing and the collision avoidance, and conducts path planning research on the aircraft that taxis in the airport surface to reduce the total taxiing time of the aircraft, save aviation fuel and reduce the amount of gaseous pollutants emitted in the taxiing stage of the aircraft. Although this method lacks robustness, it is more realistic than others. The proposed method is faster than other complex algorithms considering multiple factors, and has higher practical application value.

2. Analysis of Airport Surface Movement of the Aircraft

In this paper, we first analyzed taxiing aircrafts' motion on airport surfaces, and then established the aircraft taxiway path planning model. Figure 1 shows surveillance data covering the movement status of a typical arriving and departing aircraft [5,6], which was used to analyze the motion state of an aircraft on airport surfaces.

In Figure 1a,b, the horizontal axis is the time axis. The left and right longitudinal axes measure the speed and heading, respectively (Figure 1a,b). The aircraft taxied at a relatively high speed ~10 m/s, when there was no demand for steering and effect of conflict on airport surfaces. The aircraft with flight track number 991 arrived with relatively large steering at 80 s, 150 s, and 390 s of its taxiing time, and the aircraft's taxiing speed during the turnaround period decreased to less than 2 m/s. Around 270 s of its taxiing time, aircraft 991 stopped evacuation due to taxi conflict on the path, during which the aircraft's speed was reduced to zero. The departure aircraft with flight track number 2091 performed relatively large steering at 120 s, 370 s, and 470 s of its taxiing time. The taxiing speed was also significantly reduced during these times. For around 400 s of its taxi time, this aircraft queued before entering the runway, where its speed decreased to zero.

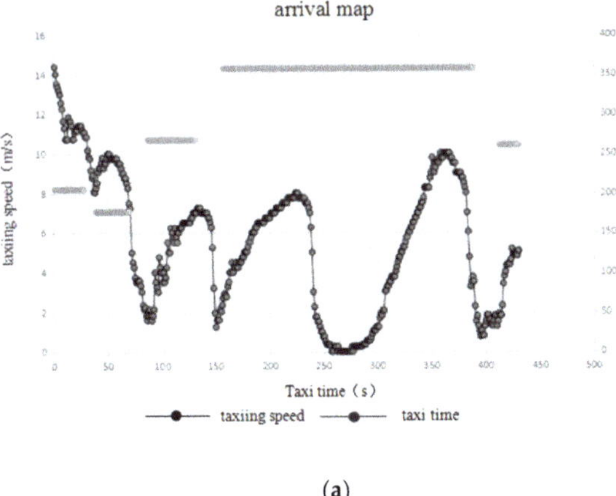

(a)

Figure 1. *Cont.*

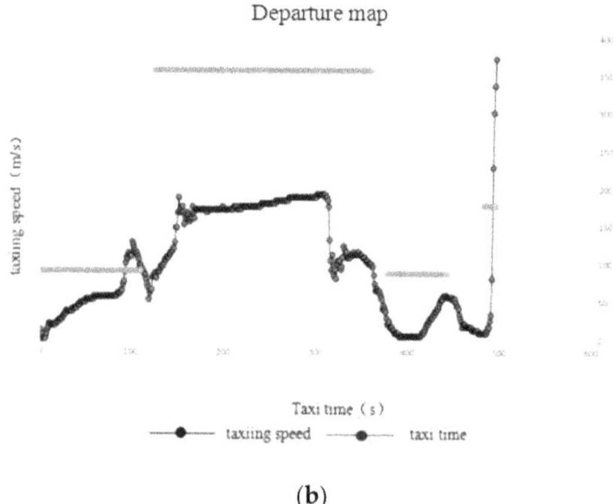

(**b**)

Figure 1. Approach and departure aircraft movement state diagram. (**a**) Arrival aircraft and (**b**) departure aircraft.

Analysis of the data showed that the taxiing time of the arrival and departure aircrafts on airport surfaces was not only directly affected by the taxiing distance factor, but was also closely related to instances of large steering and collision avoidance during the taxiing process. Therefore, this paper comprehensively considers the aircraft's taxiing distance, number of large steers during taxiing, and conflict avoidance. The study also covers path planning of aircraft taxiing in order to reduce the total aircraft taxiing time on airport surfaces, save aviation fuel, reduce gas emissions during aircraft taxiing, and improve air quality around the airport.

3. Establishing the Path Planning Model

Airport taxiway systems comprise very complicated networks. The taxiway system can be resolved into a node and edge G = (V, E) network model, when optimization is performed for aircraft taxi paths on airport surfaces. In the network model, V represents the set of nodes in the taxiway system; it consists of intersections between taxiways a intersections between taxiways and runways, parking position, runway ends, etc. E represents the set of edges connecting each node within the taxiway system, and consists of the taxiway segments between adjacent nodes in the taxiway system.

3.1. Setting Model Variables

Let $P = \{1, 2, 3, \ldots, r\}$ be the set of aircrafts for the required planning. Among them, the arrival aircraft is represented by A, and the departure aircraft is represented by D, $A, D \in P$. The set $R = \{R_1, R_2, \ldots, R_r\}$ represents the airport surface taxi path for each aircraft. The taxi path $R_k (1 \leq k \leq r)$ of any aircraft k consists of a set of ordered nodes $\{N_1^k, N_2^k, \ldots, N_q^k\}$. If the number of airport network model nodes V is M, then there are: $q \leq M$; $N_j^k \leq V, j = 1, 2, \ldots, q$; $\left(N_j^k, N_{j+1}^k\right) \in E, j = 1, 2, \ldots, q - 1$.

To avoid the complications of character expressions, when the nodes "N_i", "N_j", etc. appear as subscripts of the characters, they are simplified by "i" and "j". This replacement is only for writing format needs, and the meaning remains unaltered. The following define the variables used in the path optimization model:

$x_{ijk} = 1$: indicates that the aircraft k passes the node N_i of the taxiway, and then taxis to the next neighboring node N_j, otherwise $x_{ijk} = 0$;

s_{ij} is the length of the taxiway section between two adjacent nodes N_i and N_j in the taxiway system;

t_{jk} is the moment of entering the taxiway node N_j during aircraft k taxiing;

t_s is the minimum safety interval between two aircrafts that successively pass through the same node;

t_d is the deceleration taxi time on the landing runway after the aircraft k has landed;

$y_{ik} = 1$, indicates the taxiway node N_i in the taxi path of the aircraft k, otherwise $y_{ik} = 0$;

t_{ijk} is the taxi time used by the aircraft k to taxi from the taxiway node N_i to the next adjacent node N_j;

v_k is the taxi speed when the aircraft k is taxiing on the airport surface;

t_{ok} is the initial taxiing moment of aircraft k in the taxiway system. The node N_o represents the initial taxiing point of the aircraft within the taxiway system. For the arriving aircraft, N_o represents the initial point of quick vacated taxiway, for the departing aircraft, N_o represents gates;

t_{ek} is the termination taxiing time of aircraft k in the taxiway system. Node N_e indicates the aircraft's termination taxiing point in the taxiway system. For the arriving aircraft, N_e denotes the parking position. For the departing aircraft, N_e denotes end of the departure runway.

3.2. Building Objective Functions

The purpose of this paper is to reduce aircraft emissions during its taxi stage. The aircraft's thrust during taxiing is described by the taxiing thrust state ($7\%F_\infty$) [9]. Hence, aircraft emissions positively correlate with taxiing time; thus, the objective of path planning is to minimize the total taxiing time of all aircrafts on airport surfaces. The established objective function is:

$$\min T = \sum_{k=1}^{r} \sum_{i=N_1}^{N_M} \sum_{j=N_1}^{N_M} x_{ijk} t_{ijk} + \sum_{k=1}^{r} n_k t_n \tag{1}$$

The above formula contains two items: the first item is the time it takes for the aircraft to taxi either way. Where x_{ijk} is a decision variable defined in Section 3.1, which is 1 or 0 depending on whether the edge (N_i, N_j) is the taxi path for aircraft k, and t_{ijk} is the time at which the aircraft k taxis from the taxiway node N_i to the next adjacent node N_j, and can be expressed as:

$$t_{ijk} = s_{ij}/v_k \tag{2}$$

The second term is the time it takes the aircraft to make large steering at each node where a turn is needed. Where n_k denotes the total number of turns aircraft k has accumulated on the taxi path, and t_n denotes the time it takes for aircraft k to average each turn. The number of turns n_k for aircraft k during taxiing is determined by the following formula:

$$n_k = \sum_{j=N_2^k}^{N_{q-1}^k} num_j^k \tag{3}$$

where node $N_j \in \left\{ N_2^k, N_3^k, \ldots, N_{q-1}^k \right\}$, $num_j^k = 1$ or 0. When aircraft k passes through the steering angle θ_j^k of node N_j and reaches the preset threshold cri, i.e., $\theta_j^k \geq cri$, it is considered that when the aircraft passes through the node N_j, a large steering is performed to make the recorded value $num_j^k = 1$, otherwise, $num_j^k = 0$, n_k is the cumulative value of the num_j^k values of the nodes on aircraft k's taxi path.

To solve the aircraft steering angle θ^k at each node, the cosine theorem can be used for calculation. Assume that aircraft k on the airport's surface starts taxiing at node N_u, and continues taxiing through nodes N_v and N_w. The airport surface coordinates of the three nodes are currently (x_u, y_u), (x_v, y_v), and (x_w, y_w). θ_v^k is used to indicate the angle by which aircraft k turns through node N_v. By applying the cosine theorem, we can see that angle θ_v^k turns at the node N_v:

$$\theta_v^k = \pi - \arccos \frac{(x_w - x_v)^2 + (y_w - y_v)^2 + (x_v - x_u)^2 + (y_v - y_u)^2 - (x_w - x_u)^2 - (y_w - y_u)^2}{2\sqrt{(x_w - x_v)^2 + (y_w - y_v)^2}\sqrt{(x_v - x_u)^2 + (y_v - y_u)^2}} \tag{4}$$

3.3. Construction of Aircraft Taxiing Dynamic Model

For arriving aircrafts, t_d indicates the time needed to decelerate and vacate from the runway after landing, enter the quick vacated taxiway, or by-pass taxiway. The moment the high-speed taxiway or the by-pass runway is entered, is the moment when taxiing begins, which is represented by the Equation (5):

$$t_{ok} = ETOA_k + t_d \tag{5}$$

For departing aircrafts, the moment of pushback from the apron is the moment taxi begins, which is represented by the Equation (6):

$$t_{ok} = EOBT_k \tag{6}$$

If aircraft k's taxi path is $R_k = \left\{ N_1^k, N_2^k, \ldots, N_q^k \right\}$, the initial taxiing point is $N_o = N_1^k$, and the initial taxiing time is t_{ok}, then the time R_k when aircraft k taxis to any node N_{kj} on its path t_{jk} is:

$$t_{jk} = t_{ok} + \sum_{(N_1^k, N_2^k)}^{(N_{j-1}^k, N_j^k)} t_{ijk} + n_k^* t_n \tag{7}$$

where $\sum_{(N_1^k, N_2^k)}^{(N_{j-1}^k, N_j^k)} t_{ijk}$ represents the total time spent by aircraft k from the node N_1^k taxi to node N_j^k accumulated on each taxi segment; $n_k^* t_n$ indicates the time taken by the nodes that requiring a large steer when the aircraft k is taxied from node N_1^k to node N_j^k; n_k^* indicates the number of accumulated steers by aircraft k from N_1^k taxi to N_j^k; t_n indicates the time taken by each aircraft k for steering. n_k^* can refer to Equation (3), which is represented by the Equation (8):

$$n_k^* = \sum_{j=N_2^k}^{N_{j-1}^k} num_j^k \tag{8}$$

The moment when any aircraft taxis to any node N_j^k, on path R_k can be obtained by Equation (7). Having determined the aircraft's taxiing time through each node, we can use the taxiway network system nodes as the object of operation to construct the constraint function, and add constraints to the path planning model in Section 3.2 to avoid the problem of taxi conflict in the path planned by the model.

3.4. Building Constraint Functions

Aircraft taxi collision on airport surfaces can be divided into three types: intersection conflict, head-on conflict, and tail conflict. During aircraft taxiing, in order to prevent the front plane's engine vortex from affecting the rear plane, both aircrafts must meet the requirements of vortex separation. Table 1 shows the vortex separation standards.

Table 1. Vortex separation Standards for aircrafts on airport surfaces (m).

The Front Plane	The Rear Plane		
	Light	Medium	Heavy
Light	200	200	200
Medium	200	200	200
Heavy	300	300	300

During the aircraft's taxi path planning, to avoid taxi conflicts and meet the vortex separation requirements, constraints must be placed on the objective function to ensure the safety and feasibility of the planned taxi path.

The distance parameter is inconvenient when the taxiway network node is the operating object for the path planning model constraint. Therefore, vortex separation standards in Table 1, given in the form of distance, must be converted into the time separation between aircraft. According to relevant regulations by the Civil Aviation Administration, the maximum speed of an aircraft during its taxi on airport surfaces should not exceed 50 km/h (13.8 m/s). According to the data analyzed in Section 1, the linear taxiing speed of aircrafts on airport surfaces was conservatively taken as 10 m/s. The vortex separation t_s in the form of time is as shown in Table 2.

Table 2. Surface taxi, aircraft time vortex separation (s).

The Front Plane	The Rear Plane		
	Light	Medium	Heavy
Light	20	20	20
Medium	20	20	20
Heavy	30	30	30

3.4.1. Intersection Conflict and Vortex Separation Constraints

Intersection conflicts occur when two aircraft cross over the same taxiway network node, with/without meeting safety separations. The intersection conflict and vortex separation problems can be constrained at the same time. This is achieved by controlling the time separation t of two consecutive aircrafts k_1 and k_2, passing through the same taxiway network node t_s. Intersection conflict between aircrafts can be prevented and the vortex separation requirement can be met. The constraints are:

$$t_{jk_2} - t_{jk_1} \geq t_s \tag{9}$$

In the formula: $\forall k_1, k_2 \in P$, $k_1 \neq k_2$, $\forall N_j \in R_{k1} \cap R_{k2}$. Among them, t_{jk1}, t_{jk2} can be obtained by Equation (7).

3.4.2. Head-on Conflict Constraint

Head-on collision is the confrontation between two opposing aircrafts occupying the same taxiing path, in the taxiway network at the same time. The head-on conflict caused by two aircrafts is more difficult to resolve. This type of conflict should be avoided as much as possible. Therefore, the two opposing aircrafts k_1 and k_2 should satisfy the following formula:

$$\left(x_{ijk_1} t_{ik_1} - x_{jik_2} t_{ik_2} \right)\left(x_{ijk_1} t_{jk_1} - x_{jik_2} t_{jk_2} \right) > 0 \tag{10}$$

In the formula: $\forall k_1, k_2 \in P$, $k_1 \neq k_2$, $\forall (N_i, N_j) \in R_{k_1}$, $\forall (N_j, N_i) \in R_{k_2}$. Equation (10) indicates that if aircraft k_1 passes the taxiing path (N_i, N_j), and aircraft k_2 passes the taxiing path (N_j, N_i), then aircraft k_1 must first pass through both nodes N_i and N_j; otherwise, if aircraft k_2 passes first through node N_j, then it must also pass first through node N_i.

3.4.3. Tail Conflict

Tail conflict refers to two aircraft traveling in the same direction, while occupying the same taxi path, in the taxiway network at the same time. These aircrafts thus surpass the phenomenon caused by the rear and front planes catching up, or the problem of small safety separations. Two aircrafts k_1 and k_2, traveling in the same direction, should satisfy the following formula:

$$\left(x_{ijk_1}t_{ik_1} - x_{ijk_2}t_{ik_2}\right)\left(x_{ijk_1}t_{jk_1} - x_{ijk_2}t_{jk_2}\right) > 0 \tag{11}$$

In the formula: $\forall k_1, k_2 \in P$, $k_1 \neq k_2$, $\forall\left(N_i, N_j\right) \in R_{k_1}$, $\forall\left(N_j, N_i\right) \in R_{k_2}$. Equation (11) shows that if both aircrafts pass through taxiway $\left(N_i, N_j\right)$, then the aircraft passing first through node N_i, must also pass first through node N_j.

4. Solving the Path Planning Model

The airport surface taxi path planning model for aircrafts, established in this paper, has many complex variables and formulas, with some nonlinear equations. The traditional algorithm is difficult to solve. Genetic algorithms have high searchability and strong robustness, and are widely applied to path planning problems such as highway traffic and drones [14,15]. Genetic algorithms are based on populations and not single-point searches, and can generate multiple extreme values from different points at the same time. Obtaining a local optimal solution of the path planning model is not easy; therefore, the genetic algorithm was used to solve it.

4.1. Chromosome Coding and Initial Population Generation

When genetic algorithms are used to solve the taxi path planning model, chromosome encoding is firstly required. Genetic algorithms generally have binary encoding, floating-point encoding, symbol encoding, and other encoding methods. Based on the specific requirements of this model, floating-point encoding was used (real value encoding method).

Figure 2 is a simplified illustration of a taxiway network, and an example illustrating chromosome encoding. First, the nodes in the taxiway system were given real values such as 1, 2, 3, ... , 10, 11, etc. The order of these real values indicated the genetic sequence of a chromosome, and also corresponded to the representation of an aircraft's taxi path on airport surfaces.

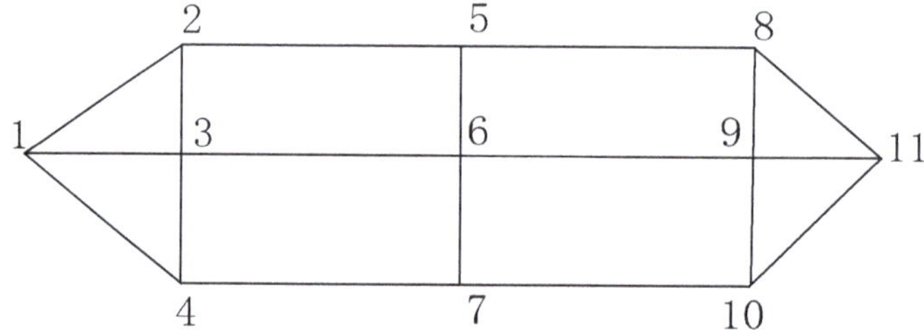

Figure 2. Simple taxiway network diagram.

If the length of the chromosome is set to 10, and the aircraft taxis from node "1" to node "11", then chromosome A shown in Figure 3 is an effective chromosome, and chromosome A indicates that the aircraft's taxi path is $1 \rightarrow 4 \rightarrow 7 \rightarrow 10 \rightarrow 11$, less than 10 bits are filled with "0" in the back.

Chromosome A	1	4	7	10	11	0	0	0	0	0	0

Figure 3. Chromosome A.

Iterative optimization of genetic algorithms is not based on a single chromosome, it is based on the population of multiple chromosomes. This paper used computer-generated random generation methods to form the initial population. First, sequence numbers of all nodes in the taxiway system were randomized; then, the illegal chromosomal sequences formed in the random sorting were eliminated. That is, the chromosomal sequences that were subject to taxiing path jumps were eliminated, and a certain number of initial feasible solutions were generated.

4.2. Selecting the Operation and Fitness Function

In this paper, the roulette method was used to select the operator performing chromosome selections. This method generates the individual population according to the fitness value as the selected variable, where individuals with large fitness values have a high probability of being selected. However, this does not only select individuals with large fitness values, as this would generate a unitary population and limit the algorithm's rapid convergence to the local optimal solution.

The fitness function is a measure to evaluate an individual's fitness value. The evaluation function that was established in this paper, did not only require chromosome encoding fitness values of the aircraft with a smaller taxiing time, but also effectively eliminated chromosome coding in those aircrafts with taxiing conflicts. According to the above requirements, the fitness function z was constructed as follows.

Construct an equivalent taxi time function z first:

$$z = \sum_{k=1}^{r} \sum_{i=N_1}^{N_M} \sum_{j=N1}^{N_M} x_{ijk}t_{ijk} + \sum_{k=1}^{r} n_k t_n + M\max\left[-\left(t_{jk_2} - t_{jk_1}\right), 0\right] +$$
$$M\max\left[-\left(x_{ijk_1}t_{ik_1} - x_{jik_2}t_{ik_2}\right) \cdot \left(x_{ijk_1}t_{jk_1} - x_{jik_2}t_{jk_2}\right), 0\right] +$$
$$M\max\left[-\left(x_{ijk_1}t_{ik_1} - x_{ijk_2}t_{ik_2}\right) \cdot \left(x_{ijk_1}t_{jk_1} - x_{ijk_2}t_{jk_2}\right), 0\right] \tag{12}$$

where M is the penalty coefficient, which is a value much larger than the normal taxiing time $T = \sum_{k=1}^{r} \sum_{i=N_1}^{N_M} \sum_{j=N_1}^{N_M} x_{ijk}t_{ijk} + \sum_{k=1}^{r} n_k t_n$. The calculation result z of Equation (12) is the total aircraft taxiing time of airport surface taxiing; if intersection conflicts, head-on conflicts, or tail conflicts existed in the taxi paths represented by this population of chromosomes, a very large z-value would be obtained due to the penalty coefficient M.

Let the fitness function be: $f_i = 1/z_i$, so that the fitness value of chromosome i is obtained, and the larger the taxiing time, the smaller the fitness value of a population of chromosomes. According to the proportion of fitness values, the distribution probability of each chromosome population was selected. The formula is:

$$p_i = f_i / \sum f_i \tag{13}$$

This type of equation shows that the probability of a highly-adapted genome is selected. If the population is N, then randomly selected N times will form the next generation of new populations.

4.3. Crossover Operations

Commonly-used crossover operators include single-point, two-point, and multi-point crossings. Chromosomes in this article represent the taxi path of an aircraft on the airport's surface. Chromosome order represents the order of nodes passing through the aircraft's taxi path. To ensure that cross-replicated chromosomes still have practical significance, this paper used a special single-point crossover operator. The crossover steps are as follows:

Randomly select a pair of chromosomes from the population as the chromosomes to be crossed with a certain crossover probability p_c (generally a large probability).

Compare two chromosomal genes to investigate the possibility of identifying identical taxiing node genes in the two chromosomes, except for the start node and termination node. If there are no identical taxiing node genes, then select a pair of new chromosomes and compare them. However, if the same taxiing node gene exists, one of the same taxiing nodes is randomly selected first, and the gene sequence behind this node is exchanged.

This article uses the same chromosome taxiing node rather than the same length node as the allele, thus the length of the two chromosomes after the crossover may not be equal. After the crossover, if the progeny chromosome has more digits than the parent chromosome, the extra "0" is removed from the end of the progeny chromosome; if the progeny chromosome has fewer cipher digits than the parent, the progeny chromosome occupies the "0" at the end.

This allowed us to detect whether there were repeated taxiing nodes before and after the newly generated chromosome pair cross points. If no repeated node were detected, then it must be an effective chromosome, and the crossover operation ends. If a repeating node exists, then the aircraft would repeat the round-trip taxiing on the taxiing path, indicating that the progeny chromosome generated by the crossover operation had no practical meaning, the crossover operation is invalid, and the "1" operation is performed again.

Still referring to the simplified taxiway network in Figure 2, the two chromosomes A and B were randomly selected, as shown in Figure 4a, except that the starting and terminating nodes had the same taxiing node "6". Then the gene sequence was exchanged after node "6", thus changing A and B into two new taxi paths, A* and B*, as shown in Figure 4b. Then, "1" and "0" bits were added at the end of the new individual A* and B*, respectively, as shown in Figure 4c. The two chromosomes had no identical nodes before and after node "6", thus this crossover operation was effective. Chromosomes A* and B* reserve the basic characteristics of their parent chromosomes; however, they are new individuals representing the new taxiing path.

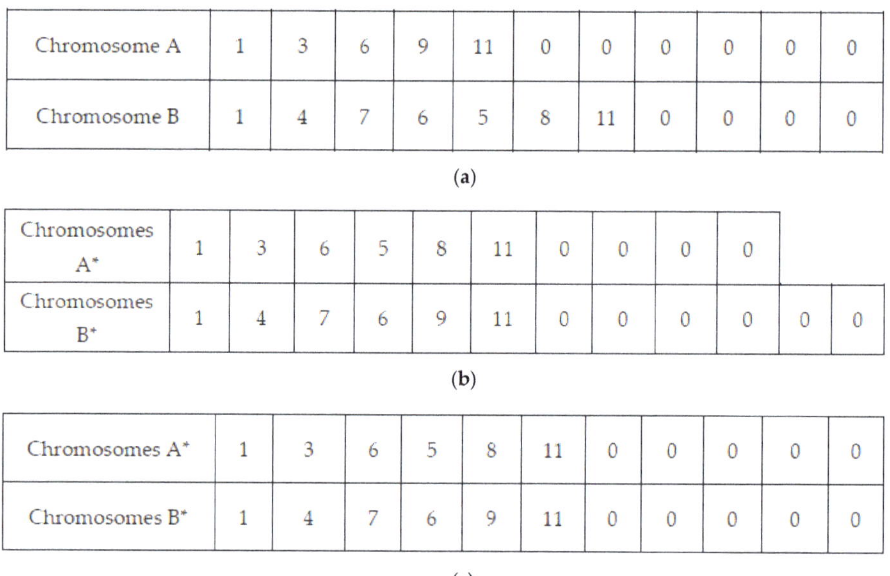

Chromosome A	1	3	6	9	11	0	0	0	0	0	0
Chromosome B	1	4	7	6	5	8	11	0	0	0	0

(**a**)

Chromosomes A*	1	3	6	5	8	11	0	0	0	0		
Chromosomes B*	1	4	7	6	9	11	0	0	0	0	0	0

(**b**)

Chromosomes A*	1	3	6	5	8	11	0	0	0	0	0
Chromosomes B*	1	4	7	6	9	11	0	0	0	0	0

(**c**)

Figure 4. Operation diagram of chromosome crossover. (**a**) Chromosome A and B before crossing; (**b**) chromosomes A* and B* after crossing; and (**c**) chromosome A* and B* after complement and delocation.

4.4. Mutation Operation

After population cross-selection, in order to improve the local algorithm's searchability and maintain the population's diversity, it was necessary to perform chromosome mutation operations on the population. Common mutation operators included single-point and multi-point mutations.

In this model, chromosome sequences represent the aircraft's taxiing path; therefore, their ordering sequence characterized the sequence of the nodes passing through the taxiing path. The traditional mutation method changes single or several nodes within a chromosome, which cannot improve the local algorithm's searchability and maintain the population's diversity. It also destroys the generated taxi path and affects population convergence.

Therefore, in the solution of this model, the mutation operation selects chromosomes for mutation from the population based on the mutation probability p_m (generally less), and then replaces it with a randomly generated reasonable taxi path as a new chromosome (Figure 5). This can also be understood as a multi-point variation in the broad sense, although the new chromosome produced by this variation method is quite different from the original chromosome, it can supplement new gene sequences, effectively avoiding the loss of certain genes due to the selection of crossover operations and maintaining the population diversity.

chromosome A	1	3	6	9	11	0	0	0	0	0	0

(a)

chromosome A*	1	4	7	10	11	0	0	0	0	0	0

(b)

Figure 5. Operation diagram of chromosome mutation. (**a**) Chromosome A before mutation; and (**b**) chromosome A* after mutation.

5. Emulation Experiment

5.1. Data Description

In this article, we used the Shanghai Hongqiao Airport as an example for simulation analysis. Figure 6 shows the network diagram of the taxiway at Hongqiao Airport. The airport is a two-runway airport, 18R/36L is mainly used for departure. It is 3300 m in length and 60 m in width. For arrivals, 18L/36R was used. It is 3400 m in length and 45 m in width. The airport has four main taxiways parallel to the runways A, B, C, and D. It also includes quick-vacate taxiways, by-pass taxiways, and apron taxiways.

According to the airport surveillance radar data at Hongqiao Airport, the 1 h traffic during peak hours includes 40–50 aircrafts. We took 20 real flight sequences within 30 min of a busy period to record monitoring data as an example; thereby, aircraft taxiway planning analysis was performed. Due to wind direction, the 18R and 18L runways were used for departure and arrival, respectively. The starting and ending moments of the flights are shown in Table 3.

5.2. Parameter Settings

During path planning, the initial taxiing time t_{ok} of each aircraft k in the taxiway system remained unchanged at the starting time (Table 3); according to the relevant regulations of the Civil Aviation Administration and the airport operational data analysis, the linear taxi speed of the aircraft was set to 10 m/s. Since the data did not have aircraft type items, assuming that flights in Table 3 were all medium aircraft, then the vortex separation was $t_s = 20s$.

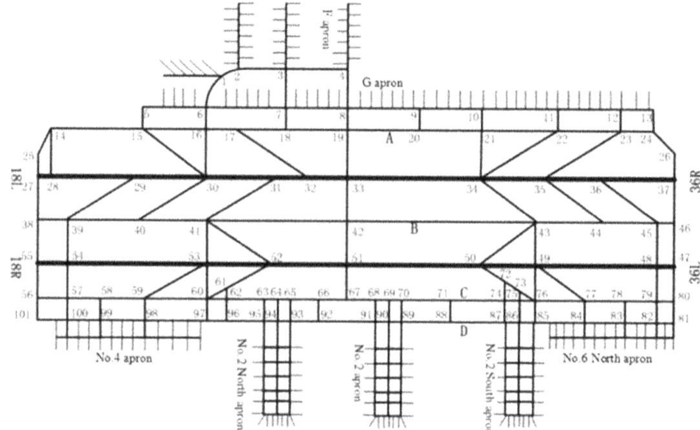

Figure 6. Network diagram of taxiway at Hongqiao Airport.

Table 3. Airport Flight Schedules.

Flight	Flight Tracking Number	Arrival and Departure Type	Apron Location	Taxiing Start and End Time (s)
1	617	departure	No. 2 apron	49,837~50,135
2	834	departure	F apron	49,916~50,333
3	2587	arrival	No. 4 apron	50039~50,455
4	98	departure	G apron	50,055~50,533
5	132	arrival	No. 4 apron	50,173~50,621
6	80	departure	No. 2 apron	50,276~50,665
7	2894	arrival	No. 6 apron	50,323~50,543
8	3433	departure	No. 2 North Apron	50,433~50,743
9	1368	arrival	No. 2 North Apron	50,443~50,740
10	339	departure	No. 2 apron	50,467~50,852
11	1704	arrival	No. 2 North Apron	50,565~51,010
12	1527	arrival	No. 2 South Apron	50,704~50,855
13	3676	departure	No. 2 South Apron	50,752~51,128
14	1727	arrival	No. 6 apron	50,825~51,030
15	2712	departure	No. 2 North Apron	50,893~51,078
16	925	arrival	No. 2 North Apron	50,940~51,216
17	3570	departure	No. 2 apron	50,956~51,348
18	1639	departure	F apron	51,065~51,465
19	438	arrival	No. 6 apron	51,075~51,280
20	1510	arrival	No. 2 South Apron	51,214~51,344

High-speed taxiways vacating and runway angles are 25° to 45°, typically 30°. When the aircraft turns to a high-speed vacate taxiway, taxiing speed is high; therefore, it is not considered as large steering, and a large steering critical angle $cri = 45°$ was set. For the Airbus A320, for example, it takes 30–60 s to achieve 90° steering during taxiing. Therefore, it is assumed that the time taken by an aircraft to make a large steer is $t_n = 30s$.

Departure flights still originate from their respective original aprons in Table 3, and terminate at the 18R runway end; the flight's arrival begins at the original high-speed vacate taxiway of the 18L runway, and ends at the original apron in Table 3. Without changing the above conditions, only taxiing path optimization for each aircraft on the airport surface is now performed.

The genetic algorithm was used to solve the established taxi path optimization model. Eight feasible taxiing paths were randomly generated for each aircraft, and the aircraft taxiing paths were randomly combined. The initial population number was set to 100, the crossover and mutation probabilities were set to Pc = 0.6 and Pm = 0.1, respectively, and the number of iterations was 100 generations. We then used the MATLAB7.9.0 software for programming.

5.3. Analysis of Results

Figure 7 shows optimal solution curves for each generation in the genetic algorithm. After optimizing the 100-generation iterative genetic algorithm, the total taxiing time of 20 aircrafts on the airports surface converged to 5343 s, and the average aircraft taxiing time was about 267 s. Among them, the total taxiing distance was 38,728 m, and the number of turns was 49.

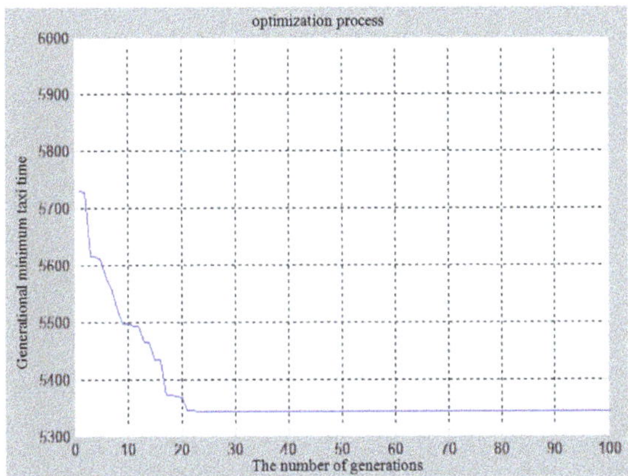

Figure 7. The minimum taxiing time curve generated using the genetic algorithm.

The taxiing path for each aircraft and the time it took the aircraft to pass through each node after path optimization are shown in Appendix A. Appendix A recorded the nodes that passed through the aircraft taxi path planning, and the time of entry into the node. For the non-steering node on the aircraft's taxi path, it was considered that aircrafts taxi into the node and then directly out, occupying the node for only a certain moment. For the steering node on the aircraft's taxi path, it was considered that after the aircraft taxied into the node, it took 30 s to perform the steering operation, after which the aircraft taxis out the node.

Taxi data before and after optimization of 20 aircraft paths listed in Table 3 were satisfied. The comparison results are shown in Table 4.

Table 4. Comparison of important data before and after aircraft path optimization.

	Before Optimization	**After Optimization**
Total taxiing distance (m)	39,650	38,728
Total number of turns	52	49
Total taxi time (s)	6423	5343
Number of flights using the C main taxiway	1	7
Number of flights using the D main taxiway	16	10
The number of slow down or stop to avoidance conflict	6	0

Comparing statistical data from Table 4 shows that the total taxiing distance and number of aircraft turns after path optimization were reduced. However, the impact on total taxiing time reduction was not significant and taxi collision avoidance was the major factor for reducing total taxiing time.

In Table 3, there are 17 aircrafts whose parking positions were the 2nd, 4th, and 6th aprons on the west side of runway 18R/36L. Among them, 16 aircraft's arrival and departure taxiing was in the D main taxiways, that is, the taxi paths were between nodes 81–101, and only one aircraft with flight track number 3676 used the C master taxiway, the taxiing route segment between nodes 56–80. This mode of airport operation caused aircrafts on the D main taxiway to run slowly, with frequent conflicts, and the average taxi speed was low, while the C main taxiway was obviously underutilized.

The model built in this paper was based on a conflict avoidance path planning model with the shortest taxi time. Therefore, the planned taxi path was different from the actual running path of the airport, and the airport taxiway system resources were fully utilized. Among the taxiing paths planned by the model, 10 of the 17 arrival and departure aircrafts on the north apron were assigned D main taxiways, and seven were assigned C main taxiways, in order to avoid the problem of taxi path conflict between aircrafts, and so that aircrafts on the planned path could meet the vortex separation requirements. Figure 8 shows the number of taxi aircraft on each taxiway before and after path optimization. The data show that the airport taxiway resource allocation was more reasonable post taxi path optimization.

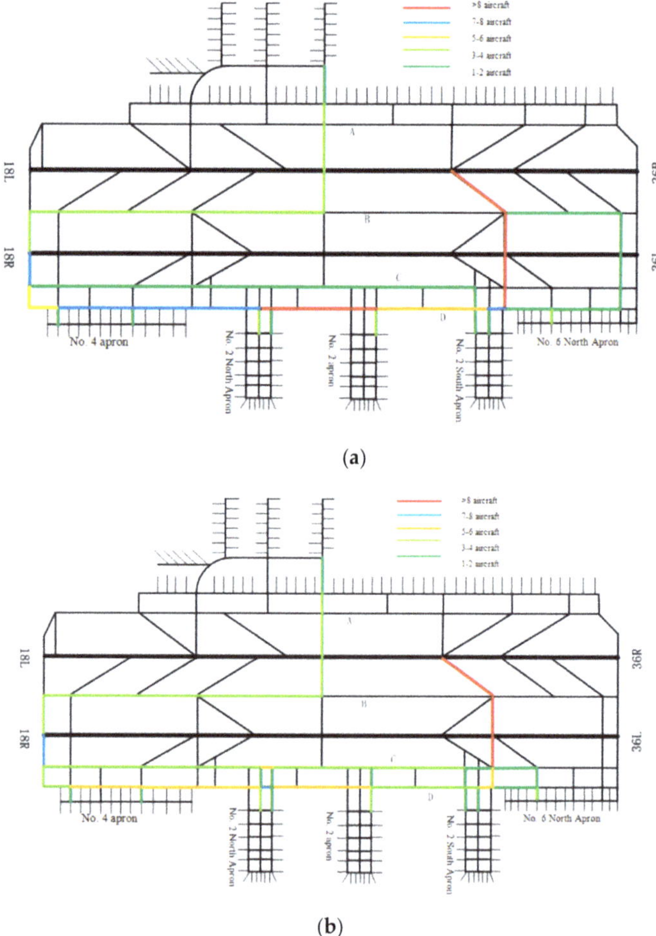

Figure 8. Number of aircraft taxiing on various taxi paths before and after path optimization. (**a**) Before optimization; and (**b**) after optimization.

6. Comprehensive Optimization Strategy and Fuel Consumption Calculation

Various aircraft taxiing modes have been studied within the local research environment. In addition to two-engine taxiing, aircrafts have three low-emission airport surface taxiing modes. Among them, single-engine and electric taxiing speeds are the same as the all-engine taxiing mode [7]. Therefore, while path planning for aircrafts on airport surfaces is carried out, aircrafts can also implement a comprehensive optimization strategy using a low-emission taxiing mode, such as single-engine taxiing or electric-driven taxiing.

To plan the taxiing path of the aircraft on the airport surface, we then used the 20 aircrafts mentioned in Section 5 to apply the path planning model and the solution algorithm established in this chapter. At the same time, the aircrafts were considered to adopt the conditions of full engine taxiing, single-engine taxiing, and electric taxiing. Using the fuel consumption and emissions calculation model under various taxiing methods [7], we calculated the total fuel consumption and emissions of all aircrafts in the airport taxiing stage before and after optimization of different combination strategies. The fuel flow rate and pollutant emission factor of the aircrafts were based on the B736-700 model. The calculation results for total fuel consumption and pollutant emissions from the 20 aircrafts are shown in Figure 9a,b.

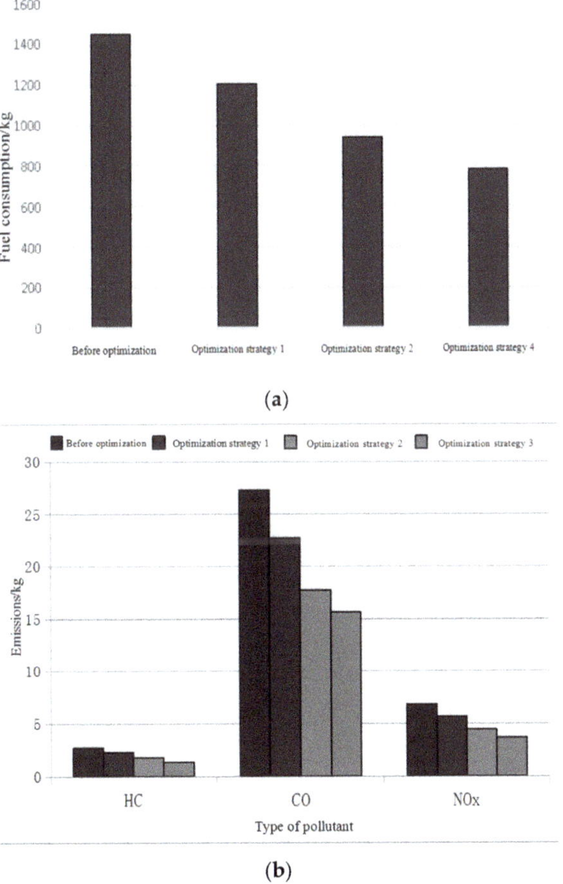

Figure 9. Aircraft fuel consumption (**a**), and emissions (**b**) under different optimization strategies.

In Figure 9, "Before Optimization" indicates that no optimization strategy was implemented, at this time, total fuel consumption and emissions of all aircraft were the largest. "Optimization Strategy 1" indicates that the aircrafts had conducted the taxi path planning under the all-engine taxiing mode. The data show that after implementing taxi path planning

For aircrafts on the airport surface, the total fuel consumption and emissions of all aircrafts were reduced by about 17% compared with before optimization. "Optimization Strategy 2 and Optimization Strategy 3" indicate that the aircraft had planned the taxi path in the single-engine taxi mode and the electric drive taxi mode, respectively. This model is a comprehensive optimization strategy for low-emission taxiing and taxi path planning. The data also indicated that total fuel consumption and emissions of all aircraft were reduced by 35% and 46%, respectively, compared with before optimization. The calculations show that the comprehensive optimization strategy is more effective in energy saving and emission reductions during the aircrafts taxiing stage.

7. Conclusions

In this paper, the motion data of aircrafts were analyzed using monitoring data. Factors such as the taxiing distance of the aircraft, the number of large turns in the taxiing, and conflict avoidance were comprehensively considered. The aircraft path optimization model, with the shortest total time of aircraft taxiing on airport surfaces, was established—the aircraft taxiing dynamic model. According to this model, possible conflict problems in the path optimization model were constrained. Finally, the genetic algorithm was used to solve the model, and the crossover and mutation operators of the traditional genetic algorithm were improved. The simulation was carried out with real flight data, and the energy-saving and emission-reducing effects of the aircraft's path planning during the taxiing stage were analyzed and calculated. The results showed that, after aircraft's taxi path optimization, the aircraft's total taxiing distance and steering times were reduced, taxiing conflict was avoided, taxiing time was relatively reduced, and the overall speed of the aircraft's taxiing in airport surface was increased, which improved the airport's operating efficiency, which in turn reduced fuel consumption and pollutant emissions during the taxiing phase. Compared with traditional optimization methods that do not consider turning factors, energy saving and emission reduction effects were more significant. The proposed method is faster than other complex algorithms considering multiple factors, and has higher practical application value.

Aircraft taxi path planning requires a high degree of rationality, accuracy, real-time, and reliability. Due to factors such as development time constraints, this paper has certain limitations. Mainly reflected in the proposed method, although it can give the path of the airport surface with free collision and the shortest taxi time, but in the actual taxiing process, the aircraft cannot ensure that the route is strictly followed, and it is possible to generate new conflicts. How to dynamically adjust the path of the aircraft under the premise of ensuring real-time performance needs further study. The future work needs to be improved by increasing the actual interference factors of the airport, comprehensively analyzing the actual situation of the airport's arrival and departure flights, dynamically adjusting the taxiing path of the aircraft and maintaining the real-time performance of the system, and further optimizing the algorithm to improve the algorithm's operation speed and accuracy and greatly improve the performance of the algorithm.

Author Contributions: Writing: N.L.; Providing case and idea: Y.S., J.Y. and J.-C.L.; Revising and editing: H.-f.Z. and S.T.

Funding: This research was funded by the national key R&D program, grant number "No.2016YFB0502405"; the National Natural Science Foundation, grant number "No.71801215"; the National Social Science Foundation, grant number "No.13CGL005".

Acknowledgments: We are grateful to Zhongshan City Science and Technology Bureau Project (No. 2017B1015) and 2018 Zhongshan Innovation and Development Research Center.

Conflicts of Interest: The authors declare no conflict of interest.

Appendix A

Table A1. The taxiing path of each aircraft, and the moment when the aircraft passes through each node.

Flight Track Number	Path Node (Passing Node Time (s))
617	89 (49,842 s)– > 90 (49,880 s)– > 91 (49,888 s)– > 92 (49,907 s)– > 93 (49,925 s)– > 94 (49,933 s)– > 95 (49,941 s)– > 96 (49,956 s)– > 97 (49,973 s)– > 98 (49,997 s)– > 99 (50,017 s)– > 100 (50,048 s)– > 10 (50,061 s)– > 56 (50,099 s) – > 55 (50,117 s)
834	4 (49,916 s)– > 8 (49,931 s)– > 19 (49,940 s)– > 33 (49,964 s)– > 42 (49,982 s)– > 41 (50,084 s)– > 40 (50,114 s)– > 39 (50,160 s)– > 38 (50,173 s)–>55 (50,222 s)
2587	34 (50,039 s)– > 43 (50,072 s)– > 49 (50,120 s)– > 76 (50,138 s)– > 85 (50,146 s)– > 86 (50,184 s)– > 87 (50,192 s)– > 88 (50,212 s)– > 89 (50,230 s)– > 90 (50,238 s)– > 91 (50,246 s)– > 92 (50,265 s)– > 93 (50,283 s)– > 94 (50,291 s)– > 95 (50,299 s)– > 96 (50,314 s)– > 97 (50,331 s)– > 98 (50,355 s)– > 99 (50,375 s)– > 100 (50,407 s)
98	8 (5,0055 s)– > 19 (50,064 s)– > 33 (50,088 s)– > 42 (50,106 s)– > 41 (50,208 s)– > 40 (50,238 s)– > 39 (50,284 s)– > 38 (50,297 s)– > 55 (50,346 s)
132	34 (50,173 s)– > 43 (50,206 s)– > 49 (50,254 s)– > 76 (50,272 s)– > 85 (50,280 s)– > 86 (50,318 s)– > 87 (50,326 s)– > 88 (50,346 s)– > 89 (50,364 s)– > 90 (50,372 s)– > 91 (50,380 s)– > 92 (50,399 s)– > 93 (50,417 s)– > 94 (50,425 s)– > 95 (50,433 s)– > 96 (50,448 s)– > 97 (50,465 s)– > 98 (50,489 s)
80	88 (50,276 s)– > 89 (50,294 s)– > 90 (50,302 s)– > 91 (50,310 s)– > 92 (50,329 s)– > 93 (50,348 s)– > 94 (50,356 s)– > 95 (50,364 s)– > 96 (50,379 s)– > 97(50,396 s)– > 98(50,419 s)– > 99(50,439 s)– > 100(50,471 s)– > 101 (50,483 s)– > 56 (50,521 s)– > 55 (50,539 s)
2894	34 (50,323 s)– > 43 (50,356 s)– > 49 (50,404 s)– > 76 (50,422 s)– > 77 (50,479 s)– > 84 (50,517 s)
3433	93 (50,438 s)– > 94 (50,476 s)– > 95 (50,484 s)– > 96 (50,498 s)– > 97 (50,516 s)– > 98 (50,539 s)– > 99 (50,559 s)– > 100 (50,591 s) – > 101 (50,603 s)– > 56 (50,641 s)– > 55 (50,659 s)
1368	34 (50,443 s)– > 43(50,476 s)– > 49(50,524 s)– > 76(50,542 s)– > 75 (50,578 s)– > 74 (50,587 s)– > 71 (50,605 s)– > 70 (50,624 s) – > 69 (50,632 s)– > 68 (50,640 s)– > 67 (50,650 s)– > 66 (50,659 s)– > 65 (50,677 s)– > 64 (50,685 s)– > 94 (50,723 s)
339	89 (50,472 s)– > 90 (50,510 s)– > 91 (50,518 s)– > 92 (50,537 s)– > 93 (50,555 s)– > 94 (50,563 s)– > 95 (50,571 s)– > 96 (50,586 s) – > 97 (50,603 s)– > 98 (50,627 s)– > 99 (50,647 s)– > 100 (50,678 s)– > 101 (50,691 s)– > 56 (50,729 s)– > 55 (50,747 s)
1704	34 (50,565 s)– > 43 (50,598 s)– > 49 (50,646 s)– > 76 (50,664 s)– > 75 (50,700 s)– > 74 (50,709 s)– > 71 (50,727 s)– > 70 (50,746 s) – > 69 (50,754 s)– > 68 (50,762 s)– > 67 (50,772 s)– > 66 (50,781 s)– > 65 (50,799 s)– > 64 (50,807 s)– > 94 (50,845 s)
1527	34 (50,704 s)– > 43 (50,737 s)– > 49 (50,785 s)– > 76 (50,803 s)– > 85 (50,811 s)– > 86 (50,849 s)
3676	87 (50,757 s)– > 74 (50,765 s)– > 71 (50,813 s)– > 70 (50,832 s)– > 69 (50,840 s)– > 68 (50,848 s)– > 67 (50,858 s)– > 66 (50,867 s) – > 65 (50,886 s)– > 64 (50,894 s)– > 63 (50,902 s)– > 62 (50,916 s)– > 60 (50,933 s)– > 59 (50,957 s)– > 58 (50,977 s)– > 57 (51,008 s)– > 56 (51,021 s)– > 55 (51,069 s)
1727	34 (50,825 s)– > 43 (50,858 s)– > 49 (50,906 s)– > 76 (50,924 s)– > 85 (50,932 s)– > 84 (50,987 s)
2712	93 (50,898 s)– > 65 (50,906 s)– > 64 (50,944 s)– > 63 (50,952 s)– > 62 (50,966 s)– > 60 (50,983 s)– > 59 (51,007 s)– > 58 (51,027 s) – > 57 (51,059 s)– > 56 (51,071 s)– > 55 (51,119 s)
925	34 (50,940 s)– > 43 (50,973 s)– > 49 (51,021 s)– > 76 (51,039 s)– > 85 (51,047 s)– > 86 (51,085 s)– > 87 (51,093 s)– > 88 (51,113 s) – > 89 (51,131 s)– > 90 (51,139 s)– > 91 (51,147 s)– > 92 (51,166 s)– > 93 (51,184 s)– > 94 (51,192 s)
3570	89 (50,961 s)– > 70 (50,969 s)– > 69(51,007 s)– > 68(51,015 s)– > 67 (51,025 s)– > 66 (51,033 s)– > 65 (51,052 s)– > 64 (51,060 s) – > 63(51,068 s)– > 62(51,083 s)– > 60 (51,099 s)– > 59 (51,123 s)– > 58 (51,144 s)– > 57 (51,175 s)– > 56 (51,188 s)– > 55(51236 s)
1639	4 (51,065 s)– > 8 (51,080 s)– > 19 (51,089 s)– > 33 (51,113 s)– > 42 (51,131 s)– > 41 (51,233 s)– > 40 (51,263 s)– > 39 (51,309 s) – > 38 (51,322 s)– > 55 (51,371 s)
438	34 (51,075 s)– > 43 (51,108 s)– > 49 (51,156 s)– > 76 (51,174 s)– > 77 (51,231 s)– > 84 (51,269 s)
1510	34 (51,214 s)– > 43 (51,247 s)– > 49 (51,295 s)– > 76 (51,313 s)– > 85 (51,321 s)– > 86 (51,359 s)

References

1. Landry, S.J.; Chen, X.W.; Nof, S.Y. A decision support methodology for dynamic taxiway and runway conflict prevention. *Decis. Support Syst.* **2013**, *55*, 165–174. [CrossRef]
2. Ravizza, S.; Atkin, J.; Burke, E.K. A more realistic approach for airport ground movement optimisation with stand holding. *J. Sched.* **2014**, *17*, 507–520. [CrossRef]
3. Weiszer, M.; Chen, J.; Stewart, P. Preference-Based Evolutionary Algorithm for Airport Runway Scheduling and Ground Movement Optimisation. In Proceedings of the IEEE International Conference on Intelligent Transportation Systems, Las Palmas, Spain, 15–18 September 2015.
4. Ortega Alba, S.; Manana, M. Energy Research in Airports: A Review. *Energies* **2016**, *9*, 349. [CrossRef]
5. Li, N.; Lv, H. Method to Find out Path of Aircraft on Surface Based on Surface Movement Radar Data. *Aeronaut. Comput. Tech.* **2016**, *46*, 6–9.
6. Lv, H. The Analysis of Aircraft Taxiing and Prediction Study Based on Surveillance Data. Master's Thesis, Civil Aviation University of China, Tianjing, China, 2016.
7. Przemysław, P.; Skorupski, J. Aircraft Taxi Route Choice in Case of Conflict Points Existence. In Proceedings of the 16th International Conference on Transport Systems Telematics, Warsaw, Poland, 16–19 March 2016.
8. Li, N.; Zhang, H. Calculating aircraft pollutant emissions during taxiing at the airport. *Acta Sci. Circumst.* **2017**, *37*, 1872–1876.
9. Chen, J.; Weiszer, M.; Zareian, E. Multi-Objective Fuzzy Rule-Based Prediction and Uncertainty Quantification of Aircraft Taxi Time. In Proceedings of the 20th International Conference on Intelligent Transportation Systems, Yokohama, Japan, 16–19 October 2017.
10. Zhu, X.; Li, N.; Sun, Y.; Zhang, H.; Wang, K.; Tsai, S.-B. A Study on the Strategy for Departure Aircraft Pushback Control from the Perspective of Reducing Carbon Emissions. *Energies* **2018**, *11*, 2473. [CrossRef]
11. Ding, J.; Li, X.; Li, Q. Optimal Scheduling Model for Hub Airport Taxi Based on Improved Ant Colony Collaborative Algorithm. *J. Comput. Appl.* **2010**, *30*, 1000–1003. [CrossRef]
12. Chiba, R.; Arai, T.; Ota, J. Integrated Design for Automated Guided Vehicle Systems Using Cooperative Co-evolution. *Adv. Robot.* **2010**, *24*, 25–45. [CrossRef]
13. Ren, C. Driverless Vehicle Path Planning Based on Genetic Algorithm Technology Research. Master's Thesis, Tianjin University, Tianjing, China, 2015.
14. Tang, J.; Shi, W.; Meng, L. Time-Dependent Dynamic Vehicle Routing Based on Genetic Algorithm. *Geomat. Inf. Sci. Wuhan Univ.* **2008**, *8*, 875–879.
15. Tsai, S.-B. Using the DEMATEL Model to Explore the Job Satisfaction of Research and Development Professionals in China's Photovoltaic Cell Industry. *Renew. Sustain. Energy Rev.* **2018**, *81*, 62–68. [CrossRef]
16. Lee, Y.C.; Hsiao, Y.C.; Peng, C.F.; Tsai, S.B.; Wu, C.H.; Chen, Q. Using Mahalanobis-Taguchi System, Logistic Regression and Neural Network Method to Evaluate Purchasing Audit Quality. *Proc. Inst. Mech. Eng. Part B J. Eng. Manuf.* **2015**, *229*, 3–12. [CrossRef]
17. Lee, Y.C.; Chen, C.Y.; Tsai, S.B.; Wang, C.T. Discussing green environmental performance and competitive strategies. *Pensee* **2014**, *76*, 190–198.
18. Liu, B.; Li, T.; Tsai, S.B. Low carbon strategy analysis of competing supply chains with different power structures. *Sustainability* **2017**, *9*, 835. [CrossRef]
19. Qu, Q.; Tsai, S.B.; Tang, M.; Xu, C.; Dong, W. Marine ecological environment management based on ecological compensation mechanisms. *Sustainability* **2016**, *8*, 1267. [CrossRef]
20. Lee, Y.C.; Wang, Y.C.; Chien, C.H.; Wu, C.H.; Lu, S.C.; Tsai, S.B.; Dong, W. Applying revised gap analysis model in measuring hotel service quality. *SpringerPlus* **2016**, *5*, 1191. [CrossRef]
21. Wang, J.; Yang, J.M.; Chen, Q.; Tsai, S.B. Collaborative Production Structure of Knowledge Sharing Behavior in Internet Communities. *Mob. Inf. Syst.* **2016**, *2016*, 8269474. [CrossRef]
22. Tsai, S.B.; Lee, Y.C.; Guo, J.J. Using modified grey forecasting models to forecast the growth trends of green materials. *Proc. Inst. Mech. Eng. Part B J. Eng. Manuf.* **2014**, *228*, 931–940. [CrossRef]
23. Tsai, S.B.; Zhou, J.; Gao, Y.; Wang, J.; Li, G.; Zheng, Y.; Ren, P.; Xu, W. Combining FMEA with DEMATEL Models to Solve Production Process Problems. *PLoS ONE* **2017**, *12*, 0167710. [CrossRef]
24. Ge, B.; Jiang, D.; Gao, Y.; Tsai, S.B. The influence of legitimacy on a proactive green orientation and green performance: A study based on transitional economy scenarios in China. *Sustainability* **2016**, *8*, 1344. [CrossRef]

25. Wang, J.; Yang, J.; Chen, Q.; Tsai, S.B. Creating the sustainable conditions for knowledge information sharing in virtual community. *SpringerPlus* **2016**, *5*, 1019. [CrossRef]
26. Tang, Z.; Zhu, X.; Xia, Z. Taxi Routing Update Algorithm for A-SMGCS with Taxi Time Delay. *Sci. Technol. Eng.* **2017**, *17*, 326–332.

MDPI

St. Alban-Anlage 66

4052 Basel

Switzerland

Tel. +41 61 683 77 34

Fax +41 61 302 89 18

www.mdpi.com

Energies Editorial Office

E-mail: energies@mdpi.com

www.mdpi.com/journal/energies